DC MOTORS SPEED CONTROLS SERVO SYSTEMS

AN ENGINEERING HANDBOOK

prepared by

ELECTRO-CRAFT CORPORATION, USA

PERGAMON PRESS

Oxford • New York • Toronto • Sydney • Paris • Frankfurt

U.K.	Pergamon Press Ltd., Headington Hill Hall, Oxford OX3 OBW, England
U.S.A.	Pergamon Press Inc., Maxwell House, Fairview Park, Elmsford, New York 10523, U.S.A.
CANADA	Pergamon of Canada Ltd., 75 The East Mall, Toronto, Ontario, Canada
AUSTRALIA	Pergamon Press (Aust.) Pty. Ltd., 19a Boundary Street, Rushcutters Bay, N.S.W. 2011, Australia
FRANCE	Pergamon Press SARL, 24 rue des Ecoles, 75240 Paris, Cedex 05, France
WEST GERMANY	Pergamon Press GmbH, 6242 Kronberg-Taunus, Pferdstrasse 1, Frankfurt-am-Main, West Germany

First published by Pergamon Press Ltd., 1977

Library of Congress Catalog Card No. 76-56647

ISBN 0 08 021715 X F
0 08 021714 1 H

Printed in the United States of America

Table of Contents

APPENDIX

□

Foreword to the Third Edition

Since its first publication back in 1972 Electro-Craft's engineering handbook, "DC MOTORS - SPEED CONTROLS - SERVO SYSTEMS", has become a standard reference in the field of motor control. In all 35,000 copies have been sold averaging about 1,000 per month. Many of these books are used by professional engineers to provide day to day assistance in their system design work, and many are playing a useful role in colleges and universities helping to train the engineers of the future.

As a result of the continuous feedback received from users of this book, we have prepared a new, third edition, which has been expanded by about 100 pages to a total of little over 500. Apart from the correction of errors this new edition includes the following changes:

Chapter 5 – The application section has been expanded to include more applications and notes on the application of the new P6000 series servomotor control systems.

Chapter 6 – An entirely new chapter discussing the new and exciting developments made by Electro-Craft in the field of DC Brushless Motors.

Chapter 7 – Data sheets on Electro-Craft products have been revised, updated, and expanded. The new models now available in the MOTOMATIC range of Speed Controls, Servomotor Controls, and Digital Positioning Systems are described in full.

Once again Electro-Craft Corporation cordially invites your comments and criticisms so that the next edition of this book will remain one of the foremost reference works in its field.

ELECTRO-CRAFT CORPORATION
August 1975

Foreword to the Second Edition

Since October 1972, when the first edition of this book was published, we have distributed 15000 copies to interested engineers and designers. The overwhelming majority of the comments which we received from our readers have been favorable and complimentary to our effort for creating this book. We have also received some constructive criticism and requests for expanding the treatment and discussion on certain subjects. In preparing this expanded second edition, we have taken into account all these requests which we could accommodate. All the changes and additions enlarge this book by about 30 percent from the first edition to approximately 400 pages.

Listed below are the nature of the various changes and additions by chapter:

Chapter 1 - Recognizing the need for inclusion of the metric system, we revised the Terminology and Symbols sections and have added the Systems of Units section.

Chapter 2 - The treatment of Motor Equations, Transfer Functions, Power Dissipation and Thermal Characteristics has been rewritten and expanded.

Chapter 3 - Minor revisions.

Chapter 4 - Servo Components treatment has been revised and expanded. The section on Servo Amplifiers has been rewritten, and a new section Phase-Locked Servo Systems has been added.

Chapter 5 - The Applications section has been expanded, sections of Application Classification and Specification, and Motor Selection Criteria have been added.

Chapter 6 - Data sheets on motors and amplifiers have been revised, updated and expanded. Electro-Craft's new line of servo amplifiers is also presented in this chapter.

Two new Appendices have been added; a description of the SI System of Units (Metric System) and a Units' Conversion Factors and Tables.

It is the continuing interest of the authors and editors of this engineering handbook to make it a most useful, relevant and informative source of knowledge regarding all aspects of DC motors, speed controls and servo systems. To this end we again wish to solicit the constructive comments of our readers. These comments, together with the results of our continually expanding search for valuable, pertinent technical information, will be incorporated into future editions of this book.

ELECTRO-CRAFT CORPORATION

September 1973 □

Foreword to the First Edition

THE ELECTRO-CRAFT STORY

Electro-Craft is a growing, publicly held corporation started in 1960. For close to ten of these twelve years, the company has specialized in the development and manufacture of fractional horsepower permanent magnet DC motors, generators and control amplifiers.

In the early years, business centered around patented, integrally constructed motor-generators and matching transistorized "Class A" control amplifiers, marketed under the MOTOMATIC® trademark. These smaller size, transistorized motor speed control systems found their first applications mostly in office equipment, medical electronics, and instrumentation products.

Over the years, the product line has expanded into larger sizes, higher performance servomotors, low ripple tachometer-generators, and increasingly sophisticated servo control amplifiers to meet the growing demands of our customers.

During the past decade, Electro-Craft has sold over half a million precision DC motors and controls to more than 1000 customers for approximately 300 different applications.

There are many motor manufacturers who do not make matching controllers, and vice versa; Electro-Craft is one of the very few companies which make both. It is this special "system-oriented" approach, coupled with constant technological innovation, that is perhaps most responsible for the company's success in the field.

WHY THIS BOOK

Around the turn of this century, AC motors won out over DC machines as the generally accepted sources of electric motive power. The deciding factor then was the state of technology and materials, rather than the inherent characteristics of each device.

Automation and the need for controlled speed and torque, rather than just raw motive power, have brought about a resurgent interest in the use of DC motors. High energy permanent magnets, epoxy resins, improved brushes and other new materials and technologies have also made possible DC machines which are more reliable, smaller and less expensive to produce.

The combination of the above factors produced a great deal of change during the last decade in the variety and performance of DC motors. The greatest innovation took place in the fractional horsepower sizes. The concurrent rapid advances in semiconductor technology, especially in low cost, high power SCR's and transistors, facilitated the emergence of a variety of matching control amplifiers, from economy models to the very sophisticated ones.

Due to this fast growth of technology, much of what is available today is not described in the standard textbooks. The purpose of this book is to offer a basic background in DC motors, speed controls and servo systems, and to describe new developments in the field. The presentation is aimed at the designer and user in industry, with emphasis on many practical problems from testing to applications engineering.

The selection of the power drive system for an application is greatly facilitated by proper communication: a specific understanding of the problem between the user and the manufacturer. Another aim of this book is to make this task simpler by attempting to share the accumulated technical knowledge of our company, which has specialized in this field for the past decade.

HOW TO USE THE BOOK

This book was prepared as a compendium of articles on many related subjects. The editors made no attempt to homogenize the styles of the different authors. An effort was made, however, to make the chapters self-contained and to organize the subject matter in a meaningful way. The index and the definitions in Chapter 1 should help the reader to decide how much background he needs to cover.

Some of the information presented is standard reference material, and some probably will not be found elsewhere. Some chapters are quite theoretical, while others focus on the more practical "how to" aspects.

This is a handbook which will probably be consulted as a reference as problems arise, rather than being read cover to cover as a student's textbook.

ABOUT THE AUTHORS

Approximately twenty people contributed to the preparation of this book. They comprise the technical staff of Electro-Craft, with backgrounds in design engineering, research and development, testing, teaching, systems development, sales, application engineering, and - most of all - practical problem solving. The authors collectively have over 200 man-years of experience in DC motor and control technology.

In addition to their regular daily work, the authors gave their time generously over the past year to make this book possible.

ELECTRO-CRAFT CORPORATION □

October 1972

Chapter 1

Terminology, Symbols, Systems of Units

1.1. TERMINOLOGY

Acceleration, Maximum from Stall - the angular acceleration of an unloaded motor, initially at rest, when the peak armature current I_{pk} is applied.

Ambient Temperature - the temperature of the cooling medium immediately surrounding the motor or another device.

Armature Reaction - the production of a magnetic field shifted by 90 electrical degrees with respect to the direction of stator magnetic field. This magnetic field is produced by armature current.

Bandwidth - the frequency range in which the magnitude of system gain expressed in **dB** is within the **3 dB** band.

Block Diagram - a simplified representation of a system, with each component represented by a block, and each block positioned in the order of signal flow through the system.

Bode Plot - a plot of the magnitude of system gain in **dB** and the phase of system gain in degrees versus the sinusoidal input signal frequency in logarithmic scale.

Break Frequency - the frequency(-ies) at which the asymptotes of the gain curve in the Bode plot intersect.

Characteristic Equation - the characteristic equation of a servo system is $1 + GH = 0$ where G is the transfer function of the forward signal path and H is the transfer function of the feedback signal path.

Circulating Current - the current in armature conductors which are short-circuited during commutation.

Cogging - the non-uniform rotation of a motor armature caused by the tendency of the armature to prefer certain discrete angular positions.

Coupling Ratio - a general term to define the relative motion between motor armature and the driven load.

Critical Damping - a critically damped system's response to a step disturbance is the return to its equilibrium state without overshooting the equilibrium state in minimum possible time.

Crossover Frequency - the frequency at which the magnitude of the product of the forward path gain and the feedback path gain is unity.

Damping (in servo theory) - a term to describe the amplitude decay of an oscillatory signal.

Damping Ratio - a measure of system damping expressed as the ratio between the actual damping and the critical damping.

Dead Band - a range of input signals for which there is no system response.

Decibel (dB) - a measure of system gain (A).

$$A_{dB} = 20 \log_{10} A$$

Dielectric Test - a high voltage test of the motor insulation ability to withstand an AC voltage. Test criterion limits the leakage current to a specified maximum at test voltage of specified magnitude and frequency, applied between the motor case and winding.

Dynamic Braking - a method for braking a DC servomotor by controlling armature current during deceleration.

Efficiency - the ratio of output power to input power.

$$\eta = \frac{P_o}{P_i}$$

Electrical Time Constant - the electrical time constant of a DC servomotor is the ratio of armature inductance to armature resistance.

Fall Time - the time for the amplitude of system response to decay to 37% of its steady-state value after the removal of steady-state forcing signal.

Field Weakening - a method of increasing the speed of a wound field motor by reducing stator magnetic field intensity by reducing magnet winding current.

Flux Biasing - a method for controlling the torque constant of a servomotor by varying the magnetic field intensity of a separate wound field assembly.

Form Factor - the form factor of a harmonic signal is the ratio of its RMS value to its average value in one half-wave.

Full Load Current - the armature current of a motor operated at its full load torque and speed with rated voltage applied.

Full Load Speed - the speed of a motor operated with rated voltage and full load torque.

Gain - the ratio of system output signal to system input signal.

Gain Margin - the magnitude of the system gain at the frequency for which the phase angle of the product of the forward path and feedback path gains is -180°.

Hunting - the oscillation of system response about theoretical steady-state value due to insufficient damping.

Incremental Motion System - a control system which changes the load position in discrete steps rapidly and repetitively.

Inertial Match - an inertial match between motor and load is obtained by selecting the coupling ratio such that the load moment of inertia referred to the motor shaft is equal to the motor moment of inertia.

Lag Network - an electrical network which

increases the delay between system input signal and system output signal.

Lead Network - an electrical network which decreases the delay between system input signal and system output signal.

Lead Screw - a device for translating rotary motion into linear motion consisting of an externally threaded screw and an internally threaded carriage (nut).

Lead Screw, Ball - a lead screw which has its threads formed as a ball bearing race and the carriage contains a circulating supply of balls for increased efficiency.

Linearity - for a speed control system it is the maximum deviation between actual and set speed expressed as a percentage of *set* speed.

Loop Gain - the product of the forward path and feedback path gains.

Mechanical Time Constant - the time for an unloaded motor to reach 63.2% of its final velocity after applying a DC armature voltage.

Motor Constant - the ratio of motor torque to motor input power.

No Load Speed - motor speed with no external load.

Phase-Locked Servo System - a hybrid control system in which the output of an optical tachometer is compared to a reference square wave signal to generate a system error signal proportional to both shaft velocity and position errors.

Phase Margin - the phase angle of the loop gain minus 180° at the crossover frequency.

Pole - a term of root locus plotting for the frequency(-ies) at which system gain goes to infinity.

Safe Operating Area Curve (SOAC) - the boundary on the speed-torque characteristic of a motor inside of which the motor may be operated continuously without exceeding its thermal rating.

Speed Regulation - for a speed control system, speed regulation is the variance in actual speed expressed as a percentage of set speed.

Speed Regulation Constant - the slope of the motor speed-torque characteristic.

System Order - the degree of the system characteristic equation.

System Stiffness - a measure of system accuracy when subjected to disturbance signals.

System Type - is given by the number of poles located at the origin of the loop gain characteristic in the complex plane.

Torque Ripple - the cyclical variation of generated torque at a frequency given by the product of motor angular velocity and number of commutator segments.

Transfer Function - the ratio of the Laplace transforms of system output signal and system input signal.

Zero - a term of root locus plotting for the frequency(-ies) at which system gain goes to zero.

1.2. SYMBOLS

Symbol	Definition	Unit	
		SI	British
a	linear acceleration $a = \frac{dv}{dt}$	m/s^2	in/s^2
A	gain	–	–
A_{dB}	gain expressed in dB	–	–
AC	alternating current (symbol)		
B	magnetic flux density	T	$line/in^2$
C	capacitance	F	F
CW	clockwise rotation		
CCW	counterclockwise rotation		
D	viscous damping factor $D = \frac{T_D}{\omega}$	$Nm/rad\ s^{-1}$	$oz\text{-}in/rad\ s^{-1}$
dB	decibel	–	–
DC	direct current (symbol)		
deg	temperature degree	^oC, K	not used
e	(2.71828) base of natural logarithms	–	–
E	internal voltage (emf)	V	V
E_g	internally generated voltage (counter emf) $E_g = K_E \cdot n$	V	V
f	frequency	Hz	Hz, c/s, cps

Symbol	Definition	Unit	
		SI	British
$f(t)$	real function of time		
$f(s)$	Laplace transform of $f(t)$		
F	force	N, kp	oz, lb
F_f	friction force	N, kp	oz, lb
g	(9.80665 m/s^2 or 386.09 in/s^2) standard gravitational acceleration	m/s^2	in/s^2
$G_m(s)$	motor transfer function		
H	magnetic field intensity (magnetizing force)	A/m	A-turn/in
$i(t)$	current (instantaneous value)	A	A
I	current	A	A
I_a	armature current	A	A
I_{ad}	maximum armature pulse current to avoid demagnetization	A	A
I_{ao}	armature current at no load	A	A
I_{pk}	peak current	A	A
j	($\sqrt{-1}$) imaginary operator	–	–
J	moment of inertia	kg m^2	oz-in-s^2
J_c	capstan moment of inertia	kg m^2	oz-in-s^2
J_L	load moment of inertia	kg m^2	oz-in-s^2

Symbol	Definition	Unit	
		SI	British
J_m	armature (rotating part of motor) moment of inertia	kg m^2	oz-in-s^2
K_D	damping constant (not equal to D) $K_D = \frac{1}{R_m}$	Nm/krpm	oz-in/krpm
K_E	voltage constant	V/krpm, V/rad s^{-1}	V/krpm, V/rad s^{-1}
K_M	motor constant	Nm/W	oz-in/W
K_T	torque constant	Nm/A	oz-in/A
l	length	m	in
L	inductance	H	H
L_a	armature inductance	H	H
m	mass	kg	oz-s^2/in
n	rotational speed, shaft speed	rpm, krpm	rpm, krpm
n_o	shaft speed at no load $n_o = \frac{V}{K_E}$	rpm, krpm	rpm, krpm
N	a) gear ratio (coupling ratio) b) number of coil turns	— —	— —
p	a) number of poles b) pole of transfer function	— s^{-1}	— s^{-1}
P	a) power b) lead screw pitch	W not used	W turns/in

Symbol	Definition	Unit	
		SI	British
P_i	input power	W	W
P_L	power loss $P_L = P_i - P_o$	W	W
P_o	output power	W	W
P_{th}	thermal flux	W	W
PM	permanent magnet (symbol)		
PR	power rate $PR = \frac{T_{os}^2}{J_a}$	$kg\ m^2 s^{-4}$	$oz\text{-}in/s^2$
r	radius	m	in
rad	$(57.29578^o = \frac{360^o}{2\pi})$ radian - unit of plane angle	—	—
R	a) resistance b) motor terminal resistance $R = R_a + R_b$	Ω Ω	Ω Ω
R_a	armature winding resistance	Ω	Ω
R_b	brush resistance	Ω	Ω
R_m	speed regulation constant $R_m = \frac{R}{K_E K_T}$	krpm/Nm	krpm/oz-in
R_{th}	thermal resistance	$^oC/W$	$^oC/W$
s	Laplace operator $s = \sigma + j\omega$	s^{-1}	s^{-1}
SI	international system of units (metric system)		
t	time	s	s
t_i	time interval	s	s

Symbol	Definition	Unit	
		SI	British
t_f	fall time	s	s
t_r	rise time	s	s
t_s	settling time	s	s
t_R	$(1.69\tau_m)$ theoretical reversing time - time interval from full speed in one direction to 63.2% of full speed in reverse direction, when motor supply voltage is stepwise reversed	s	s
T	torque	Nm, kpm	oz-in, lb-in
T_a	accelerating torque	Nm	oz-in
T_D	damping torque $T_D = D \cdot \omega$	Nm	oz-in
T_f	internal friction torque	Nm	oz-in
T_g	internally generated torque $T_g = T_o + T_f + T_D$	Nm	oz-in
T_{gs}	internally generated torque at stall $T_{gs} = T_{os} + T_f$	Nm	oz-in
T_L	load torque	Nm	oz-in
T_{LR}	rated load torque	Nm	oz-in
T_m	motor opposing torque $T_m = T_f + T_D$	Nm	oz-in
T_o	motor output torque $T_o = T_g - T_f - T_D = T_L + T_a$	Nm	oz-in

Symbol	Definition	Unit	
		SI	British
T_{os}	motor output torque at stall $T_{os} = T_{gs} - T_f$	Nm	oz-in
T_{pk}	peak torque $T_{pk} = K_T I_{pk}$	Nm	oz-in
v	linear velocity $v = \frac{dx}{dt}$	m/s	in/s
v(t)	voltage (instantaneous value)	V	V
V	a) voltage b) motor terminal voltage	V V	V V
V_{ad}	maximum terminal voltage a motor can withstand at stall to avoid demagnetization $V_{ad} = I_{ad} R$	V	V
V_b	voltage drop across brushes	V	V
V_{pk}	peak voltage	V	V
V_r	ripple voltage	V	V
V_S	supply voltage	V	V
w	weight (heaviness)	N, kp	oz, lb
W	energy (work)	J	lb-ft
x	linear displacement	m	in
x,y,z	rectangular coordinates	m	in
X	reactance	Ω	Ω
z	zero of transfer function	s^{-1}	s^{-1}
Z	impedance	Ω	Ω

Symbol	Definition	Unit	
		SI	British
a	angular acceleration $a = \frac{d\omega}{dt}$	rad/s^2	rad/s^2
a_m	maximum theoretical acceleration from stall $a_m = \frac{K_T I_{pk} - T_f}{J_a}$	rad/s^2	rad/s^2
ϵ	permittivity (dielectrical constant) $\epsilon = \epsilon_r \epsilon_o$ in SI system	F/m	—
ϵ_o	(8.85416×10^{-12} F/m) permittivity of vacuum	F/m	not used
ϵ_r	relative permittivity	—	—
ζ	damping ratio	—	—
η	efficiency	—	—
θ	angular displacement	rad	rad
Θ	temperature	^oC, K	^oC, ^oF
Θ_a	armature temperature $\Theta_a = \Theta_A + \Theta_r$	^oC	^oC
Θ_A	ambient temperature	^oC	^oC, ^oF
Θ_c	case temperature	^oC	^oC
Θ_h	motor housing temperature	^oC	^oC
Θ_r	temperature rise	^oC	^oC
μ	permeability $\mu = \mu_r \mu_o$ in SI system	H/m	line/A-turn-in
μ_o	(1.25664×10^{-6} H/m) permeability of vacuum	H/m	not used

Symbol	Definition	Unit	
		SI	British
μ_r	relative permeability	—	—
ρ	density	kg/m^3	$oz(mass)/in^3$
σ	real part of s	s^{-1}	s^{-1}
τ	time constant	s	s
τ_e	electrical time constant	s	s
τ_m	mechanical time constant	s	s
τ_{th}	thermal time constant	s	s
φ	phase angle	rad, ^{o}	rad, ^{o}
Φ	magnetic flux	Wb	line
ψ	temperature coefficient of resistance ψ_{al} = 0.00415 (aluminum) ψ_{cu} = 0.00393 (copper)	$1/^oC$, deg^{-1}	$1/^oC$
ω	a) angular velocity $\omega = \frac{d\theta}{dt}$	rad/s	rad/s
	b) angular frequency	s^{-1}	s^{-1}
	c) imaginary part of s	s^{-1}	s^{-1}
o	a) angular degree	—	—
	b) temperature degree	^oC, K	^oF, ^oC

1.3. SYSTEMS OF UNITS

The U.S.A. has been a member nation of the Metric Convention since 1875, but today, except for sporadic use of metric units in science and part of industry, the United States is the only industrialized nation which has not established a national policy committing itself to use of the metric system.

Since 1972, however, the situation has been dramatically changed. The Senate Commerce Committee passed legislation which will convert the United States to the use of international system (SI) of metric measurement. The bill does not establish a mandatory date for the change, but does set the conversion to the metric system within ten years as a national goal.

The editors felt the necessity to bring this second edition up-to-date in this respect. Therefore, in the Appendix A.1. the comprehensive description of SI system is introduced and its basic philosophy is explained.

Further, all equations in the book are presented in SI units and, where necessary, in British units at the same time. Finally, the reader can in Appendix A.2. find exact conversion factors between SI and British units.

The British system of units is used in this book in its technical form, briefly explained in Appendix A.2. Therefore, the units *oz* and *lb* throughout the book are units of force unless otherwise specified.

To help the reader keep both SI and British units in order we have provided all important equations in the book with a listing of units to be used. The following example:

$$P = T\omega \quad [W; Nm, rad/s]$$

shows that this equation is correct only if the torque is expressed in newtonmeters, angular velocity in radians per second, and power in watts. If other units are used, a conversion factor must be introduced. □

Chapter 2
DC Motors and Generators

2.1. BASIC THEORY

HISTORICAL BACKGROUND

The direct current (DC) motor is one of the first machines devised to convert electrical power into mechanical power. Its origin can be traced to disc-type machines conceived and tested by Michael Faraday, the experimenter who formulated the fundamental concepts of electromagnetism.

Faraday's primitive design was quickly improved upon; and many DC machines were built in the 1880's when DC was the principal form of electric power generation. With the advent of 60 Hz AC as the electric power standard in the United States and 50 Hz in Europe, and invention of the induction motor with its lower manufacturing costs, the DC machine became less important. In recent years, the use of DC machines has become almost exclusively associated with applications where the unique characteristics of the DC motor (e.g., high starting torque for traction motor application) justify its cost, or where portable equipment must be run from a DC (or battery) power supply.

The ease with which the DC motor lends itself to speed control has long been recognized. Compatibility with the new thyristor (SCR) and transistor amplifiers, plus better performance due to the availability of new improved materials in magnets, brushes and epoxies, has also revitalized interest in DC machines. The need for new high-performance motors with highly sophisticated capabilities has produced a superabundance of new shapes and sizes quite unlike DC machines of 10 years ago.

CONCEPT OF TORQUE AND POWER

An understanding of electro-mechanical energy conversion, as exemplified by a motor, is based upon acquaintance with several fundamental concepts from the field of mechanics.

The first concept is that of *torque.* If a force is applied to a lever which is free to pivot about a fixed point, the lever, unless restrained, will rotate. The motive action for this rotation, termed torque, is defined as *the product of force and the perpendicular distance from the pivot to the force vector.* This concept is illustrated in Fig. 2.1.1.

Referring to Fig. 2.1.1, it can be seen that by applying this definition we can write an equation for torque (**T**) so that:

$$\mathbf{T} = \mathbf{F} \cdot \mathbf{r} \qquad [\text{Nm; N, m}] \quad \text{or} \quad [\text{oz-in; oz,in}] \qquad (2.1.1)$$

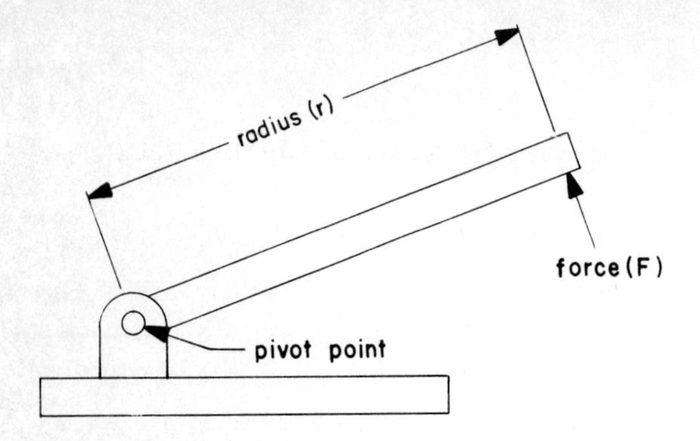

Fig. 2.1.1 Applying a force F to a lever with radius r will produce a torque equal to F·r at the pivot point.

If several forces of different magnitudes are applied at different points along the lever, a resultant torque can be calculated. A counterclockwise torque direction will be adopted as the positive one, clockwise will be negative, and resultant torque will then be the sum of all the torques applied to the lever with the appropriate plus or minus signs assigned to each of the $\mathbf{F}_i \cdot \mathbf{r}_i$ products. Then if we look at Fig. 2.1.2, the resultant torque can be written as:

$$\mathbf{T} = \Sigma \ (\mathbf{F}_i \cdot \mathbf{r}_i) \qquad (2.1.2)$$

$$\mathbf{T} = \mathbf{F}_1\mathbf{r}_1 - \mathbf{F}_2\mathbf{r}_2 - \mathbf{F}_3\mathbf{r}_3 \qquad (2.1.3)$$

This definition of torque, while elementary, nevertheless correctly describes the turning moment which can be observed at the output shaft of an electric motor at stall. The measurable shaft output torque is the resultant one that occurs after the various negative torques (bearing and brush friction torque, gear friction torque, etc.) are subtracted from the developed electromagnetic torque.

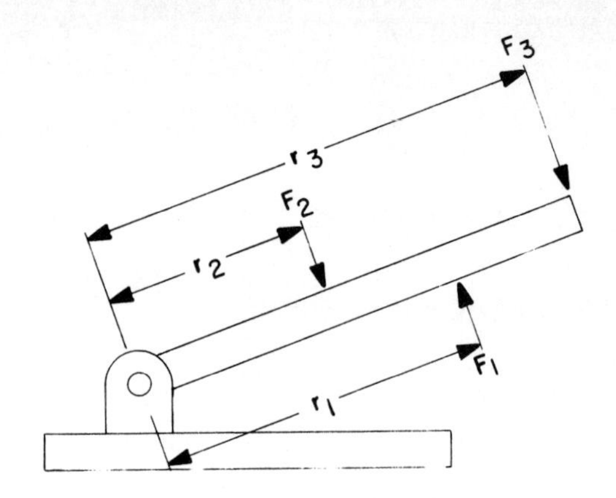

Fig. 2.1.2. The resultant torque of several applied forces at various radii is the sum of all torques applied with appropriate plus and minus signs.

Having established the concept of torque it is then necessary to develop an understanding of *power* which is defined as *the rate at which work is done or energy is expended.*

*Work (*or *energy)* can be also defined in terms of torque acting through a given displacement. If a torque **T** acts through an angle θ, as shown in Fig. 2.1.3, the work performed will be:

$$\mathbf{W} = \mathbf{T} \cdot \theta \qquad [\text{J; Nm, rad}] \text{ or } [\text{oz-in; oz-in, rad}] \qquad (2.1.4)$$

If the above work **W** is performed in the time interval **t**, we can describe the power expended as:

$$\mathbf{P} = \frac{\mathbf{W}}{\mathbf{t}} = \frac{\mathbf{T} \cdot \theta}{\mathbf{t}} \qquad [\text{W;Nm,rad,s}] \qquad (2.1.5)$$

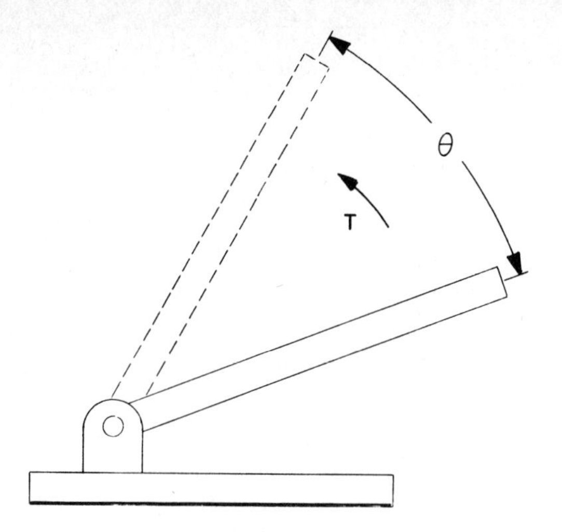

Fig. 2.1.3. Work, defined as a torque acting through a given angular displacement.

LAW OF ELECTROMAGNETIC INDUCTION

If a current-carrying conductor is placed in a magnetic field, a force will act upon it. The magnitude of this force is a function of the magnetic flux density, **B**, the current, **I**, and the orientation of the field and current vectors. Referring to Fig. 2.1.4, the force acting upon the conductor can be written as:

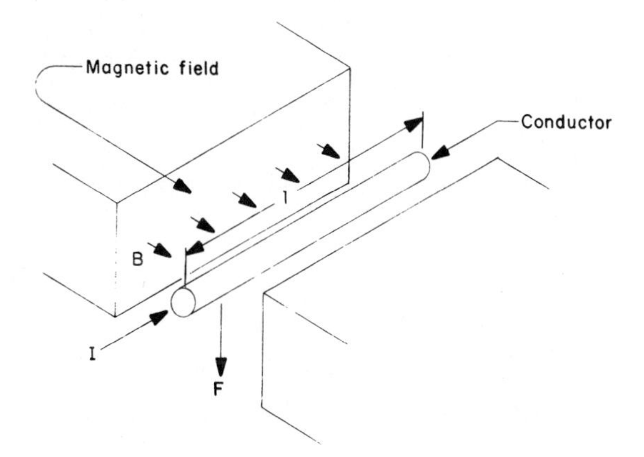

Fig. 2.1.4. A current-carrying conductor in a magnetic field experiences a force acting upon it.

$$\mathbf{F} = \mathbf{B}L\mathbf{I} \qquad [\text{N; T, m, A}] \qquad (2.1.6)$$

where L is the length of the conductor.

The direction in which force on a conductor will act is a function of the direction of current and magnetic field vectors.

Vector notation is a convenient means of including the directional information in an expression for force. If the conductor is considered to be of unit length, electro magnetic force can be written as the cross product of two vectors:

$$\mathbf{F} = \mathbf{I} \times \mathbf{B} \qquad (2.1.7)$$

or

$$|\mathbf{F}| = |\mathbf{I}|\ |\mathbf{B}|\ \sin\theta \qquad (2.1.8)$$

Since two vectors, **B** and **I**, will determine a plane, their vector product will be a vector, perpendicular to the plane, and will be proportional to the sine of the angle between them. This is shown in Fig. 2.1.5. The analogy of the "screw" can also be used to determine the direction. The force vector will point in the direction that a right hand screw advances when its head is rotated from the current vector, **I**, towards the magnetic flux density vector, **B**.

The exact opposite effect on the foregoing discussion is also observed: if a conductor was moved through a magnetic field, an electric voltage appeared (or was generated) across the conductor.

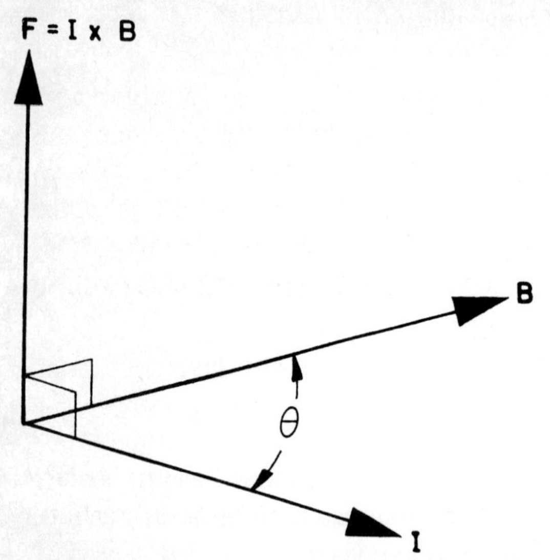

Fig. 2.1.5. The direction of the force on a current carrying conductor is perpendicular to the plane determined by the current and field vectors.

The magnitude of this generated voltage, **E**, was found to depend upon magnetic flux density, **B**, length of conductor, L, and speed of the conductor, **v**, as it moved through the field. An expression for the generated voltage is:

$$\mathbf{E} = \mathbf{B}L\mathbf{v} \qquad [\mathrm{V;T,m,m/s}] \qquad (2.1.9)$$

It occured to Faraday that the best way to obtain a useful function from the electromagnetic effect was to make the conductors a part of a rotational body by fixing them to a shaft which could turn continuously through a magnetic field. The problem of introducing current (motor action) or picking off a voltage (generator action) was solved by the use of sliding contacts called brushes. The type of machine built by Faraday (Fig. 2.1.6) would today be called a homo-polar (single pole) machine, and with other design improvements is used to provide large currents at low voltages.

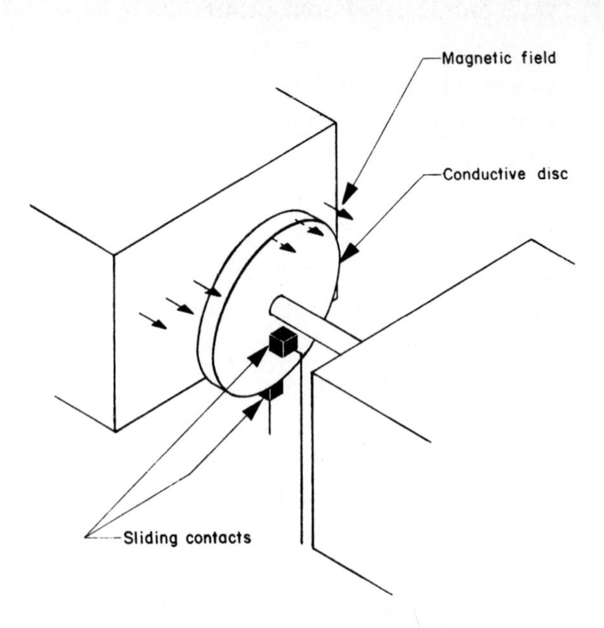

Fig. 2.1.6. An example of the type of machine built by Faraday in the 1880's would be this single pole, or homo-polar machine as it would be called today.

A rudimentary DC motor can be seen to be comprised of three main parts: current-carrying conductors called an *armature;* a *circuit for magnetic field* to provide the means for energy conversion, which can be provided either by permanent magnets or by electromagnets (for the "wound-field" motors); some type of *sliding contact arrangement* for introducing current to moving conductors (usually carbon brushes and commutator).

MAGNETIC CIRCUIT PRINCIPLES

Since electric motor design is based upon the placement of conductors in magnetic field, a discussion of magnetic circuit principles will help understand motor action.

Magnetomotive Force (mmf)

If a conductor were wound into a coil with many turns, the magnetic contribution of each individual turn would add to the magnetic field intensity which exists in the space enclosed by the coil. In this way, extremely strong magnetic fields can be developed. The force which acts to push the magnetic flux through a space is called variously magnetomotance, magnetomotive force, or simply *mmf.* The unit of *mmf* is A-turn and its dimension is [A]. Thus, coils with 10 turns and 2 A, or 5 turns and 4 A will each develop a magnetomotance of 20 A-turn.

Magnetic Flux (Φ)

The term *magnetic flux* is used to describe "how much magnetism" there is in the space around a coil or permanent magnet, or in the air gap of a motor. The unit of magnetic flux is the *line* (or 1 kilo-line = 10^3 lines) in the British system or the *weber* in the SI system. One weber is equal to 10^8 lines.

Magnetic Flux Density (B)

Magnetic flux density is a measure of concentration of magnetic flux in an area. It is defined as the ratio of total flux to the given area and is specified in terms of *lines per square inch* in the British system. The SI unit for magnetic flux density is the *tesla* and is equal to one weber per square meter.

Permeability (μ)

Permeability is the degree to which a medium will support a magnetic field, also stated as the ease with which flux will penetrate a medium. The unit of permeability is *line per A-turn/in* in the British system. In the SI system the permeabiltiy is given as the product

$$\mu = \mu_o \mu_r$$

where μ_o is the permeability of vacuum [H/m] and μ_r is the relative permeability. The relative permeability is used to describe the ability of different materials to support magnetic fields. Vacuum or air has a relative permeability of one, while ferromagnetic materials such as cold-rolled steel will have a relative permeability of several hundred. This means that for a given value of *mmf*, a magnetic circuit which contains iron may have several hundred times the flux of one constructed with non-ferrous materials. This is the reason most motors are built with steel materials.

Magnetic Field Intensity (H)

Magnetic field intensity has the properties of a gradient and gives an indication of how a *mmf* is used up around a magnetic circuit. The unit of magnetic field intensity is *A-turn/in* in the British system and *A/m* in the SI system.

This discussion of electromagnetism is portrayed in Fig. 2.1.7a and 2.1.7b. In Fig. 2.1.7a a coil of N turns is excited by the

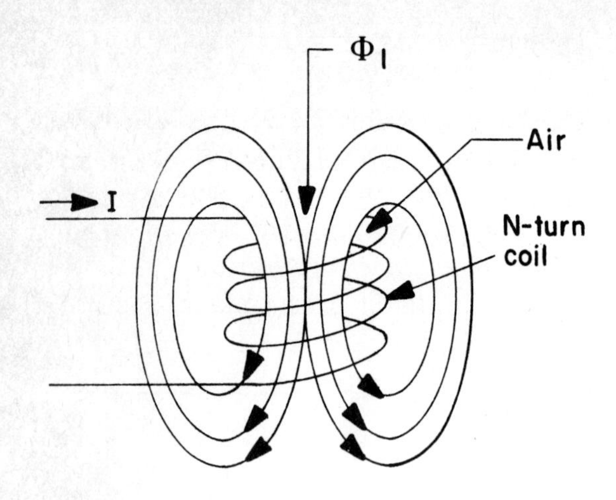

Fig. 2.1.7a. With air surrounding N turns of wire, the magnetic flux is Φ_1.

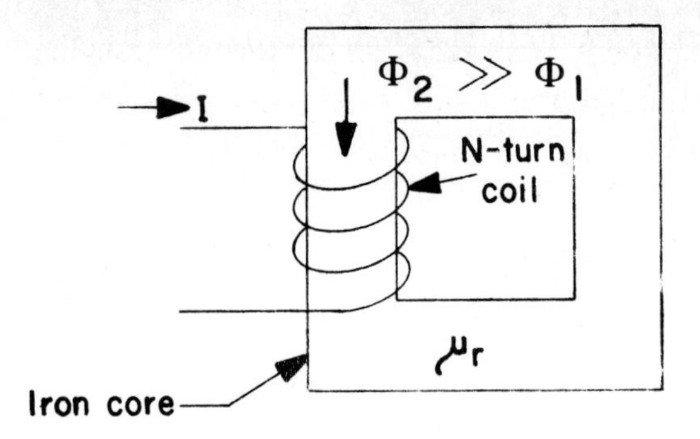

Fig. 2.1.7b. If N turns of wire are placed around an iron core with relative permeability μ_r, the magnetic flux is increased μ_r - times.

current **I**. The resultant flux is Φ_1. If the same coil is now placed around an iron core with relative permeability μ_r, as shown in Fig. 2.1.7b, the magnetic field conditions will change greatly, with flux increasing to $\Phi_2 = \mu_r \Phi_1$.

WORKING EQUATIONS FOR VOLTAGE AND TORQUE

Voltage

Using the laws outlined previously useful expressions can be derived to describe such machine characteristics as torque and generated voltage in terms of design parameters (e.g., number of conductors, magnet size, etc.). We will first derive an expression for generated voltage using the equation **E = Blv** (the voltage generated in a single conductor) as a starting point.

Velocity **v** of an armature conductor can be rewritten as:

$$\mathbf{v} = \mathbf{r}\omega \qquad [\text{m/s; m, rad/s}] \qquad (2.1.10)$$

where ω is angular velocity, and **r** is radius of rotation.

The magnetic flux density **B** can be rewritten in terms of pole flux and area:

$$\mathbf{B} = \frac{\Phi}{\mathbf{A}} \qquad [\text{T; Wb, m}^2] \qquad (2.1.11)$$

$$\mathbf{A} = \frac{2\pi \mathbf{r} L}{\mathbf{p}} \qquad [\text{m}^2\text{; m, m, –}] \qquad (2.1.12)$$

where

r is the radius of rotation of the armature

L is the length of the conductor

p is the number of poles.

Then, the generated voltage per conductor will be:

$$\mathbf{E}' = \frac{\Phi \mathbf{p} \omega}{2\pi} \qquad [\text{V; Wb, –, rad/s}] \qquad (2.1.13)$$

If there are many conductors (z) moving through the field, their cumulative effect will be z-times the above voltage.

$$E = \frac{z\Phi p\omega}{2\pi} \quad [V; -, Wb, -, rad/s] \quad (2.1.14)$$

or

$$E = \frac{z\Phi pn}{60} \quad [V; -, Wb, -, rpm] \quad (2.1.15)$$

Finally, if we introduce the factor **n′**, the number of parallel conductor paths in the armature, the average generated voltage will be:

$$E = \left[\frac{z\Phi p}{60\, n'}\right] n \quad [V; Wb, rpm] \quad (2.1.16)$$

in the SI system, or

$$E = \left[\frac{z\Phi p}{60 \cdot 10^8 n'}\right] n \quad [V; lines, rpm] \quad (2.1.17)$$

in the British system.

The quantity enclosed in brackets in (2.1.16) and (2.1.17) will be a design constant and is commonly called the voltage constant, K_E, of the motor. It is an important motor (or generator) characteristic since it will determine the speed of the motor at a given value of applied voltage. Using the new motor parameter the voltage equation becomes:

$$E_g = K_E n \quad [V; V/krpm, krpm] \quad (2.1.18)$$

Torque

The basic equation for torque is derived from the equation for force (**F = B*L*I**) in a manner similar to that used in obtaining the voltage equation. After the appropriate substitutions are made, we obtain the following expression for torque:

$$T = \left[\frac{z\Phi p}{2\pi n'}\right] I \quad [Nm; Wb, A] \quad (2.1.19)$$

If it is desired to express the torque and magnetic flux in British units, a conversion factor must be introduced and (2.1.19) then becomes:

$$T = \left[2.254 \times 10^{-7} \frac{z\Phi p}{n'}\right] I \quad (2.1.20)$$

[oz–in; lines, A]

The quantity in brackets in (2.1.19) and (2.1.20) will also be a constant for a given motor design and is called the torque constant, K_T, of the motor. Substituting, the torque equation becomes:

$$T = K_T I \quad [Nm; Nm/A, A] \quad (2.1.21)$$

or

[oz–in; oz–in/A, A]

If the torque and voltage equation are each solved for a common factor, the pole flux, Φ, and the results equated, we obtain other useful relationships:

$$K_T = K_E \quad [Nm/A; V/rad\ s^{-1}] \quad (2.1.22)$$

$$K_T = 9.5493 \times 10^{-3}\ K_E \qquad (2.1.23)$$

[Nm/A; V/krpm]

$$K_T = 1.3524\ K_E \qquad (2.1.24)$$

[oz–in/A; V/krpm]

COMMUTATION

A practical machine based upon the principles discussed earlier, for maximum performance and minimum cost, must get many conductors into the magnetic field existing around its armature. This allows the cumulative effect of many conductors to add to the output torque and/or voltage characteristic of the machine. A practical machine will therefore have the entire armature surface covered with conductors. Or, as in the case with conventionally constructed motors, the conductors are placed in deep slots of an iron laminated rotor structure. In either case, there are many conductors enveloping the entire periphery of the rotor structure. This multiplicity of conductors leads to another constructional feature of the motor: the *commutator.*

Recalling the vector diagram of Fig. 2.1.5, it was seen that in order to have the force vector **F** remain constant in magnitude and direction, it is necessary to have the vectors **B** and **I** remain fixed in space with respect to each other. This condition is not met if direction of current through the conductors were to remain as the rotor turns on its axis of rotation. In Fig. 2.1.8, current flowing into the paper is indicated by a circled **x**, while current flow out of the paper is indicated by a circled dot. The rotor member shown in Fig. 2.1.8 has six full pitch coils equally spaced around its periphery.

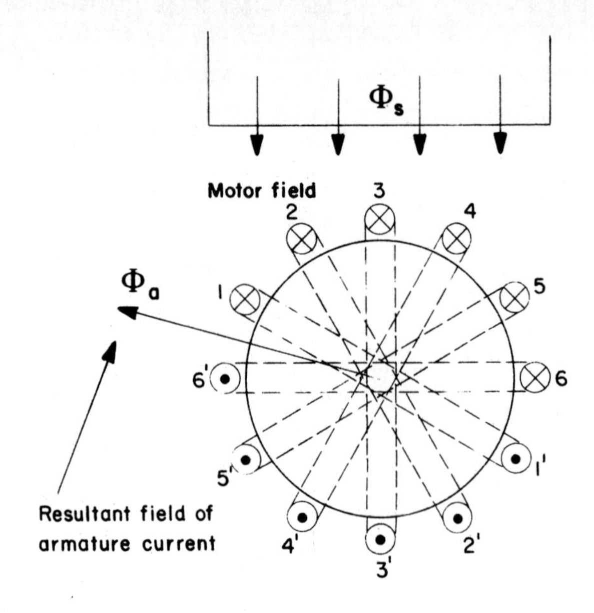

Fig. 2.1.8. By means of a commutator switching current in the rotor, the vector of resultant electromagnetic field remains fixed in space.

As the rotor turns, the resultant electromagnetic field will rotate with it unless some provision is made to switch the direction of current in individual coils as they pass a fixed point in space. This switching is accomplished by means of a commutator. The most commonly used commutator is a cylinder, consisting of segments of conductive material (usually copper) interspersed with, and insulated from one another, by an insulating material, as shown in Fig. 2.1.9.

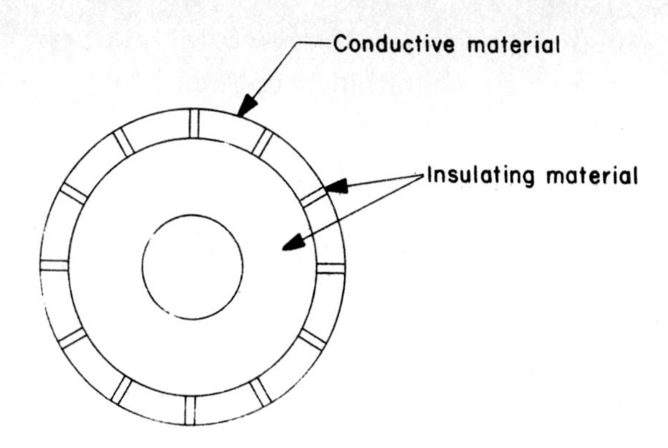

Fig. 2.1.9. The end view of a commutator showing the interspersing of conductive and insulating materials.

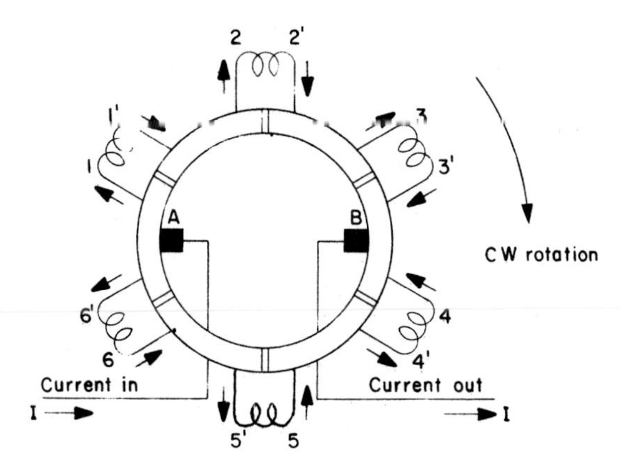

Fig. 2.1.10. A representation of current flow through a motor armature.

The machine conductors (which constitute the armature) are connected in sequence to segments of the commutator. The connection of the coils to the commutator is shown symbolically in Fig. 2.1.10.

Current flow is in at brush A and out at brush B. The small radially oriented arrows show current direction in the individual coil sides. If the rotation is clockwise, it can be seen that 1/6 of a revolution after the instant shown, the current in coils 3-3′ and 6-6′ will have changed directions. As successive commutator segments pass under the brushes, their current directions will also change. As a result of this switching (commutation), current flow in the armature occupies a fixed position in space independent of rotation and a steady unidirectional torque will result.

Referring again to Fig. 2.1.10, it can be seen that at an instant of time, each of the brushes may be contacting two adjacent commutator segments. At this condition the coil connected to those two segments will be "shorted" through the brush. The result is a short circuit current through the coil. In some cases this short circuit current can produce undesirable vibration.

By action of the commutator, the wound armature of the DC motor can be regarded, as shown in Fig. 2.1.11, as a wound core with an axis of magnetization fixed in space.

The axis of magnetization is determined by the position of the brushes and for this reason is also called the brush axis. In order for a motor to be bidirectional, i.e., have equal characteristics for both directions of rotation, the brush axis must be at an angle of 90 electrical degrees with the main field, in a position called the neutral position or neutral axis. Most DC motors are built so that the structure holding the brushes can be rotated over a limited range. Then after the motor is assembled, the brush axis can be adjusted to the neutral position by actual observance of machine characteristics. If a machine is not required to function in a bidirectional role, it is possible to move the brush axis sightly off the neutral position so that commutation is optimum for one direction of rotation. □

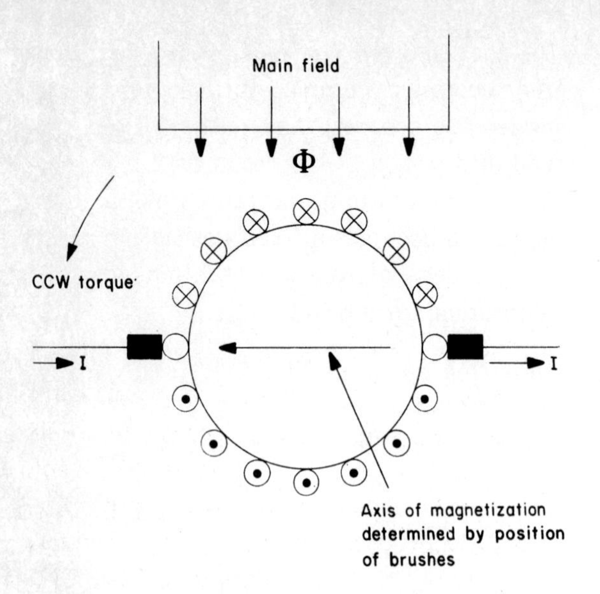

Fig. 2.1.11. The brush axis, or neutral axis of a motor is determined by the position of the brushes.

2.2. MOTOR COMPARISON

There are two large families of DC motors, the *integral horsepower* types having power ratings of one horsepower or more, and the *fractional horsepower* motors, with power ratings of less than one horsepower. For the purposes of this manual, we will confine our considerations to the latter classification.

Fractional horsepower DC motors can be subdivided into types based on the type of *motor magnetic field* employed (permanent magnet, or wound in straight series, split series, shunt, compound), with further delineation based on *duty cycle* (continuous or intermittent), *cooling method* or *enclosure* (fan-cooled, no cooling, closed frame, open frame), and by *hot-spot temperature* (the hottest part of the motor insulation).

GENERAL DESCRIPTION

The class of fractional horsepower DC motors which utilize *electromagnets* to generate the stator magnetic field are called *wound-field motors.* The electromagnets can be energized individually or in conjunction with the armature. When the electromagnets are energized in conjunction with the armature, the motors are known as *self-excited* motors. Various configurations of electromagnet windings for the self-excited motors are possible, such as: series, shunt, or compound.

The class of fractional horsepower DC motors which utilize *permanent magnets* differs from wound-field types in that no external power is required in the stator structure. These various motors are described and compared below.

Wound-Field Motors

Straight-series motors provide very large torque at start-up due to their use of coils in series with the armature to produce the stator magnetic flux. Because the field winding carries the full armature current, it consists of a few turns of heavy gage wire. As motor speed increases, current reduces and so does the stator magnetic flux. This in turn causes another increase in the motor speed. Under no-load conditions this type of motor could theoretically run away because of the steep speed curve, but internal friction and losses in the windings provide sufficient load to hold motor speed within safe operating limits.

The straight-series motor is usually employed where large starting torques are required. A typical straight-series motor is shown in Fig. 2.2.1, along with its speed-torque curve.

Split-series motors are quite similar to straight-series motors, except that they have two *field coils,* oppositely *connected.* This feature accommodates applications where rapid polarity switching is desired for rapid changes of direction. A representative application for motors of this type is their use in aircraft actuators. A typical split-series motor is shown in Fig. 2.2.2.

Shunt motors have armature and field coils connected in parallel. Line current in the

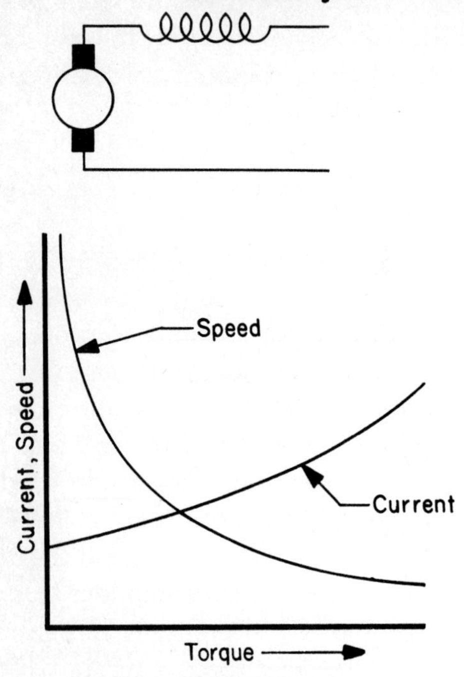

Fig. 2.2.1. Straight-series motor.

armature is a function of the load configuration. The shunt motor has in the past been popular for both fixed speed and

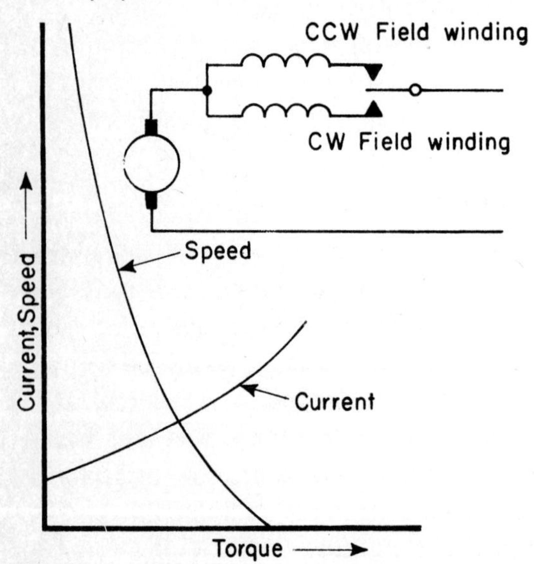

Fig. 2.2.2. Split-series motor.

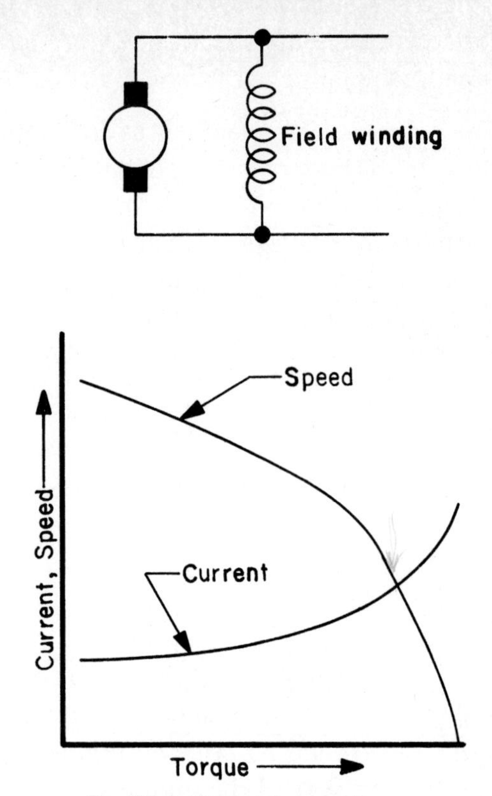

Fig. 2.2.3. Typical shunt motor.

variable speed applications. The torque-speed characteristics are non-linear at higher current levels, as shown in Fig. 2.2.3.

Compound motors have both series and shunt field windings. When the series winding aids the shunt winding, the motor is termed a "cumulative compound" motor, and when the series winding opposes the shunt winding, the motor is termed a "differential compound" motor. In general, small compound motors have a strong shunt field and a weak series field to help start the motor.

The compound motor exhibits high starting torque and relatively flat speed-torque characteristic at rated load. In reversing applications, the polarity of both fields, or of the armature, must be switched. Because rather complex circuits are required for control, most compound motors built are large bidirectional types. A typical compound motor is shown in Fig. 2.2.4.

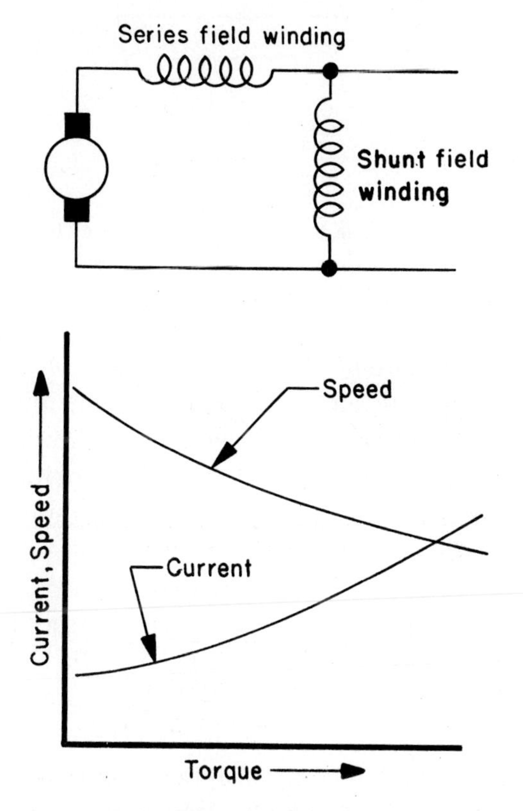

Fig. 2.2.4. Typical compound motor.

Permanent Magnet Motors

Since the stator magnetic field of Permanent Magnet (PM) motors is generated by permanent magnets, no power is used in the field structure. The stator magnetic flux remains essentially constant at all levels of armature current and, therefore, the speed-torque curve of the PM motor is linear over an extended range (see Fig. 2.2.5). With

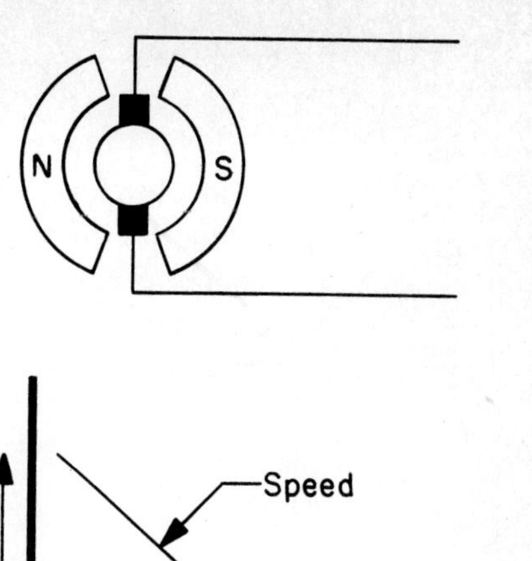

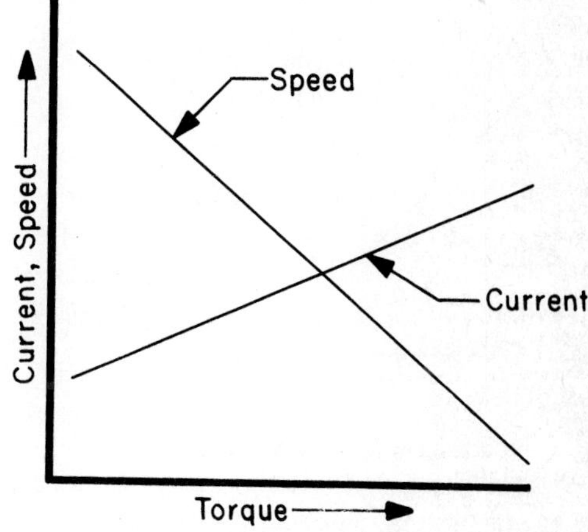

Fig. 2.2.5. Typical permanent magnet motor.

modern ceramic magnets, the stalled torque will tend to be higher and the speed-torque curve will tend to be more linear than for a comparable wound-field motor.

A comparison of a permanent magnet motor and a shunt motor is shown in Fig. 2.2.6. Note the non-linear characteristic of the wound-field motor at higher torque levels.

The reason for the non-linear torque-speed curve in the case of the shunt motor is that the armature reaction flux (which always is orthogonal to the main stator flux in any DC motor) tends to follow the low-reluctance path through the pole shoe, and at higher current levels causes a net effect of an angular shift in pole location and a lower effective flux level. This is illustrated in Fig. 2.2.7.

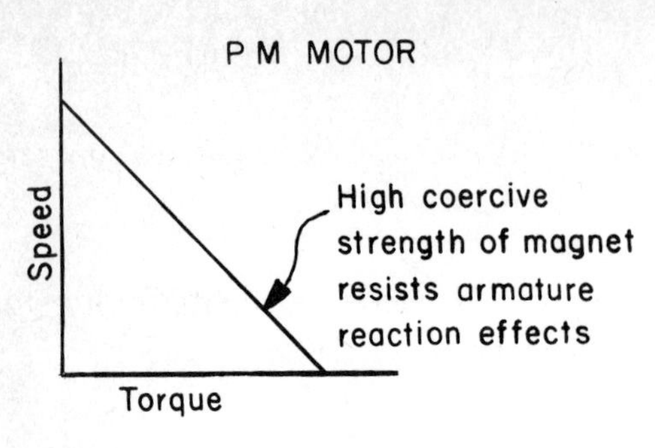

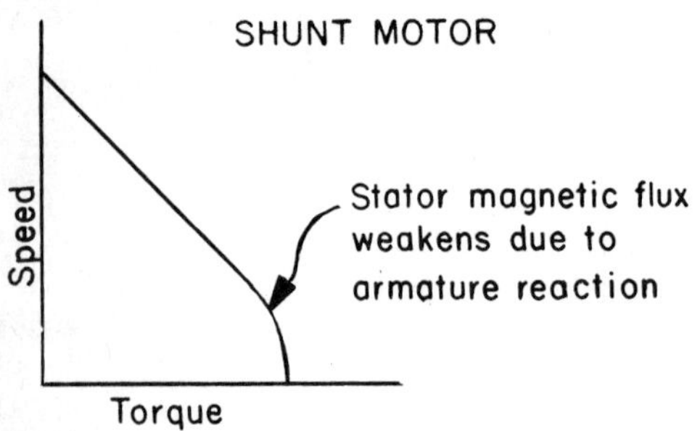

Fig. 2.2.6. Stalled torque tends to be higher, and speed-torque curves more linear for the permanent magnet motors, compared to a shunt motor.

In the case of the ceramic permanent magnet (Fig. 2.2.8.), the armature reaction flux stays orthogonal to the permanent magnet flux, since the permeability of ceramic

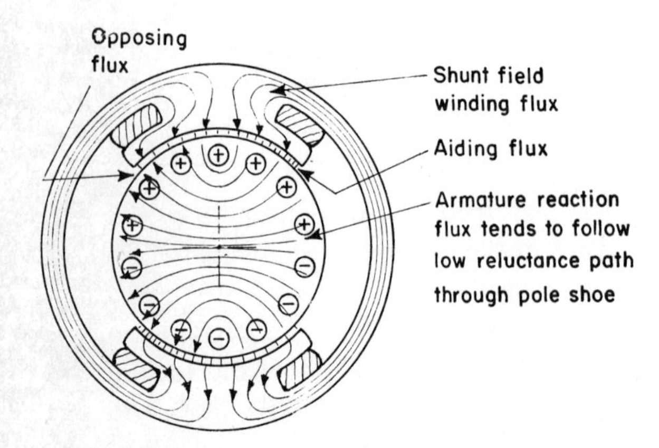

Fig. 2.2.7. Demagnetization effect in wound-field motors is caused by a component of the armature reaction field.

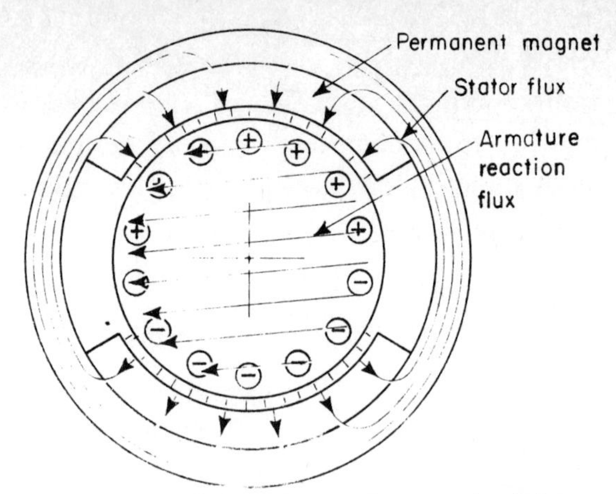

Fig. 2.2.8. Due to the coercive strength of the permanent magnet motor, it is virtually insensitive to demagnetization effects.

magnet material is very low (almost equal to that of air). In addition, the high coercive force of the magnet material resists any change in flux whenever the armature reaction field enters. The result is a linear torque-speed characteristic.

The PM motor offers several advantages in addition to those discussed above. Perhaps the most obvious advantage is that electrical power need not be supplied to generate the stator magnetic flux. Since the conversion of electrical power to mechanical power takes place in the armature winding, the power supplied to the field winding results mostly in an RI^2 loss (heat loss) in the winding itself. The PM motor thus simplifies power supply requirements, while at the same time it requires less cooling.

Another benefit of the PM motor is a reduced frame size for a given output power. This characteristic is also portrayed in Figs. 2.2.7 and 2.2.8. Because of the high coercive strength of permanent magnets, their radial dimension is typically one-fourth that of the wound-field motor for a given air gap.

The significant advantages of PM motors over wound-field types are summarized as follows:

1. Linear torque-speed characteristic.
2. High stall (accelerating) torque.
3. No need for electric power to generate the magnetic flux.
4. A smaller frame and lighter motor for a given output power. □

2.3. MOTOR EQUATIONS AND TRANSFER FUNCTION

This section presents the equations of a DC motor and the derivation of its transfer function. It is assumed that the reader is familiar with the Laplace transformation and with the concept of the transfer function. Readers who wish to review this material are directed to the first section of Chapter 4. For simplicity in notations, the Laplace transform of a time variable is denoted by the same letter, followed by the argument **s**. For example, the Laplace transform of **x(t)** is **x(s)**.

The description of motor equations starts with the relationships between electrical variables. These are known as the *electrical equations.*

2.3.1. ELECTRICAL EQUATIONS

The first characteristic to be considered is the motor impedance. The best way to determine this is by direct measurements. Locking the motor shaft at fixed position, and applying sinusoidal voltage at varying frequencies, one can measure the resulting current and evaluate the impedance as the ratio of the voltage to the current.

When this is done, it appears that the *motor impedance at stall* is equal to a resistance, **R**, in series with a parallel combination of an inductance, L_a, and another resistance, R_L. When the motor rotates, the armature coils move in the stator magnetic field. The induced emf appears across the armature terminals as internally generated voltage (counter emf), E_g. Therefore, the equivalent electric circuit of the motor is the impedance at stall, connected in series to a voltage source, E_g. This equivalent circuit is shown in Fig. 2.3.1. The physical explanation for

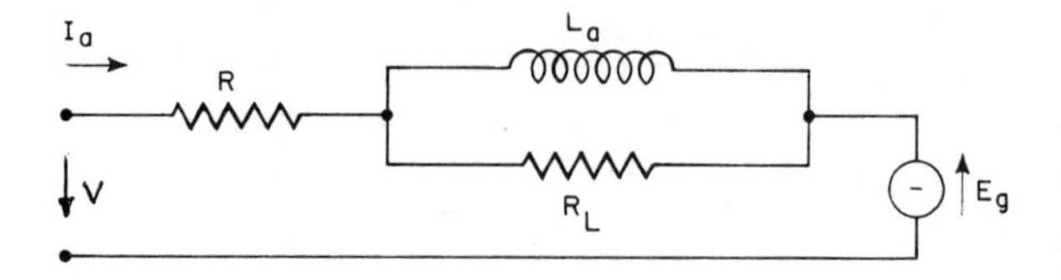

Fig. 2.3.1. Motor equivalent circuit.

this model is that R_L represents the losses in the magnetic circuit. This model was found to be accurate for the motor. However, the resistance R_L is usually larger than **R** (typically, about 5-10 times), and hence, the effect of R_L on the motor operation is insignificant. Therefore, it is possible to ignore this resistance in most practical applications, and approximate the motor equivalent circuit by **R**, L_a, and E_g. This assumption will be made throughout the book. However, in moving coil motors, the resistance R_L affects the measurement of the inductance L_a, and must be considered. Let the motor voltage and current be **V** and I_a, respectively. The relation between these variables is given by

$$V = L_a \frac{dI_a}{dt} + RI_a + E_g \tag{2.3.1}$$

where E_g, the internally generated voltage, is proportional to the motor velocity, ω.

$$E_g = K_E \omega \tag{2.3.2}$$

The above equations can be combined into

$$V = L_a \frac{dI_a}{dt} + RI_a + K_E\omega \qquad (2.3.3)$$

Equation (2.3.3) is known as the *electrical equation of the motor.*

2.3.2. DYNAMIC EQUATIONS

Since the magnetic field in the motor is constant, the current produces a proportional torque

$$T_g = K_T I_a \qquad (2.3.4)$$

Let us denote the moment of inertia of the motor by J_m, and let T_f represent the constant friction torque in the motor. Also, denote all the viscous friction torques and other torques which are proportional to the velocity by $D\omega$. Then, the opposing torque in the motor, T_m, is given by

$$T_m = T_f + D\omega \qquad (2.3.5)$$

Now assume that the motor is coupled to a load. Denote the load moment of inertia by J_L and the load opposing torque by T_L. The relation between the torques and velocity is the following:

$$T_g = (J_m + J_L)\frac{d\omega}{dt} + D\omega + T_f + T_L \qquad (2.3.6)$$

Equation (2.3.6) is the dynamic equation of the motor, and along with (2.3.3) and (2.3.4), it describes the relations between the electrical and mechanical variables. Note that equation (2.3.6) is based on a tacit assumption that motor velocity is the same as that of the load. While this assumption holds in most cases, for high-performance servo systems one has to investigate the case where the deflections of the motor shaft and other elastic parts lead to *torsional resonance.* The following discussion of the transfer function is based on the equal velocity assumption, and section 2.3.4. describes the torsional resonance.

2.3.3. MOTOR TRANSFER FUNCTION

Equation (2.3.3), (2.3.4), and (2.3.6) provide a model for the motor which describes the relation between its variables. However, when the motor is used as a component in a system, it is desired to describe it by the appropriate *transfer function* between the motor voltage and its velocity. For this purpose we can assume that $T_L = 0$ and $T_f = 0$, since neither affects the transfer function. If we now apply the Laplace transformation to the motor equations, we get:

$$V(s) = (sL_a + R)\, I_a(s) + K_E\omega(s) \qquad (2.3.7)$$

$$T_g(s) = K_T I_a(s) \qquad (2.3.8)$$

$$T_g(s) = (J_m + J_L)\, s\omega(s) + D\omega(s) \qquad (2.3.9)$$

Let us define the total moment of inertia, J, by

$$J = J_m + J_L \qquad (2.3.10)$$

Now combine (2.3.8) with (2.3.9) to obtain an expression for the current:

$$I_a(s) = \frac{1}{K_T} (sJ + D)\, \omega(s) \tag{2.3.11}$$

Next, combine (2.3.11) with (2.3.7) to form

$$V(s) = \frac{1}{K_T} (sL_a + R)(sJ + D)\, \omega(s) + $$

$$+ K_E \omega(s) \tag{2.3.12}$$

and the corresponding transfer function is

$$G_m(s) = \frac{\omega(s)}{V(s)}$$

$$= \frac{K_T}{(sL_a + R)(sJ + D) + K_E K_T} \tag{2.3.13}$$

The transfer function has two poles, which, in all practical cases, are negative and real. Therefore, it may be written as

$$G_m(s) = \frac{1/K_E}{(s\tau_1 + 1)(s\tau_2 + 1)}$$

$$= \frac{K_T}{JL_a} \frac{1}{(s - p_1)(s - p_2)} \tag{2.3.14}$$

τ_1 and τ_2 are the time constants, whereas p_1 and p_2 are the poles of the transfer function. The poles are the roots of the characteristic equation

$$s^2 L_a J + s(L_a D + RJ) + RD + K_E K_T = 0 \tag{2.3.15}$$

The time constants are related to the poles by

$$\tau_1 = -\frac{1}{p_1} \qquad \tau_2 = -\frac{1}{p_2} \tag{2.3.16}$$

Clearly, the time constants τ_1 and τ_2 are functions of all the motor parameters.

In the case where some of the parameters are negligible, the transfer function may be simplified considerably. For example, the following discussion considers a common case where the damping factor, **D**, is negligible and the inductance, L_a, is small.

Low Inductance and No Damping

If we assume that the motor damping factor is zero, **D = 0**, the transfer function (2.3.13) becomes

$$G_m(s) = \frac{\omega(s)}{V(s)} = \frac{K_T}{s^2 L_a J + sRJ + K_E K_T} \tag{2.3.17}$$

The poles in this case are the roots of

$$s^2 L_a J + sRJ + K_E K_T = 0 \tag{2.3.18}$$

These are

$$p_{1,2} = \frac{-RJ \pm \sqrt{(RJ)^2 - 4 L_a J K_E K_T}}{2L_a J} \tag{2.3.19}$$

In all practical motors the inductance, L_a, is small so that

$$R^2J^2 - 4L_aJ\,K_EK_T > 0 \tag{2.3.20}$$

and therefore the two poles are negative real. They are defined as

$$p_1 = \frac{-RJ + \sqrt{(RJ)^2 - 4L_aJ\,K_EK_T}}{2L_aJ} \tag{2.3.21}$$

and

$$p_2 = \frac{-RJ - \sqrt{(RJ)^2 - 4L_aJ\,K_EK_T}}{2L_aJ} \tag{2.3.22}$$

When the inductance L_a is much smaller than the term $\frac{R^2J}{K_EK_T}$, one can use the approximation

$$\sqrt{1-x} \approx 1 - \frac{x}{2} \tag{2.3.23}$$

which is valid for small values of **x**.

The approximation (2.3.23) gives

$$\sqrt{(RJ)^2 - 4L_aJ\,K_EK_T}$$

$$= RJ\sqrt{1 - \frac{4L_a\,K_EK_T}{R^2J}}$$

$$= RJ\left(1 - \frac{2L_a\,K_EK_T}{R^2J}\right) \tag{2.3.24}$$

Now, when (2.3.24) is substituted in (2.3.21) and (2.3.22), it is found that

$$p_1 = \frac{-RJ + RJ\left(1 - \frac{2L_a\,K_EK_T}{R^2J}\right)}{2L_aJ}$$

$$= \frac{-K_EK_T}{RJ} \tag{2.3.25}$$

and

$$p_2 = \frac{-RJ - RJ\left(1 - \frac{2L_aK_EK_T}{R^2J}\right)}{2L_aJ} \approx \frac{-2RJ}{2L_aJ}$$

$$= \frac{-R}{L_a} \tag{2.3.26}$$

In view of (2.3.25) and (2.3.26), the transfer function can be written as

$$G_m(s) = \frac{K_T/L_aJ}{\left(s + \frac{K_EK_T}{RJ}\right)\left(s + \frac{R}{L_a}\right)} \tag{2.3.27}$$

or

$$G_m(s) = \frac{1/K_E}{(s\tau_m + 1)(s\tau_e + 1)} \tag{2.3.28}$$

where

$$\tau_m = \frac{RJ}{K_EK_T} \quad \text{[Metric Units]} \tag{2.3.29}$$

$$\tau_m = \frac{10_4\,RJ}{K_EK_T} \quad \text{[English Units]}$$

is the *mechanical time constant,* and

$$\tau_e = \frac{L_a}{R} \tag{2.3.30}$$

is the *electrical time constant.*

Note, however, that the poles of the transfer function will be negative reciprocals of τ_m and τ_e only if the two time constants are sufficiently different, or if

$$\tau_m > 10\tau_e \qquad (2.3.31)$$

However, if the two time constants are close (as for the Electro-Craft ET-4000 motor), the poles will be functions of the motor parameters, and equation (2.3.18) is to be used to determine them.

Furthermore, if the motor inductance is large enough so that

$$4L_a J K_E K_T > R^2 J^2 \qquad (2.3.32)$$

the two poles of the transfer function become complex. However, this is not the case for the majority of DC motors.

2.3.4. TORSIONAL RESONANCE

In the previous discussion of the motor equation it was assumed that the velocity of all the parts in the motor-load system is identical; i.e., the motor and load may be approximated by a single body. This assumption is not accurate for high-performance servo systems, since the mechanical parts of the system are elastic, and they deflect under torque. Consequently, the instantaneous velocities of various parts are different, and at some frequencies will be in opposite directions. This condition allows the system to store a large amount of mechanical energy, which results in noticeable angular vibrations. This phenomenon is called *torsional resonance.*

Since many motors are coupled to both an inertial load and a tachometer, it is reasonable to approximate this mechanical system by three solid bodies, coupled by inertialess shafts. The approximate system is shown in Fig. 2.3.2. Let us denote the

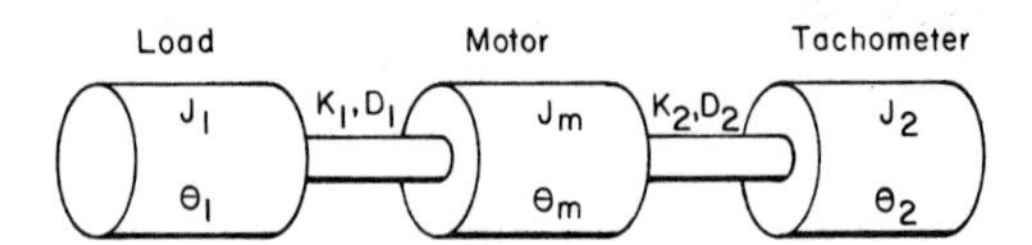

Fig. 2.3.2. Three-body model for load-motor-tachometer system.

moments of inertia and angular positions of the load, motor, and tachometer by J_1, J_m, J_2, and θ_1, θ_m, θ_2, respectively. Also, denote the stiffness and damping factors of the equivalent shafts by K_1, D_1 and K_2, D_2. Note that this model is an overall approximation and, hence, K_1 and D_1 may be influenced by characteristics of the motor armature, shaft, coupling, or the load.

In order to derive the dynamic equations for this system, let T_1 and T_2 be the torques delivered from the motor to J_1 and J_2, respectively. The dynamic equations of the three bodies are:

$$T_g = J_m\ddot{\theta}_m + D\dot{\theta}_m + T_1 + T_2 \qquad (2.3.33)$$

$$T_1 = J_1\ddot{\theta}_1 \qquad (2.3.34)$$

$$T_2 = J_2\ddot{\theta}_2 \qquad (2.3.35)$$

The deflections of the shafts are described by the following equations:

$$T_1 = K_1 (\theta_m - \theta_1) + D_1 (\dot{\theta}_m - \dot{\theta}_1) \quad (2.3.36)$$

$$T_2 = K_2 (\theta_m - \theta_2) + D_2 (\dot{\theta}_m - \dot{\theta}_2) \quad (2.3.37)$$

It may be useful to consider the electrical analog of this mechanical system. The analogy between the variables is the following:

A torque T_g is equivalent to a proportional current.

A velocity $\dot{\theta}$ is equivalent to a proportional voltage.

A moment of inertia **J** is equivalent to a proportional capacitance.

A damping factor **D** is equivalent to resistance of **1/D**.

A stiffness factor **K** is equivalent to inductance of **1/K**.

Then the electrical circuit analogous to the three-body mechanical system of Fig. 2.3.2 is shown in Fig. 2.3.3. The analogy may be helpful in understanding the model. However, it is not necessary for the derivation of the equations.

If we substitute (2.3.34) and (2.3.35) for the torques T_1 and T_2, the dynamic equations become

$$T_g = J_m \ddot{\theta}_m + D\dot{\theta}_m + J_1 \ddot{\theta}_1 + J_2 \ddot{\theta}_2 \quad (2.3.38)$$

$$J_1 \ddot{\theta}_1 + D_1 \dot{\theta}_1 + K_1 \theta_1 = D_1 \dot{\theta}_m + K_1 \theta_m \quad (2.3.39)$$

$$J_2 \ddot{\theta}_2 + D_2 \dot{\theta}_2 + K_2 \theta_2 = D_2 \dot{\theta}_m + K_2 \theta_m \quad (2.3.40)$$

Now consider the electrical equations,

$$V = L_a \frac{dI_a}{dt} + RI_a + K_E \dot{\theta}_m \quad (2.3.41)$$

$$T_g = K_T I_a \quad (2.3.42)$$

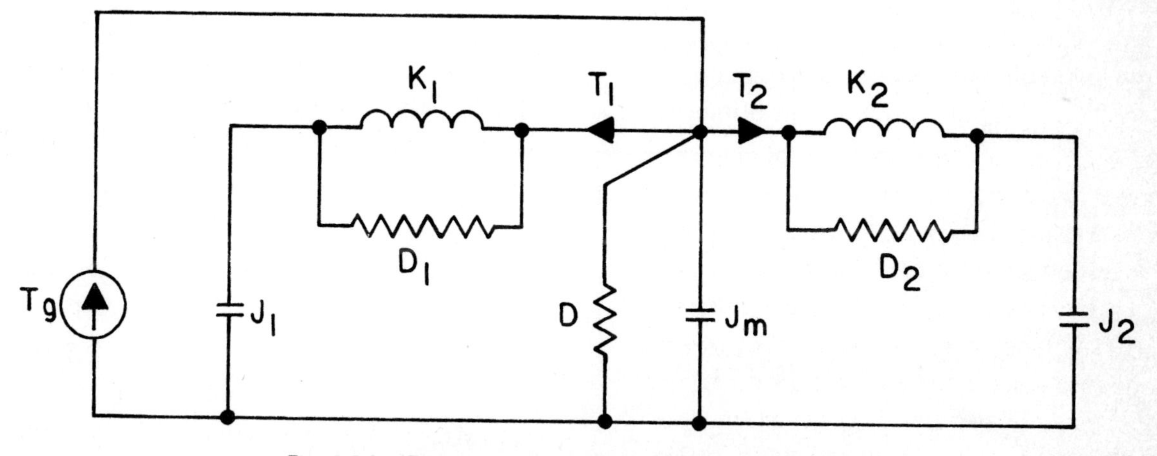

Fig. 2.3.3. Electrical analogy of three-body mechanical system.

The above equations describe the behavior of the system. In order to simplify the analysis, substitute (2.3.42) in (2.3.38) and apply the Laplace transformation to all the equations.

$$(sL_a+R)\,I_a(s) + K_E s\theta_m(s) = V(s)$$

$$-K_T\,I_a(s) + (s^2J_m+sD)\theta_m(s) + $$

$$+ s^2J_1\theta_1(s) + s^2J_2\theta_2(s) = 0$$

$$(sD_1+K_1)\theta_m(s) -$$

$$- (s^2J_1+sD_1+K_1)\theta_1(s) = 0$$

$$(sD_2+K_2)\theta_m(s) -$$

$$- (s^2J_2+sD_2+K_2)\theta_2(s) = 0$$

(2.3.43)

The system equations (2.3.43) may be written in matrix form for simplification.

(See Below)

The matrix equation (2.3.44) is a complete model and includes (2.3.13) as a special case. Note also that if the mechanical system is simple and can be approximated by two bodies J_m and J_1, we can obtain the corresponding model by crossing the fourth row and column in the matrix of (2.3.44).

Equation (2.3.44) enables us to derive the transfer function between the input voltage, V, and the velocities of the three bodies. The transfer function of the motor velocity is given by (2.3.45), whereas (2.3.46) is the transfer function for either the tachometer or the load velocity (when subscripts 1 and 2 are interchanged).

$$G_m(s) = \frac{s\theta_m(s)}{V(s)}$$

$$= \frac{1}{\Delta}\Big[K_T(s^2J_1+sD_1+K_1)\cdot$$

$$\cdot(s^2J_2+sD_2+K_2)\Big] \quad (2.3.45)$$

$$G_2(s) = \frac{s\theta_2(s)}{V(s)}$$

$$= \frac{1}{\Delta}\Big[K_T(s^2J_1+sD_1+K_1)(sD_2+K_2)\Big]$$

(2.3.46)

where Δ is the determinant of the matrix in (2.3.44), divided by s.

$$\begin{bmatrix} sL_a+R & sK_E & 0 & 0 \\ -K_T & s^2J_m+sD & s^2J_1 & s^2J_2 \\ 0 & sD_1+K_1 & -(s^2J_1+sD_1+K_1) & 0 \\ 0 & sD_2+K_2 & 0 & -(s^2J_2+sD_2+K_2) \end{bmatrix} \begin{bmatrix} I_a(s) \\ \theta_m(s) \\ \theta_1(s) \\ \theta_2(s) \end{bmatrix} = \begin{bmatrix} V(s) \\ 0 \\ 0 \\ 0 \end{bmatrix} \quad (2.3.44)$$

$$\Delta = (sL_a+R)\Big[(sJ_m+D)(s^2J_1+sD_1+K_1)(s^2J_2+sD_2+K_2)+$$
$$+ sJ_2\,(s^2J_1+sD_1+K_1)(sD_2+K_2)+$$
$$+ sJ_1\,(sD_1+K_1)(s^2J_2+sD_2+K_2)\Big]+$$
$$+ K_E K_T (s^2J_1+sD_1+K_1)(s^2J_2+sD_2+K_2) \quad (2.3.47)$$

Using the notations,

$$R_i(s) = s^2J_i + sD_i + K_i \qquad i = 1, 2$$
$$(2.3.48)$$
$$Q_i(s) = sD_i + K_i \qquad i = 1, 2$$

The transfer functions may be written as

$$G_m(s) = \frac{s\theta_m(s)}{V(s)}$$

$$= \frac{1}{\Delta}\Big[K_T\, R_1(s)\, R_2(s)\Big] \quad (2.3.49)$$

$$G_2(s) = \frac{s\theta_2(s)}{V(s)}$$

$$= \frac{1}{\Delta}\Big[K_T\, R_1(s)\, Q_2(s)\Big] \quad (2.3.50)$$

where

$$\Delta = (sL_a+R)\Big[(sJ_m+D)R_1R_2+sJ_2R_1Q_2+sJ_1Q_1R_2\Big]+$$
$$+ K_E K_T R_1 R_2 \quad (2.3.51)$$

Note that the two transfer functions, (2.3. 49) and (2.3.50), have the same poles, but the zeros may be different. Let us consider the zeros and the poles of the tachometer velocity transfer function.

The zeros are the roots of the equation

$$(s^2J_1 + sD_1 + K_1)(sD_2 + K_2) = 0$$
$$(2.3.52)$$

This equation has one real root and two complex ones.

$$z_1 = -\frac{K_2}{D_2} \quad (2.3.53)$$

and

$$z_{2,3} = \frac{-D_1 \pm j\sqrt{4K_1J_1 - D_1^2}}{2J_1} \quad (2.3.54)$$

The poles of the transfer function are the roots of the characteristic equation,

$$\Delta = 0 \quad (2.3.55)$$

Since Δ is a polynomial of sixth order, there will be six roots. It would be impossible to determine the roots exactly, but we can find reasonable approximations.
Note that at low frequencies we may perform the following approximations:

$$R_i(s) \approx K_i$$

$$Q_i(s) \approx K_i \quad (2.3.56)$$

$$sL_a + R \approx R$$

Under these conditions equation (2.3.55) becomes

$$sR\,K_1K_2\,(J_m + J_1 + J_2) +$$

$$+ RD\,K_1K_2 + K_E K_T\,K_1K_2 = 0 \quad (2.3.57)$$

Define J, the total moment of inertia, as

$$J = J_m + J_1 + J_2 \qquad (2.3.58)$$

Then (2.3.57) becomes

$$(sRJ + RD + K_E K_T)\, K_1 K_2 = 0 \qquad (2.3.59)$$

and the low frequency pole is

$$p_1 = -\frac{K_E K_T + RD}{RJ} \qquad (2.3.60)$$

If the motor damping factor, D, is small and can be ignored, the pole p_1 becomes

$$p_1 = -\frac{K_E K_T}{RJ} \qquad (2.3.61)$$

which corresponds to the mechanical time constant of the motor.

In order to determine the other poles of the motor transfer function, we assume that the other poles correspond to frequencies much higher than p_1. This assumption is realistic, especially for moving coil motors, where the other poles are about 50 times larger than p_1. Thus, assume that for these poles,

$$|s| \gg \frac{J}{J_m} |p_1| = \frac{K_E K_T + RD}{R J_m} \qquad (2.3.62)$$

It follows that

$$|s|\, R\, J_m \gg K_E K_T \qquad (2.3.63)$$

and if we multiply both sides by $R_1(s)\, R_2(s)$, it becomes

$$|sR\, J_m\, R_1(s)\, R_2(s)| \gg$$

$$\gg |K_E K_T\, R_1(s)\, R_2(s)| \qquad (2.3.64)$$

Note, however, that the two terms in (2.3.64) are parts of Δ, as given by equation (2.3.51). This suggests that the term on the right-hand side is small and, hence, it may be ignored.

Similarly, we may conclude from (2.3.62) that

$$|s| \gg \frac{RD}{RJ_m} \qquad (2.3.65)$$

and therefore

$$|J_m\, s| \gg D \qquad (2.3.66)$$

Thus, we may approximate the term $(sJ_m + D)$ by sJ_m.

In view of the above simplifications, equation (2.3.55) becomes

$$\Delta = s(sL_a + R) \times$$

$$\times \left[J_m R_1 R_2 + J_2 R_1 Q_2 + J_1 Q_1 R_2\right] = 0$$

(2.3.67)

The root at zero represents the low frequency pole, p_1. The second pole is the root of

$$(sL_a + R) = 0 \qquad (2.3.68)$$

or

$$p_2 = -\frac{R}{L_a} \qquad (2.3.69)$$

which corresponds to the electrical time constant of the motor. The remaining four poles are the roots of the equation

$$J_m R_1 R_2 + J_2 R_1 Q_2 + J_1 Q_1 R_2 = 0 \tag{2.3.70}$$

In order to approximate them, assume that the tachometer moment of inertia, J_2, is much smaller than that of the motor or the load. This assumption is reasonable in most cases. Consequently, we choose to ignore the term $J_2 R_1 Q_2$ in equation (2.3.70). The remaining terms in the equation are

$$(J_m R_1 + J_1 Q_1) R_2 = 0 \tag{2.3.71}$$

or, explicitly,

$$\left[J_m (s^2 J_1 + sD_1 + K_1) + + J_1 (sD_1 + K_1)\right]\left[s^2 J_2 + sD_2 + K_2\right] = 0$$

Note that one pair of poles is the solution of

$$s^2 J_2 + sD_2 + K_2 = 0 \tag{2.3.72}$$

and it is determined mostly by the tachometer parameters. Therefore, this pair is called the tachometer resonance poles:

$$p_{3,4} = \frac{-D_2 \pm j\sqrt{4J_2 K_2 - D_2^2}}{2J_2} \tag{2.3.73}$$

The frequency of the tachometer poles is

$$\omega_T \approx \sqrt{\frac{K_2}{J_2}} \quad \text{[rad/s]} \tag{2.3.74}$$

Similarly, the other pair of poles is the solution of

$$s^2 J_m J_1 + s(J_m + J_1) D_1 + (J_m + J_1) K_1 = 0 \tag{2.3.75}$$

or, if we define the equivalent moment of inertia, J_e, as

$$J_e = \frac{J_m J_1}{J_m + J_1} \tag{2.3.76}$$

equation (2.3.75) becomes

$$s^2 J_e + sD_1 + K_1 = 0 \tag{2.3.77}$$

The poles are

$$p_{5,6} = \frac{-D_1 \pm j\sqrt{4K_1 J_e - D_1^2}}{2J_e} \tag{2.3.78}$$

and the frequency of the poles is approximately

$$\omega_L \approx \sqrt{\frac{K_1}{J_e}} \quad \text{[rad/s]} \tag{2.3.79}$$

Clearly, these poles are formed by the interaction between the motor and the load, and they are called the load resonance poles. In summary, the transfer function between the motor voltage and the tachometer velocity has three zeros and six poles.

$$G_2(s) = \frac{s\theta_2(s)}{V(s)}$$

$$= \frac{K_T D_2}{J_m J_2 L_a} \frac{(s-z_1)(s-z_2)(s-z_3)}{(s-p_1)(s-p_2)(s-p_3)(s-p_4)(s-p_5)(s-p_6)} \tag{2.3.80}$$

The first zero, z_1, is negative real and the other two are complex. They are given by equations (2.3.53) and (2.3.54). On the other hand, the pole p_1 is at low frequency and is caused by the mechanical load. The second pole, p_2, corresponds to the electrical time constant of the motor. The remaining poles are two complex pairs, where the first one is determined mostly by the tachometer and the other pair is caused by the motor and the load.

Note that the zero z_1 and the pole p_2 are negative real, and correspond to high frequencies; therefore, their contribution to the motor transfer function is insignificant and they may be ignored. Thus, the transfer function may be reduced to the form

$$G_2(s) = \frac{s\theta_2(s)}{V(s)}$$

$$= \frac{K_T K_2}{R\, J_m J_2} \frac{(s-z_2)(s-z_3)}{(s-p_1)(s-p_3)(s-p_4)(s-p_5)(s-p_6)} \tag{2.3.81}$$

The transfer function was examined for several motors, under various load conditions, and was found to be in excellent agreement with the theoretical analysis given above. For example, when the Electro-Craft M-1030 motor with the M-110 tachometer was loaded by an inertial load, $J_L = 0.0006$ oz-in-s^2 (approximately 42.4 gcm^2), the following poles and zeros were measured:

$$p_1 = -20 \text{ Hz}$$

$$p_{3,4} = -200 \pm j\,4900 \text{ Hz}$$

$$p_{5,6} = -177 \pm j\,4000 \text{ Hz}$$

$$z_{2,3} \cong \pm j\,2900 \text{ Hz}$$

The pole-zero distribution of the transfer function is shown in Fig. 2.3.4, and the motor-tachometer frequency response is shown in Fig. 2.3.5.

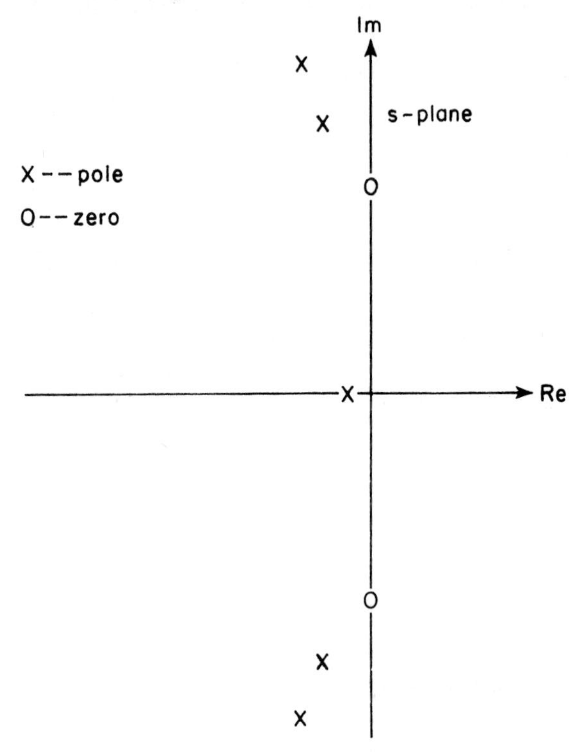

Fig. 2.3.4. Poles and zeros of motor-tachometer transfer function.

The analysis of the resonance phenomenon was based on the assumption model of the three bodies with inertialess shafts, where the stiffness and damping were constant for each shaft. However, as the motor temperature rises due to power dissipation, the mechanical properties of the armature and coupling change and, consequently, the locations of the poles and zeros change

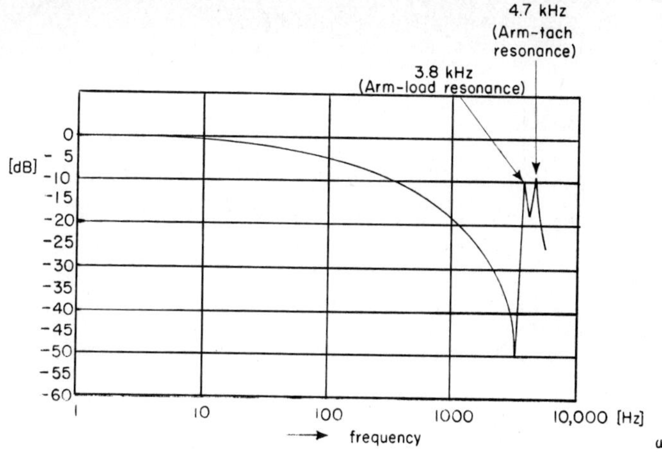

Fig. 2.3.5. Frequency response of motor-tachometer with inertial load.

along with it. In general, the damping and the shaft compliance tend to increase with rising temperature, resulting in increase in the real part of poles and zeros, and decrease in their imaginary part.

2.3.5. SPEED-TORQUE CURVE

When a constant voltage, V, is applied to the motor terminals, the motor shaft will accelerate according to (2.3.3.) and 2.3.6.) and attain a final steady state velocity.

Under steady state conditions the current is constant and the motor equations become

$$V = RI_a + K_E\omega \tag{2.3.82}$$

$$T_g = K_T I_a \tag{2.3.83}$$

When the two equations are combined they become

$$V = \frac{T_g R}{K_T} + K_E\omega \tag{2.3.84}$$

Equation (2.3.84) shows the relation between the velocity and the generated torque at steady state conditions. Fig. 2.3.6 shows that the equation forms a straight line in the speed-torque plane.

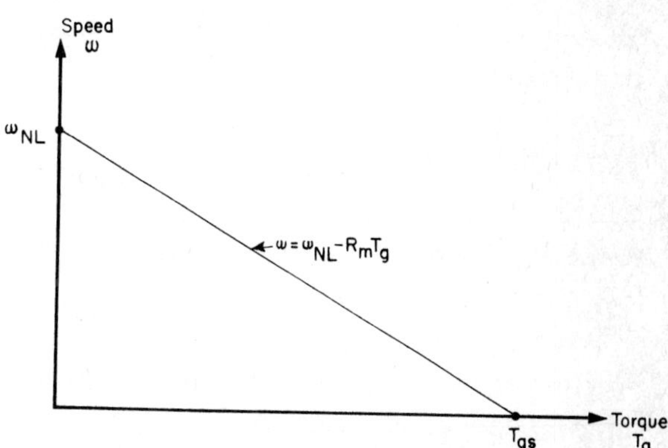

Fig. 2.3.6. DC motor speed-torque curve.

If the torque is zero, the no-load velocity ω_{NL}, is given by

$$\omega_{NL} = \frac{V}{K_E} \tag{2.3.85}$$

On the other hand, the generated torque at stall, T_{gs}, equals

$$T_{gs} = \frac{VK_T}{R} \tag{2.3.86}$$

Accordingly, the relation between the speed and the torque can be expressed by the equation

$$\omega = \omega_{NL} - R_m T_g \tag{2.3.87}$$

where

$$R_m = \frac{R}{K_T K_E} \tag{2.3.88}$$

R_m, the slope constant of the speed-torque curve, is called the *speed regulation constant.* For a given motor, we can plot a family of speed-torque lines that correspond to various terminal voltages. For example, if the motor parameters are:

$K_T = 13.5$ oz-in/A

$K_E = 10$ V/krpm

$R = 1\ \Omega$

and the applied voltages are 10, 20, and 30 V, the no-load velocity will be 1000, 2000, and 3000 rpm, respectively. The complete family of speed-torque lines is shown in Fig. 2.3.7.

This family of speed-torque characteristics used to be important to users of DC motors in the past, so that they could graphically trace motor speed regulation in response to load and voltage changes. However, since the permanent magnet DC motors have linear characteristics, the designer can calculate all needed information from the knowledge of the motor most basic parameters, namely the K_T, R, J and R_{th}. The value of R_m can be calculated from R and K_T, and static and dynamic characteristics can be derived, as we have just seen in this chapter. □

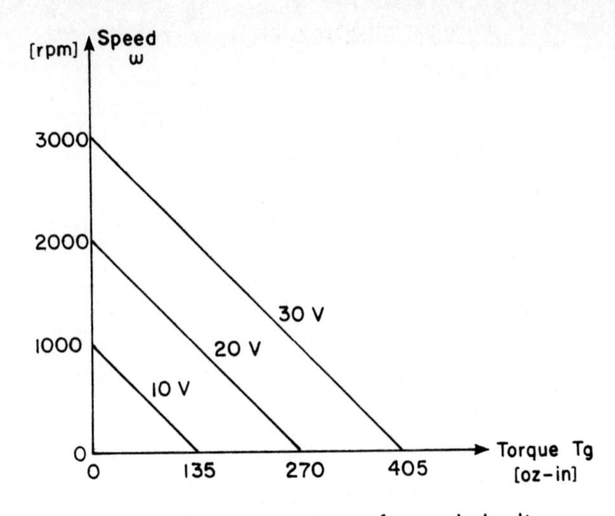

Fig. 2.3.7. Speed-torque curves for terminal voltages of 10, 20, and 30 V.

2.4. POWER DISSIPATION IN DC MOTORS

2.4.1. ORIGINS OF POWER DISSIPATION

The DC motor is used to produce mechanical power from electric power, and as such it may be viewed as an energy converter. It converts electric power into mechanical power during acceleration and steady run, and back to electric form during deceleration. However, the motor is not an ideal converter, due to the armature resistance and other losses. Therefore, it has heat losses as by-products of the energy conversion. The power flow through the motor and the distribution of losses are illustrated in Fig. 2.4.1. The figure shows that the motor losses can be grouped into two divisions. The *load sensitive* losses are dependent upon the generated torque, and the *speed sensitive* losses are determined by rotational speed. Since those losses are important in establishing the limits of motor application, let us consider them in detail.

Winding Losses $I_a^2 R$

These losses are caused by the resistance of the motor and equal $I_a^2 R$. Clearly, they depend on the generated torque, as the armature current, I_a, is proportional to it.

Brush Contact Losses

As the carbon brush slides over the conductive surface of the commutator, a film is formed which is necessary in providing lubrication for proper brush function. It has some of the properties of a dielectric material. Thus, it is not uncommon for this apparent contact resistance to be of sizable value with respect to the actual winding resistance. Current flow through the brush and the film creates heat with resulting loss of power.

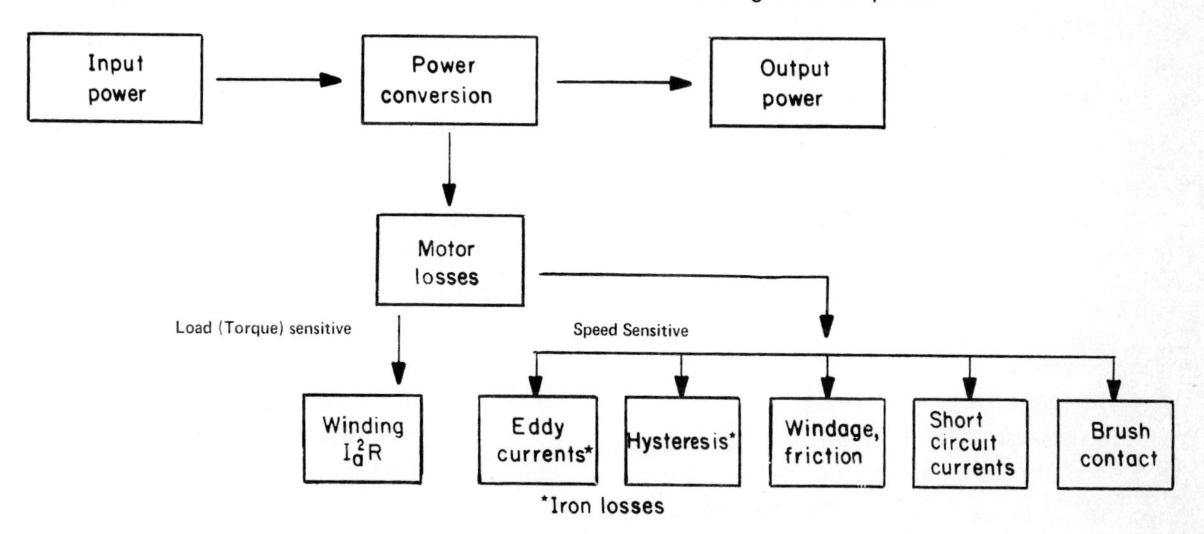

Fig. 2.4.1. Power losses in a permanent magnet motor.

The problem of brush and contact losses is complex, owing to three factors: (1) the carbon has a negative temperature coefficient of resistivity which causes the resistance of the brush itself to decrease with temperature; (2) the contact film between brush and commutator breaks down as current density increases, thereby also decreasing the apparent resistance of the brush-commutator interface; and (3) at increasing rotational speeds a cushion of air is swept along by the commutator, producing an aerodynamic lift under the brush. This brush lift effect tends to increase the apparent contact resistance. To summarize the characteristics of the losses it can be said that:

(1) *They are unpredictable, depending upon factors of current density, rotational speed, and even atmospheric conditions (humidity promotes the film-forming process).*

(2) *Increasing load (current) tends to decrease the percentage of machine loss at the brush-commutator interface.*

(3) *Increasing speed tends to increase the proportional loss at brush-commutator interface.*

Eddy Current Losses

Eddy currents are phenomena associated with a change of magnetic field in a medium that can also support a flow of electric current. In the case of the PM motor, the medium that experiences the change of magnetic field and allows current to flow is the iron of the armature. Fig. 2.4.2. shows how all parts of the armature iron undergo a change of magnetization as it rotates in the magnetic field.

The circulating currents (eddy currents) which are produced in the iron are proportional to speed, and can have significant heating effect on the motor, particularly when the motor operates at high speed. Eddy currents, along with hysteresis losses, usually determine the maximum speed that may be obtained from an iron core armature.

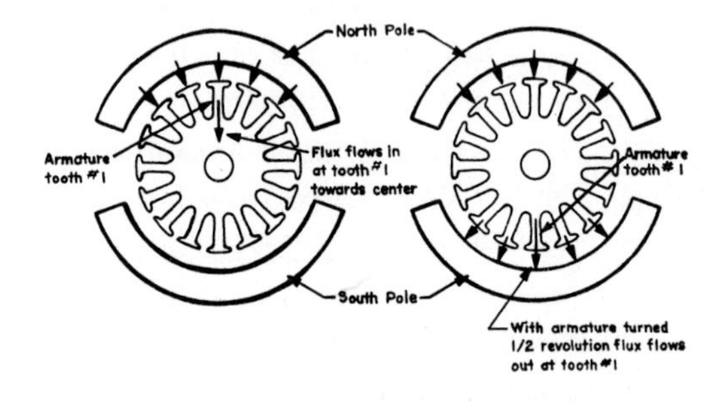

Fig. 2.4.2. All parts of the armature undergo a change of magnetization as it rotates in the magnetic field, thus causing the magnetic domain boundaries to shift.

Eddy current effect is not present in all PM motors. There is a type of low inertia motor, designed for maximum acceleration capability, which does not have moving iron in its magnetic field. These "non-ferrous armature" motors have neither iron eddy currents nor hysteresis effects, and consequently require lower power inputs to obtain high rotational speeds. Typical of this type are Electro-Craft's MCM®* Moving Coil Motors.

*MCM is a registered trade mark for Electro-Craft Moving Coil Motors.

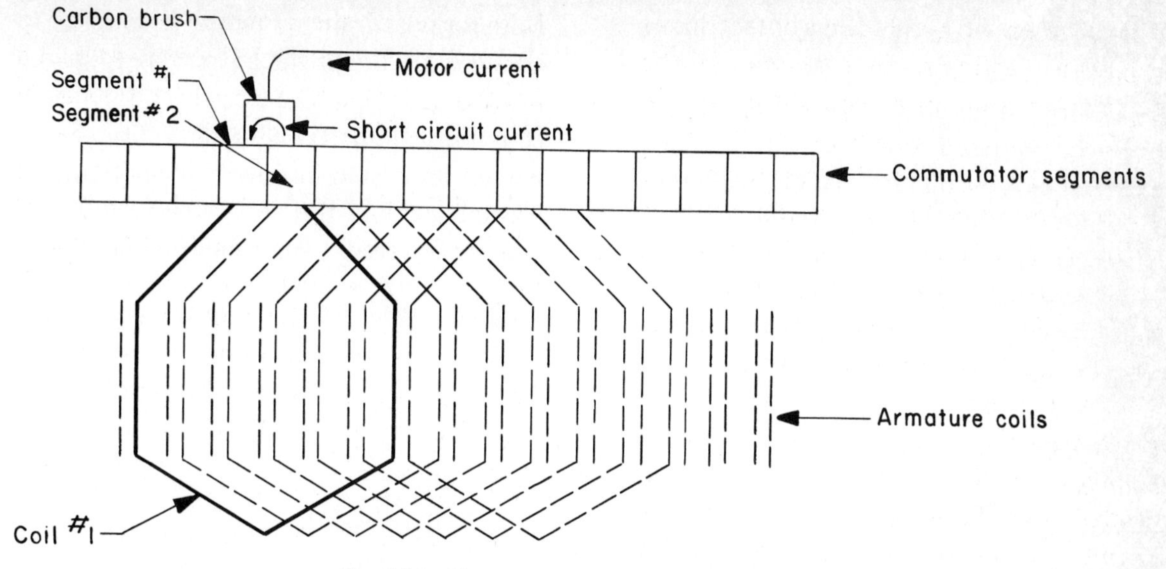

Fig. 2.4.3. Short circuit current in a commutated coil.

Eddy current effects can be minimized in iron core armatures by using laminated sheet steel for the core. The use of thinner laminations results in reduced eddy currents. Silicon steel with high electrical resistance will further reduce eddy current heating.

Hysteresis Effect

Fig. 2.4.2. shows how the iron core of the armature experiences changing magnetization as it rotates in the stator field. As the armature rotates, magnetic domain boundaries shift. The resistance to this magnetic domain shifting also causes heat generation that increases with motor speed. Hysteresis loss is a characteristic of the type of steel used. It is usually combined with eddy current effect and called "iron loss."

Windage and Friction

This category of motor heat generation includes all those factors which produce a mechanical drag on the machine, such as bearing friction, friction of the brushes on the commutator, and air resistance. The last item is usually insignificant in smaller sized servomotors. The magnitude of windage and the friction losses will depend upon the mechanical features of design; i.e., coefficient of friction between brush and commutator and size of bearing and lubricant.

Short Circuit Currents

This term does not imply a fault in a machine, but rather describes a normal current which exists briefly in the armature coil that is commutated.

As shown in Fig. 2.4.3, at the instant the brush is riding on two commutator segments, it actually provides a short circuit connection between the ends of the coil that is connected to those two segments.

It can be seen that the start of coil #1 is connected to segment #1, while the end of coil #1 is connected to segment #2. It can easily be surmised that any voltage which exists across the coil terminations will cause a current to flow through the coil. This current can not be measured directly, but in cases where it becomes excessive, it will cause arcing at the trailing edge of the motor brush.

The effect of short circuit current is to contribute a component of loss which increases with motor speed. It appears as a viscous drag on the armature. It will also impose a maximum speed limitation on motors which are not otherwise limited by their iron losses.

Short circuit current effects can be minimized by proper motor design so that the brushes only commutate coils which are in regions of low magnetic flux. This tends to minimize the voltage that will exist across the ends of the coil.

2.4.2. POWER DISSIPATION

The speed sensitive losses, when combined together, act as a velocity dependent opposing torque, T_m, which contains a constant term, T_f, and a velocity dependent term, $D\omega$.

$$T_m = T_f + D\omega \tag{2.4.1}$$

The terms T_f and D may be interpreted as constant friction and viscous damping, but they represent all the velocity sensitive losses.

Now suppose that the motor is coupled to a load with a moment of inertia, J_L, and a load torque, T_L. The electric and dynamic equations will be (2.3.3) and (2.3.6). Since we are interested in power dissipation, the inductance, L_a, has very small effect and may be ignored. The simplified equations for the motor would be:

$$V = RI_a + K_E\omega \tag{2.4.2}$$

$$I_a = \frac{1}{K_T}\left[(J_m+J_L)\frac{d\omega}{dt} + T_m + T_L\right] \tag{2.4.3}$$

The input power to the motor, P_i, is given by

$$P_i = VI_a = I_a(RI_a+K_E\omega)$$

$$= I_a^2R + K_E\omega I_a \tag{2.4.4}$$

In view of (2.4.3), P_i becomes

$$P_i = I_a^2R + \frac{K_E\omega}{K_T}\left[T_m + T_L + (J_m+J_L)\frac{d\omega}{dt}\right] \tag{2.4.5}$$

or

$$P_i = I_a^2 R + \frac{K_E\omega}{K_T} T_m + \frac{K_E\omega}{K_T} T_L + \frac{K_E(J_m+J_L)}{K_T}\omega\frac{d\omega}{dt} \tag{2.4.6}$$

Note that the last term of (2.4.6) is equal to

$$\frac{K_E(J_m + J_L)}{K_T} \omega \frac{d\omega}{dt}$$

$$= \frac{K_E (J_m + J_L)}{2K_T} \frac{d(\omega^2)}{dt} \qquad (2.4.7)$$

This quantity will have zero average value when the velocity is constant or periodic, or over any time interval where the final velocity is equal to the initial velocity. Consequently, the contribution of this term to the power dissipation is zero and it can be ignored. Of the remaining terms we can identify the last one as the output power which is delivered to the load,

$$P_o = \frac{K_E \omega T_L}{K_T} \qquad (2.4.8)$$

Note that the ratio of the voltage and torque constants, K_E/K_T, forms the conversion factor in (2.4.8), whose value depends on the units used in the equation. If all the quantities are expressed in basic or main SI units, i.e., $[P_o] = W$, $[\omega] = rad \cdot s^{-1}$, $[T_L] = Nm$, $[K_E] = V/rad \cdot s^{-1}$, $[K_T] = Nm/A$, then $K_E/K_T = 1$ and (2.4.8) becomes

$$P_o = \omega T_L \qquad (2.4.8a)$$

The same applies for other quantities in this section.

The remaining two terms are the losses in the motor.

$$P_L = I_a^2 R + \frac{K_E \omega T_m}{K_T} \qquad (2.4.9)$$

Note that the first term, $I_a^2 R$, is the winding resistive loss and the second term corresponds to the velocity sensitive losses. Moreover, by substituting equation (2.4.1) into (2.4.9), the losses become

$$P_L = I_a^2 R + \frac{K_E \omega T_f}{K_T} + \frac{K_E D \omega^2}{K_T} \qquad (2.4.10)$$

Equation (2.4.10) describes the *instantaneous power dissipation* in general, and it can be used to determine the losses under specific velocity profiles. In the following discussion the power dissipation is evaluated for two common types of angular motion: constant speed and incremental motion.

2.4.3. DISSIPATION AT CONSTANT VELOCITY

When the motor runs at a constant velocity, ω_m, the power dissipation may be found directly from (2.4.10). This is given by

$$P_L = I_a^2 R + \frac{K_E T_f \omega_m}{K_T} + \frac{K_E D \omega_m^2}{K_T} \qquad (2.4.11)$$

This is illustrated by the following example.

Example

Given are the following motor and load parameters:

$$R = 1\,\Omega$$

$$K_T = 6 \text{ oz-in/A} = 4.24 \cdot 10^{-2} \text{ Nm/A}$$

$$K_E = 4.45 \text{ V/krpm}$$
$$= 4.24 \cdot 10^{-2} \text{ V/rad} \cdot s^{-1}$$

$$J_m = 0.5 \cdot 10^{-3} \text{ oz-in-s}^2$$
$$= 3.53 \cdot 10^{-6} \text{ kg m}^2$$

$$T_f = 6 \text{ oz-in} = 4.24 \cdot 10^{-2} \text{ Nm}$$

$$D = 0.04 \frac{\text{oz-in}}{\text{rad/s}} = 2.825 \cdot 10^{-4} \frac{\text{Nm}}{\text{rad} \cdot s^{-1}}$$

$$J_L = 2 \cdot 10^{-3} \text{ oz-in-s}^2$$
$$= 1.412 \cdot 10^{-5} \text{ kg m}^2$$

$$T_L = 12 \text{ oz-in} = 8.474 \cdot 10^{-2} \text{ Nm}$$

The motor will run at constant velocity ω_m = **300 rad/s** (2860 rpm). The power dissipation is found as follows.

The armature current is found from (2.4.3) and (2.4.1)

$$I_a = \frac{1}{K_T} (T_f + D\omega_m + T_L)$$
$$= \frac{1}{6} [6 + 0.04 \cdot 300 + 12] = 5 \text{ A}$$

The power dissipation can be found from (2.4.11). It can be calculated by using either the SI or the British system of units. In this example we will use the SI system.

Note that in this system K_T and K_E are numerically equal, when expressed in basic and main SI units.

$$P_L = 5^2 \cdot 1 + \frac{4.24 \cdot 10^{-2} \cdot 300 \cdot 4.24 \cdot 10^{-2}}{4.24 \cdot 10^{-2}} + \frac{4.24 \cdot 10^{-2} \cdot 2.825 \cdot 10^{-4} \cdot 300^2}{4.24 \cdot 10^{-2}}$$

$$P_L = 25 + 12.72 + 25.42 = 63.14 \text{ W}$$

2.4.4. DISSIPATION DURING INCREMENTAL MOTION

Equation (2.4.10) for power dissipation can be applied to the case of incremental motion where the motor steps periodically following a trapezoidal velocity profile, as shown in Fig. 2.4.4.

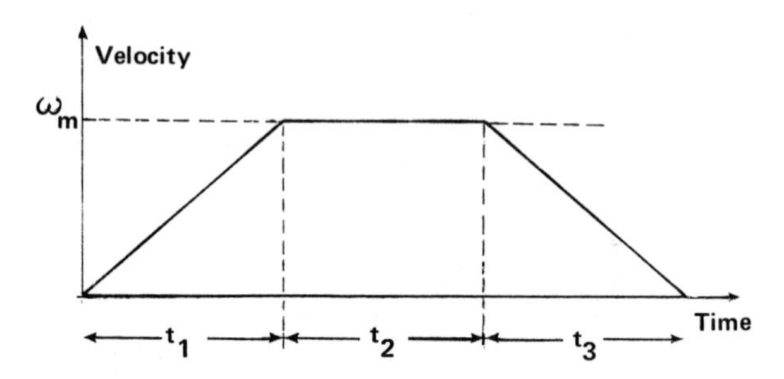

Fig. 2.4.4. Motor velocity profile.

The motor accelerates over t_1 to the velocity ω_m and runs at that speed for t_2. Later, it decelerates to zero velocity over t_3. The stepping is periodic with a repetition rate of **f** steps per second. Equation (2.4.10) describes the instantaneous power dissipation in the motor. However, it is of greater importance to determine the average power loss in the motor over one cycle by averaging the power. The analysis was found long and laborious, and we present here the final results for this case.

The average power dissipation, $\bar{P}$, was found to have five terms.

$$\bar{P} = P_1 + P_2 + P_3 + P_4 + P_5 \qquad (2.4.12a)$$

where

$$P_1 = \frac{f R J^2 \omega_m^2}{K_T^2} \left(\frac{1}{t_1} + \frac{1}{t_3} \right) \qquad (2.4.12b)$$

$$P_2 = \frac{fR}{K_T^2} (T_f + T_L + D\omega_m)^2 t_2 \qquad (2.4.12c)$$

$$P_3 = \frac{fR}{K_T^2} \left[(T_f + T_L)^2 + (T_f + T_L) D\omega_m + \frac{1}{3} D^2 \omega_m^2 \right] (t_1 + t_3) \qquad (2.4.12d)$$

$$P_4 = \frac{f K_E T_f \omega_m}{2K_T} (t_1 + 2t_2 + t_3) \qquad (2.4.12e)$$

$$P_5 = \frac{f K_E D \omega_m^2}{3K_T} (t_1 + 3t_2 + t_3) \qquad (2.4.12f)$$

Clearly, the various terms can be interpreted physically. P_1 describes the power dissipation due to the acceleration and deceleration of the motor and load. P_2 and P_3 describe the winding losses due to the opposing torques during the various intervals of motion. Finally, P_4 and P_5 represent the rotational losses due to the constant opposing torque, T_f, and the damping, D, respectively.

The following example illustrates the use of the equations (2.4.12). Special care should be taken to use the right units.

Example

Consider a loaded motor with the following parameters:

K_T = 25 oz-in/A = 0.176 Nm/A

K_E = 18.4 V/krpm = 0.176 V/rad $\cdot$ s^{-1}

R = 1.2 Ω

T_f = 6 oz-in = $4.24 \cdot 10^{-2}$ Nm

D = 4 oz-in/krpm = $2.7 \cdot 10^{-4}$ Nm/rad $\cdot$ s^{-1}

T_L = 50 oz-in = 0.353 Nm

J = 0.02 oz-in-s^2 = 1412 g cm^2
= $1.412 \cdot 10^{-4}$ kg m^2

The motion requirements are given by

t_1 = 50 ms = 0.05 s

t_2 = 200 ms = 0.2 s

t_3 = 40 ms = 0.04 s

ω_m = 3000 rpm = 314 rad/s

and the repetition rate is

$$f = 2 \text{ steps/s}$$

We may use any system of units to evaluate the power. In this example, we use both systems to illustrate their use. To evaluate P_1, we use the SI system, whereas the British system is used for the other terms.

$$P_1 = \frac{2 \cdot 1.2 \, (1.412 \cdot 10^{-4})^2 \cdot 314^2}{0.176^2} \times$$

$$\times \left(\frac{1}{0.05} + \frac{1}{0.04} \right) = 6.82 \text{ W}$$

$$P_2 = \frac{2 \cdot 1.2}{25^2} (6 + 50 + 4 \cdot 3)^2 \cdot 0.2 = 3.55 \text{ W}$$

$$P_3 = \frac{2 \cdot 1.2}{25^2} \left[56^2 + 56 \cdot 4 \cdot 3 + \frac{1}{3} 4^2 \cdot 3^2 \right] \times$$

$$\times (0.05+0.04) = 1.33 \text{ W}$$

$$P_4 = \frac{2 \cdot 18.4 \cdot 6 \cdot 3}{2 \cdot 25} (0.05 + 0.4 + 0.04) = 6.49 \text{ W}$$

$$P_5 = \frac{2 \cdot 18.4 \cdot 4 \cdot 3^2}{3 \cdot 25} (0.05 + 0.6 + 0.04)$$

$$= 12.19 \text{ W}$$

and the total average power dissipation is

$$\bar{P} = 30.38 \text{ W}$$

In the following section we will discuss the ability of a motor to handle armature power dissipation. □

2.5. THERMAL CHARACTERISTICS OF DC MOTORS

As the motor is operated, power losses are dissipated in the armature, resulting in temperature rise. The increase of temperature is important since it may limit the motor performance and, therefore, it deserves a careful study.

In this section we consider the temperature rise in the armature resulting from the power losses. We start with the simple case of continuous operation and determine the boundaries for working conditions. Later, we study the case of intermittent operation and develop a thermal model for the motor in order to analyze the temperature response.

2.5.1. CONTINUOUS OPERATION

When the motor runs at a constant or near constant velocity and the operation time is long (30 min or more), the motor temperature reaches a steady-state level, depending on the power dissipation.

In order to limit the temperature rise, we have to limit the losses, $\mathbf{P_L}$, in the motor. Since the losses depend on the velocity ω_m, and the load torque $\mathbf{T_L}$ [see (2.4.10)], we can limit the values of the speed and the torque so that the resulting temperature is acceptable.

This can be done by constructing a boundary in the speed-torque plane which defines the safe operation area. The boundary curve is called Safe Operation Area boundary for Continuous operation (SOAC). An example for such SOAC curve is shown in Fig. 2.5.1. The area to the left of the curve is safe for operation, whereas points to the right of the SOAC will result in excessive losses and heating.

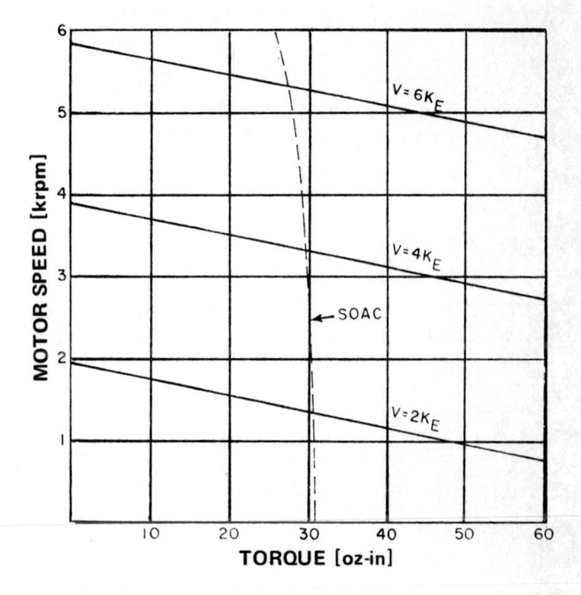

Fig. 2.5.1. Speed-torque diagram and the Safe Operation Area boundary for Continuous operation (SOAC).

The speed-torque diagram may include additional information on the voltage required to drive the motor under those conditions, or on rating points. This, however, does not affect the SOAC.

When a motor is forced-air cooled, the thermal resistance decreases with the cooling, allowing a larger amount of power losses to be dissipated. This can be described by two SOAC curves for the uncooled and the cooled motors. An example is shown in Fig. 2.5.2.

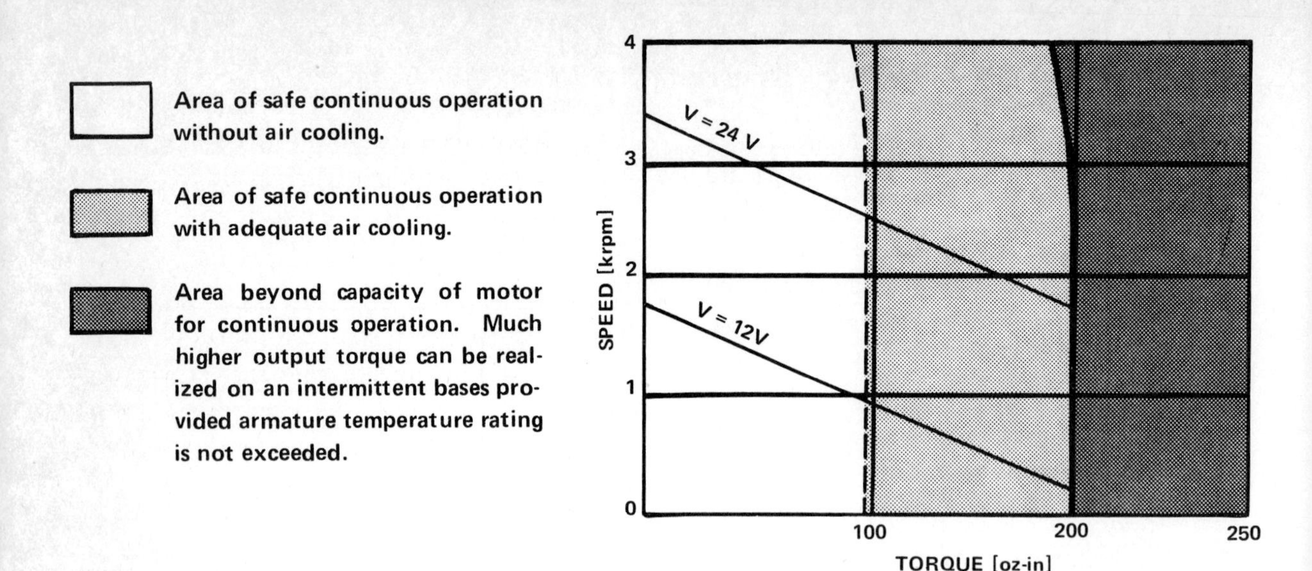
Area of safe continuous operation without air cooling.
Area of safe continuous operation with adequate air cooling.
Area beyond capacity of motor for continuous operation. Much higher output torque can be realized on an intermittent bases provided armature temperature rating is not exceeded.
V = 24 V
V = 12V
SPEED [krpm]
4
3
2
1
0
100
200
250
TORQUE [oz-in]

Fig. 2.5.2. The SOAC for a motor with cooling capability.

2.5.2. INTERMITTENT OPERATION

When the motor is operated continuously for a short time, or when the operation is periodic, the power losses in the motor vary with the time and the analysis of the resulting temperature is more complex than for the continuous case.

In order to determine the temperature rise, a thermal model is developed for the motor. The model describes the relation between the power and the temperature, and makes a thermal analysis for various power dissipation functions possible.

2.5.3. THERMAL MODEL

The motor consists of two parts, the armature (rotor) and the housing (stator). It is assumed here that the temperature over the two bodies are uniform. In other words, the armature temperature, *relative to ambient,* Θ_a, is the same all over the armature. Similarly, the relative temperature of the housing, Θ_h, is also uniform.

The heat capacity is the ratio between the heating energy and the temperature rise. For the given model the heat capacities of the armature and the housing are denoted by $\mathbf{C_a}$ and $\mathbf{C_h}$, respectively.

The power losses, $\mathbf{P_L}$, are assumed to be dissipated in the armature, since this is the common case. However, the final results of the following analysis are valid also for the case where some of the losses are dissipated in the housing.

The third assumption about the motor is that the heat transfer is proportional to temperature difference. Therefore, the power transfer from the armature to the housing, $\mathbf{P_a}$, equals

$$P_a = G_a (\Theta_a - \Theta_h) \quad (2.5.1)$$

where G_a is the thermal conductance between the two parts. Similarly, the heat transfer from the housing to the ambient, P_h, is given by

$$P_h = G_h \Theta_h \quad (2.5.2)$$

with G_h being the heat conductance between the housing and the ambient.

If the motor is cooled by forced air, with the temperature of the air inflow being equal to ambient temperature, the heat transfer by cooling, P_c, equals

$$P_c = G_c \Theta_a \quad (2.5.3)$$

where G_c is the heat conductance and depends on the air flow.

2.5.4. THERMAL EQUATIONS

The thermal equations for the motor armature and housing are:

$$\left. \begin{aligned} C_a \frac{d\Theta_a}{dt} &= P_L - P_a - P_c \\ C_h \frac{d\Theta_h}{dt} &= P_a - P_h \end{aligned} \right\} \quad (2.5.4)$$

In view of (2.5.1), (2.5.2), and (2.5.3) the thermal equations become

$$C_a \frac{d\Theta_a}{dt} = P_L - G_a (\Theta_a - \Theta_h) - G_c \Theta_a \quad (2.5.5a)$$

$$C_h \frac{d\Theta_h}{dt} = G_a (\Theta_a - \Theta_h) - G_h \Theta_h \quad (2.5.5b)$$

The heat flow in the motor is illustrated in Fig. 2.5.3.

The electrical analog of the thermal system can be constructed by noting the equivalence between temperature and power on one hand, with voltage and current on the other hand. The resulting equivalent network is shown in Fig. 2.5.4.

Since we are interested in Θ_a, which is higher than the housing temperature, it is desired to develop the transfer function between P_L and Θ_a.

Upon application of the Laplace transformation to (2.5.5), it becomes

$$sC_a \Theta_a(s) = P_L(s) - (G_a + G_c) \Theta_a(s) + + G_a \Theta_h(s) \quad (2.5.6a)$$

$$sC_h \Theta_h(s) = G_a \Theta_a(s) - (G_a + G_h) \Theta_h(s) \quad (2.5.6b)$$

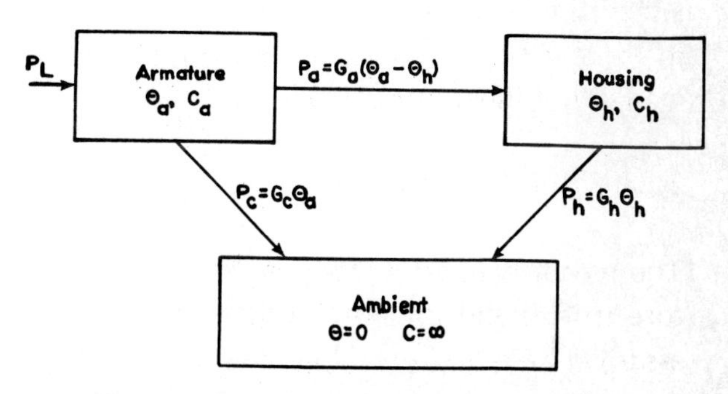

Fig. 2.5.3. Thermal model and heat flow in a DC motor.

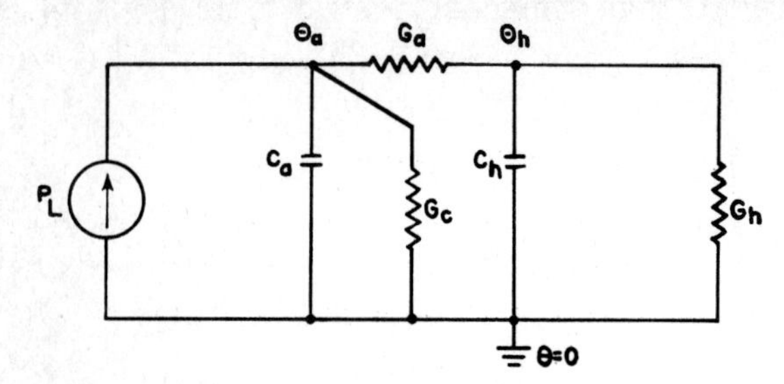

Fig. 2.5.4. Electrical analog of thermal system.

Equation (2.5.6b) may be used to substitute for $\Theta_h(s)$.

$$\Theta_h(s)\ (sC_h + G_a + G_h) = G_a\ \Theta_a(s) \quad (2.5.7)$$

Combining (2.5.7) with (2.5.6a) results in

$$\Theta_a(s)\left[sC_a + G_a + G_c - \frac{G_a^2}{sC_h + G_a + G_h}\right] = P_L(s) \quad (2.5.8)$$

and the transfer function between P_L and Θ_a is

$$\frac{\Theta_a(s)}{P_L(s)} = \frac{sC_h + G_a + G_h}{s^2 C_h C_a + s\left[C_a(G_a+G_h) + C_h(G_a+G_c)\right] + G_aG_c + G_aG_h + G_cG_h} \quad (2.5.9)$$

Note that (2.5.9) may be written in the form

$$\frac{\Theta_a}{P_L}(s) = \frac{A\left(s\,\tau_{th_3} + 1\right)}{\left(s\,\tau_{th_1} + 1\right)\left(s\,\tau_{th_2} + 1\right)} \quad (2.5.10)$$

where

$$A = \frac{G_a + G_h}{G_aG_c + G_aG_h + G_cG_h} \quad (2.5.11)$$

and

$$\tau_{th_3} = \frac{C_h}{G_a + G_h} \quad (2.5.12)$$

The time constants, τ_{th_1} and τ_{th_2}, correspond to the roots of the thermal characteristic equation:

$$s^2C_hC_a + s\left[C_a(G_a+G_h) + C_h(G_a+G_c)\right] + G_aG_c + G_aG_h + G_cG_h = 0 \quad (2.5.13)$$

Note that the characteristic roots of (2.5.13) depend on the air cooling conductance, G_c. Therefore, the thermal time constants, τ_{th_1} and τ_{th_2}, will vary with the introduction of forced-air cooling.

The transfer function (2.5.10) may be written as the sum of partial fractions:

$$\frac{\Theta_a}{P_L}(s) = \frac{R_{th_1}}{s\,\tau_{th_1} + 1} + \frac{R_{th_2}}{s\,\tau_{th_2} + 1} \quad (2.5.14)$$

where

$$R_{th_1} = \frac{A\left(\tau_{th_3} - \tau_{th_1}\right)}{\tau_{th_2} - \tau_{th_1}} \quad (2.5.15a)$$

and

$$R_{th_2} = \frac{A\left(\tau_{th_2} - \tau_{th_3}\right)}{\tau_{th_2} - \tau_{th_1}} \quad (2.5.15b)$$

Now define $\Theta_1(s)$ and $\Theta_2(s)$ as

$$\Theta_1(s) = \frac{R_{th_1}}{s\,\tau_{th_1} + 1} P_L(s) \qquad (2.5.16a)$$

$$\Theta_2(s) = \frac{R_{th_2}}{s\,\tau_{th_2} + 1} P_L(s) \qquad (2.5.16b)$$

Then the armature temperature, Θ_a, equals

$$\Theta_a(s) = \Theta_1(s) + \Theta_2(s) \qquad (2.5.17)$$

Equations (2.5.16) and (2.5.17) describe the transfer function as a parallel combination of two partial fractions, as shown in Fig. 2.5.5. This model is easier to analyze due to its simplicity; it is therefore preferred over (2.5.9).

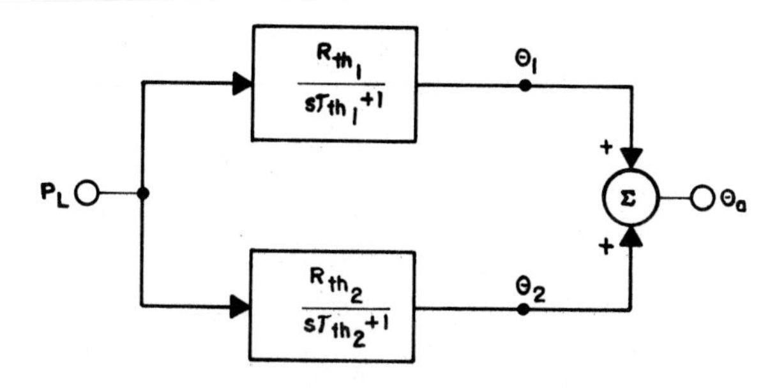

Fig. 2.5.5. Parallel model for thermal system.

2.5.5. THERMAL ANALYSIS

The temperature rise in a motor due to power losses may be determined from the transfer function model of Eq. (2.5.9). However, this method may be unnecessarily complicated in most cases, and an easier procedure has been developed, based on the parallel model of (2.5.16) and (2.5.17), and is presented below.

The basic idea of this approach is that the temperatures, Θ_1 and Θ_2, should be found separately. Later, they can be combined to form Θ_a. Since the transfer function of either Θ_1 or Θ_2 is of the form

$$\frac{\Theta(s)}{P_L(s)} = \frac{R_{th}}{s\,\tau_{th} + 1} \qquad (2.5.18)$$

the method for determining Θ_1 and Θ_2 is the same.

The engineer has the freedom to solve (2.5.18) in any way. However, in the following discussion, the solution for most common cases is presented.

Case 1 – Constant Power

When a constant power loss, P_L, is dissipated in the motor from **t = 0**, the resulting temperature rise is

$$\Theta = R_{th} P_L \left(1 - e^{-\frac{t}{\tau_{th}}}\right) \qquad (2.5.19)$$

If the power is applied for a long time,

$$t \geqslant 4\,\tau_{th} \qquad (2.5.20)$$

the temperature reaches the steady-state value of

$$\Theta = R_{th} P_L \qquad (2.5.21)$$

Therefore, the steady-state temperature rise due to constant power dissipation is the product of the power and the thermal resistance.

Case 2 – Short Term Power Variations

When the power loss variations are much faster than the thermal time constant, e.g., power variation time is less than $\tau_{th}/5$, the power variations will be filtered and only the average power will affect the temperature. To determine the temperature rise due to the average power, use the results of Case 1.

Case 3 – Low Frequency Power Variations

When the power dissipation varies periodically and the period is neither too short nor too long with respect to the thermal time constant, we can not use any of the results of Case 1 or 2, and the temperature has to be found from the transfer function. In the following discussion the case where the applied power is of square wave form is analyzed and the temperature rise is determined.

Pulsed Power Case

If P_L is as shown in Fig. 2.5.6, with an "ON" time t_1 and a period t_c, the temperature can be found as follows:

At time $t = 0$, temperature equals Θ_0; then power is applied and the resulting temperature is found:

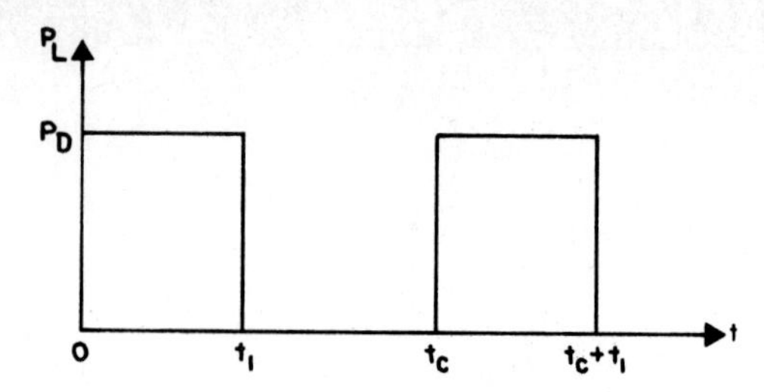

Fig. 2.5.6. An example of "pulsed power".

$$\Theta_m = \Theta_0 + (R_{th} P_D - \Theta_0)\left(1 - e^{-t_1/\tau_{th}}\right) \tag{2.5.22}$$

At time t_1, power is removed and temperature drops according to:

$$\Theta_0 = \Theta_m e^{-\left(\frac{t_c - t_1}{\tau_{th}}\right)} \tag{2.5.23}$$

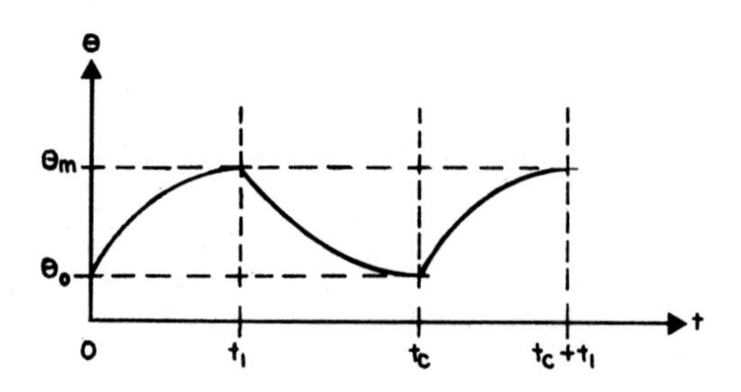

Fig. 2.5.7. "Pulsed power" is applied for a long time period so that thermal equilibrium is attained.

Fig. 2.5.7 shows the time display of (2.5.22) and (2.5.23) as the temperature variation with respect to time.

Suppose the system is operated long enough to achieve a steady-state condition as shown in Fig. 2.5.7. Then for the maximum temperature rise, we can substitute (2.5.23) into (2.5.22) to obtain

$$\Theta_m = \frac{R_{th} P_D \left(1-e^{-t_1/\tau_{th}}\right)}{1-e^{-t_c/\tau_{th}}} \qquad (2.5.24)$$

In order to illustrate the method of thermal analysis, consider the following example.

Example

The thermal behavior of a motor is described by the following parameters:

$R_{th_1} = 1\ ^oC/W$

$R_{th_2} = 0.8\ ^oC/W$

$\tau_{th_1} = 15\ s$

$\tau_{th_2} = 15\ min = 900\ s$

The power loss is of pulse type with

$P_D = 150\ W$

$t_1 = 5\ s$

$t_c = 15\ s$

and the operation lasts for two hours. First, we will calculate Θ_2, which corresponds to the larger time constant. Note that the period $t_c = 15$ s is much shorter than $\tau_{th_2} = 900$ s, and only the average power will be considered (Case 2).

The average power is

$$P_{av} = \frac{150 \cdot 5}{15} = 50\ W$$

and the resulting steady-state temperature rise is

$$\Theta_2 = R_{th_2} P_{av} = 0.8 \cdot 50 = 40\ ^oC$$

To determine Θ_1, note that $t_c = \tau_{th_1}$, and therefore, neither Case 1 or 2 will apply. The temperature rise is found from (2.5.24), which becomes

$$\Theta_1 = \frac{1 \cdot 150 \cdot \left(1-e^{-\frac{5}{15}}\right)}{1-e^{-\frac{15}{15}}} = 90\ ^oC$$

Then the total temperature rise, Θ_a, equals

$$\Theta_a = \Theta_1 + \Theta_2 = 130\ ^oC$$

If the ambient temperature is 25 ^oC, the temperature of the armature will be 155 ^oC.

□

2.6. MOTOR CHARACTERISTICS AND TEMPERATURE

Motor constants such as K_T, K_E, and R are parameters used to define motor characteristics. But it is well to remember that they are not true constants in the strictest sense, since these parameters will be temperature sensitive to some degree. The following describes the effects of temperature variations, and presents a method for calculating the magnitude of parameters' change.

ARMATURE RESISTANCE

The majority of PM motor armatures are wound with copper conductors, and some of the new low inertia types use aluminum conductors. With either metal, electrical resistance increases with temperature, each at a different rate. There are several effects of an increase in armature resistance as the motor is heated.

1) Motor I_a^2R *losses will increase for a given value of current.* If a servomotor is being driven with a constant value of current to achieve a constant acceleration, the winding loss will increase in proportion to the amount of resistance increase. For this reason the "hot" resistance of the motor armature winding should be used when assessing a motor's capability for performing a prescribed duty cycle.

2) Speed regulation constant will increase. This constant was defined as the slope of the speed versus torque curve and was found to be proportional to **R**. Thus a motor loaded close to rated torque will slow up somewhat as it warms up due to the higher value of speed regulation constant.

3) The frequency response characteristic of the motor will be different from "cold" to "hot". This is a consideration in high-performance servo applications, where the motor is required to respond to rapidly varying input signals, and occurs since the mechanical time constant used in the transfer function is proportional to **R**.

The change of **R** can be calculated if the anticipated temperature rise can be estimated. The resistance of copper as a function of temperature is shown in Fig. 2.6.1.

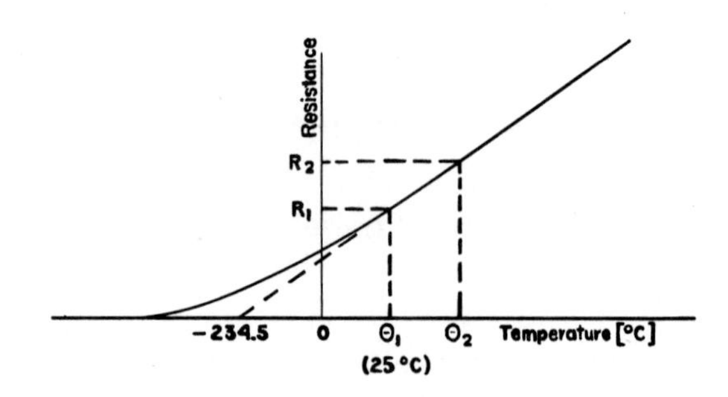

Fig. 2.6.1. Resistance versus temperature characteristic for copper conductor.

As can be seen from the curve, the characteristic is linear in the region which is of most interest, from 25 °C to 200 °C. If this straight portion is extended, as shown by the broken line, it intersects the temperature axis at −234.5 °C. Then by similar triangles, we can say that

$$\frac{R_2}{R_1} = \frac{234.5 + \Theta_2}{234.5 + \Theta_1} \qquad (2.6.1)$$

or,

$$R_2 = \frac{234.5 + \Theta_2}{234.5 + \Theta_1} R_1 \qquad (2.6.2)$$

Example

If an armature resistance $\mathbf{R = 2.5\ \Omega}$ at a standard 25 ^oC temperature, calculate the expected resistance when the winding is at its rated 155 ^oC operating temperature.

$\Theta_2 = 155\ ^oC$

$\Theta_1 = 25\ ^oC$

$R_1 = 2.5\ \Omega$

then,

$$R_2 = \frac{(234.5 + 155)}{(234.5 + 25)} \times 2.5 = 3.75\ \Omega \qquad (2.6.3)$$

It can be seen that over the operating temperature range a 50% variation in resistance is observed, a very significant change, indeed.

An alternate formula for calculating resistance with temperature change is:

$$R_2 = R_1 [1 + \psi (\Theta_2 - \Theta_1)]$$

$$[\Omega; \Omega, \Omega/^oC, ^oC] \qquad (2.6.4)$$

where

R_2 is resistance at Θ_2

R_1 is resistance Θ_1

and

$\psi_{cu} = 0.00393$ for copper

$\psi_{al} = 0.00415$ for aluminum

TORQUE CONSTANT K_T AND VOLTAGE CONSTANT K_E

The effects of temperature variation in K_T and K_E are:

1) An increase in I_a^2R *losses.* If a given torque is required to accelerate a load, as the motor heats up, K_T and K_E decrease and more current will be required to provide the same torque.

2) Speed regulation constant will increase. The decrease in K_T and K_E would cause an additional fall off in speed of a motor carrying rated load torque.

3) Change in frequency response characteristics. Because of the reduced torque capability, the frequency response of the motor is reduced.

Of the two main types of magnets commonly used, the temperature coefficient of magnetic flux density for ceramics is −0.002; Alnico magnets have coefficients that range from −0.0001 to −0.0005.

Using the above temperature coefficients of magnetic flux density, it is possible to develop an expression to calculate a derated torque constant to be used where the motor magnets will experience a significant temperature rise.

$$K_{TD} = K_T \; [1 + \psi_B \, \Theta_{rm}] \qquad (2.6.5)$$

where

K_{TD} is the derated torque constant

K_T is the rated torque constant (at 25 ^oC)

ψ_B is the temperature coefficient of magnetic flux density

Θ_{rm} is the temperature rise of the magnet material.

This equation points out a great advantage in using forced-air cooling. Air circulating through the motor will greatly reduce magnet temperature rise, and will minimize temperature variation.

DERATING MOTOR TORQUE

The above discussion of temperature effects should make it very obvious that a motor's torque capability is degraded as operating temperature increases. An expression can be developed to compare the maximum torque that can be produced by "hot" and "cold" motors. If a motor at stall is excited with a current I_a to its maximum dissipation level, the power will be dissipated as I_a^2R loss. Then torque developed will be:

$$T_g = K_T I_a \qquad (2.3.4)$$

As the armature heats up the current must be reduced to avoid exceeding the rated armature dissipation. The "hot" current is found to be:

$$I_{ah} = \frac{I_a}{\sqrt{1 + \psi_{cu} \, \Theta_{rw}}} \qquad (2.6.6)$$

where Θ_{rw} is the winding temperature rise

Torque can be calculated as the product $K_{TD} I_{ah}$:

$$T_{gh} = \frac{K_T \; [1 + \psi_B \, \Theta_{rm}] \; I_a}{\sqrt{1 + \psi_{cu} \, \Theta_{rw}}} \qquad (2.6.7)$$

The ratio of maximum "hot" torque to "cold" torque will be:

$$\frac{T_{gh}}{T_g} = \frac{1 + \psi_B \, \Theta_{rm}}{\sqrt{1 + \psi_{cu} \, \Theta_{rw}}} \qquad (2.6.8)$$

Example

A motor with copper armature winding and ceramic magnets has the following ratings:

K_T = 80 oz-in/A

R = 2.5 Ω

$(I_a^2R)_{max}$ = 105 W

If the motor is used with a controller that limits current so that the maximum armature dissipation cannot be exceeded, calculate the torques at 25 ^{o}C and at 155 ^{o}C operating temperature, and torque derating factor.

Allowable armature current at moment of excitation (i.e., at 25 ^{o}C) is:

$$I_a = \sqrt{\frac{105}{2.5}} = 6.48 \text{ A}$$

"Cold" torque at 25 ^{o}C is:

$$T_g = 6.48 \times 80 = 518 \text{ oz-in}$$

With armature at 155 ^{o}C and ambient of 25 ^{o}C, we estimate magnet temperature will be 90 ^{o}C. Then

$$\Theta_{rm} = 90 - 25 = 65 \ ^{o}C$$

$$\Theta_{rw} = 155 - 25 = 130 \ ^{o}C$$

Use (2.6.7) to calculate allowable "hot" torque.

$$T_{gh} = \frac{80\,(1 - 0.002 \times 65) \times 6.48}{\sqrt{1 + 0.00393 \times 130}}$$

$$T_{gh} = 367 \text{ oz-in}$$

The derating factor is found by use of (2.6.8):

$$\frac{T_{gh}}{T_g} = \frac{1 - 0.002 \times 65}{\sqrt{1 + 0.00393 \times 130}} = 0.708$$

or, using the calculated T_{gh} and T_g:

$$\frac{T_{gh}}{T_g} = \frac{367}{518} = 0.708$$

□

2.7. OTHER CHARACTERISTICS

Other mechanical motor characteristic besides those previously discussed (speed and torque output, power dissipation, and elasticity) are also important. These are: mounting accuracies; noise problems (such as mechanically generated audible noise, noise and dynamic balancing, noise caused by side loading, bearing and brush noise); environmental considerations; brush wear; and motor life. Each of these characteristics will be covered separately in the following discussion.

Further, we will discuss the problems of demagnetization.

MOUNTING

As an illustration of the construction accuracies built into a servomotor, we will look at an Electro-Craft ET-4000 motor integrally mounted with a M-110 Tachometer, as shown in Fig. 2.7.1.

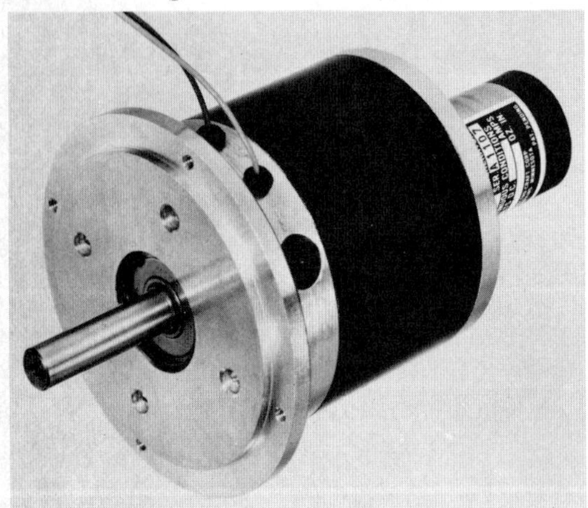

Fig. 2.7.1. Electro-Craft ET-4000 Motor and M-110 Tachometer, integrally mounted.

The outline drawing (Fig. 2.7.2) shows that the mounting surface is held perpendicular to the shaft within 0.005 in , which means that when the shaft is rotated about the mounting surface it can rise and fall no more than 0.005 in total when measured at a radius of 1.75 in.

Also note that shaft end play and radial play are closely controlled. Radial play is especially significant in applications which require a stable shaft position when radially loaded.

Projecting above the level of the mounting surface is a close-tolerance, circular pilot boss which, when matched with a pilot hole in the mating mounting structure, facilitates interchangeability and minimizes the need for machine calibration adjustments. The mounting surface in our illustration is fitted with four tapped holes to accept screws which clamp it to the mounting panel.

The mounting panel, in addition to containing the pilot hole and the attachment screw clearance holes, must have certain other characteristics. For example, it should be stiff enough so that it does not deflect significantly when radial loading is applied to the motor shaft. Also, it should have good thermal conductivity, which is particularly necessary if peak performance is being demanded of the motor.

All applications, of course, are not identical; but all must adequately consider the factors described above. Foot-mounted motors or

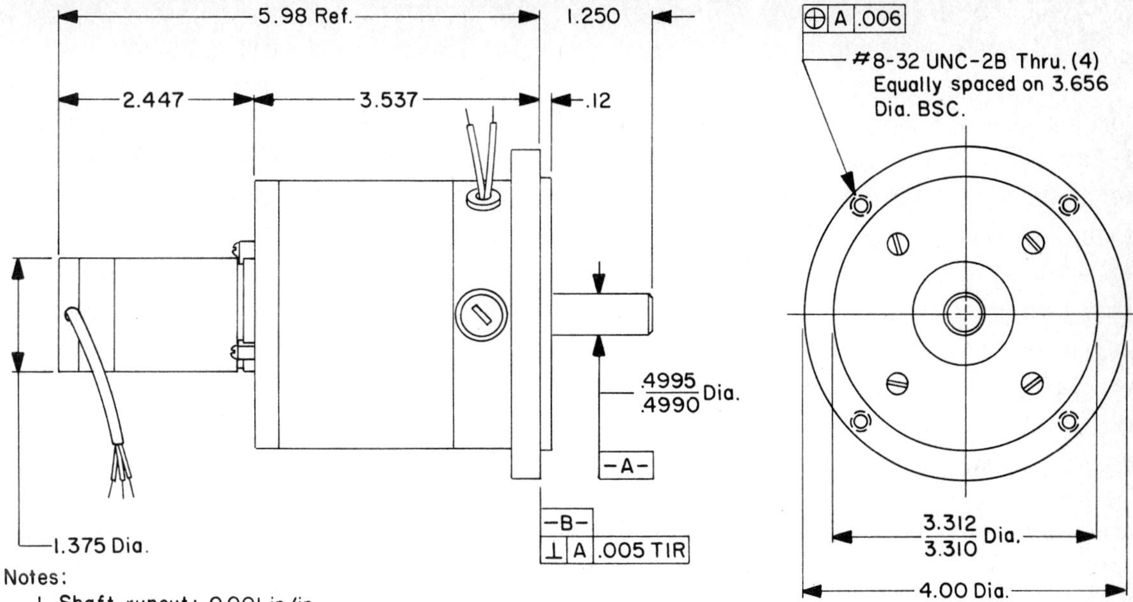

Notes:

1. Shaft runout: 0.001 in/in.
2. Shaft endplay: 0.004 in max. 0.005 in min. when measured with a 20 lb reversing load.
3. Shaft radial play: 0.0015 in max. when measured with a 3 lb load.
4. Runout of mtg. surface -B- within 0.005 T.I.R. measured at 1.75 in radius.

Fig. 2.7.2. Typcial outline drawing of a motor.

those mounted to a gearhead may utilize slotted mounting holes and/or flexible couplings, for example, instead of pilot mounting.

NOISE

Mechanically generated audible noise is usually an unwelcome output from any machine, and DC servomotors are no exception. Unfortunately, however, acceptable noise limits are difficult to define and even more difficult to verify, since quantitative noise measurements are costly. Consequently, in the interest of economics, noise is seldom specified as a measurable parameter. In those applications which require it, noise is controlled qualitatively by means of some readily measured output, such as a vibration level, which can be attributed to rotor imbalance, excessive side loading, excessive bearing noise, and improper bedding of brushes to the commutator.

NOISE AND DYNAMIC BALANCING

Critical speed of a motor is the speed at which vibrational resonance occurs. This critical speed should be as far above the maximum operating speed of the motor as practical to avoid undue amplification of the vibration amplitude. For this reason, and since servomotors are intended for rotational services, the rotors should be dynamically balanced; i.e., balanced under running speeds.

The higher the operating speed, the more critical dynamic balancing becomes. For very high speed operation, the rotor should be balanced at the operating speed rather than at the normally used standard balancing speed. The amount of rotor imbalance which can be tolerated depends on the particular motor and applications. In general, however, it can be said that for satisfactorily, smooth and quiet operation, rotor vibration must not result in motor accelerations greater than a fraction of standard gravitational acceleration, **g**.

NOISE CAUSED BY SIDE LOADING

Excessive side loading of the output shaft can result in shaft bowing and internal misalignments, thus causing noisy operation due to overloading of the bearings. To avoid this, a particular motor design must be matched to the application.

BEARING NOISE

Excessive bearing noise can be caused by a mismatched ball complement, dented races, ring distortion, misalignment, insufficient lubrication, or overloading. Bearings which exhibit rough, noisy operation should be replaced. Sleeve bearings, while they are usually quieter than ball bearings, are not as suitable as ball bearings for many applications. They require more length than ball bearings and use a lower viscosity lubricant, which is more likely to migrate to unwanted locations in the motor. Moreover, they are not as well suited for carrying axial loads.

BRUSH NOISE

Smooth operation of the brushes on the commutator enhance the quietness of the motor. This is often one of the most difficult factors to control, since it is dependent on the initial surface finish, cylindricity, concentricity, subsequent wear and filming of the commutator, bedding of the brushes to the commutator, and fitting of the brushes to their holders. Material content of the brushes will also contribute to the noise factor. Brushes having high metal content (which is introduced to reduce resistance) are far noisier than nonmetallic brushes. Therefore, it may be necessary to sacrifice low brush resistance in order to achieve quieter motor operation.

ENVIRONMENTAL CONSIDERATIONS

Motors which operate in environments which are more severe than a standard room temperature environment usually require special design attention. For example, at high altitudes, motors must be provided with specially treated brushes; in elevated temperatures, motors must be either derated or provided with cooling; and when exposed to chemical liquids or fumes, dust or dirt, motors require additional sealing. All of these affect the useful life of the servomotor.

BRUSH WEAR

The key to proper brush operation is the commutator surface film formation. The film prevents undue wear, and yet permits a low resistance current path between brush and the commutator. A variety of conditions affect the film maintenance, but two basic factors are dominant: brush *pressure* and the *current density* of the brush.

The inter-relationship of brush pressure and brush wear is shown in Fig. 2.7.3, where the optimum wear region is shown as an operational area between the electrical wear region (insufficient pressure to insure proper conductance) and the mechanical wear region (high pressure which destroys the film). Thus, the brush spring must be designed so that the brush pressure will stay in the optimum wear region during its wear life.

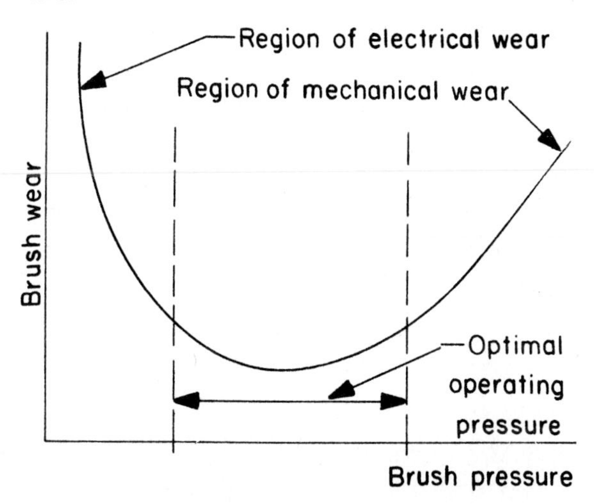

Fig. 2.7.3. Generalized picture of brush wear at a constant speed under varying brush pressure.

The brush must be designed with sufficient cross-sectional area with respect to its conductance value to insure cool operation under worst-case current conditions.

Brush life, then, depends on load conditions, commutator surface velocity, operating temperature, humidity level, vibration, reversing cycles, choice of brush and spring material, commutator design, commutator finish and runout, to mention the most important factors.

Today's DC permanent magnet motor has brush designs which exceed 5 000 hours life in normal uses. Electro-Craft Corporation has Moving Coil Motors which have qualified for use in incremental motion systems, measuring the life span in hundreds of millions of start-stop cycles of pulse currents of 20-30 A. Brush selections in Electro-Craft motors have been made after many years of life testing under various conditions.

MOTOR LIFE

In a typical application, the service-free life of a motor may exceed 10 000 operating hours. The useful life of a DC servomotor is dependent on the severity of its application. Operating a motor continuously at rated output will result in shorter life in terms of operating hours when compared with operation at a lower average power level. Life is limited by brushes, bearings, and winding insulation.

Brush life is influenced by the brush material, geometry, brush spring pressure, commutator, motor duty cycle, and operating environment. Brush life is reduced by electrical erosion, which will cause rapid wear of the brushes and commutator, and also by excessive spring pressure, rough commutators, reduced atmospherical pressure, or by the presence of abrasive particles.

Bearing life is dependent on loading, speed, and lubrication. Most servomotors have their bearings loaded to only a small portion of their rated capacity, but are operating at a relatively high speed. Lubrication must then be considered.

Ball bearings utilize a grease to inhibit corrosion, transfer heat, and to prevent the entry of contaminating particles. For each bearing application there is an optimum fill of grease. Deviating from the optimum by using either too much or too little grease, or perhaps a grease of different composition, can shorten the life of the bearing.

Sleeve bearings are usually made from a porous material (typically sintered bronze) which contains oil to lubricate the bearings. Sometimes additional oil is stored in a reservoir.

The insulation of the motor winding can break down with time due to the influence of high temperature, rotational stresses, oil migration, or contamination with chemical fluids or fumes, thus limiting motor life.

DEMAGNETIZATION OF PM MOTORS

It was shown that energy conversion is accomplished in a PM motor by means of the magnetic field provided by the permanent magnets. These magnets, which provide the field, must be activated or magnetized by placing them into a strong external field. This initial magnetization of the motor is usually accomplished by magnetizing equipment at a stage in the production process prior to the assembly of the motor. Even as the magnets must be magnetized to make them functional, so they can be demagnetized to lose their torque-developing capability. Such a demagnetizing effect occurs whenever a current flows in the motor armature. The armature in effect becomes an electromagnet which tends to oppose the main stator magnetic flux.

In most cases this armature demagnetization, which is called *armature reaction*, has only a reversible effect; i.e., when the current goes to zero the magnets return to full strength. This is the effect that causes an apparent loss in the torque constant, K_T, of the motor under heavy load conditions. However, there is a definite value of current which will cause a permanent demagnetization of the motor. When this occurs the magnetic field intensity, developed by the armature reaction, is great enough to overcome the magnetic domains of the magnet structure.

Fig. 2.7.4. shows how armature reaction and main field relate to each other. It can be seen that if the armature reaction is represented by a resultant vector, **A**, lying along the neutral axis there will be successively weaker components acting radially outward as the angle θ increases. As those component vectors at angles $\theta > \theta_n$ reach the intrinsic coercive strength of the magnet material, they will cause the portion of magnet opposite them to be permanently demagnetized.

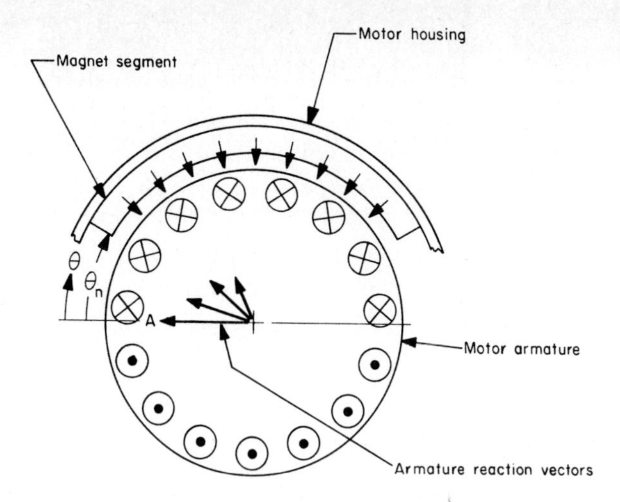

Fig. 2.7.4. The armature demagnetization, or armature reaction effect causes an apparent loss in K_T.

It can be seen that demagnetization is not an "all or nothing" proposition. Rather, the amount of demagnetization is a function of the magnitude of armature current. For purpose of a standard definition, it is common to speak of demagnetizing current as the current that will cause a permanent decrease of 5% of the torque constant, K_T. Of course greater values of current will cause a greater decrease of K_T. The partially demagnetized motor will always be stabilized at the reduced value of K_T. That means that having been partially demagnetized, it can withstand subsequent applications of demagnetizing current without additional degradation of K_T. If a motor is demagnetized, it must be disassembled and remagnetized.

A major reason for the popularity of PM motors in recent years is the high intrinsic coercivity of the new ceramic magnets. This characteristic permits relatively thin magnet segments (for economical motor design) to be used without experiencing bothersome demagnetizing problems.

It is common practice in motor design to provide the magnets strong enough to withstand armature reaction developed by up to seven or eight times the rated current. Thus if the rated current for a motor is 5 A, it is usually safe to assume that current pulses up to 35 A can be applied safely. If some applications create currents in excess of the demagnetizing level, it is necessary for the controller to incorporate some type of current limiter.

Example 1: A motor has the following ratings:

rated current $I_a = 16$ A

armature resistance (at 25°C) $R = 0.25\Omega$

What is the maximum voltage that can be applied suddenly with the motor at standstill without running the risk of demagnetization?

Using the 7:1 criterion to determine the demagnetizing current for the subject motor we obtain:

$$I_d = 7 \times 16 = 112 \text{ A}$$

At standstill the armature current is limited only by the armature resistance; so then,

$$V = 112 \times 0.25 = 28 \text{ V}$$

We can assume that applying 28 V is likely to cause some demagnetization. Then a safe voltage would be 10% less; or, a safe voltage is:

$$V = 28 \times 0.9 = 25.2 \text{ V}$$

Motor demagnetization commonly occurs when the motors are used in an application requiring a "plug reversal". A plug reversal means a sudden switching of the polarity of the applied voltage to achieve a rapid reversal of motor rotation, or perhaps a quick stop. When a motor is plug-reversed, it sees not only the voltage applied to its terminals, but also the armature generated voltage as voltage sources.

This condition is shown in Fig. 2.7.5. It can readily be seen that when this type of application is encountered a proper investigation of the motor demagnetization characteristics must be made.

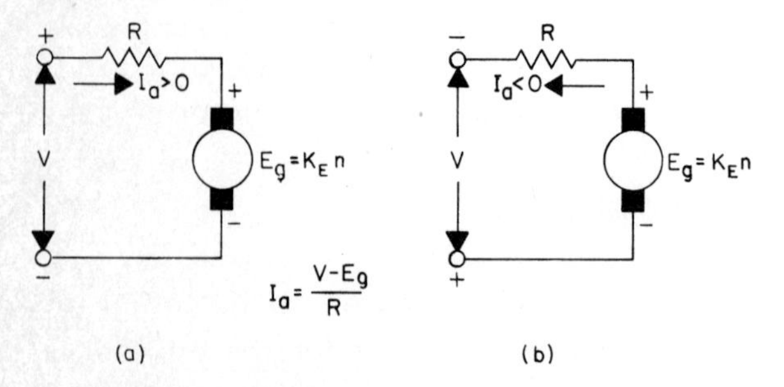

Fig. 2.7.5. In (a) the motor is running in a forward manner. In (b) the motor is plug-reversed. The generated voltage, E_g, is that voltage at the instant of reversal.

Example 2: The same motor used above in example 1 is to be used in an application where it runs under a no-load condition at 1200 rpm and then must be stopped as quickly as possible. If the voltage constant of the motor **K_E = 18 V/krpm**, what is the maximum plugging voltage that can be used to help stop the motor?

At 1200 rpm, the generated armature voltage will be:

$$E_g = 18 \times 1.2 = 21.6 \text{ V}$$

From example 1, demagnetizing current is 112 A. Then the plugging voltage can be calculated by the current formula shown in Fig. 2.7.5.

$$V = R I_a + E_g$$

$$= 0.25\,(-112) + 21.6$$

$$V = -6.4 \text{ V}$$

Again, since the demagnetizing current was calculated from an assumed 7:1 ratio of demagnetizing current to rated current, a prudent designer would downgrade the plugging voltage that will be used, or provide current limiting.

In PM motors with Alnico magnets, it is often necessary to provide some type of protection against demagnetization. The most common form of this protection is the use of a soft iron pole shoe between the armature and magnet. This arrangement is shown in Fig. 2.7.6. The high permeability of the pole show provides a low reluctance shunt for armature reaction flux around the permanent magnet, thus protecting it.

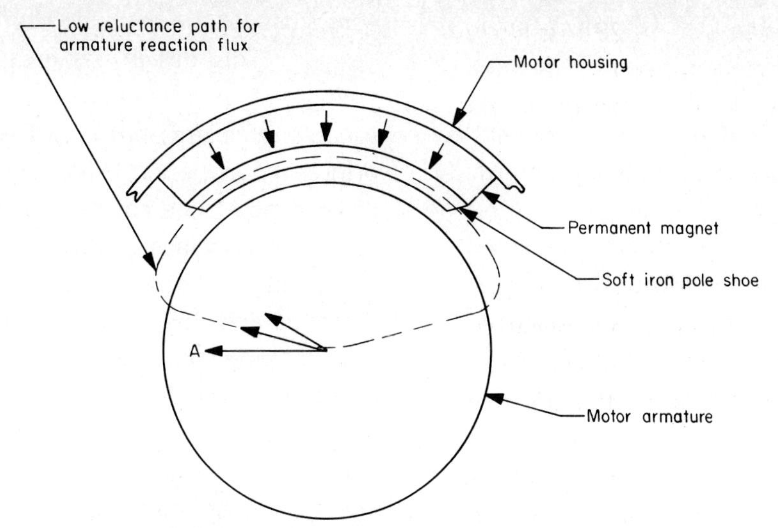

Fig. 2.7.6. A soft iron pole shoe is used for portection against demagnetization.

□

2.8. MOVING COIL MOTORS (MCM®)*

A description of the "iron-less" or "moving coil" concept, covered in this section, will introduce the reader to the MCM* line of permanent magnet DC motors — motors with the highest torque to moment of inertia ratio of any motor type. Motor ratings and their relationship to this unique design is discussed, as well as thermal properties, resonance, and demagnetization.

MOVING COIL MOTORS

In their efforts to improve the performance of servo mechanism actuators, designers have continually sought to produce motors with higher torque and lower moment of inertia. The trend in the design of AC servomotors in the late 1940's and early 1950's was to make motors with small armature diameter and maximum practicable magnetic flux density. A high-performance, low inertia AC servomotor of two decades past was capable of a maximum acceleration of 15 000 rad/s^2.

Dramatic improvements in permanent magnet motor characteristics accompanied the introduction of high-performance Alnico materials, and some basic trends in the design of DC low inertia motors appeared. One was the small diameter iron core armature, operating in very high magnetic flux levels. The more refined style of this design featured conductors bonded to the armature core surface: the so-called *slotless armature* design. This design has great mechanical strength, high torsional rigidity and is reasonably efficient; but it suffers from two principal drawbacks: 1) inductance — hence, its electrical time constant is great; and 2) armature resistance and the torque constant are small — requiring very high armature currents and low power supply voltages, and making the transistor amplifier unnecessarily complex and expensive.

The other design trend, which has since gained almost universal acceptance, is the *moving coil* concept. This principle basically consists of a multiple d'Arsonval movement with a commutation arrangement. This working concept is not new; patents granted at the turn of the century described moving coil motors — not necessarily of the low inertia variety — but supposedly offering high efficiency and good commutation. The adhesive and polymer technology being what is was then, the inventor had the rotor structure held together with iron bands, continuing around the circumference, and undoubtedly contributing enormous eddy current losses.

The moving coil structure design of the present era has followed two general paths: the flat disc armature and the "shell" or "cup" armature. The two structures are shown in Fig. 2.8.1 and 2.8.2. Both units have a multitude of conductors which move in a magnetic field, connected in a manner explained in section 2.1. The armature structure is supported mainly by non-magnetic materials and the active conductors

*Moving Coil Motor, or MCM is a registered trademark of Electro-Craft Corporation.

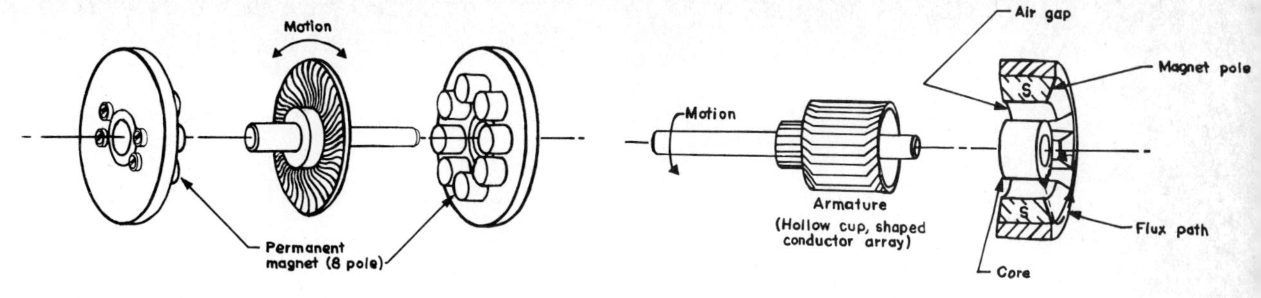

Fig. 2.8.1. Disc armature motor ("printed" motor).

Fig. 2.8.2. Shell armature motor.

are therefore moving in an air gap with a high magnetic flux density.

The moving coil motor is characterized by the absence of rotating iron in the armature. Since the conductors are operating in an air gap, the armature features low inductance, hence low electrical time constant – typically less than 0.1 ms. The absence of iron in the armature also brings about another benefit: there is no reluctance torque effect, i.e., no magnetic cogging.

The *disc* armature motor shown in Fig. 2.8.1 is often called the "printed motor". This is in reference to its early production process in which the armature was fabricated by means of photoetching – similar to the technique used in making printed circuits. Now printed motors are made from stamped segments, which are arranged and joined to form a continuous conductor pattern and a commutating surface. Fig. 2.8.1 shows an 8-pole configuration which, when assembled, provides flux across an air gap of about 0.1 in. Current flow is radial across the disc surface, with rotary forces acting on these conductors tangentially. The *end turns* (the path from an active conductor under the "north" pole to the corresponding conductor under the adjoining "south" pole) are at a relatively large radius, and since the moment of inertia of a disc increases by the fourth power of its diameter, the outer end turns therefore contribute a large amount of inertia to this type of armature.

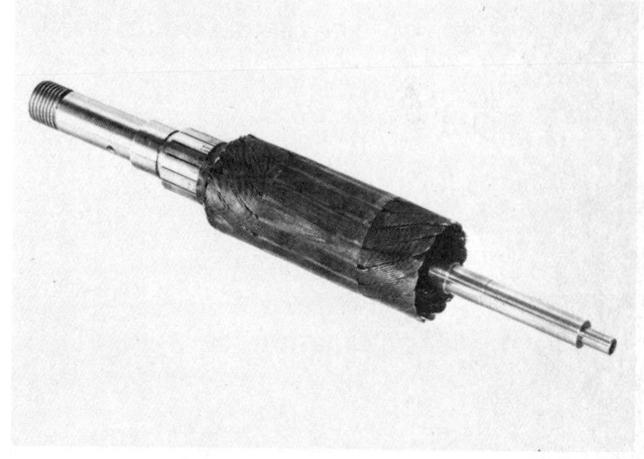

Fig. 2.8.3. Shell armature.

The *shell* type armature (Fig. 2.8.2) on the other hand consists of a cylindrical, hollow rotor which is fabricated to form a rigid shell structure by bonding copper or aluminum coils or *skeins* by the use of polymer resins and fiberglass and other structural members (Fig. 2.8.3). This method offers

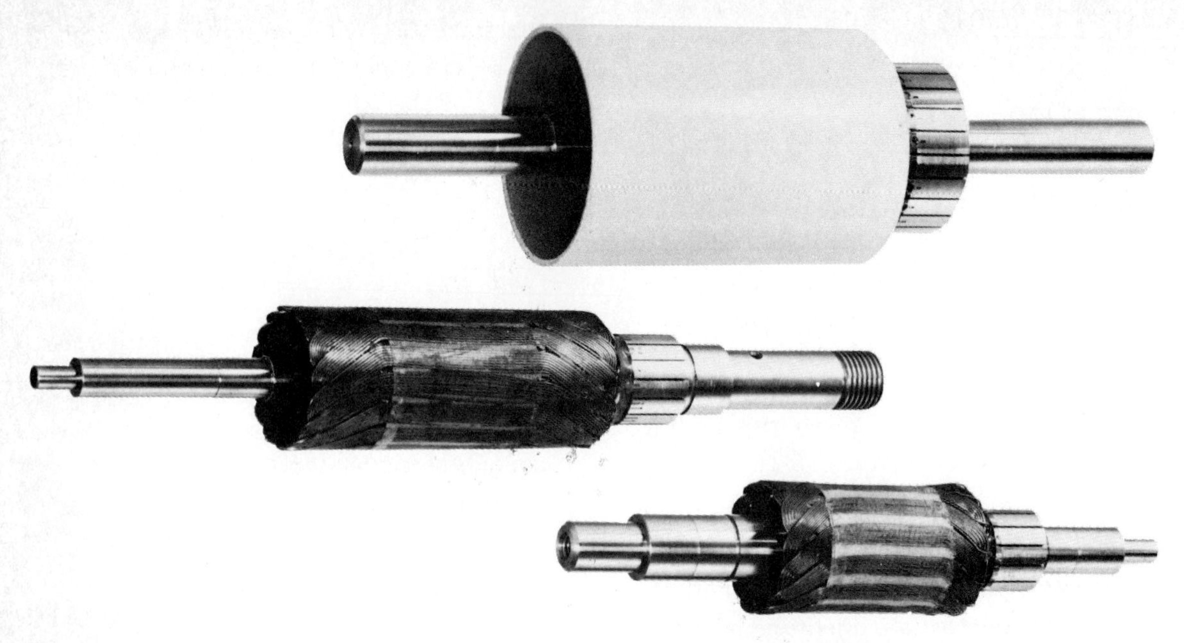

Fig. 2.8.4. Typical Electro-Craft MCM motor armatures.

considerable flexibility in design since the manufacturer can offer a variety of wire sizes, turns per coil, and diameter and length options.

A sample of the variety of armatures manufactured by Electro-Craft is shown in Fig. 2.8.4. Because of the cylindrical shape of the shell-type armature, the end turns do not burden the armature by inequitable contributions of inertia.

Consequently, the shell-type armature has the highest torque-to-moment of inertia ratio, providing acceleration capabilities of up to 1 000 000 rad/s^2. Fig. 2.8.5 shows the complete Electro-Craft MCM motor, capstan, optical encoder and vacuum manifold for an incremental tape transport operating at speed of 200 in/s, capable of 250 start-stop cycles per second.

Electro-Craft moving coil motors are used in incremental tape transports, line printers, optical character readers, incremental motion drives, phase-locked servos, computer printers, machine tool drives and video recorders.

Fig. 2.8.5. Entire motor, capstan, optical encoder and vacuum manifold for tape transport.

MOVING COIL MOTORS AND MOTOR RATINGS

The moving coil motor has the highest acceleration capability of the DC motor family; it can handle a start-stop duty cycle of several thousand cycles per second. On the other hand, it can be destroyed in seconds if improperly handled. Therefore, a good understanding of the physical make-up of this exceedingly fast actuator is important to the user.

Moving coil servomotors obey all the fundamental concepts and equations which were developed in section 2.1. However, some of their features are unique. How moving coil motors differ from conventional servomotors is discussed in the following paragraphs.

One common misunderstanding arises from the use of conventional motor ratings among the makers of moving coil motors. For example, the specification sheet for a high performance moving coil motor may state: *Rated Voltage = 24 V; Rated Current = 8 A.* This may be a motor which the manufacturer has proposed to be used in a system having an amplifier supply voltage of 45 V and working at peak currents of 24 A. The designer may rightfully ask what meaning does the ratings have if there is no need to adhere to it.

The use of motor rating numbers originated with the specifications for fixed speed motors, illustrated in Fig. 2.8.6, where we see a typical induction motor torque-speed characteristic with the motor rating point shown at its maximum output power point.

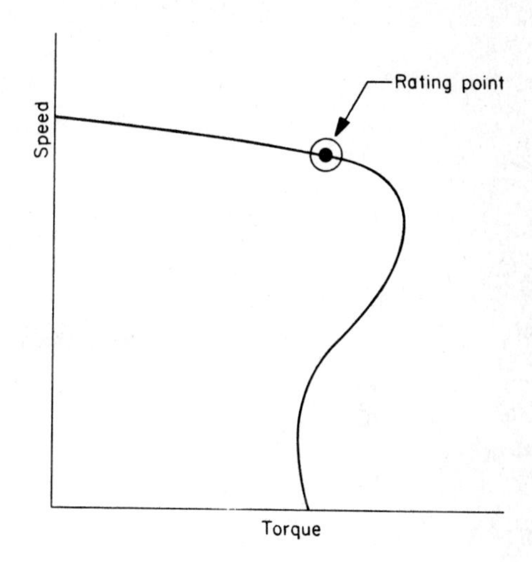

Fig. 2.8.6. Typical rating point on an AC induction motor.

In Fig. 2.8.7 we see similar characteristics for a conventional DC motor intended for use in a drive system. The maximum allowable internal power dissipation locus is shown as a line emerging from stall speed (zero speed) conditions vertically, then leaning toward the speed axis as speed increases. The rating point could then be chosen to be at the one location along this line which would give the maximum output power, thus giving the motor a most favorable rating (if the point would be at a reasonable, utilizable speed).

The moving coil motor speed torque characteristic shown in Fig. 2.8.8 has a maximum power dissipation locus which is essentially vertical for the useful speed range of the motor. One is therefore tempted, in setting

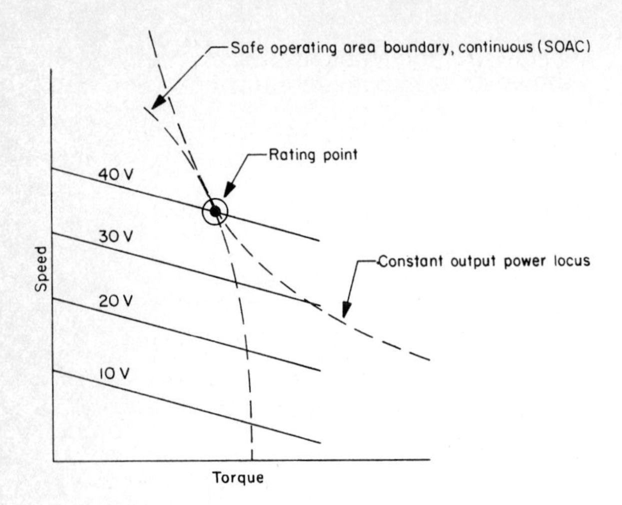

Fig. 2.8.7. Rating point on a typical DC motor.

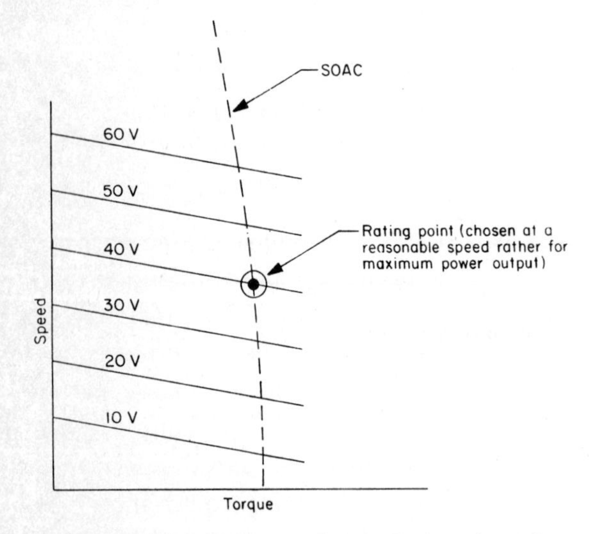

Fig. 2.8.8. Rating point for a typical moving coil DC motor.

a rating point for such motors, to choose the highest possible speed in such a way as to obtain a high rated output power to place on the specification sheet.

Since moving coil motors are used primarily for their incrementing ability rather than ultra high speed performance, such a rating would be unrealistic. It therefore happens that the moving coil motor designer may pick a point along the maximum uncooled power dissipation locus for the motor at a speed which represents a good compromise between speed and long life. Then why this discrepancy between ratings and actual intended use? *The rating numbers for moving coil motors are really obsolete methods of describing the performance capability of a motor intended for use in incremental motion applications, but are still given in specifications merely as a concession to those who like to see these numbers in the rare cases where moving coil motors are used in velocity control applications.*

The really meaningful parameters which properly describe the performance capability of a moving coil motor are the torque constant K_T, motor resistance R, motor moment of inertia J_m, thermal resistance, R_{th} and maximum armature temperature, $\Theta_{a\ max}$. From these parameters most all essential characteristics can be found or derived. Other factors of importance (but often not discussed until the customer discovers them) are shaft and armature torsional resonance and motor shaft critical speed vibrations.

Going back to the essential parameters we recall the following:

Speed regulation constant

$$R_m = \frac{R}{K_E K_T} \tag{2.8.1}$$

[krpm/Nm; Ω, V/krpm, Nm/A]

or

[krpm/oz-in; Ω, V/krpm, oz-in/A]

where

$$K_T = 9.5493 \times 10^{-3} K_E \qquad (2.1.23)$$

[Nm/A; V/krpm]

or

$$K_T = 1.3524 K_E \qquad (2.1.24)$$

[oz-in/A; V/krpm]

Motor mechanical time constant:

$$\tau_m = \frac{RJ_m}{K_E K_T} \qquad (2.8.2)$$

[s;Ω, kg m^2, V/rad s^{-1}, Nm/A]

or

$$\tau_m = 104.72 \frac{RJ_m}{K_E K_T} \qquad (2.8.3)$$

[s; Ω, oz-in-s^2, V/krpm, oz-in/A]

Maximum armature power dissipation (steady state):

$$P_{Dmax} = \frac{\Theta_{a\,max} - \Theta_A}{R_{th}} \qquad (2.8.4)$$

[W; °C, °C/W]

where Θ_A is the ambient temperature.

As we see, all these essential parameters can be derived from the basic ratings given earlier.

THERMAL PROPERTIES

The thermal resistance of a moving coil motor should not be looked on as a single number (as given above for simplicity's sake) but as at least two separate numbers: an armature-to-housing thermal resistance, and a housing-to-ambient thermal resistance. In conjunction with this, the thermal time constant of each case should be considered. This is essential when considering the transient temperature characteristics under severe load conditions, since typical moving coil motors may have a thermal time constant of 20-30 s for armature-to-housing and a thermal time constant of 30-60 min for housing-to-ambient. It is easy to see that the armature could be heated to destructive temperatures in less than a minute without it giving any warning because of the long thermal time constant between housing and ambient.

An analysis of this problem was given in section 2.5.

In order to prevent thermal destruction, air cooling can be provided for the motor. To show the effects of varying amounts of air cooling on motor power dissipation capability, a graphical illustration of cooling air flow aerodynamical impedance and total motor thermal resistance, as shown in Fig. 2.8.9, is usually provided.

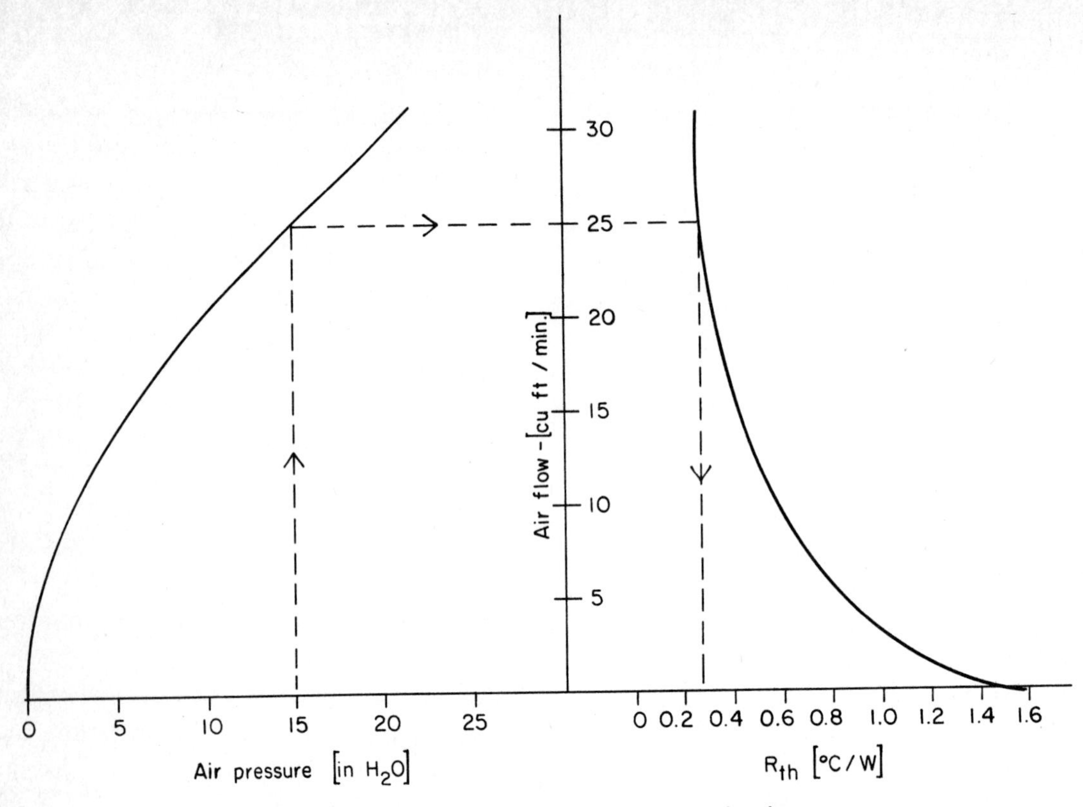

Fig. 2.8.9. The effect of air cooling on thermal resistance.

We can see from this figure that a pressure in 15 in of H_2O will supply 25 cubic feet of air per minute, giving a thermal resistance of 0.28°C/W. By such impedance charts a system designer can select the most economical cooling system to fulfill his needs, within the thermal resistance range of a high of 1.5°C/W to an asymptotic limiting value of 0.25°C/W. Note that this chart shows total motor thermal resistance; it cannot be used for transient studies where the short term power dissipation may exceed the heat capacity of the motor.

Almost all moving coil motors are equipped with Alnico 5 or 5–7 magnets which have a magnetic flux temperature coefficient of –0.05%/°C. This means that the armature magnetic flux will decrease by that amount as the magnet temperature increases, giving a change in torque constant of –0.05%/°C. The motor torque constant, K_T, is directly proportional to the air gap magnetic flux. A common error is to calculate the decline in the torque constant using the proper temperature coefficient, but using the armature temperature as parameter. This gives the wrong answer, since the magnet temperature never reaches the maximum armature temperature. For calculation purposes the housing-to-ambient thermal resistance can be used as a guideline for the magnet temperature. It generally amounts to a worst-case decline of torque constant by a few percent.

RESONANT PHENOMENA IN MOVING COIL MOTORS

Due to the unique construction of moving coil motors and their use in high-performance (wide bandwidth) servo systems, the resonant characteristics may cause problems to the user. The most prominent resonant mode is the torsional resonance between the motor armature and the load, and between the motor armature and the tachometer. Theory of this torsional resonance is described in full detail in section 2.3.4. Fig. 2.8.10 shows a plot of tachometer output with the motor armature excited by a constant amplitude, variable frequency input signal. At lower frequencies, up to 10 Hz, the tachometer output is essentially constant. At the mechanical "break" frequency, 48 Hz, the tachometer response declines −6 dB per octave up to 3 kHz, then a sharp resonant peak occurs at 4.7 kHz, rapidly declines, and again peaks at 5.9 kHz.

This example shows the behavior of the motor-tachometer unit without being connected to a load. When a typical inertial load is direct-coupled to the motor shaft, the plot takes the appearance shown in Fig. 2.8.11. Note that the resonant peak between armature and shaft with load has decreased to 3.8 kHz, and this is what the servo system designer should look at in designing servo system bandwidth and compensation. The amplitude of the resonances and respective frequencies will dominate the servo characteristics in high-performance systems, and it is very important to have these characteristics measured and established before the system design is completed. Thus, if a given motor model and shaft configuration would prove to be difficult for the designer to reconcile, the motor manufacturer has means of suggesting and providing alternatives to solve these problems.

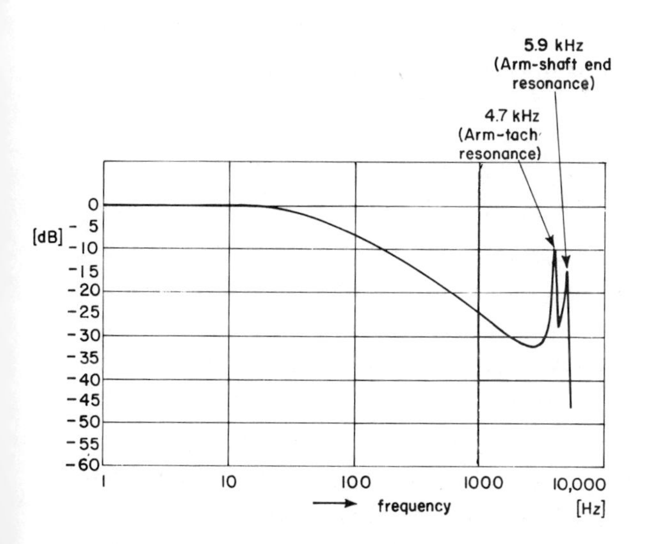

Fig. 2.8.10. Bode plot of an unloaded motor-tachometer.

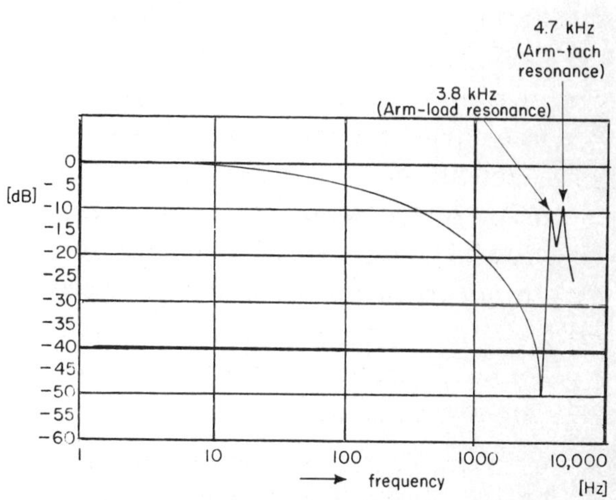

Fig. 2.8.11. Bode plot of motor-tachometer with inertial load.

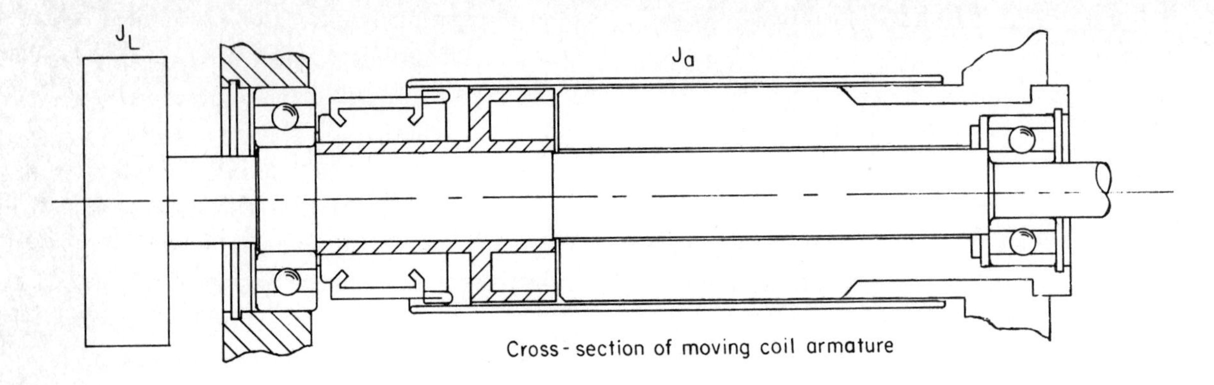

Fig. 2.8.13. Mechancial diagram of a moving coil motor armature and shaft.

Another phenomenon which has been puzzling to many motor users is *critical speed resonance.* The term critical speed is used here advisedly, since it is not the classical critical speed phenomenon described in various mechanical textbooks. The classical critical speed problem is generally defined as the speed at which a shaft will resonate in a bending mode equivalent to that of the "free bar" bending as shown in Fig. 2.8.12.

Note the similarity between the "free bar" and the moving coil shaft-armature arrangement in Fig. 2.8.13.

The true critical speed is that at which the time (period) of one revolution will be equivalent to the period of free bar resonance oscillations. *For the machine sizes we are discussing, the true critical speed would occur in the range of 20 000 to 50 000 rpm — much higher than any practical useful speed for devices under discussion.* However, the bending resonance oscillations can be excited at shaft speeds lower than the true critical speed, such as, for instance, by the commutation frequency.

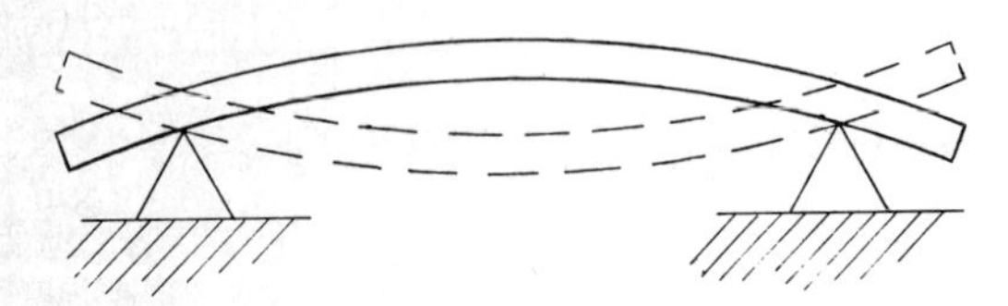

Fig. 2.8.12. Classical example of a "free bar" bending.

Thus it turns out that critical shaft speeds of a moving coil motor are defined by the following equation:

$$n_c = \frac{60\, f_c K_1}{N\, K_2} \qquad [\text{rpm; Hz}, -] \qquad (2.8.5)$$

where

f_c is basic "free bar" resonance frequency of armature shaft and load

N is number of commutator bars

K_1 = 1, 2, 4, 8

K_2 = 1, 2 and sometimes 4

At these critical speeds the problem may manifest itself as a peak in audio noise with tendency for the output shaft to exhibit "whip". The resonance is induced by the commutation current flowing in the winding, generating a torque vector which disturbs the shaft radially. This excites the free bar resonance oscillations of the shaft and armature assembly, and may transmit forces into end caps and other structural members, causing audible noise.

The factors K_1 and K_2 contribute harmonics of the fundamental resonant frequency with amplitudes dependent on specific internal component configurations.

Electro-Craft has developed several means of coping with this problem, and can supply moving coil motors which will meet the most rigid critical speed resonance standards.

DEMAGNETIZING CURRENT

Moving coil motors are generally not as susceptible to demagnetizing peak currents as iron core motors because of their significant airgap and special pole shoe design. Experience at Electro-Craft with moving coil motors shows that peak current limits are set not by demagnetizing effects, but by structural and conductance limits in the armature structure and brush assembly. □

2.9. SPECIALTY MOTORS – Permanent Magnet Motors With Variable K_T

2.9.1. INTRODUCTION

While the permanent magnet DC motor has many advantages over other motor types for speed control and servo applications, in some instances its inherent characteristics cause problems. Such applications are, typically, digital tape transport drive systems, where high torque forward drive and high speed rewind requirements exist; and numerically controlled production machinery, where close position control and rapid traverse conditions prevail. Typically, these applications require the motor to run intermittently at light load and high speed, and other times at heavy load and low speed.

The problem is caused by the inherent correlation between the torque constant K_T and voltage constant K_E (see section 2.1.). Thus, when the designer desires large torque constant to achieve large torque without unduly high current, he then underwrites the need for a high voltage from the amplifier for the high speed condition. On the other hand, if he limits the amplifier voltage, he needs a smaller voltage constant

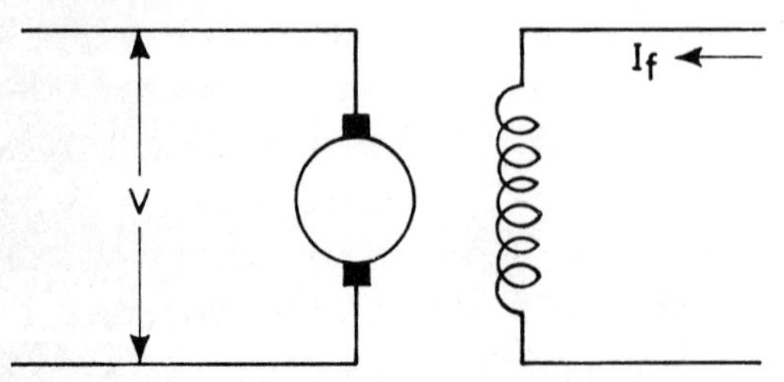

Fig. 2.9.1. Schematic of wound-field motor showing separately excited field winding.

and torque constant, and he must now supply a higher current for the torque application. Since the ratio of optimum K_T for the two modes can easily be 3:1, we can see that the servo controller cost can be a big factor in making design decisions.

2.9.2. THE WOUND-FIELD MOTOR

The shortcoming of the PM motor is due to its constant permanent magnet field; or, in terms of its transfer function characteristic, it is the invariance of the torque constant, K_T. A solution to the problem of prohibitive amplifier costs is to use a motor with an adjustable K_T. The small K_T can then be employed when high speed is required and the large K_T for the high torque mode. A variable K_T motor permits an economical amplifier design.

As is often the case in going to the wound-field (WF) motor, a number of undesirable things also develop. The motor will be substantially larger and heavier (see Fig. 2.9.2) because of volume required by the field windings. This could cause problems with the installation and maintenance of the motors. The wound-field motor is usually more expensive than the PM motor of comparable rating. This is also due to the added costs of the field windings and larger housing. A third factor to consider is the loss of magnetic detent. While a PM motor can be designed so that it will require a measurable torque to turn it from a position of minimum reluctance, when power is removed from the field winding of the WF motor there is nothing to prevent the shaft from turning. This can be very annoying if the application requires that a certain shaft position is to be held after power is removed. In some cases, it is necessary to add a fail-safe brake to prevent shaft rotation, as in the case of a tape reel unwinding due to the weight of tape.

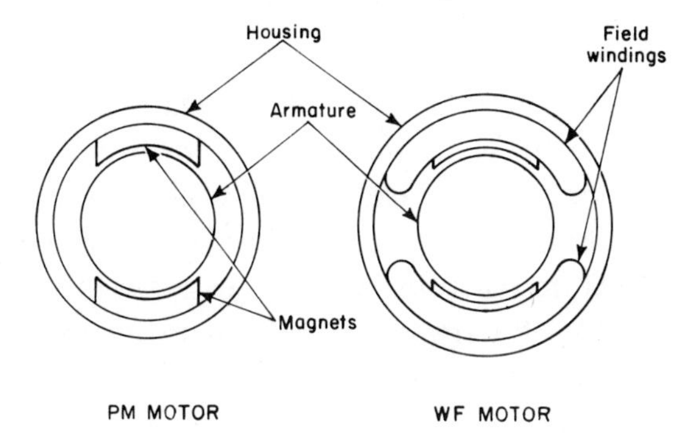

Fig. 2.9.2. A comparison of the relative sizes of wound-field and permanent magnet motor diameters.

2.9.3. NEW MOTOR TYPES

From the foregoing discussion it can readily be seen that the ideal DC servomotor would have cost and performance characteristics of the PM motor and also the versatility of speed control as exemplified by the wound-field motor. This ideal motor has been the goal of DC servomotor manufacturers for a number of years. The efforts in this direction have resulted in several new types of special purpose motors which offer some attractive characteristics. The motors described in the following paragraphs are the subjects of patent applications action by Electro-Craft Corporation.

The Hybrid Motor

An early approach to the design of a motor that would incorporate desirable features of the "ideal motor" was the hybrid motor shown in cross-section in Fig. 2.9.3. The hybrid motor simply took a "wound" field and put it in series with a PM field with both encompassing a common armature.

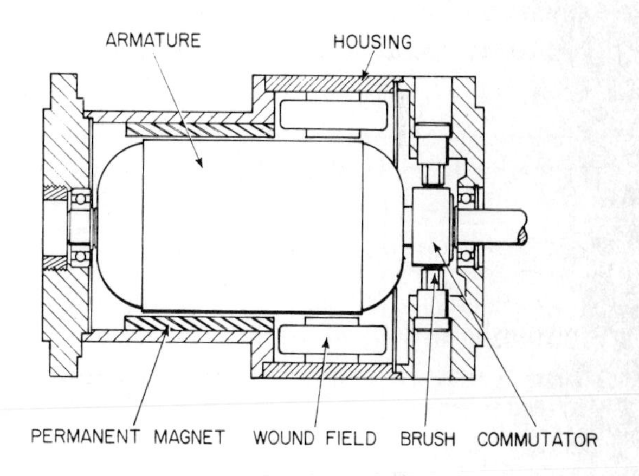

Fig. 2.9.3. The hybrid motor showing permanent magnet and wound-field functions in one armature.

The portion of the armature rotating in the PM field generates a torque that is proportional to the magnetic flux density and the length of conductors therein. This part of the armature is designed also to provide magnetic detent. Another portion of the armature rotates within the "wound" field and generates a torque that can either add to or subtract from that due to the PM. By properly proportioning the relative lengths of WF and PM portions of the armature, the motor K_T can be changed by a fixed ratio merely by switching the polarity of the voltage applied to the field winding.

Utilizing the hybrid motor approach, a 3:1 variation in torque constant can be achieved very easily with a relatively small field winding power. The field winding can be equipped with a center tap so that the winding polarity can be simply switched by means of a flip-flop when triggered by an appropriate signal. The dual mode characteristics of a hybrid motor (the Electro-Craft 5260) are shown in Tab. 2.9.1. It can be seen that the 3:1 ratio is obtained with only 50 W of field power required.

	PM field only	+ voltage on WF	– voltage on WF
Torque constant [oz-in/A]	28	42	14
No-load speed at 30 V [rpm]	1450	965	2850
Field winding power - 50 W			

Tab. 2.9.1. The hybrid motor characteristics.

Although the hybrid approach offers some desirable application characteristics it does not achieve the full cost savings potential and efficiency of the PM motor. A portion of it does have the cost and bulkiness of the WF motor associated with it.

The DAARC Motor

The not completely satisfactory costs and bulk of the hybrid motor led to the DAARC concept. DAARC is a handy acronym which derives from the words "Direct Axis Armature Reaction Control". The acronymous phrase provides a very accurate description of the principle of operation of the motor, which seems to offer the most potential at this time.

The principle of operation of the DAARC motor can most easily be understood by considering Fig. 2.9.4. It shows the arrangement of magnets and armature conductors on either side of the air gap in the PM motor. It can be seen that armature current developes a triangular mmf distribution around the air gap. The peak absolute values of the armature mmf (also called armature reaction) coincide with the placement of the brush axis. If the brush axis is shifted, the armature reaction axis will shift with it. Now in a normal motor the brush axis is usually placed close to the field quadrature axis since this is the most efficient area in developing torque. In contrast current introduced through brushes on the direct axis would produce zero torque. Getting back to Fig. 2.9.4, it can be seen that the armature mmf tends to increase magnet flux at one corner of the magnet and decrease it at the other corner.

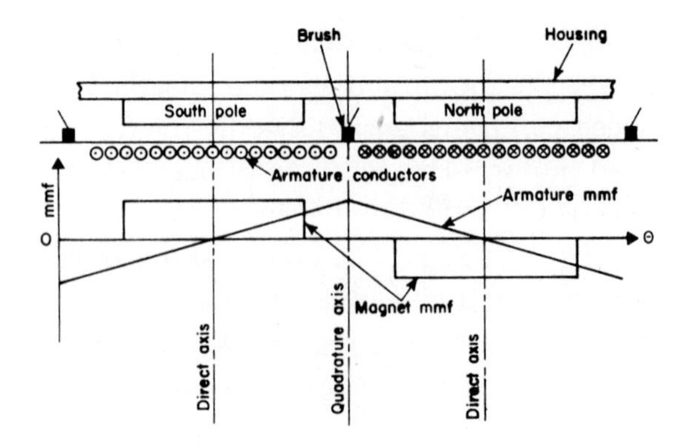

Fig. 2.9.4. Armature mmf in a DC permanent magnet motor.

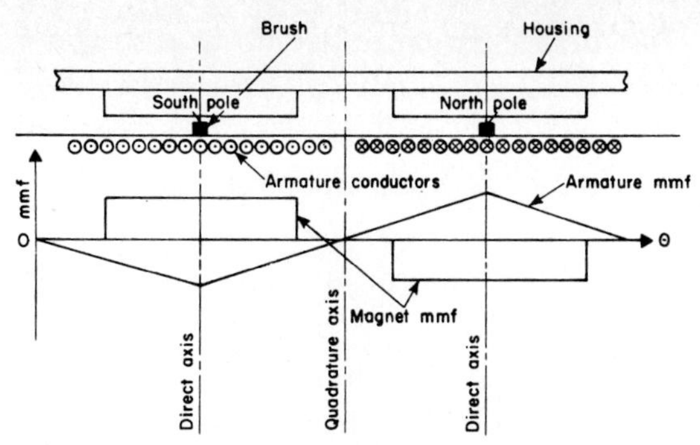

Fig. 2.9.5. Armature mmf in a DC permanent magnet motor with brushes shifted 90°el.

In a normal motor, the two effects will approximately offset and there is little effect on motor action. Now suppose that the brush axis in Fig. 2.9.4 is shifted 90°el so that the positive peak is aligned with the north pole axis. This arrangement is shown in Fig. 2.9.5. It can be seen that the armature mmf is now working in opposition to the permanent magnet, and will act to decrease the air gap flux. Thus, by controlling the current to the direct axis brushes, the air gap flux (and hence torque constant) can be reduced to a desired lower level. This is the principle of the DAARC motor operation.

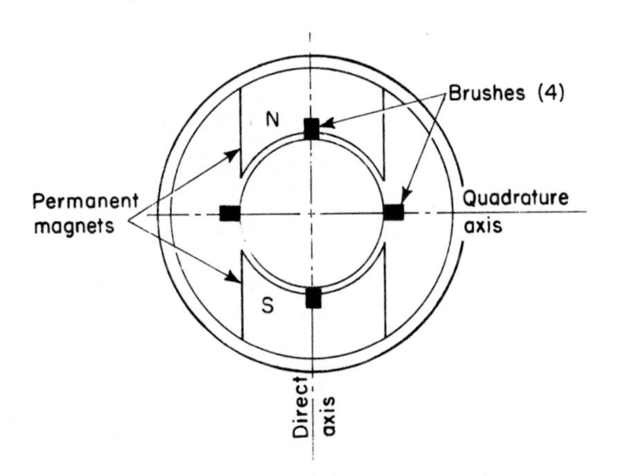

Fig. 2.9.6. Location of brushes in a DAARC motor.

Of course, current going to brushes on the direct axis will not develop torque or motor rotation. It is therefore necessary to add another set of brushes on the quadrature axis for motor operation. The arrangement of brushes is shown for a two-pole configuration in Fig. 2.9.6. The direct axis brushes are used to control air gap flux. They can be excited by a predetermined fixed voltage to drop the motor K_T to the desired level for high speed operation. The quadrature axis brushes are excited by the servo amplifier in the normal manner to develop torque and achieve the desired speed.

The DAARC approach has several advantages over other approaches. Since a zero net voltage is generated across the direct-axis brushes, armature resistance is theoretically the only limiting parameter on direct axis current. This means that a low voltage source is sufficient to supply the current to these brushes. By actually reducing the magnetic flux in the air gap the rotational losses of the motor due to iron loss and circulating currents are made small. These losses are otherwise proportional to speed and can become excessive at high speeds.

Perhaps the greatest advantage of the DAARC motor approach is the potential for cost savings. The basic cost structure is that of a PM motor with the addition of an extra set of brushes. In the practical case some additional design changes to optimize DAARC operation do add back cost but overall cost is very favorable compared to wound-field motors.

	Without direct axis current	With direct axis current
Torque constant [oz-in/A]	43	13
No-load speed at 30 V [rpm]	940	2600
Direct axis input power - 120 W		

Tab. 2.9.2. The DAARC motor characteristics.

Test data from an actual motor is shown in Tab. 2.9.2. It is seen that the motor characteristics are comparable to those of the hybrid motor as shown in Tab. 2.9.1.

The DAARC motor is a response to a requirement for a very special kind of motor. It is truly a good example of the saying that "necessity is the mother of invention".

□

2.10. MOTOR TESTING

This section will enlighten the reader as to how a motor manufacturer checks motor parameters. Incoming inspection and troubleshooting charts are included to completely describe tests performed on motors. The following tests are described:

End play
Radial play
Shaft runout
Moment of inertia
Resistance
Inductance
Friction torque and starting current
No-load current and rotational losses, no-load voltage, no-load speed
Demagnetization current
Torque constant
Voltage constant
Electrical time constant
Mechanical time constant
Torque ripple
Speed regulation constant
Efficiency
Frequency response
Thermal resistance
Thermal time constant
Air flow impedance

END PLAY

A movement of a motor shaft in the axial direction, caused when an axial force is applied on either end of the motor shaft, is termed *end play.* To measure the amount of end play, a hub is placed on the motor shaft, the motor is locked in a test fixture, and an indicator point is placed touching the hub. A predetermined force is applied on one end of the shaft such as $\mathbf{F_1}$ in Fig. 2.10.1, and the deflection is noted on the indicator. The indicator displays the end play directly.

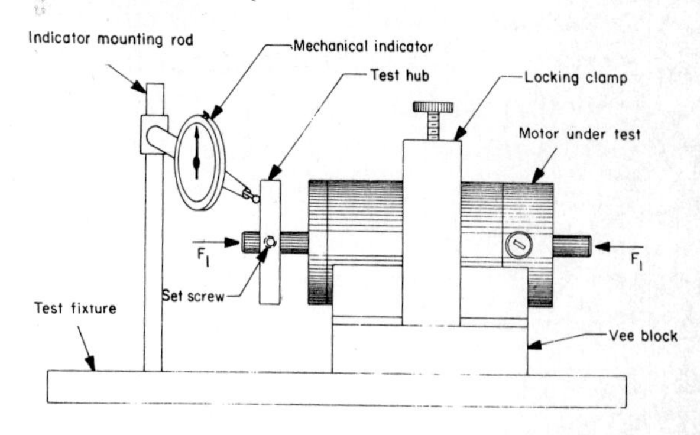

Fig. 2.10.1. End play test setup.

RADIAL PLAY

The amount of deflection of the motor shaft in response to a radial force in the radial direction is termed *radial play*. The amount of radial play is determined by placing a hub on the motor shaft at a specific distance from the mounting surface, locking the motor in a test fixture and placing an indicator point on the hub. A predetermined force, such as $\mathbf{F_2}$ in Fig. 2.10.2, is applied at a specific distance from the mounting surface, first in one direction, then in the other. The total displacement measured is the radial play.

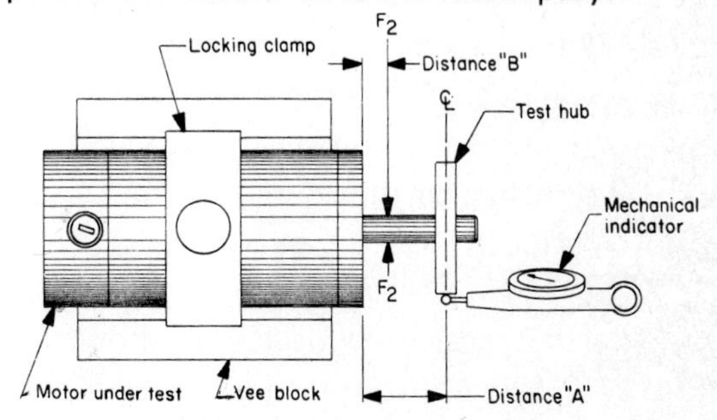

Fig. 2.10.2. Radial play test setup.

SHAFT RUNOUT

As a motor shaft rotates it may exhibit non-concentric characteristics. The amount of wobble is termed *runout*. To measure runout, an indicator point is placed on the motor shaft at a specific distance from the mounting surface, see Fig. 2.10.3. The shaft is rotated through one revolution and deflection is measured on the indicator which reads shaft runout directly.

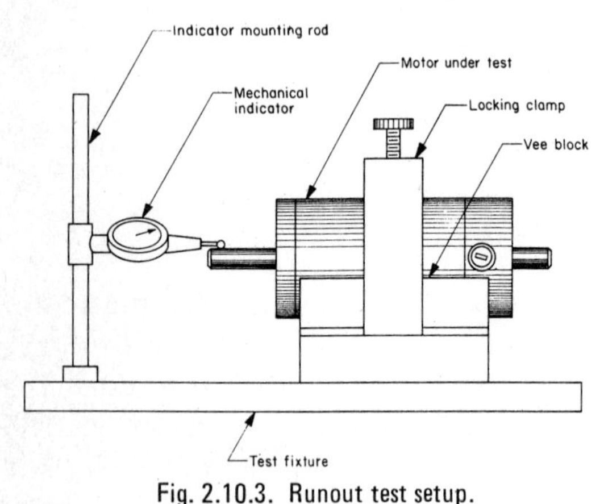

Fig. 2.10.3. Runout test setup.

MOMENT OF INERTIA

Inertia is the inherent property of bodies which resist any change in their state. A measure of this property is the *moment of inertia.* There are two means of measuring moment of inertia. The first involves an inertial measuring device, while the second method makes use of the torsional pendulum technique. Both methods relate moment of inertia through a mathematical proportionality to the square of the period of oscillations.

An inertial measuring device as shown in Fig. 2.10.4 will quickly and accurately provide the desired result.

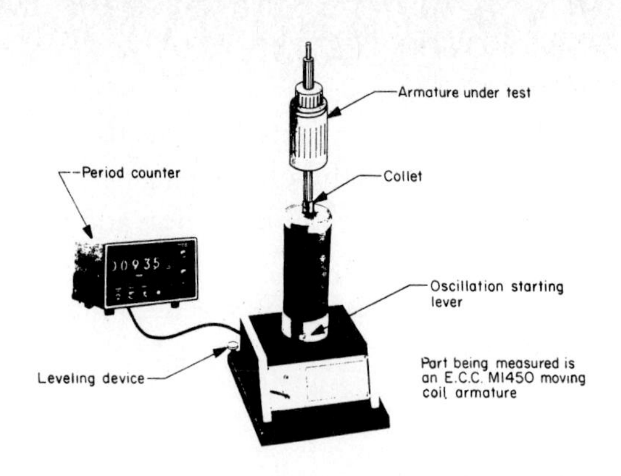

Fig. 2.10.4. Inertial measuring device.

First attach a collet to the shaft, then mount this into the test fixture, turn on the oscillation lever, and read the period of oscillations. Remove the armature from the collet and read the period of oscillations of the collet only. Moment of inertia, **J**, can be calculated from:

$$J = C\,[t_1^2 - t_2^2] \qquad (2.10.1)$$

where

C is the calibrated constant of the instrument

t_1 is the period of oscillations of armature and collet

t_2 is the period of oscillations of collet only

The second method of measurement requires a collet and master cylinder of known moments of inertia and approximately the same size as the armature under test.

Attach the collet to the armature under test and connect as shown in Fig. 2.10.5.

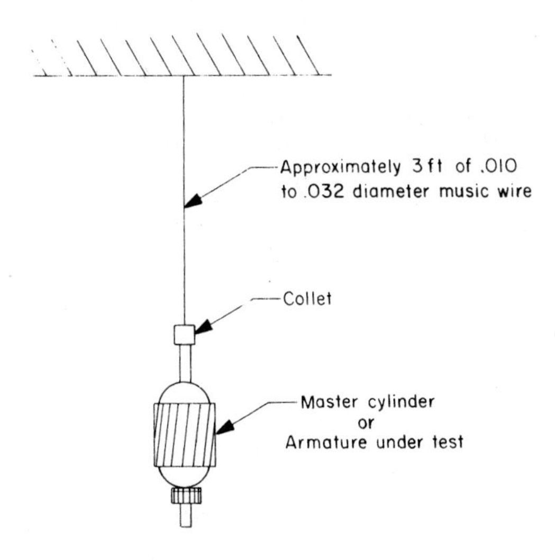

Fig. 2.10.5. Alternate method of measuring the moment of inertia.

Give the armature a half twist and allow it to oscillate freely while recording the time for 20 oscillations. Attach the collet to the master cylinder and repeat the above procedure. Moment of inertia of the armature under test can be calculated from:

$$J = J_1\left(\frac{t_2}{t_1}\right)^2 - J_2 \qquad (2.10.2)$$

where

J_1 is moment of inertia of master cylinder and collet

J_2 is moment of inertia of collet

t_1 is time of 20 oscillations of master cylinder and collet

t_2 is time of 20 oscillations of armature under test and collet

RESISTANCE

In this procedure, the motor under test is driven and the *resistance* of brushes, commutator, and armature winding is measured dynamically. Resistance values are specified at 25 °C; therefore, readings should be made quickly to avoid heating effects; the current should be about 1/4 of the rated current.

A power supply is connected across the terminals of the motor under test while it is being driven with a low-speed motor, as shown in Fig. 2.10.6.

Current is adjusted for a predetermined level and voltage is measured. Resistance is calculated from:

$$R = \frac{V}{I_a} \qquad (2.10.3)$$

INDUCTANCE

In measuring *inductance,* the motor under test is connected to the impedance bridge as shown in Fig. 2.10.7. The **Q** and **L** settings are alternately adjusted for minimum voltmeter readings. When the minimum reading is obtained, the **L** setting indicates the motor inductance. Repeat above procedure for three other shaft positions 90° apart, and take the average of the four readings to determine the motor inductance, L_a.

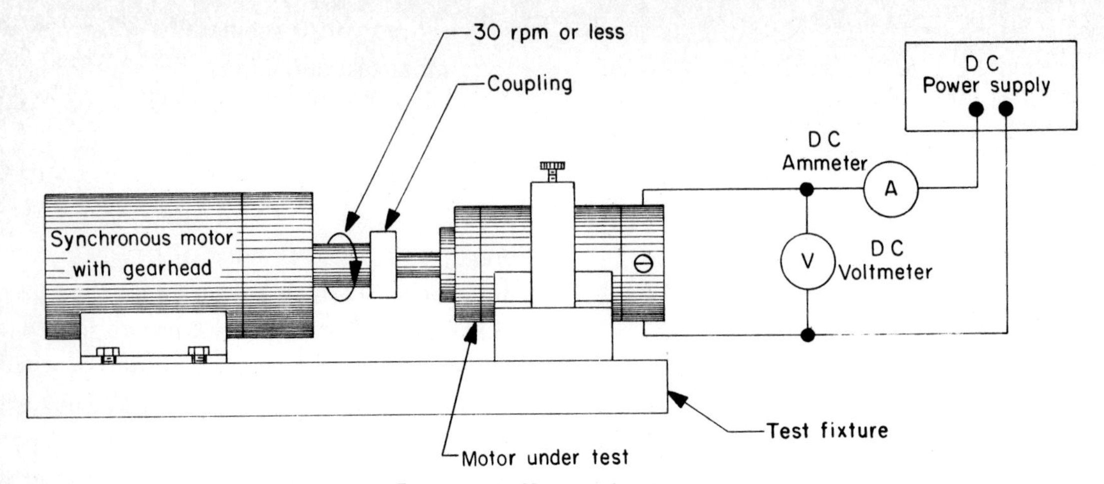

Fig. 2.10.6. Motor resistance test.

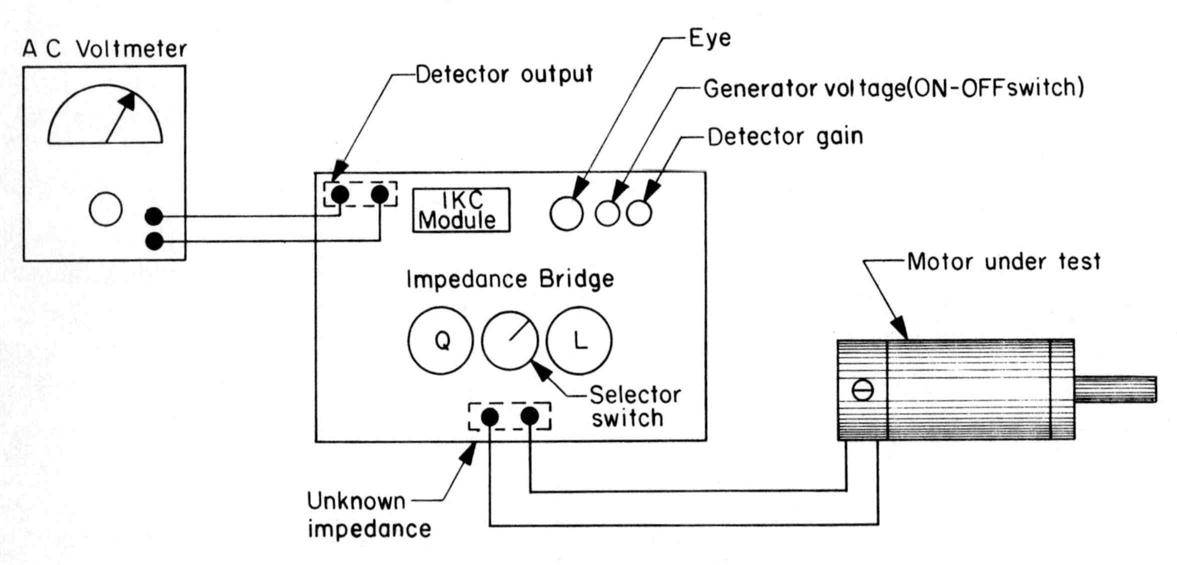

Fig. 2.10.7. Impedance bridge measurements of inductance.

FRICTION TORQUE AND STARTING CURRENT

When current becomes sufficient to overcome torque caused by static friction, motor rotation will start. The minimum current is called *starting current,* and the static friction termed *friction torque.*

Friction torque can be measured by two methods. The first utilizes a torque watch, and the second involves current and voltage measurements. In using a torque watch, attach the watch to the motor shaft as indicated in Fig. 2.10.8, and start to rotate the torque watch slowly. Torque as indicated on the watch will be friction torque.

Repeat this procedure at three other positions 90 degrees apart in both clockwise and counterclockwise rotation directions. The highest reading is the maximum friction torque, T_f.

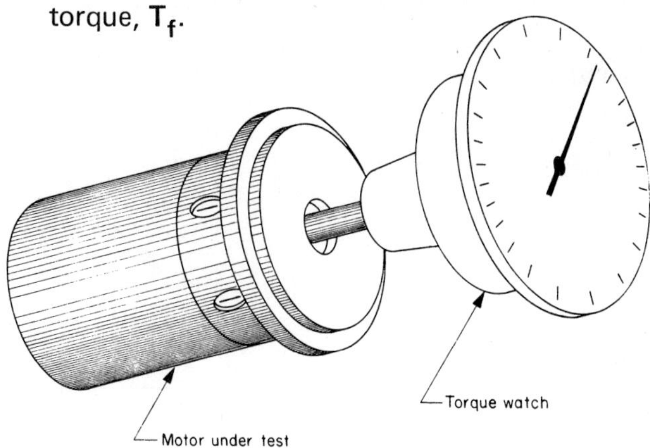

Fig. 2.10.8. Torque watch method of measuring friction torque.

The second method is based on starting current measurements as an indicator of starting torque.

To measure the minimum current required to start rotation, the motor under test is connected as illustrated in Fig. 2.10.9. Voltage is slowly increased until the motor shaft barely turns. This current value is the starting current, I_{as}.

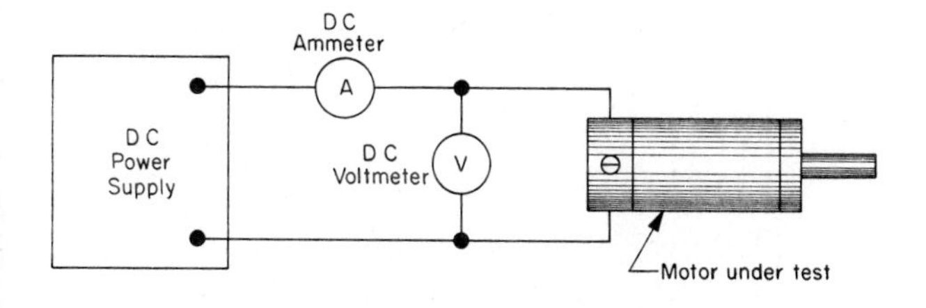

Fig. 2.10.9. Test setup for measuring minimum current required to overcome friction in order to start motor rotation.

The friction torque may be calculated by using the starting current value which was just measured, and the formula:

$$T_f = K_T\, I_{as} \tag{2.10.4}$$

NO-LOAD CURRENT AND ROTATIONAL LOSSES, NO-LOAD VOLTAGE, NO-LOAD SPEED

Not all input power supplied to a motor is converted into mechanical power. There are both mechanical and electrical losses. If a motor is rotating with no load on its shaft, a small current will still be drawn. This is no-load current, I_{ao}, directly attributed to *rotational losses.*

With the motor connected as in Fig. 2.10.-10, a predetermined (rated) voltage is applied. The measured current is termed the *no-load current,* I_{ao}; motor speed is termed *no-load speed,* n_o.

A plot is actually made for determining rotational losses. Voltage is adjusted to correspond to motor speed from 500 rpm up to the maximum allowable safe motor speed, in increments of 500 rpm, while recording currents and voltages. If losses in units of [oz-in/krpm] are desired, the $K_T I_{ao}$ values versus speed are plotted. If losses in units of [W/krpm] are desired, the current I_{ao} times the input voltage V versus speed is plotted. The slope of the line will give rotational loss in appropriate units.

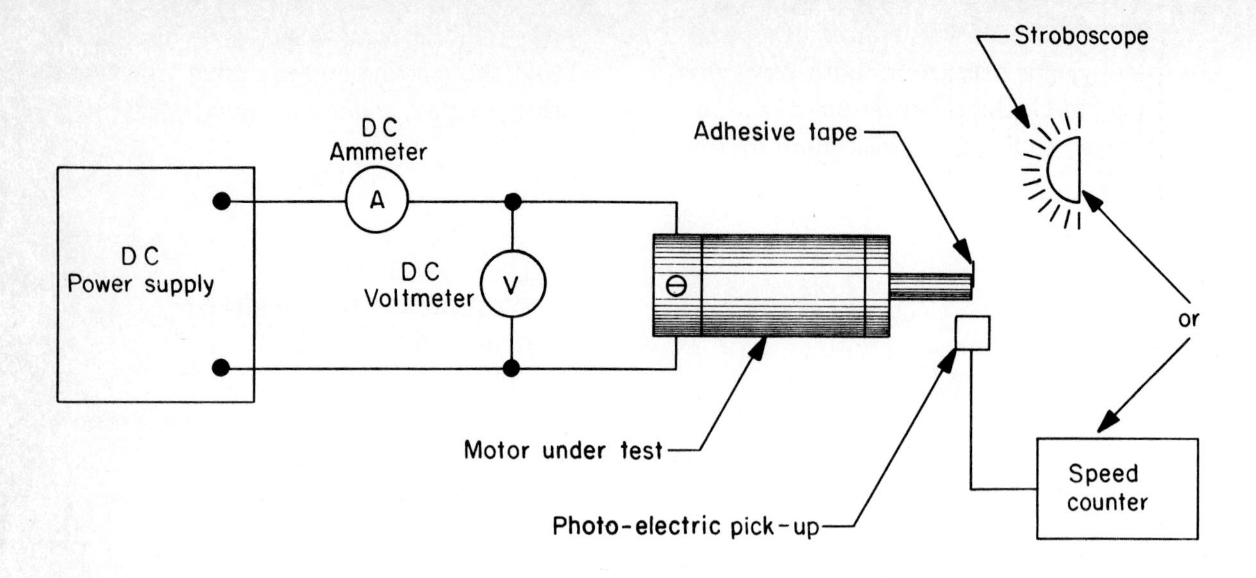

Fig. 2.10.10. Measuring no-load current and no-load speed.

DEMAGNETIZATION CURRENT

When the motor is subjected to high current pulses, the magnets may become demagnetized. This results in a lowered K_T, and requires more current to produce the same torque as before demagnetization.

The motor under test is actually demagnetized to obtain the demagnetization current. The procedure is:

1. Note voltage required for motor speed of 1000 rpm.

2. Insert motor in test circuit of Fig. 2.10.11, lock its shaft, apply a step voltage for a short period of time and measure the current.

The above procedure is repeated, always increasing the magnitude of step voltage of test #2, until test #1 requires 2% less voltage to maintain 1000 rpm. The current which causes the 2% change is the *demagnetization current.*

TORQUE CONSTANT

The measurement of the motor torque constant, K_T, will require a dynamometer or a torque indicating instrument. This test method is based on the setup shown in Fig. 2.10.18. Keep the motor running at a constant speed, while adjusting the load of the dynamometer up to the rated load of the motor. The results of the measurements can be plotted, as shown in Fig. 2.10.12, and an average slope established which will give

$$K_T = \frac{\Delta T}{\Delta I} \tag{2.10.5}$$

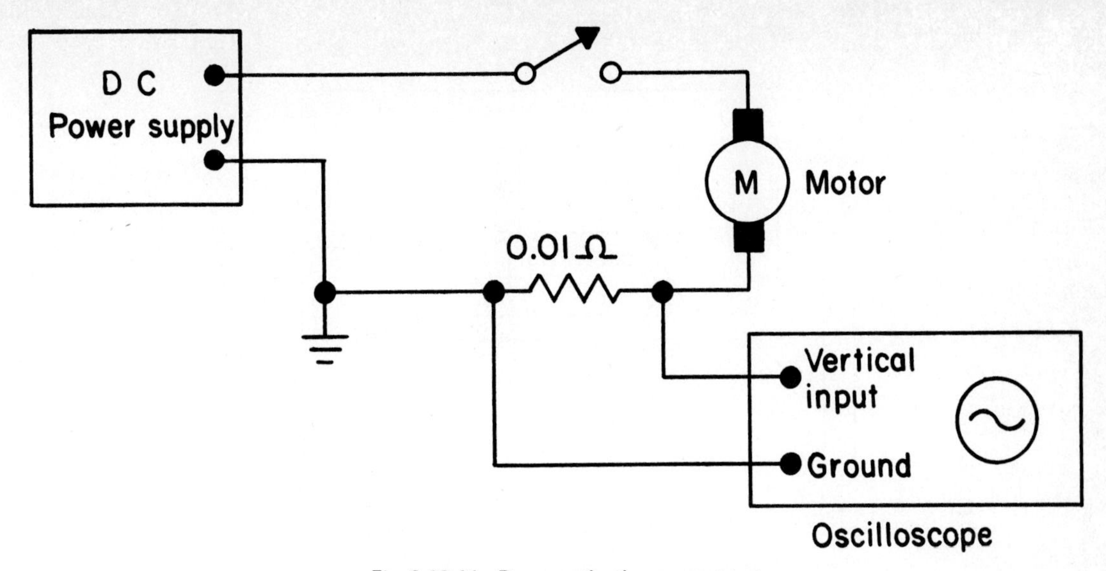

Fig. 2.10.11. Demagnetization current test.

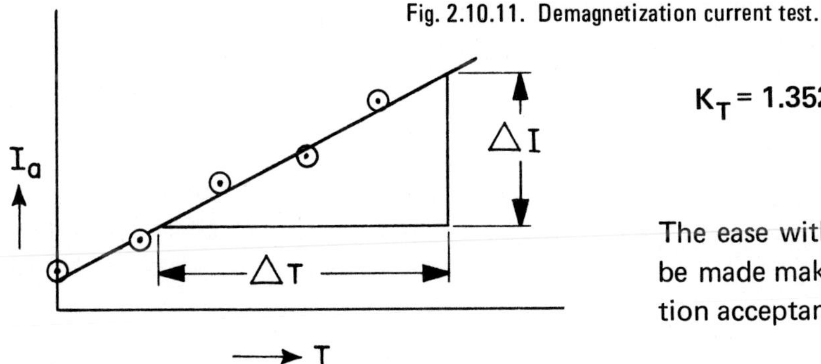

Fig. 2.10.12. Minimizing measurement errors in establishing K_T.

Another way of determining K_T is to calculate the value based on the measurement of the voltage constant K_E (see following test). The torque constant is always related to the voltage constant in the following way:

$$K_T = K_E \qquad [\text{Nm/A; V/rad s}^{-1}] \qquad (2.1.22)$$

$$K_T = 9.5493 \times 10^{-3}\ K_E \qquad (2.1.23)$$

[Nm/A; V/krpm]

$$K_T = 1.3524\ K_E \qquad [\text{oz-in/A; V/krpm}] \qquad (2.1.24)$$

The ease with which this measurement can be made makes it the most suitable production acceptance test.

VOLTAGE CONSTANT

The voltage constant, K_E, can best be tested by running the motor as a generator (driven by another motor) and measuring the generated voltage, E_g, while measuring the shaft speed, n. The voltage constant is then obtained by the following relationship:

$$K_E = \frac{E_g}{n} \qquad [\text{V/krpm; V, krpm}] \qquad (2.1.18)$$

A typical setup is shown in Fig. 2.10.13.

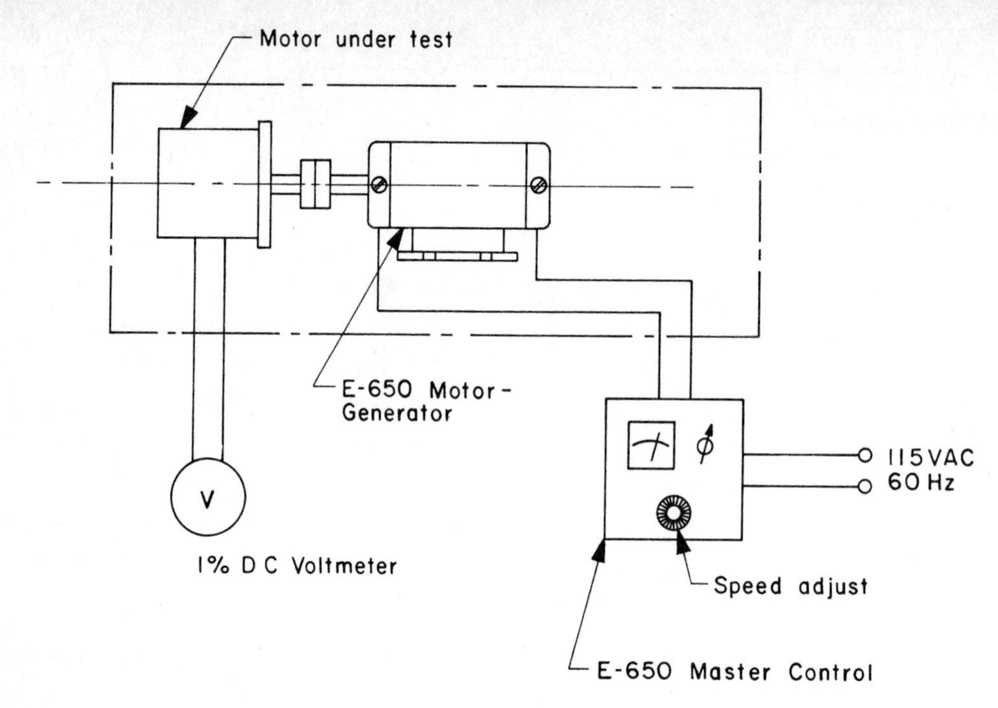

Fig. 2.10.13. Arrangement for motor back-emf test to determine K_E (voltage constant) and K_T (torque constant).

ELECTRICAL TIME CONSTANT

The electrical time constant of a motor may be calculated from the measured values of motor inductance, $\mathbf{L_a}$, and armature resistance, **R**, as follows:

$$\tau_e = \frac{L_a}{R} \qquad [s; H, \Omega] \qquad (2.3.30)$$

In an alternative method using a direct measurement, a step input voltage is applied to the motor armature with the motor shaft locked, and the exponential rise of the motor current is measured on an oscilloscope. The time required for the current to rise to 63.2% of its final value is equal to the electrical time constant, τ_e, of Fig. 2.10.14a.

The motor is connected as in Fig. 2.10.14b. A step input voltage is applied and current through a resistor of low value ($\mathbf{R_1 < 0.1\ R}$, where **R** is the armature resistance) is monitored on the scope in the form of the voltage $\mathbf{v_2}$. The time required for current to reach 63.2% of its final value is measured and the value of τ_e determined.

The present discussion is based on the assumption that the armature impedance is a series combination of a resistance and an inductance. Hence, it can be written as

$$Z(s) = R + sL_a \qquad (2.10.6)$$

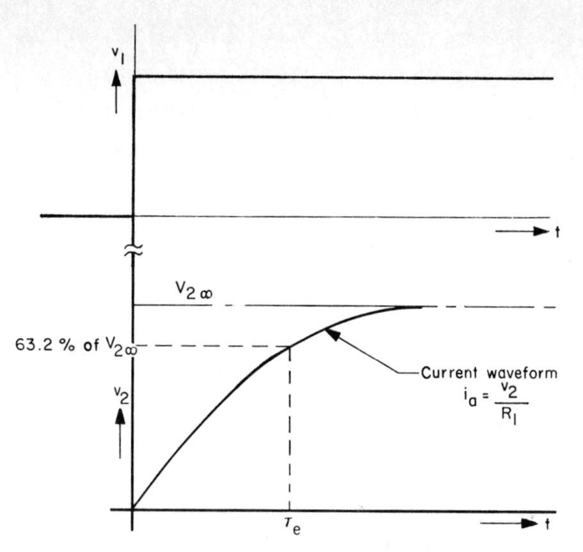

Fig. 2.10.14a. Measuring electrical time constant.

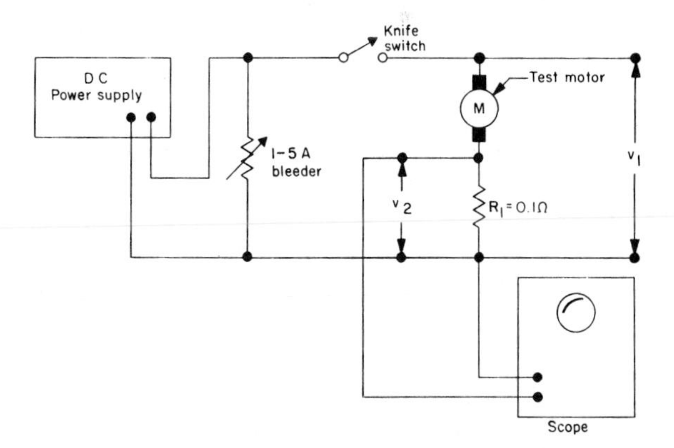

Fig. 2.10.14b. Electrical time constant test setup.

While this description is valid for "conventional" iron core motors, it is found that the impedance of moving coil motors is approximately [see Fig. 2.3.1. and Eq. (2.3.1)]:

$$Z(s) = R + \frac{sL_aR_L}{sL_a + R_L} \qquad (2.10.7)$$

where R_L is an additional resistance in parallel with L_a, making the inductance term not quite orthogonal to **R**. The parallel resistance, R_L, is very large for iron core rotors; for these the approximation of the resistance-inductance series combination is valid.

MECHANICAL TIME CONSTANT

The mechanical time constant, τ_m, defined by (2.3.29) can be measured in several ways. For example, if a motor is given a step voltage, the time required for a motor to reach 63.2% of its final speed is the mechanical time constant, provided the electrical time constant of the armature does not affect the measurement.

Moving coil motors usually have $\tau_e < 0.1\tau_m$, and such measurement is possible if a tachometer is provided for speed measurements.

The mechanical time constant can also be measured by noting the mechanical break frequency, f_b, in the frequency response test (to be described under "Frequency Response" in this section). The mechanical time constant can be derived from the measured break frequency as follows:

$$\tau_m = \frac{1}{2\pi f_b} \qquad [s; Hz] \qquad (2.10.8)$$

In cases where the electrical time constant is of the same order as the mechanical time constant, the measurement is a bit more complex and is beyond the scope of this brief treatment. However, τ_m can be calculated [see Eqs. (2.3.29), (2.8.2), (2.8.3), and (4.1.26)] from known values of moment of inertia, armature resistance and torque constant as follows:

$$\tau_m = \frac{JR}{K_E K_T} \qquad (2.10.9)$$

[s; kg m^2, Ω, V/rad s^{-1}, Nm/A]

TORQUE RIPPLE

The output torque of a DC motor at low speeds appears to be constant, but under closer examination it has in fact a cyclic component as illustrated in Fig. 2.10.15. This cyclic action is referred to as *torque ripple,* and is caused by the switching action of the commutator (and sometimes by the armature reluctance torque).

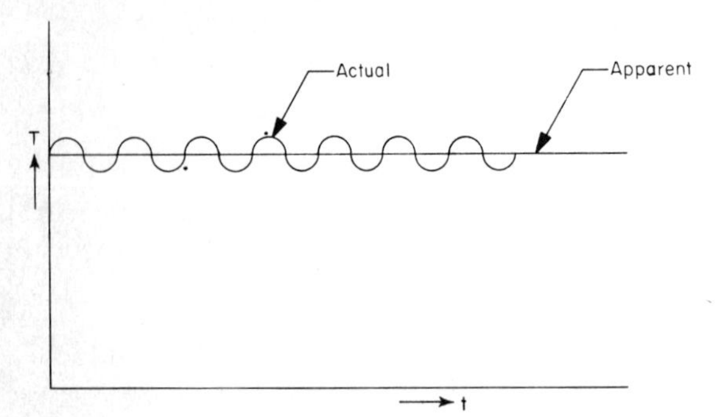

Fig. 2.10.15. Torque ripple

Although torque ripple usually is a very small percentage of the rated output torque, and in most applications can be neglected, there are times when it may be critical; it therefore becomes necessary to have a means of checking torque ripple.

By connecting the motor under test to the apparatus as illustrated in Fig. 2.10.16, measurement of torque ripple can be made, if the moment of inertia of the measuring device is much smaller than the motor moment of inertia (otherwise inertia filtering takes place and the ripple measurement will be invalid).

The percent peak-to-peak ripple torque can be calculated from:

$$T_r = \frac{\text{Peak-to-Peak Torque Ripple}}{\text{Average Output Torque}} 100[\%] \qquad (2.10.10)$$

SPEED REGULATION CONSTANT

The motor under test is operated at rated voltage from no load to 1.5 times the maximum continuous torque, while speed and torque are recorded. Readings must be taken quickly to avoid overheating. The data is plotted in a speed versus torque curve as shown in Fig. 2.10.17.

Speed regulation constant is the slope of the line and may be determined from:

$$R_m = \frac{\Delta n}{\Delta T} \qquad (2.10.11)$$

This value can be also calculated using the formula:

$$R_m = \frac{R}{K_E K_T} \qquad (2.8.1)$$

EFFICIENCY

The motor is connected in the test apparatus as shown in Fig. 2.10.18. The speed **n** and output torque $\mathbf{T_o}$, at which efficiency will be checked, should correspond closely to actual conditions under which motor will be used. Input voltage **V** and current $\mathbf{I_a}$ are measured. Efficiency can be calculated from:

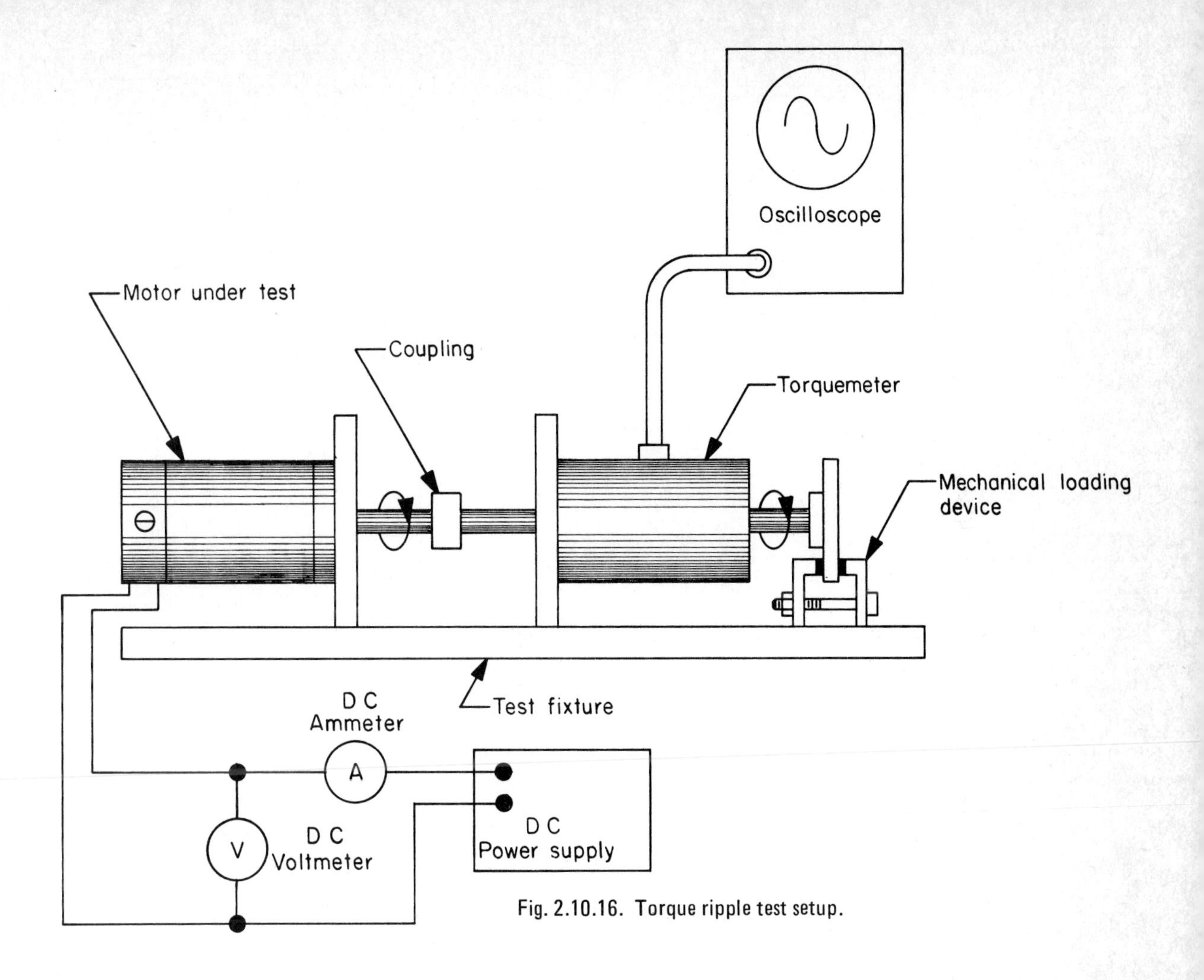

Fig. 2.10.16. Torque ripple test setup.

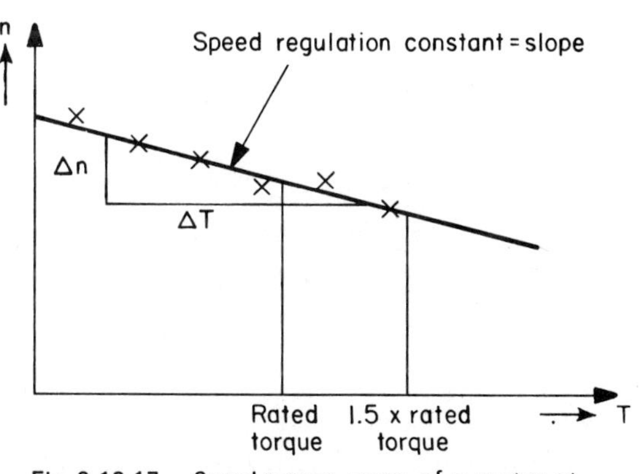

Fig. 2.10.17. Speed-torque curve of a motor at rated voltage.

$$\eta = 1.0472 \times 10^4 \frac{T_o n}{V I_a} \tag{2.10.12}$$

[%; Nm, krpm, V, A]

or

$$\eta = 73.948 \frac{T_o n}{V I_a} \tag{2.10.13}$$

[%; oz-in, krpm, V, A]

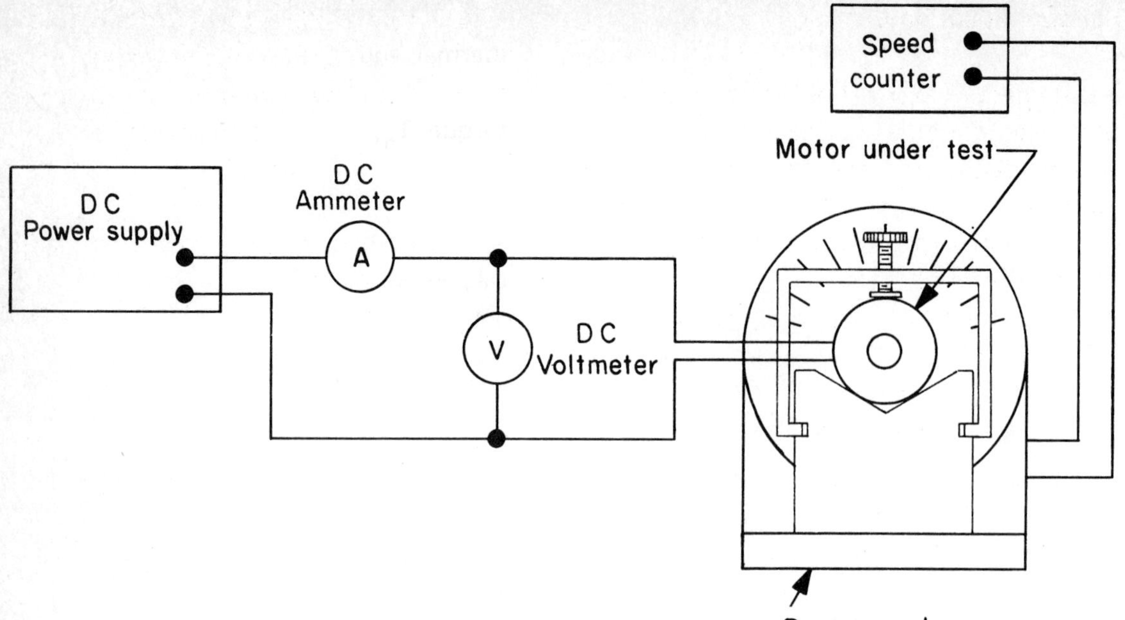

Fig. 2.10.18. Motor efficiency test setup.

FREQUENCY RESPONSE

The determination of motor-generator frequency response and resonant frequencies points can be performed in a simple test setup shown in Fig. 2.10.19. The amplifier shown is Electro-Craft's Motomatic Control System Laboratory amplifier, but any amplifier with a response from 5 Hz to 10 kHz will do. The MCSL unit is convenient in that it has adjustable gain and a current limiting output stage, preventing accidental overload of the motor or amplifier.

The test is done by first establishing a convenient reference level at a low frequency. Then the frequency region of interest is scanned and pertinent points recorded. A typical plot of frequency response of a motor-tachometer was shown earlier in Figs. 2.8.10 and 2.8.11.

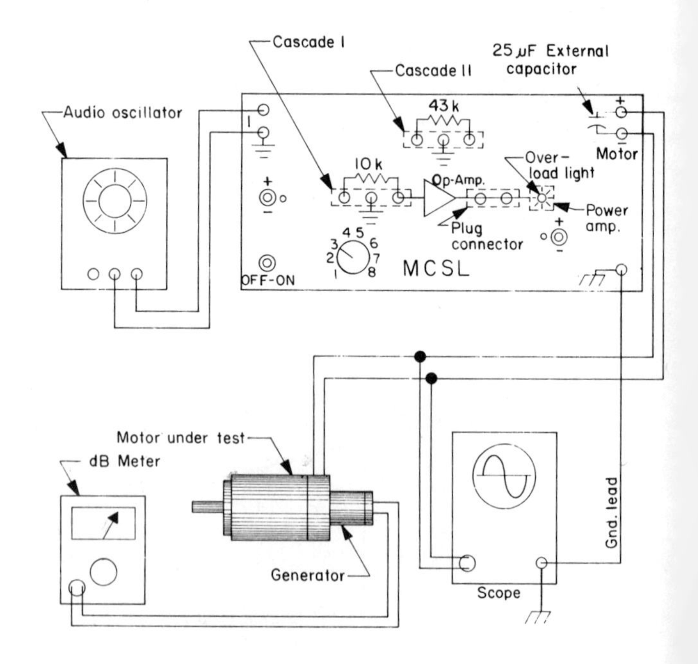

Fig. 2.10.19. Frequency response test setup using a Motomatic Control System Laboratory amplifier.

THERMAL RESISTANCE

Measurement of thermal resistance of a motor is generally done by the manufacturer; it is not usually possible for a customer to disassemble a motor and make the necessary alterations to perform this test. However, it may be of interest to the reader to understand how thermal tests are performed at Electro-Craft Corporation, and we will therefore briefly describe these tests.

A small calibrated thermistor is inserted in the expected "hot spot" as in Fig. 2.10.20 (in some motors several such thermistors are used, where the worst-case hot spot is not known). In most cases, the thermistor output leads are brought out through slip rings so that dynamic testing can be performed.

The motor is run at a power level comparable to that in an actual application (Fig. 2.10.21). Once the motor has reached thermal equilibrium, record input voltage **V**, input current $\mathbf{I_a}$, motor speed **n**, output torque $\mathbf{T_o}$, armature temperature Θ_a, and ambient temperature Θ_A.

Thermal resistance (armature-to-ambient) is calculated from:

$$R_{th} = \frac{\Theta_a - \Theta_A}{P_L} \qquad (2.10.14)$$

[°C/W; °C, W]

where the power loss in the motor is given by

$$P_L = P_i - P_o$$

$$P_L = V\,I_a - 104.72\,T_o n \qquad (2.10.15)$$

[W; V, A, Nm, krpm]

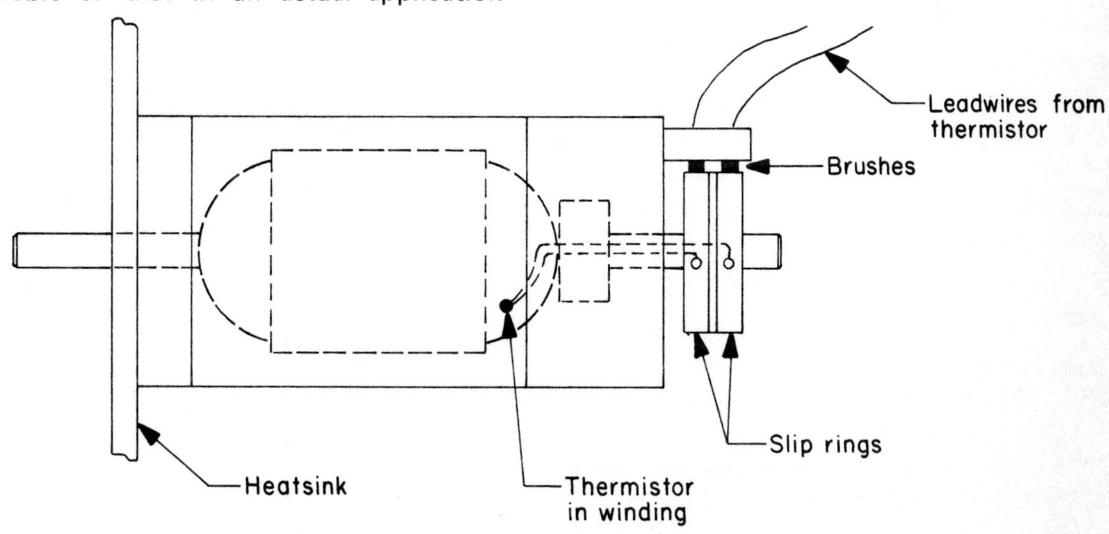

Fig. 2.10.20. Inserting a thermistor in motor armature for thermal resistance test.

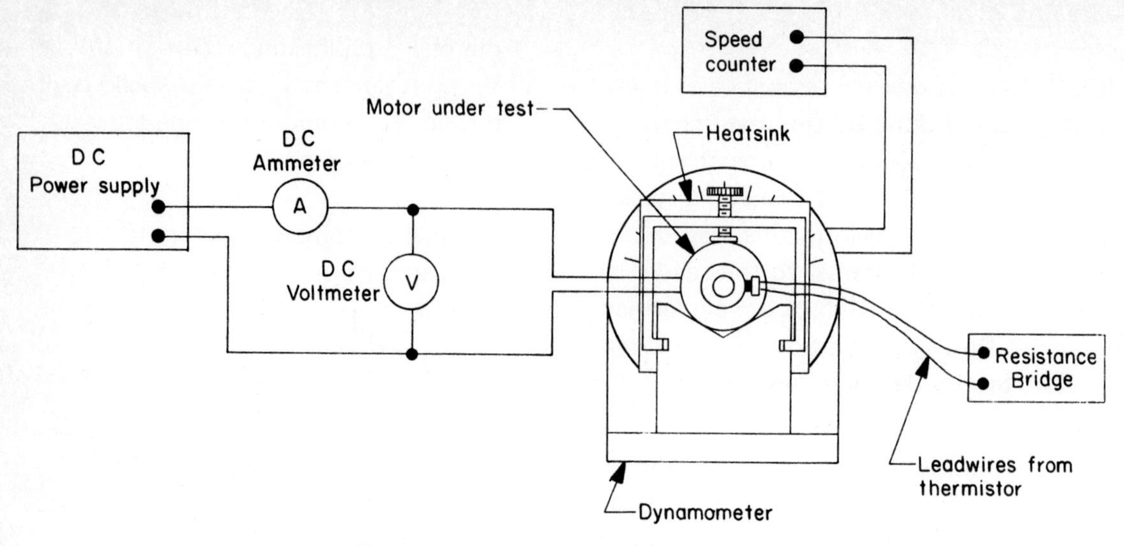

Fig. 2.10.21. Thermal resistance test setup.

or

$$P_L = V\,I_a - 0.73948\,T_o n \qquad (2.10.16)$$

[W; V, A, oz-in, krpm]

THERMAL TIME CONSTANT

Thermal time constant is the time required for temperature to attain 63.2% of its final value, as τ_{th} in Fig. 2.10.22.

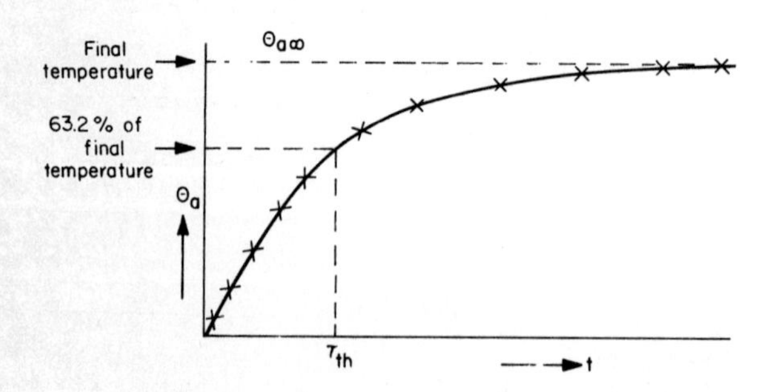

Fig. 2.10.22. Armature temperature versus time curve.

To measure the thermal time constant, the thermistor arrangement need in the thermal resistance test is used and the motor is connected as shown in Figs. 2.10.23 and 2.10.24. The motor is energized with a step function of power, which is held constant during the test. The strip chart recorder measures the armature temperature rise, and after thermal equilibrium is attained, the chart will appear as in Fig. 2.10.25.

The armature thermal time constant can be determined from the chart paper by measuring the record, as shown in Fig. 2.10.25. The thermistor calibration characteristic in circuit connection of Fig. 2.10.24 must be considered.

This example is based on a single thermal time constant for the sake of simplicity; as shown earlier in section 2.5., a motor has at least two independent thermal time constants.

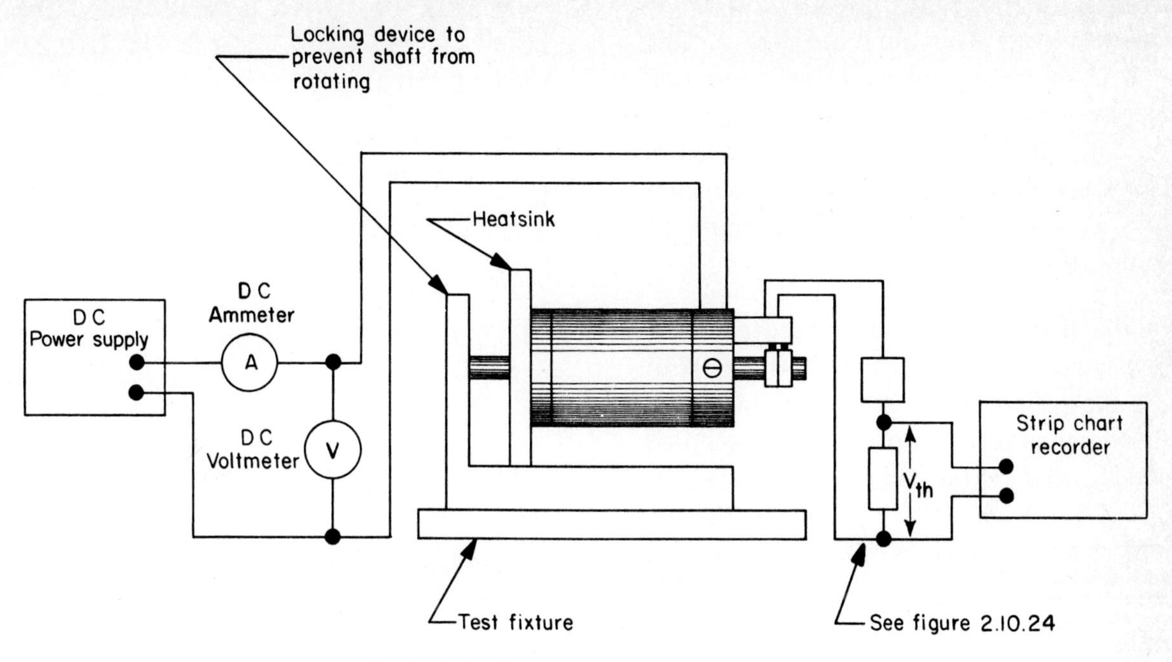

Fig. 2.10.23. Thermal time constant test setup.

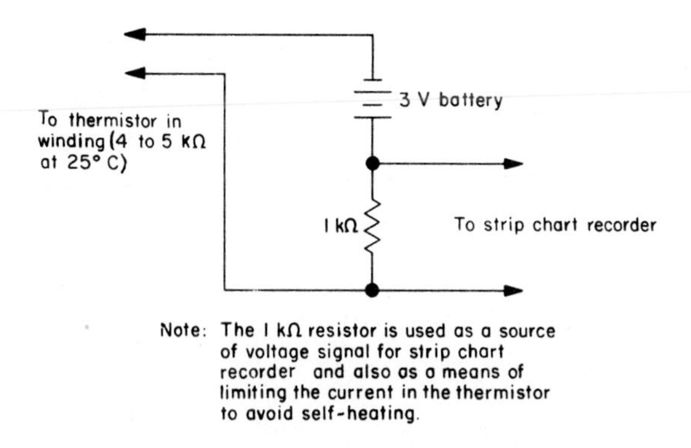

Fig. 2.10.24. Circuit used in thermal time constant test.

AIR FLOW IMPEDANCE

By increasing the air flow through a motor, the heat removal rate is increased, thereby substantially increasing the output power capability of the motor. When air is forced through a motor, the internal parts and design of the air cooling channels will influence the air flow in such a way so as to resist or impede that flow. Air flow impedance is a measure of this influence, and is given as air flow rate (volume per unit of time) divided by given pressure, e.g. in units of

$$\frac{m^3/min}{at} \text{ or } \frac{ft^3/min}{in\ of\ H_2O} = \frac{CFM}{in\ of\ H_2O}$$

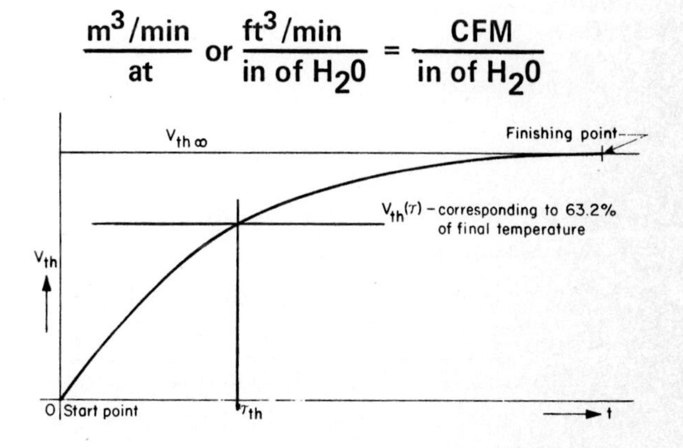

Fig. 2.10.25. Record of strip chart recorder.

Following are two alternate tests. One utilizes an air flow meter, while the other requires air pressure gages, a plastic bag, and a container of known size. The latter test is perhaps a little less accurate, but it is handy for occasional tests with a minimum of equipment outlay.

In the first test method, the motor under test is connected as in Fig. 2.10.26. With the blower on, the air flow rate on the air flow meter and the air pressure are recorded and plotted. The damper is adjusted increasing the air flow rate, while data is plotted, until a curve similar to that in Fig. 2.10.27 is obtained.

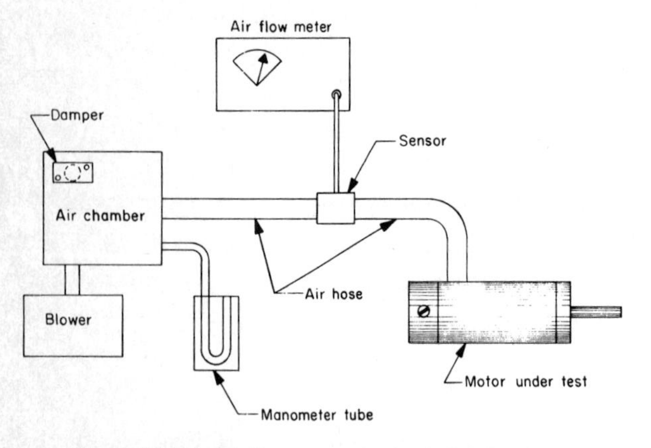

Fig. 2.10.26. Air flow meter method of measuring air flow impedance.

For the second method, the motor is connected as in Fig. 2.10.28, with air inside the plastic bag evacuated. The blower is turned on and the damper adjusted for a predetermined pressure, then the air hose is connected to the plastic bag and a stop watch is started. Time to fill the bag and air pressure are recorded. Air flow rate is calculated, and the above procedure repeated until a graph similar to Fig. 2.10.27 can be plotted.

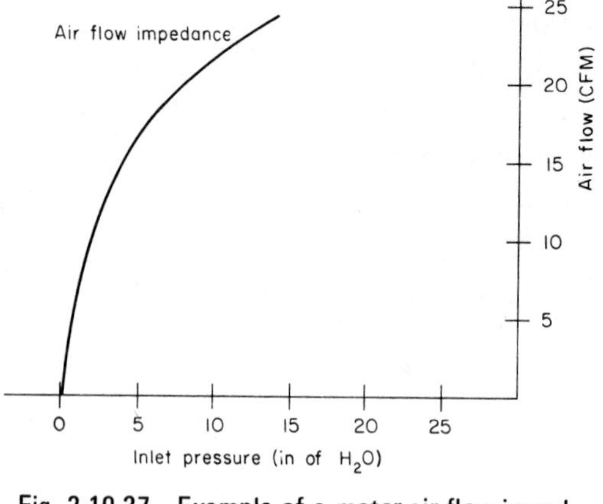

Fig. 2.10.27. Example of a motor air flow impedance.

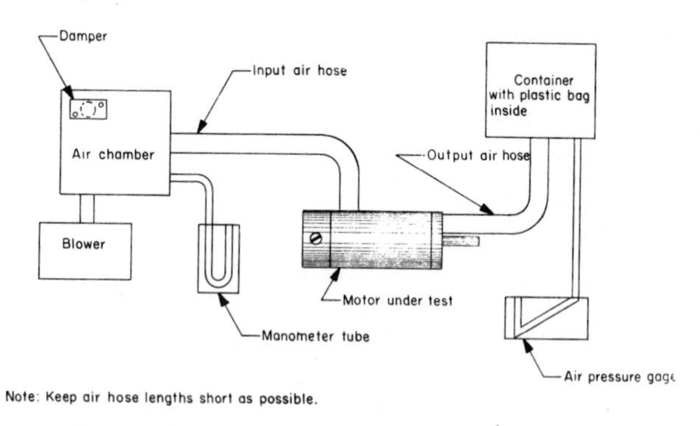

Fig. 2.10.28. Alternate scheme of measuring air flow impedance.

Example of calculating air flow rate:

Container Volume = 4.7 ft^3

Air pressure = 1 in of H_2O

Average time to fill bag = 23 s

$$\text{Air flow rate} = \frac{4.7\ ft^3}{23\ s}$$

$$= \frac{4.7}{23} \cdot 60 = 12.3\ \text{CFM @ 1 in of } H_2O$$

INCOMING INSPECTION OF MOTORS

The following list of motor parameters is intended to serve as a basic guide for incoming inspection tests. Whether this list is expanded or shortened would depend on the testing program (for example, 100% testing or sampling techniques) and on individual customer needs, as well as the type of motor under consideration.

MOTOR TESTING PARAMETERS

1. Visual Inspection

A. LEADWIRES – LOOK FOR:

NICKED, CUT OR CRACKED INSULATION WHICH EXPOSES BARE WIRE.

PROPER COLOR IDENTIFICATION.

PROPER STRIPPING OF LEADWIRE ENDS.

CONNECTORS AND TERMINALS (IF USED) ARE PROPERLY ATTACHED

B. OUTPUT SHAFT(S)

RUST OR FOREIGN SCALE.

NICKS OR UP-SETS, WHICH MAY CAUSE AN OVER-SIZE CONDITION.

IF PINIONS OR GEARS ARE USED, CHECK FOR BURRS WHICH MAY CAUSE IMPROPER MATING.

C. BALL BEARINGS

CHECK FOR DAMAGED SHIELDS.

PROPER INSTALLATION OF RETAINING RINGS.

D. PILOTS AND LOCATING SURFACES

NICKS, BURRS, OR UP-SETS WHICH MAY CAUSE IMPROPER MATING OR ALIGNMENT.

PROPER FINISH (IRRIDITE, ANODIZE, ETC., IF APPLICABLE).

E. BRUSH HOLDERS

CHECK FOR LOOSE OR CRACKED BRUSH HOLDERS.

LOOSE OR CRACKED BRUSH HOLDER CAPS.

F. NAMEPLATES

PROPER LABELING AND IDENTIFICATION.

2. Electrical and Mechanical Checks

A. END PLAY

B. RADIAL PLAY

C. SHAFT RUNOUT

D. RESISTANCE

E. NO-LOAD CURRENT

F. NO-LOAD SPEED

G. TORQUE CONSTANT

H. VOLTAGE CONSTANT

I. ABNORMAL NOISE

MOTOR TROUBLESHOOTING CHART

Problem	Probable Causes	Corrective Action
Motor Runs Hot NOTE: Don't judge the motor temperature by feel. Use an appropriate temperature measuring device.	1. Excessive load.	1. Determine if motor rating is correct for given load. Inspect coupling between motor and load for excess drag.
	2. Maximum speed of motor exceeded for long periods of time.	2. Re-check motor maximum speed rating.
	3. High ambient temperature.	3. Re-check motor rating for given ambient temperature. Reduce ambient temperature by providing ventilation.
	4. Worn bearings.	4. Replace bearings.
	5. Short-circuited coils in armature.	5. Repair or replace armature.
	6. Armature rubbing	6. Note cause of rubbing. Remove obstruction or replace with new armature.
Motor Burns Out	1. Same as "Motor Runs Hot"	1. Any of the conditions listed in "Motor Runs Hot" will cause a motor to burn up if not detected in time. Replace motor and correct problem when possible to avoid recurrences.
High No-Load Speed	1. Demagnetization of magnets. Maximum pulse current to avoid demagnetization exceeded.	1. Remagnetize magnets.
High No-Load Current	1. Worn bearings.	1. Replace bearings.
	2. Worn or sticking brushes.	2. Replace brushes. Check brush holder slots for obstructions.
	3. Magnets demagnetized.	3. Remagnetize magnets.

Problem	Probable Causes	Corrective Action
	4. Armature rubbing.	4. Note cause of rubbing. Remove obstruction or replace with new armature.
	5. Excessive pre-load on bearings.	5. Remove excess shims.
	6. Misaligned or cocked bearings.	6. Remove armature and bearing assembly and re-insert same with proper alignment.
Low Output Torque at Rated Input Power	1. Magnets demagnetized.	1. Remagnetize magnets.
	2. Open or shorted armature winding.	2. Repair or replace armature.
	3. Excessive motor drag.	3. Check for any added friction spots such as worn bearings, rubbing armature, misalignment between motor and load, etc.
High Starting Current	1. Worn bearings.	1. Replace bearings.
	2. Worn or sticking brushes.	2. Replace brushes. Check holder slots for obstructions.
	3. Obstruction in air gap.	3. Remove obstruction.
	4. Armature rubbing.	4. Note cause of rubbing, remove obstruction or replace with new armature.
	5. Magnets demagnetized.	5. Remagnetize magnets.
	6. Open or shorted armature winding.	6. Repair or replace armature.
	7. Misalignment between motor and load.	7. Re-align coupling to reduce drag.
Erratic Speed	1. Varying load.	1. Re-adjust load.
	2. Worn or sticking brushes.	2. Replace brushes. Check brush holder slots for obstructions.
	3. Obstruction in air gap.	3. Remove obstruction.
	4. Worn bearings.	4. Replace bearings.

Problem	Probable Causes	Corrective Action
	5. Open or shorted armature winding.	5. Repair or replace armature
Reversed Rotation	1. Motor leadwires to power supply reversed.	1. Reverse leadwires.
	2. Magnets have reversed polarity.	2. Rotate magnet and housing assembly to proper orientation or remagnetize.
Excessive Brush Wear	1. Incorrect spring tension.	1. Springs may lose temper due to excessive heat. Replace brush with proper spring material and tension.
	2. Dirty or rough commutator.	2. Clean or re-machine commutator. Insure a good surface finish.
	3. Incorrect brush seating (off center neutral)	3. Check to see that motor is properly neutralized.
	4. Excessive load	4. Check the value of rated current vs. the actual. If over the mfg. rating, the excessive heat and current would speed brush wear.
	5. Short circuited coils in armature.	5. Repair or replace armature.
	6. Loose-fitting brushes.	6. Check brushes and brush holder slots for proper size. Replace as needed.
	7. Vibration	7. The armature should be dynamically rebalanced. The vibration causes brush bounce, hence, arcing and excessive wear.
	8. Lack of moisture.	8. Operating in a near vacuum condition causes rapid brush wear. Use a special treated or "high altitude" brush.
Excessive Bearing Wear	1. Belt tension too great. Misalignment of belt, coupling, or drive gears. Unbalanced coupling or too closely meshed gears. Excessively heavy flywheel or other load hung on motor shaft.	1. Correct mechanical condition. Limit radial load to mfg. specification.

Problem	Probable Causes	Corrective Action
	2. Dirty bearings.	2. Clean or replace bearings. If condition is bad, provide means for shielding motor from the dirt.
	3. Insufficient or inadequate lubrication.	3. Review bearing mfr. recommendation for type and amount of lubricant to use.
	4. Excessive thrust load.	4. Reduce the load or obtain a motor designed to handle the required thrust load.
	5. Bent output shaft causing excessive vibration.	5. Remove armature and check shaft with mechanical indicator. Straighten if possible or replace armature.
Excessive Noise	1. Unbalanced armature.	1. An unbalanced armature can set up vibrations that can be easily felt. The armature should be rebalanced dynamically.
	2. Worn bearings.	2. Replace bearings.
	3. Excessive end play	3. Add washers (shims) to the motor to take up end play. Re-check friction level and end play as excessive pre-load will cause early bearing failure.
	4. Misalignment between motor and load.	4. Correct mechanical conditions of misalignment.
	5. Motor not fastened firmly to mounting.	5. Correct mechanical condition of mounting.
	6. Dirt in air gap.	6. Noise is irregular, intermittent and scratchy: dismantel and clean motor.
	7. Amplified motor noises.	7. Uncouple motor from load and allow it to run. If noise persists, loosen the mounting bolts and lift motor while it is still running. If motor is quiet, the mounting was acting as an amplifier. If possible, replace with a cushion type mounting to reduce noise.
	8. Rough commutator.	8. Re-machine commutator.

Problem	Probable Causes	Corrective Action
Excessive Radial Play	1. Shaft loose in bearing I.D.	1. Check fits, the I.D. of the bearing should be a light press fit.
	2. Worn bearings.	2. Replace bearings.
	3. Bearings loose in end cap bearing bore.	3. The bearings should be free to slide in the bearing bore. However, the clearance should not be excessive. If excessive, replace end cap.
Excessive End Play	1. Improper shimming.	1. Add shims as needed. Re-check motor for proper end play.
Excessive Vibration	1. Unbalanced armature.	1. Re-balance armature dynamically to proper level.
	2. Worn Bearings.	2. Replace bearings.
	3. Excessive radial play.	3. See corrective action for "excessive radial play".
	4. Open or shorted armature winding.	4. Repair or replace armature.
Shaft Will Not Turn	1. No input voltage	1. Check at motor leads for proper input voltage.
	2. Motor bearings tight or seized.	2. Replace bearings.
	3. Load failure.	3. Disconnect motor from load and then determine if motor will run without load. (Check load to see if it turns freely.)
	4. Dirt or foreign matter in air gap.	4. Remove the obstruction.
	5. Excessive load.	5. Check the load. It may be enough to stall the motor.
	6. Open motor winding.	6. Check for open windings. Repair or replace armature.

Problem	Probable Causes	Corrective Action
	7. Worn or sticking brushes.	7. Brushes may be worn enough so that they do not touch commutator. Clean brushes and brush holders so that brushes move freely. Replace if necessary. ☐

2.11. DC GENERATORS

INTRODUCTION

The DC tachometer-generator is a rotating electromagnetic device which, when mechanically driven, generates an output voltage proportional to speed. This characteristic can be used in applications requiring either shaft speed readout or closed-loop speed control or stabilization.

Fig. 2.11.1 illustrates an application in which a tachometer-generator, or a tachometer as it is often called, is used in conjunction with a voltmeter calibrated directly in speed units (rpm) to provide speed readout – in this case at a remote control station.

However, the most important application of tachometers is in servo systems, where they are employed to provide shaft speed feedback signal in speed control systems, as shown in Fig. 2.11.2. A tachometer providing rate feedback signal to improve stability of a position servo system is shown in Fig. 2.11.3.

THEORY

The basic theory of the DC motor in Chapter 2. applies to the DC tachometer, but while the DC tachometer can be viewed essentially as a small DC motor, it has some important distinguishing requirements:

1. The tachometer shall provide a DC voltage proportional to the shaft speed, with a controlled linearity;
2. The output voltage shall be relatively free from rotation dependent voltage ripple or random voltage

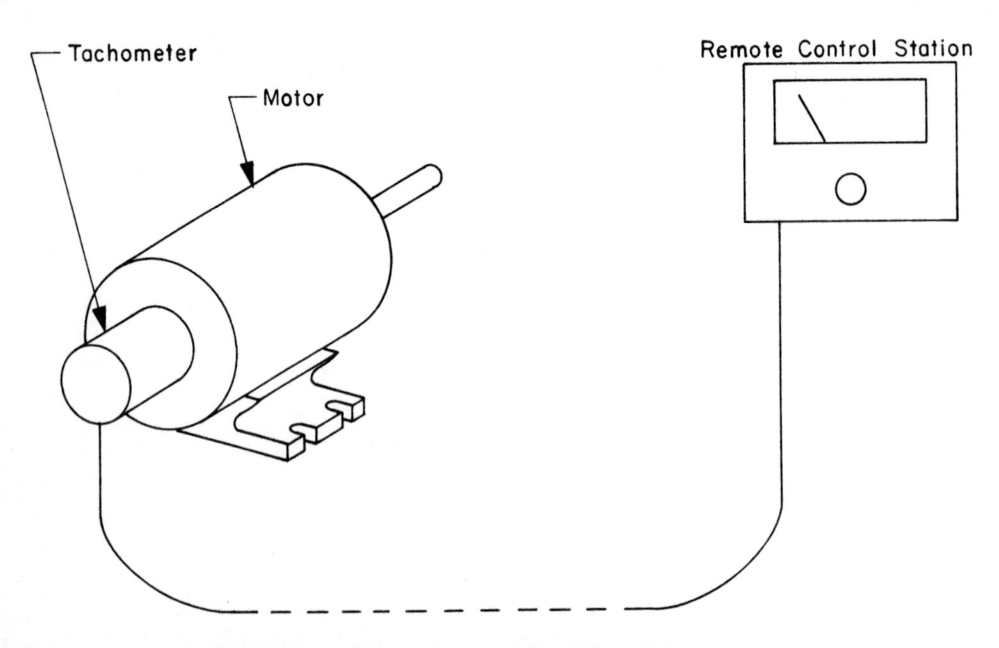

Fig. 2.11.1. Tachometer providing shaft speed readout.

changes in the frequency range in which the tachometer is to be operating;

3. The voltage gradient K_E should be stable with ambient and device temperature variations;

4. There is no requirement that the tachometer provide any significant power; hence, the design considerations are mainly directed at solving problems related to requirements 1, 2 and 3 above.

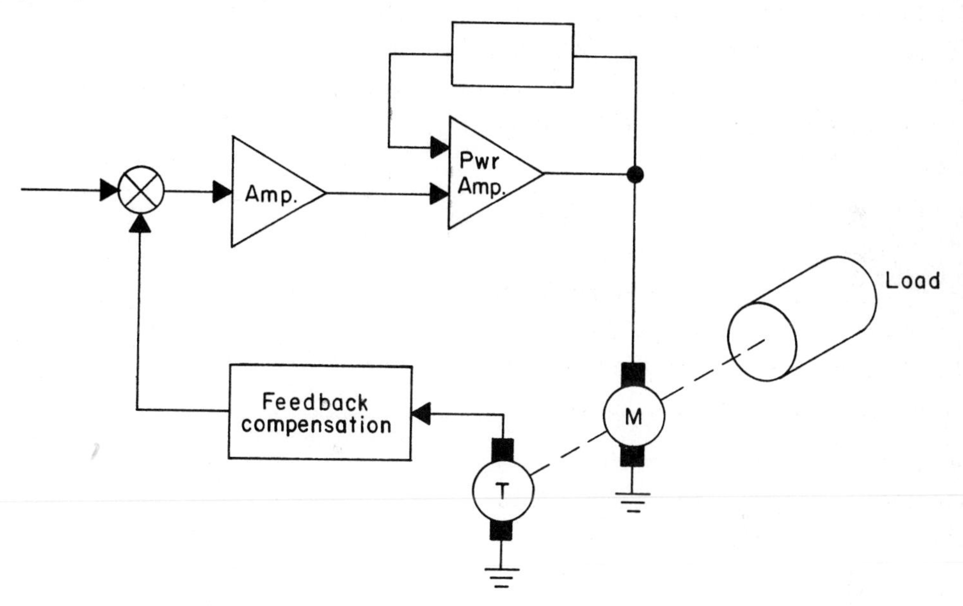

Fig. 2.11.2. Block diagram of speed servo with tachometer feedback.

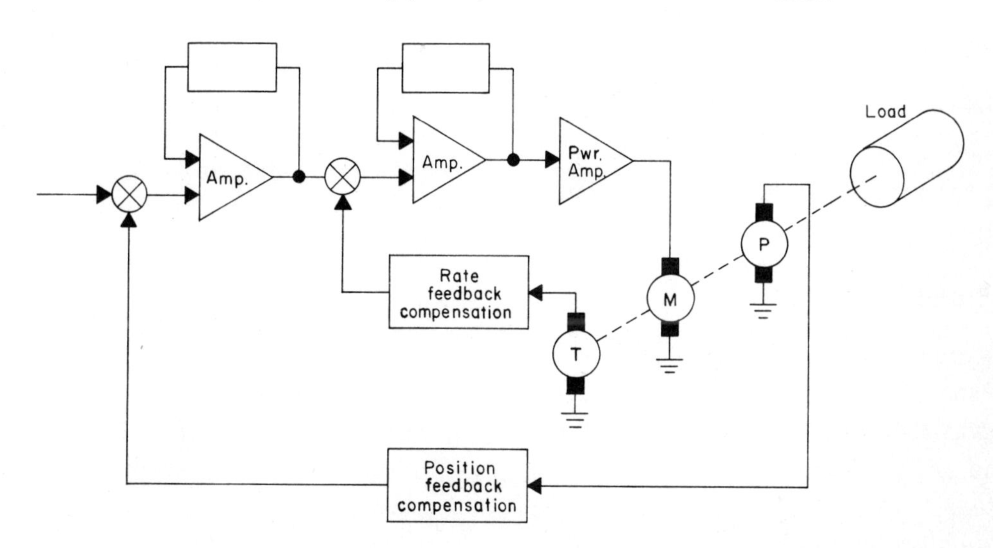

Fig. 2.11.3. Block diagram of position servo with tachometer damping.

The tachometer develops an output voltage proportional to shaft speed, with voltage polarity dependent on direction of rotation. Fig. 2.11.4 shows a relationship between tachometer speed and output voltage, which can be expressed mathematically as:

$$E_g = K_E n \qquad (2.1.18)$$

where E_g is the tachometer output voltage and n is the tachometer speed. K_E is the proportionality constant between output voltage and speed, i.e. the voltage constant of the tachometer.

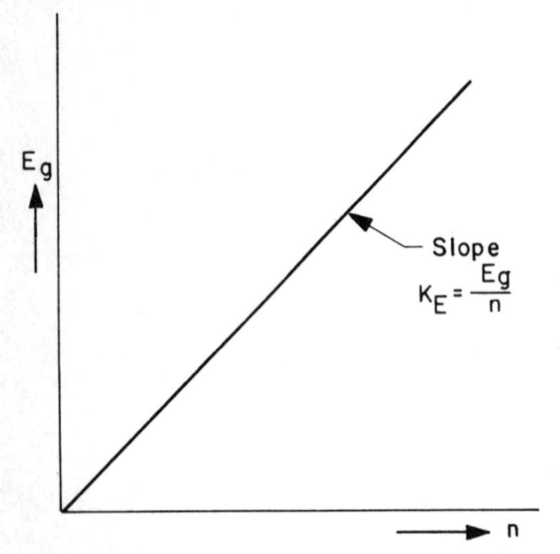

Fig. 2.11.4. Tachometer output voltage vs. shaft speed.

TYPES OF DC GENERATORS

There are basically two types of DC generators: the *shunt-wound* generator and the *permanent magnet* generator. The shunt-wound tachometer for feedback purpose is obsolete, however, and the permanent magnet units are universally accepted for their compactness, efficiency and reliability.

The majority of DC tachometers are of the conventional iron-copper armature type.

In many cases Electro-Craft provides the tachometer with coin silver commutators, which gives highly reliable commutation for cases where the tachometer is run for extended periods of time at very low speeds — a difficult operational condition for copper commutators. A variety of windings can easily be provided on iron core armatures; thus, a wide range of K_E values can be supplied.

Moving coil tachometers are, in principle, identical to moving coil motors, as described in section 2.8. with one important difference: a high number of coils per pole is desirable to minimize ripple voltage.

Essentially, the moving coil concept is characterized by a lack of iron in the armature design. The armature windings consist of a cylindrical, hollow rotor, composed of wires and held together with fiber-glass and polymer resins. Because the armature is not burdened by iron in its design, this hollow or shell-type armature has the lowest moment of inertia of any DC tachometer type. Since coil inductance is low, due to the absence of iron, moving coil tachometers have an extremely low ripple amplitude, approximately 1% peak-to-peak for Electro-Craft tachometers. This makes moving coil tachometers excellent, not only in low inertia applications, but in precision speed controls as well. The ripple content is sometimes difficult to measure, since most test methods produce inherent ripple contributions of the same order of magnitude. Suitable test procedures are discussed in detail in section 2.12.

MOUNTING FEATURES

Tachometers can, of course, be used as separate devices; but in a majority of servo applications the tachometer is closely connected to a motor.

A unique solution to the motor-tachometer combination is represented by Electro-Craft's patented Motomatic® motor-tachometer (Figs. 2.11.5 and 2.11.6). In this device the tachometer winding is wound on the motor armature alongside the motor winding. Due to the use of magnets with low permeability and high coercivity, there is no significant interaction between the two windings over a wide range of operating conditions. A thermistor-resistor network minimizes the effect of the temperature dependent magnetic flux on the tachometer voltage gradient. The "Motomatic" system — the motor-tachometer used with a transistor amplifier — has been preferred for speed control applications for many years, and its applications range from blood pumps, crystal-pulling machines and semiconductor spinners to textile machines, motion picture editing machines, and automatic welding machines.

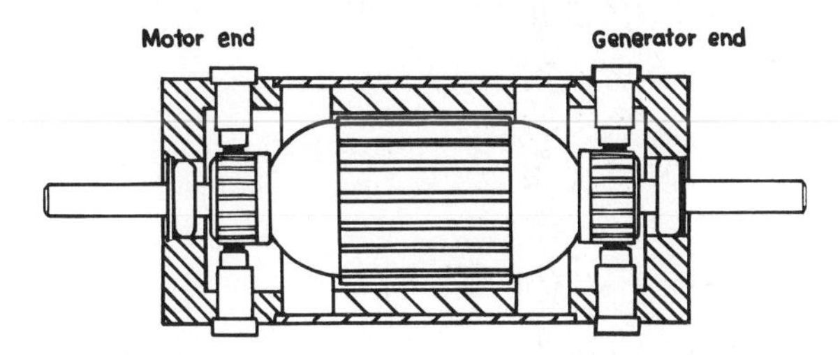

Fig. 2.11.5. Cross-sectional view of Motomatic motor-tachometer.

While the Motomatic motor-generator performs excellently for wide range speed controls, it has a limitation in its ability to respond rapidly in high-performance servo systems, due to the magnetic coupling between the two windings. Thus, for servo bandwidths beyond 15-20 Hz the tachometer must be physically separated from the motor armature.

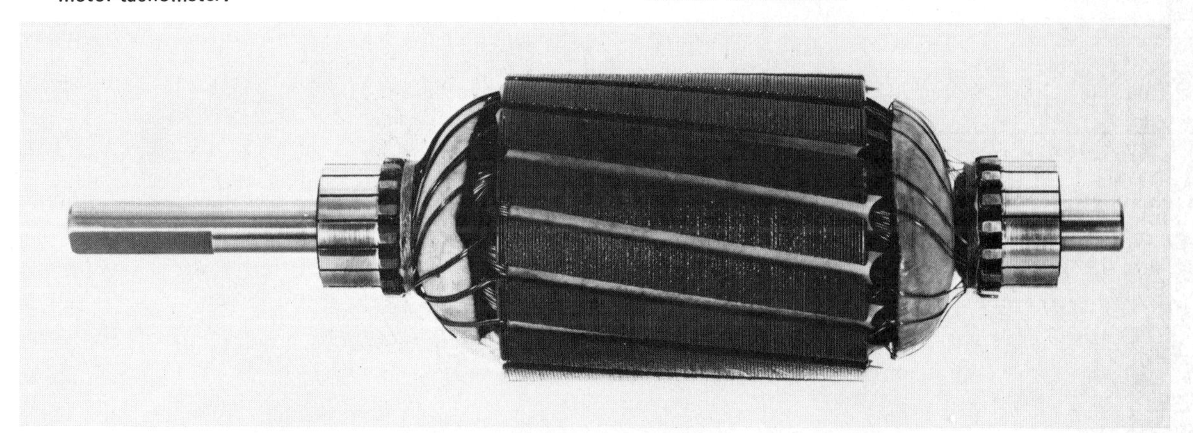

Fig. 2.11.6. Electro-Craft's Motomatic armature. Motor and tachometer windings simultaneously wound on the same armature.

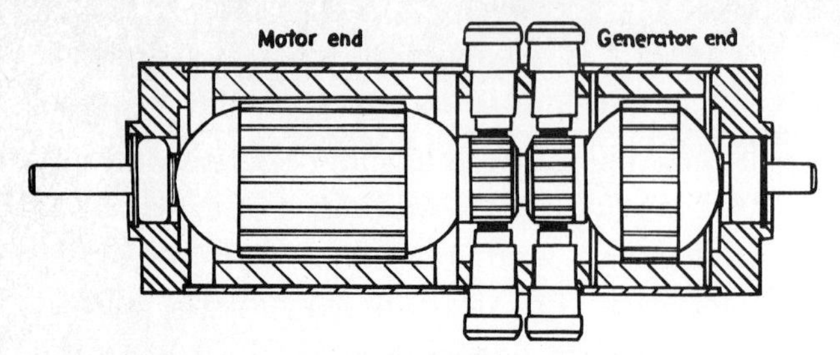

Fig. 2.11.7. Motor and tachometer on a common shaft (Electro-Craft E-576).

Fig. 2.11.7 shows a cross-sectional view of a patented Electro-Craft design of a motor-tachometer (E-576) on a common shaft. This design combines the advantage of rigidly coupled motor-tachometer armatures, which are magnetically separated, with independently adjustable magnet mounting features. The device represents a good combination of high-performance features with economical manufacturing attributes.

A photograph of the armature assembly is shown in Fig. 2.11.8, and the assembled unit is shown in Fig. 2.11.9. The physical separation of the armature enables servo bandwidths of up to 100 Hz to be achieved.

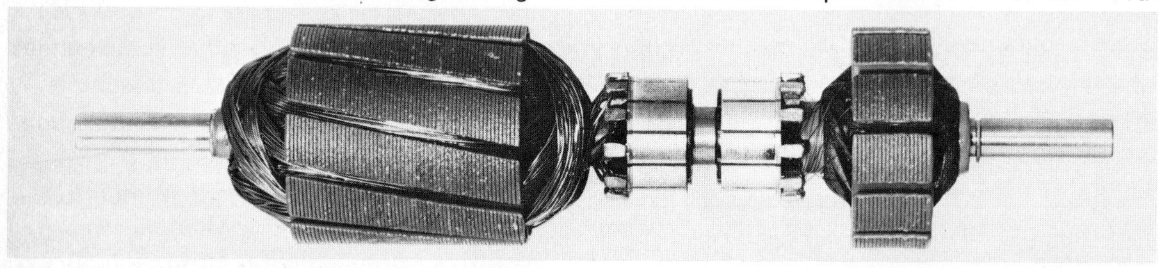

Fig. 2.11.8. The Electro-Craft E-576 is an example of motor and tachometer mounted on the same shaft in one housing.

Fig. 2.11.9. The assembled unit E-576.

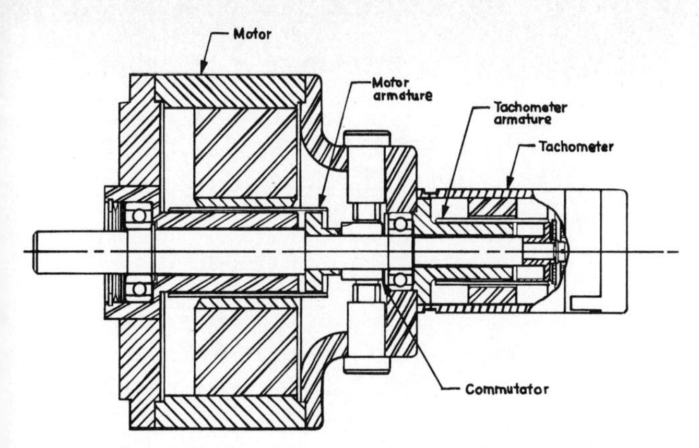

Fig. 2.11.10. Cross-section of an Electro-Craft Moving Coil Motor-Tachometer.

When moving coil motors are equipped with tachometers, it is necessary to mount the tachometer external to the motor housing, but still with the moving coil tachometer armature rigidly coupled to the motor shaft. Fig. 2.11.10 illustrates a typical moving coil motor-tachometer combination. Because of the low armature moments of inertia and rigid shaft connections, the torsional resonance frequency between motor armature and generator armature is over 3.5 kHz – well beyond most servo requirements. The assembled unit is shown in Fig. 2.11.11.

Fig. 2.11.11. Electro-Craft M1030-110 Moving Coil Motor-Tachometer. Motor and tachometer armatures are rigidly mounted on the same shaft.

In some cases a customer may want to mount a tachometer to a motor on his own, perhaps as shown in Fig. 2.11.12. Since firm alignment of two independent sets of bearings presents problems, the two shafts are usually joined by a flexible coupling. This coupling, in turn, may limit the performance of the servo system due to the torsional resonance introduced by the coupling. It is therefore advisable to use the factory-assembled, direct-coupled motor-generators whenever possible.

TEMPERATURE EFFECTS

Tachometers can be grouped into four main classes according to the accuracy of their output:

1. Ultrastable generators
2. Stable generators
3. Compensated generators
4. Uncompensated generators

The desired accuracy of each individual system design determines which generator to employ.

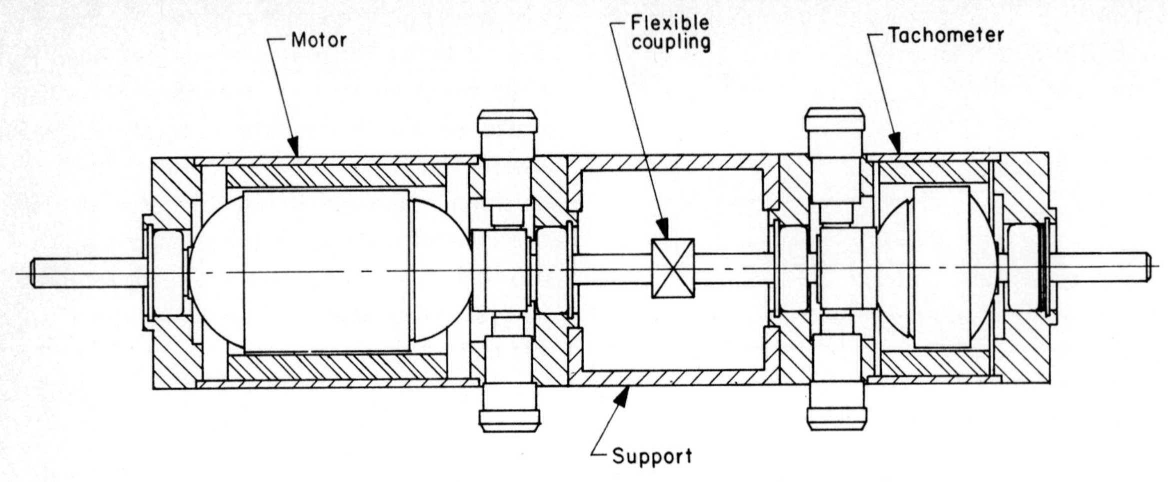

Fig. 2.11.12. Assembly of a separate motor and tachometer, using a flexible shaft coupling.

Ultrastable Generators

This class of generators will show temperature errors of less than 0.01% per °C of temperature change. To achieve this low level of temperature sensitivity, a stable magnet material is used which has a low temperature coefficient, such as Alnico 8 or Alnico 5, in combination with a compensator alloy in the magnetic circuit. Additional design features minimize heating effects in the area of the magnet. This type of unit is the most costly of the four.

Stable Generators

This group of tachometer-generators will exhibit temperature errors ranging up to 0.02% per °C of temperature change. They are usually based on uncompensated Alnico 5 magnets. This class of generators is a bit less costly to build, and will usually be quite satisfactory for many applications.

Compensated Generators

This group of tachometer-generators is characterized by temperature errors in the area of 0.05% per °C of temperature change over a restricted temperature range. This type of unit utilizes lower cost ceramic magnets in combination with thermistor temperature-compensating networks to achieve a level of temperature sensitivity that is adequate for several applications. A minor disadvantage of this type of unit is that its output impedance is generally higher than that of the previously mentioned ones. Because of the compensating circuitry and the effectiveness of the compensation, this group is limited to a temperature range of up to 75 °C.

Uncompensated Generators

This group has the lowest cost of the four types mentioned here. It is characterized by temperature errors ranging up to 0.2% per °C of magnet temperature. This type of unit is often adequate for damping purposes in position servos, where adequate phase margin is provided in the initial design.

EXAMPLE: It is desired to use a motor-tachometer combination in a constant speed drive where the speed must not vary by more than 2% of set speed over a temperature range of 20 °C to 70 °C. What type of tachometer-generator should be used?

The specified temperature range means that the tachometer-generator will see a temperature change of

$$\Delta\Theta = 70 - 20 = 50\ ^{\circ}\mathrm{C}$$

The allowable temperature coefficient of output voltage must be:

$$\psi_V \leqslant \frac{2\%}{50\ ^{\circ}\mathrm{C}} = 0.04\%/^{\circ}\mathrm{C}$$

The stable group of tachometer-generators would be the proper choice here to give this desired performance at the lowest cost.

LINEARITY AND LOAD EFFECTS

The ideal tachometer-generator would have an output voltage vs. speed characteristic as shown earlier in Fig. 2.11.4, a perfectly straight line. If a generator is run within its rated speed range (usually below 6000 rpm) one can expect such linear behavior. However, if a generator designed for low speed operation, is run at very high speed (above 10 000 rpm), aerodynamic lift or commutator runout may cause a drop in the voltage-speed characteristic such as shown in Fig. 2.11.13. This, of course, represents abnormal operating conditions, and is not recommended.

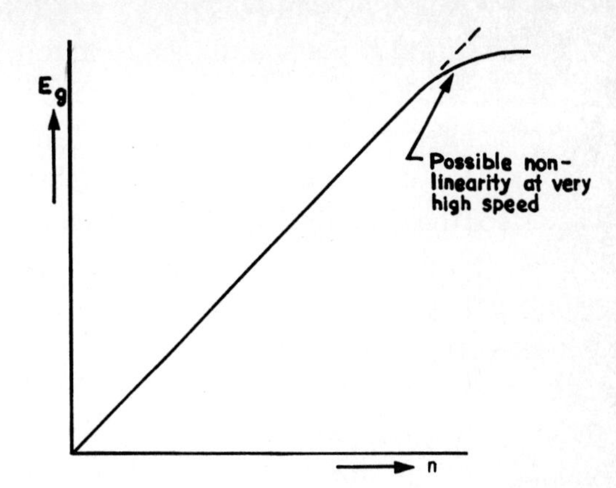

Fig. 2.11.13. Abnormal tachometer characteristic.

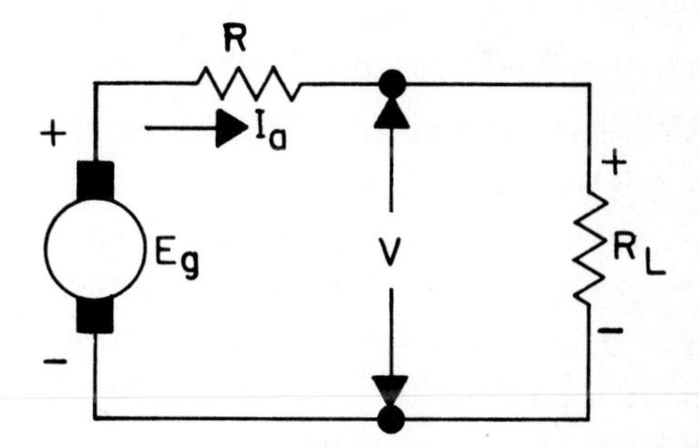

Fig. 2.11.14. Schematic of tachometer connected to resistive load R_L.

The effect of load impedance can be derived from the schematic in Fig. 2.11.14.

The loop equation can be written as:

$$E_g = RI_a + V = K_E n \qquad (2.11.1)$$

Since:

$$V = R_L I_a$$

we obtain after substituting into (2.11.1) and rearranging

$$V = \frac{R_L}{R + R_L} K_E n \qquad (2.11.2)$$

This shows output voltage to be a function of load resistance, R_L, armature resistance, R, tachometer voltage constant, K_E, and speed, n. R_L does not affect linearity, but does affect the gradient V/n. The tendency is, then, to make R_L as high as possible to get high voltage gradient. However, in order to keep a good commutation surface, the value of R_L should be in the 5 to 10 kΩ region. In cases where the tachometer-generator is run for long periods at very low speeds (less than 100 rpm), it may be necessary to use values of R_L even lower than that in order to maintain good commutation. A rule of thumb is that in such cases a current I_a = 0.1 to 1 mA should be maintained.

Linearity can be specified by limiting the maximum deviation of the voltage-speed characteristic from a straight line drawn between zero speed point and the output voltage at a specified calibration speed. This maximum deviation is usually specified as a percentage of the output voltage at the calibration speed.

The percent linearity error can be expressed mathematically as:

$$\Delta = 100 \frac{V - \dfrac{n}{n_c} V_c}{V_c} \quad [\%] \qquad (2.11.3)$$

where

Δ is the linearity error

V is output voltage

n is tachometer speed

n_c is a specified calibration speed

V_c is output voltage at n_c

The calibration speed, n_c, that is specified for a particular generator should be established at 83.3% of the maximum operating speed of the application.

EXAMPLE: If a tachometer-generator has the following characteristics:

$$K_E = 7 \text{ V/krpm}$$

and its linearity error is

$$\Delta \leqslant 0.2\% \text{ over speed range 0 to 4000 rpm,}$$

what is the maximum deviation from a perfect straight line characteristic at 1500 rpm?

For the maximum speed of 4000 rpm the calibration speed should be:

$$n_c = 0.833 \times 4000 = 3332 \text{ rpm}$$

The output voltage at the calibration speed is:

$$V_c = K_E n_c = 7 \times 3.332 = 23.324 \text{ V}$$

The output voltage at 1500 rpm should be

$$V = K_E n = 7 \times 1.5 = 10.5 \text{ V}$$

We can calculate the maximum allowable deviation at 1500 rpm by solving (2.11.3) for V and subtracting the theoretical output voltage of 10.5 V.

Then:

$$V = V_c \left(\frac{\Delta}{100} + \frac{n}{n_c} \right) \qquad (2.11.4)$$

$$V = 23.324 \left(\frac{0.2}{100} + \frac{1500}{3332} \right) = 10.542 \text{ V}$$

$$\Delta V = 10.542 - 10.5 = 0.042 \text{ V} = 42 \text{ mV}$$

□

2.12. GENERATOR TESTING

Because of the many inherent similarities between motors and generators, a good share of the device testing procedures and setups are identical. This is the case for the following characteristics and performance parameters, and the reader is directed to the discussion in section 2.10. for detailed procedure.

End play
Radial play
Shaft runout
Moment of inertia
Inductance
Friction torque

The following discussions present suggested tests and measurements which, in addition to those above, provide a complete repertoire of generator inspection routines.

OUTPUT IMPEDANCE

With a motor driven generator connected as in Fig. 2.12.1, generator no-load output voltage is recorded. Switch SW1 is then closed, applying a load to the generator, then the coarse (R1) and fine (R2) resistors are adjusted until the output voltage equals one-half of the no-load output voltage, as recorded above. The resistors' values at this point equal the generator output impedance.

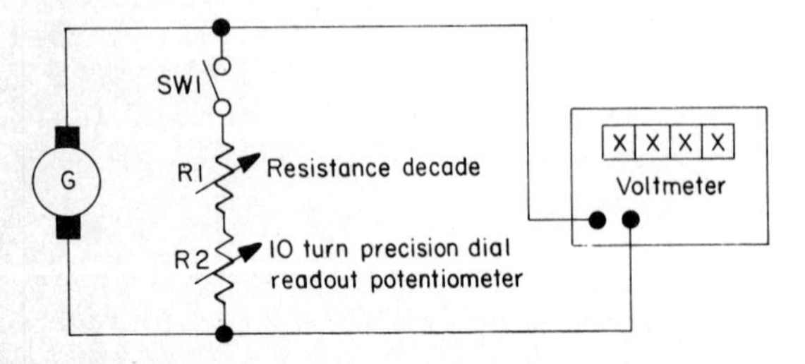

Fig. 2.12.1. Generator output impedance test.

VOLTAGE GRADIENT

The generator output voltage is proportional to the speed of armature. In testing for generator voltage gradient, the generator is driven at 1000 rpm with a voltmeter monitoring output voltage. Voltage at 1000 rpm is equal to generator voltage gradient, K_E, in units of [V/krpm].

VOLTAGE POLARITY

Generator output is checked to verify that a specific voltage polarity exists for a given direction of armature rotation.

GENERATOR RIPPLE

Iron-Core Generator Ripple

DC tachometer-generator ripple is generally specified in two alternate ways. In non-critical applications, it is given as the ratio of RMS value of AC components of the output voltage to its average DC value; in cases where good control of ripple is essential, it is the ratio of peak-to-peak value of AC components to average DC value. The latter measurement requires an oscilloscope, and the former requires true RMS and DC voltmeters.

The fundamental ripple frequency is dictated by the number of commutator segments, and can be calculated by the expression shown below:

$$f_r = \frac{100}{6} k\,u\,n \qquad [\text{Hz}; -, \text{krpm}]$$

[2.12.1)

where

u is number of commutator segments

k = 1 if u is even

k = 2 if u is odd

n is shaft speed

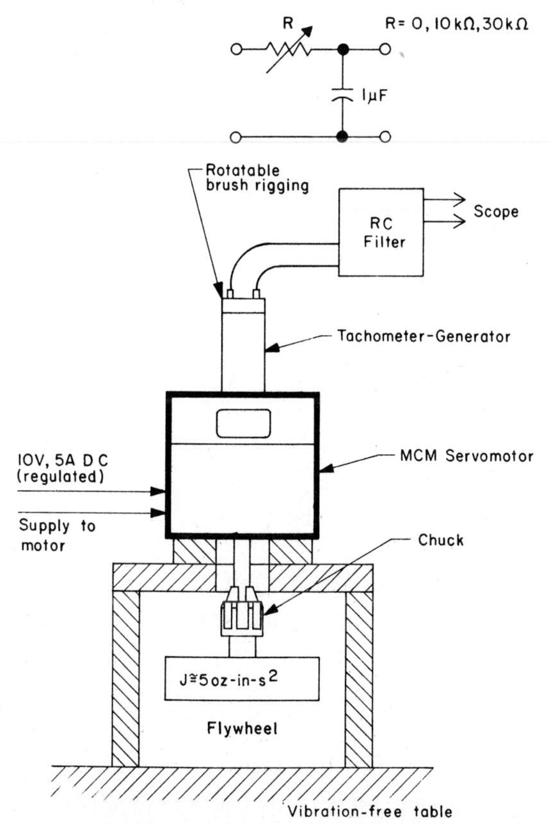

Fig. 2.12.2. Tachometer-generator test setup.

The RMS ripple measurement is simply made by connecting a DC voltmeter and a true RMS voltmeter with an AC input network to the tachometer terminals, while the tachometer is driven at a constant speed in the region of interest. The RMS percentage ripple is determined as follows:

$$\left(\bar{V}_r\right)_{RMS} = 100\,\frac{V_{RMS}}{V_{AV}} \quad [\%] \qquad (2.12.2)$$

The peak-to-peak measurement is discussed in the following section, and illustrated in Fig. 2.12.2.

Moving Coil Tachometer Ripple

Moving coil tachometers having peak-to-peak ripple amplitude of the order of 1% are difficult to test, because most test methods contribute inherent shaft velocity ripple of the same order of magnitude. The following test method was developed at Electro-Craft Corporation to test motor-generators designed for use as capstan motors for digital tape transports, but can as well be applied to any motor-tachometer or single tachometer having low ripple specifications.

The Electro-Craft test is based on the principle that the motor is run "open loop" with a heavy flywheel attached directly to the output shaft. No compliant coupling is used. A suitable collet or non-marring chuck can be used to couple the flywheel to the shaft. The flywheel, in conjunction with the motor, effectively filters out any mechanical ripple due to commutation or to motor torque ripple.

Considerable care is necessary in performing this test to insure that the motor-tachometer under test is suitably isolated from all sources of external mechanical and electrical interference.

This test should be performed only on samples which have been properly run-in with a tachometer current drain of at least 1 mA. The motor-tachometer under test is connected into a test arrangement as shown in Fig. 2.12.2.

The motor is driven from a regulated power supply, and the speed is adjusted so that the tachometer DC output voltage is 0.5 V ± 50 mV, as measured on the oscilloscope. This is a typical voltage for capstan motors; other applications may require different values.

Adjust the oscilloscope time base to 20 or 50 ms/cm and observe tachometer-generator output voltage with the oscilloscope amplifier switched to "DC". The vertical gain should be set to 0.1 V/cm. Estimate and record the AC component of generator output voltage plus output "noise".

Switch the oscilloscope amplifier to "AC" and adjust gain to suitable level (5, 10, 20 or 50 mV/cm). Adjust the waveform to the center of screen. Fig. 2.12.3 shows a typical pattern. Examine and record:

a) Amplitude (peak-to-peak) of cyclic AC component (an RC filter, typically 1μF and 10 kΩ, will remove all high frequency components);

b) Amplitude (peak-to-peak) of hf noise or "hash";

c) Amplitude (peak-to-peak) of commutator ripple;

d) Total (peak-to-peak) AC component expressed as a percentage of DC output voltage, to obtain peak-to-peak percentage ripple.

Percentage (peak-to-peak) ripple can be calculated as follows:

$$\left(\bar{V}_r\right)_{p-p} = 100\,\frac{AX}{V_{AV}} \qquad (2.12.3)$$

[%; V/cm, cm, V]

where

A is oscilloscope amplifier gain

X is amplitude (peak-to-peak) of ripple (see Fig. 2.12.3) observed on the scope screen

V_{AV} is tachometer DC average output voltage

TEMPERATURE COEFFICIENT

Temperature coefficient relates a change in generator output voltage with temperature. It is expressed as a percent change of output voltage per ^{o}C.

In testing for the temperature coefficient, the generated output voltage at a predetermined speed is recorded. The generator

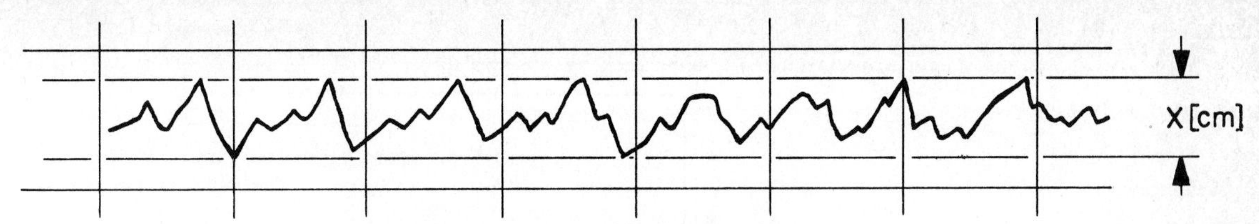

Fig. 2.12.3. Typical ripple pattern.

is placed in a temperature chamber, allowed to reach thermal stability, then driven at the same speed while voltage is recorded. This can be done for any number of temperatures. Temperature coefficient is calculated by:

$$\psi_V = 100 \frac{V_1 - V_A}{V_A (\Theta_1 - \Theta_A)} \qquad (2.12.4)$$

$$[\%/^oC; V, ^oC]$$

where

V_1 is output voltage at temperature Θ_1

V_A is output voltage at ambient temperature Θ_A

Θ_1 is adjusted chamber temperature

Θ_A is ambient temperature

DIELECTRIC TEST

This is an electrical check to see whether there will be a breakdown of dielectric materials when subjected to high voltage. The test simply calls for application of a specified high voltage between generator winding and housing while monitoring leakage current. If specified leakage is not exceeded, the unit passes successfully.

LINEARITY

The purpose of a tachometer is to generate a voltage proportional to speed. The linearity of a tachometer is defined as *the maximum deviation of generated voltage vs. speed characteristic from the straight line,* as described in section 2.11.

The generator is attached in a test fixture as illustrated in Fig. 2.12.4, and is driven at several different speeds while speed and output voltage is recorded on a chart similar to that in Tab. 2.12.1.

SPEED [rpm]	MEASURED OUTPUT VOLTAGE [V]	CALCULATED THEORETICAL OUTPUT VOLTAGE [V]	DIFFERENCE $\lvert\Delta V\rvert$ [V]	DEVIATION V [%]
1000	3.05	3.0273	0.0227	0.0074
2000	6.20	6.0546	0.1454	0.0240
3000	8.99	9.0819	0.0919	0.0100
4000	12.02	12.1092	0.0892	0.0073
5000	15.15	15.1365	0.0135	0.00089

Tab. 2.12.1. Example of generator linearity test results.

The true mean value of the tachometer voltage constant, K_E, is obtained from data in Tab. 2.12.1 by dividing the sum of measured output voltages by the sum of speeds:

$$K_E = \frac{45.41}{15} = 3.0273 \text{ V/krpm}$$

This is used to calculate theoretical output voltage (speed in [krpm] times K_E), as

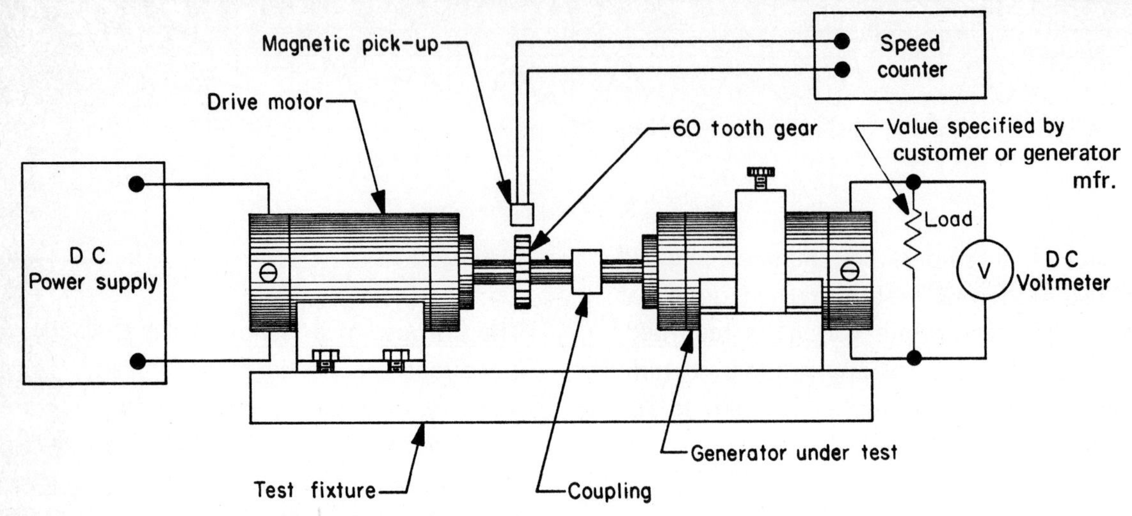

Fig. 2.12.4. Generator linearity test setup.

recorded in column 3 of Tab. 2.12.1. Differences between calculated theoretical output voltages and measured output voltages are recorded in column 4. The deviations from linearity are calculated by:

$$\Delta = 100 \frac{\Delta V}{V_{theor}} \quad [\%] \quad (2.12.5)$$

STABILITY

Output voltage variations can occur in short term and/or long term periods. Short term variations may be of a random nature. Long term voltage drift may take place over a period of hours or days. Both changes are recorded as percentage voltage variations at a specified load.

Test equipment used should have the following capabilities:

1. A recorder with stability at least four times better than the voltage stability to be measured. Full-scale range of the recorder shall be 5 to 10 times the expected voltage variations. Recorder input current should be no more than 5% of the specified generator current. The recorder may have appropriate filtering to prevent voltage ripple from recording.

2. A precision DC power supply with voltage stability at least four times better than the voltage stability to be measured.

3. Precision decade resistors, voltage dividers, and potentiometers as may be required to adjust the recorder sensitivity, zero the recorder, and match the power supply voltage to the generator voltage.

4. A room or enclosure with temperature control to maintain ambient temperature within ± 5 °C.

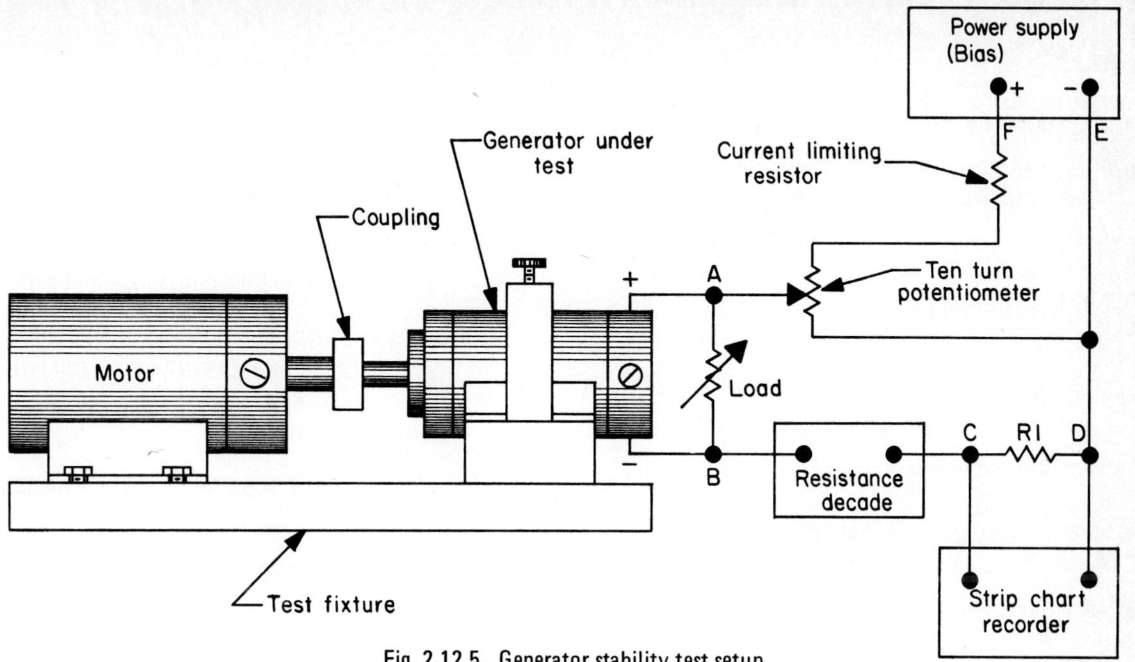

Fig. 2.12.5. Generator stability test setup.

The generator shall be loaded with a non-inductive load which will remain unchanged throughout the test. Voltage dividing networks are used to adjust the recorder sensitivity, and to match the generator and stabilized power supply voltages, as in Fig. 2.12.5.

After thermal equilibrium of the generator is attained, the stabilized DC power supply is adjusted so that the recorder is at a mid-scale position.

The recorder can be operated at intervals, or for any continuous period of time desired. Short term operation is approximately five hours. Long term drift usually is read at one-hour intervals for eight hours, then at eight-hour intervals for seventy-two hours.

The largest peak-to-peak drift amplitude in any one-hour period expressed as a percentage of average DC voltage during this hour is the voltage stability figure for short term variations. Refer to Fig. 2.12.6.

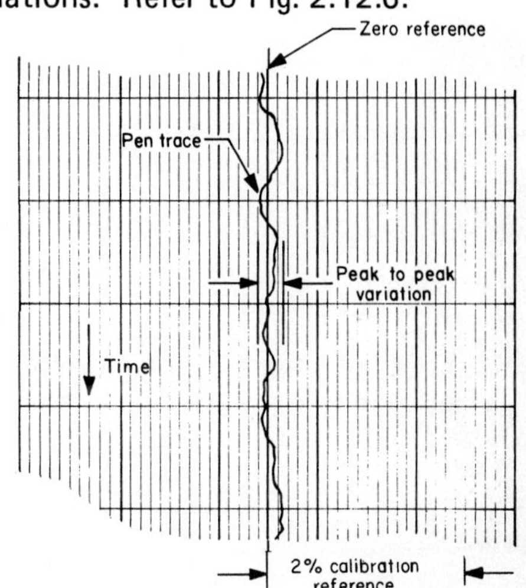

Fig. 2.12.6. Strip chart recorder results for short term generator stability testing.

For long term drift, the percentage voltage change per hour is calculated for each time interval between voltage readings. Long term voltage stability is the largest of these percentages.

INCOMING INSPECTION

The following list of electrical and mechanical tests can be employed as guidelines for incoming inspection:

GENERATOR INSPECTION

1. Visual Inspection

   A. LEADWIRES

      NICKED, CUT OR CRACKED INSULATION TO EXPOSE BARE WIRE

      PROPER COLOR IDENTIFICATION

      PROPER STRIPPING OF LEADWIRE ENDS

      IF TERMINALS OR CONNECTORS ARE USED, CHECK THAT THE RIGHT ONE IS BEING USED AND THAT IT IS ATTACHED PROPERLY.

   B. OUTPUT SHAFT(S)

      RUST OR FOREIGN SCALE

      NICKS OR UPSETS, WHICH MAY CAUSE AN OVER-SIZE CONDITION

   C. BALL BEARINGS

      CHECK FOR DAMAGED SHIELDS

      PROPER INSTALLATION OF RETAINING RINGS

   D. PILOT AND LOCATING SURFACES

      NICKS, BURRS, OR UPSETS WHICH MAY CAUSE IMPROPER MATING OR ALIGNMENT

   E. BRUSH HOLDERS

      CHECK FOR LOOSE OR CRACKED BRUSH HOLDERS

      LOOSE OR CRACKED BRUSH HOLDER CAPS

   F. NAMEPLATES

      PROPER LABELING AND IDENTIFICATION

2. Electrical and Mechanical Checks

   A. END PLAY

   B. RADIAL PLAY

   C. RUNOUT

   D. OUTPUT IMPEDANCE

   E. FRICTION TORQUE

   F. VOLTAGE GRADIENT

   G. VOLTAGE POLARITY

   H. GENERATOR RIPPLE

GENERATOR TROUBLESHOOTING CHART

Problem	Problem Causes	Corrective Action
High Generator Ripple	1. Open or shorted winding	1. An open or shorted winding will show a very high spike in the normal ripple pattern. Repair or replace armature.
	2. Improper filter	2. If filter is used, check components for proper values.
	3. Improper test speed	3. Check armature for proper speed. Generator ripple is speed sensitive. Also, filters are generally selected according to commutating frequency which is a function of speed.
	4. Speed variations in drive motor	4. Check stability of drive motor. Any variations in speed will be reflected in the generator output.
	5. Brush bounce	5. If brush bounce occurs, it's generally at high speeds. Check for rough or non-concentric commutator, weak brush springs, or worn brushes.
	6. Excessive axial or radial armature movement	6. Check coupling between drive motor and generator for misalignment.
	7. Brushes out of electrical neutral zone	7. Readjust generator for minimum ripple. See test procedure for "Generator Ripple".
No Output Voltage	1. Faulty connection	1. Check connections from generator leadwires to metering device.
	2. Open wires	2. Check leadwires and brush flex wires for breaks.
	3. Hung brushes	3. The brushes should slide freely inside the holder, otherwise should be replaced.
	4. Coupling between drive motor and generator	4. Be sure generator is rotating, check for slippage in coupling.
Intermittent Output	1. Intermittent opens	1. Check leadwires and brush flex wires for breaks.
	2. Hung brushes	2. Replace brush or brush holder.
	3. Worn brushes	3. Replace brushes.
	4. Brush bounce	4. Check for rough commutator and remachine if necessary. Check brush springs for proper tension.
Low Output Voltage	1. Incorrect resistive load value	1. Check resistive load for proper value.

Problem	Problem Causes	Corrective Action
	2. Magnets partially demagnetized	2. Remagnetize magnets.
	3. Brush out of neutral zone	3. Readjust generator for proper output voltage. See generator test procedure for "Voltage Gradient."
	4. Incorrect checking speed	4. Voltage output is proportional to speed. Be sure armature is rotating at the proper speed setting.
Reversed Polarity	1. Generator lead wires to load reversed.	1. Reverse leadwires.
	2. Magnets have reversed polarity.	2. Rotate magnet and housing assembly to proper orientation or remagnetize.
Shaft Will Not Turn	1. Generator bearings tight or seized	1. Replace bearings.
	2. Dirt or foreign matter in air gap	2. Remove the obstruction.
Excessive Vibration	1. Armature unbalanced	1. Rebalance armature dynamically to proper level.
	2. Worn bearings	2. Replace bearings
	3. Excessive radial play	3. See corrective action for "Excessive Radial Play"
Excessive Radial Play	1. Shaft loose in bearing I.D.	1. Check fits, the I.D. of the bearing should be a light press fit.
	2. Worn bearings	2. Replace bearings.
	3. Bearings loose in end cap bearing bore	3. The bearings should be free to slide in the bearing bore. However, the clearance should not be excessive. If excessive, replace end cap.
Excessive End Play	1. Improper shimming	1. Add shims as needed. Recheck generator for proper end play.
Excessive Noise	1. Armature unbalanced	1. An unbalanced armature can set up vibration that can easily be felt. The armature should be rebalanced dynamically.
	2. Worn bearings	2. Replace bearings.
	3. Excessive End Play	3. Add shims to the generator to take up end play. Re-check friction level and end play as excessive preload will cause early bearing failure.

Problem	Problem Causes	Corrective Action
	4. Misalignment between generator and drive mechanism	4. Correct mechanical condition of misalignment.
	5. Generator not firmly fastened to mounting	5. Correct mechanical condition of mounting.
	6. Dirt in air gap	6. Noise is irregular, intermittent and scratchy. Dismantle and clean generator.
	7. Rough commutator	7. Remachine commutator.
Excessive Bearing Wear	1. Dirty bearings	1. Clean or replace bearings. If condition is bad, provide means for shielding generator from the dirt.
	2. Insufficient or inadequate lubrication	2. Review bearing mfg. recommendations for type and amount of lubricant to use.
	3. Excessive thrust load	3. Reduce the load or obtain a generator designed to handle the required thrust load.
	4. Excessive vibration	4. Check coupling between drive motor and generator for misalignment.
High Friction Torque	1. Worn bearings	1. Replace bearings.
	2. Worn or sticking brushes	2. Replace brushes. Check brush holder slots for obstructions.
	3. Obstruction in air gap	3. Remove obstruction.
	4. Armature rubbing	4. Note cause of rubbing, remove obstruction or replace with new armature.
Excessive Brush Wear	1. Incorrect spring tension	1. Replace brush with proper spring material and tension.
	2. Dirty or rough commutator	2. Clean or remachine commutator. Insure a good micro-finish.
	3. Incorrect brush seating (off center neutral)	3. Check to see that generator is properly neutralized. See test procedure for "Voltage Gradient".
	4. Short circuit coils in armature.	4. Repair or replace armature.
	5. Loose-fitting brushes	5. Check brushes and brush holder slots for proper size. Replace as needed.

Problem	Problem Causes	Corrective Action
	6. Vibration	6. The armature should be dynamically rebalanced. Vibration causes brush wear.
	7. Lack of moisture	7. Operating in a new vacuum condition causes rapid brush wear. Use a special treated or "high altitude" brush. □

Chapter 3
Unidirectional Speed Controls

3.1. BASIC CONTROL METHODS

SPEED CONTROLLERS – AN INTRODUCTION

In this chapter we will discuss various speed control methods for DC motors in some detail. But first let us briefly review the various methods and their control range.

Rheostat controls for DC motors, the earliest and simplest of the control methods, have an effective controlled speed range of about 4:1 with poor regulation of motor speed against changes in load torque and line voltage. This control method is very inefficient because of the power dissipated in the rheostat.

The variable transformer, or *"Variac"*, with rectifiers, can run DC motors over a wider speed range (up to 10:1) at an improved regulation compared to the rheostat control. It is also more efficient than the former.

SCR (thyristor) controls with *half wave* operation have characteristics similar to the variable transformer control. However, SCR systems with *full wave* rectification can achieve a 20:1 speed range when used with "IR" compensation techniques (a pseudo-closed loop current sensing technique). Using such speed compensation methods, full wave SCR controls can achieve quoted regulation figures such as 3% from zero to full torque load. This regulation figure is understood to mean that the motor speed will not deviate from the *set* speed more than 3% of the *rated* speed at any speed setting. To illustrate this, say that a motor has a rated speed of 1800 rpm, and the control has a regulation of 3% of full speed; this means that the motor speed may vary 54 rpm due to load variations. Now, assume the motor is running at 180 rpm, at no load. When given rated load torque, the motor speed may decline to 180 – 54 = 126 rpm – a regulation, based on *set* speed, of 30%!

The foregoing examples of speed controls were based on "open loop" features, meaning that there is no velocity feedback element in the control loop which tells the controller what the motor speed is, so that the controller can make appropriate adjustments to keep the speed essentially constant.

True closed-loop feedback control is possible with SCR circuits, using a tachometer for feedback; with such devices a full wave or three-phase SCR control may achieve speed ranges of up to 100:1. Due to the pulsating nature of SCR control techniques, the speed stability of motors below 1 hp may not always be good in the lower speed range, due to the low moment of inertia of the motor, and due to the load. However, in integral horsepower machines excellent speed range can be achieved.

Transistor controllers such as those supplied by Electro-Craft Corporation, on the other hand, can handle closed-loop speed control with a speed range of over 1000:1, and with regulation better than 1 or 2%, based on *set speed.* For example, if an 1800 rpm, 1% regulation rated speed control system is set at 180 rpm and the load torque changes from zero to rated value, the speed will not change more than 1.8 rpm. The transistor controls have their best operational advantage for motor sizes up to 1 hp, and will presently handle high-performance servomotors up to 5 hp. For the smaller motor sizes a continuous control method is used, but for motors above 1/3 to 1/2 hp a pulse-width modulation technique is the most efficient control method.

In the following discussions we will more fully examine the various aspects of speed controllers, old and new.

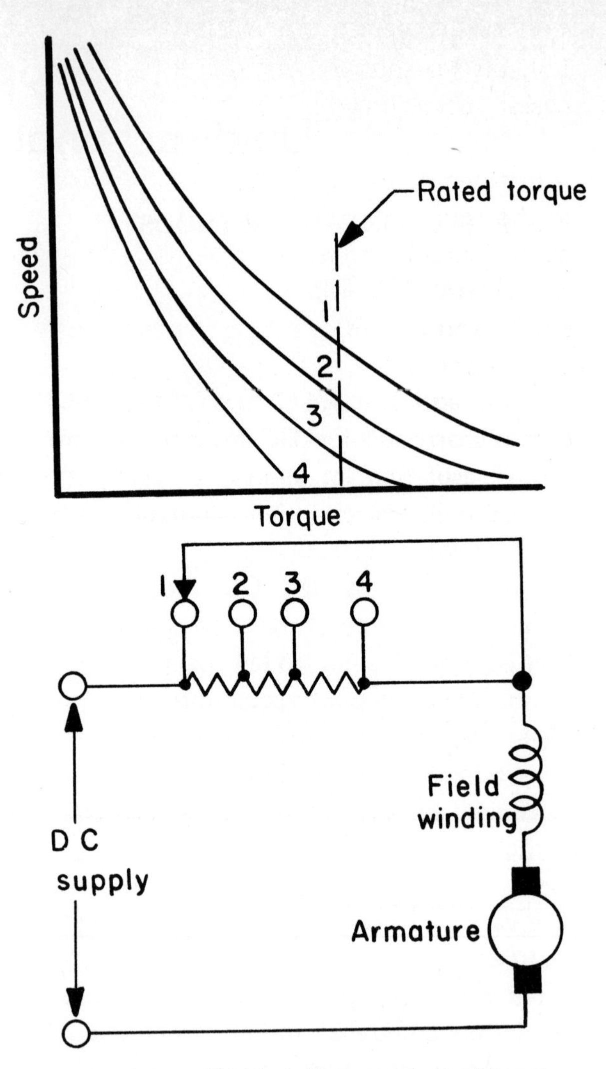

Fig. 3.1.1. Electrical diagram and speed-torque curves for series motor resistor speed control.

OPEN LOOP CONTROLS – THE TRADITIONAL APPROACH

Speed controls for DC motors have a long history. Perhaps the oldest and most widely used control for small DC motors is the series resistor speed control for series motors pictured in Fig. 3.1.1. The variable resistor is inserted in series with both the armature and field circuit, yielding performance as shown in the accompanying speed-torque diagram. This control has good starting characteristics (large torque available at low speed), but has a runaway speed tendency at small load torque conditions, making the control useful only for control applications with somewhat fixed friction conditions, such as sewing machine motors, food mixers, and the like. It is also evident that the speed regulation characteristics of the motor decline with decreasing speed, making it difficult for the operator to achieve good overall speed control.

A totally different characteristic is obtained in the shunt motor variable resistor speed control circuit shown in Fig. 3.1.2. Here, a series resistor is inserted in the armature circuit, and the field winding is excited with a constant voltage (a permanent magnet field can also be used). The resulting speed-torque characteristics show the regular shunt motor regulation for resistor setting #1 (no resistance in the circuit). With increasing insertion of resistance in the armature circuit, speed regulation degenerates. This type of control will work well for constant load torque, rather than for a widely varying torque situation.

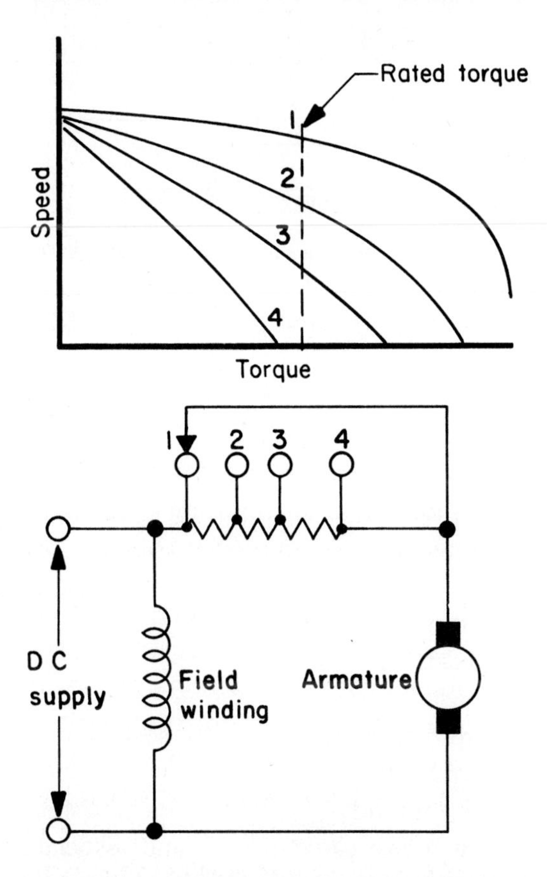

Fig. 3.1.2. Electrical diagram and speed-torque curve for shunt motor resistor speed control.

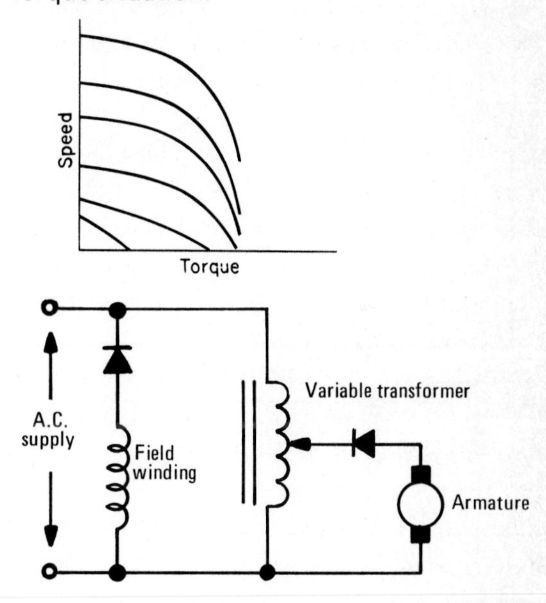

Fig. 3.1.3. Shunt motor with variable transformer to control armature voltage.

Fig. 3.1.3 shows a control utilizing a variable transformer controlling the armature voltage of a shunt motor with constant field excitation. The resulting speed-torque characteristics are much improved over variable resistor control characteristics, with a more uniform speed regulation over a wider speed range.

In Fig. 3.1.4 we see a shunt motor connected to a variable field resistor control. The action of this control circuit is unique in that the motor speed is variable *only above the speed it would have without the field resistor control.* This has an undesirable effect in that the torque constant of the motor will decrease with increasing

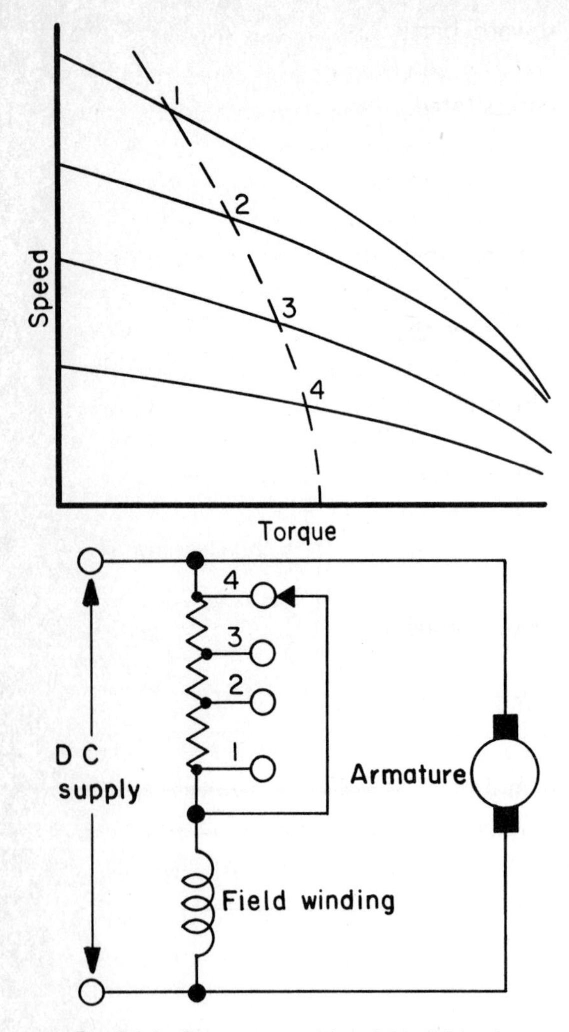

Fig. 3.1.4. Shunt motor with variable field resistor control.

resistance insertion (field weakening), the net effect is that the armature current for a given torque will increase with higher speed. The motor can easily be overloaded, and this control circuit is used only in unique cases where load conditions are both predictable and well controlled.

In some sophisticated open-loop control methods, such as the motor-generator armature control method shown in Fig. 3.1.5, a constant speed motor drives a generator with an adjustable control field voltage. The generator will produce an adjustable voltage which is delivered to the armature of the motor, and the resulting torque-speed characteristics are improved over the ones shown in Fig. 3.1.2, since in this case the regulation is essentially independent of the speed setting.

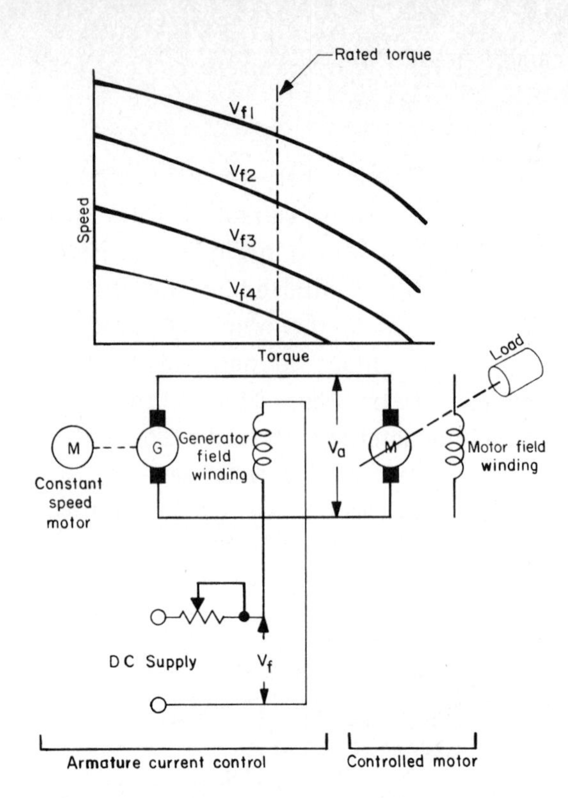

Fig. 3.1.5. An open-loop control utilizing an armature control method.

This results in superior motor speed control performance over any of the previously shown methods. However, due to the cost of the motor-generator set and associated field control, this method has not been practical for the small motor speed control applications most commonly used in home and industry. Therefore, the motor-generator control method has been mainly confined to industrial uses of large motor speed controls sets of 1 hp and above. Because of the power amplifying characteristics of the motor-generator set, this was the easiest entry into the closed-loop control systems when power vacuum tubes and magnetic amplifiers became available for control purposes.

The control methods shown above are by no means all that have been developed and used; rather they exemplify the kind of problems encountered in the open–loop, manually operated systems of the past. We can see that in each case there was some undesirable side effect that limited its use to certain applications.

CLOSED-LOOP CONTROLS

Although many of the "open-loop" controls are adequate for uses today, the trend toward better speed regulation, such as in meeting servo systems requirements, have necessitated closed-loop control.

In its elementary form, *a closed-loop control consists of an actuator (motor), a comparator, an amplifier, and a sensor (generator).* Fig. 3.1.6 illustrates these elements and how the "closed-loop" expression is derived: the amplifier drives the motor, which is coupled to the generator; the generator sends a signal to the comparator, which measures the feedback signal against the command signal; the comparator keeps the two signals in balance by giving the amplifier the proper command. Thus, the "loop" consists of the components which are shown in Fig. 3.1.6. Signal flow is indicated by the arrow.

This example is just one of many feedback techniques. Usually, a modern, high-performance servo system has several separate feedback loops, such as voltage, current, velocity and position feedback.

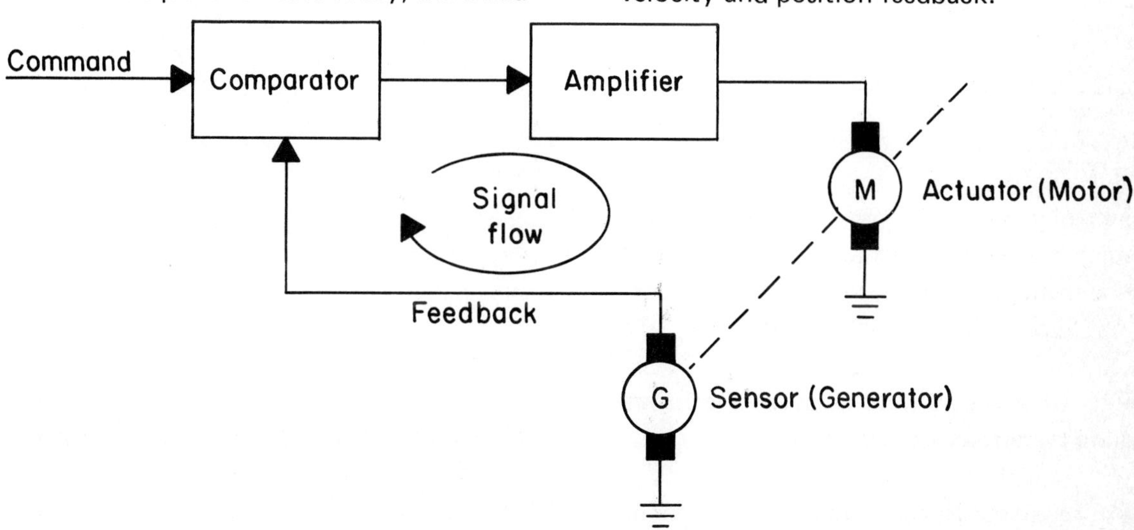

Fig. 3.1.6. A closed-loop speed control system.

In the open loop controls described previously, the human operator was the sensor or feedback element, so in a sense these systems could also qualify under certain circumstances as closed-loop controls; but human control is not an automatic operation, and that is what closed-loop controls are all about.

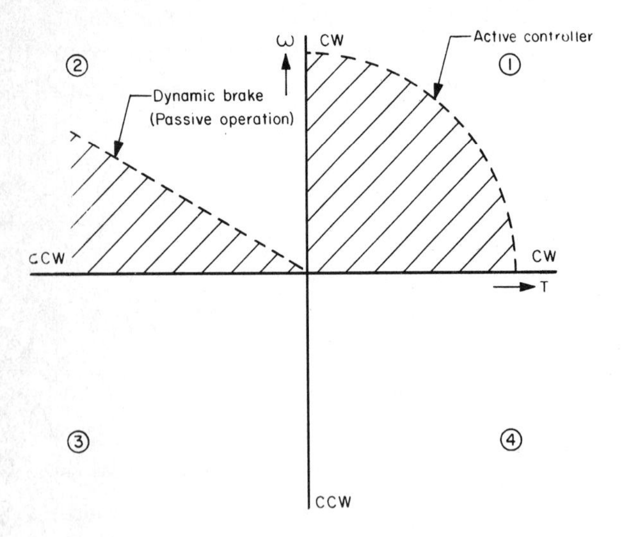

Fig. 3.1.7. Unidirectional or regulator type speed control system operates in the first quadrant.

SPEED CONTROLS VS. SERVO SYSTEMS

It is important to recognize the basic differences between the two types of motor control systems because of their individual unique advantages, and since in one case (the regulator control) linear control theory applies only in a restricted sense.

Fig. 3.1.7 shows four quadrants of shaft velocity versus torque plot.

The regulator (speed control) system operates only in the first quadrant of this diagram. The system cannot produce a negative torque; nor can it reverse direction without a reversing switch, adding operation in the third quadrant. This is generally done by *manual* control, which makes servo considerations invalid. With the addition of a dynamic brake, we can achieve limited negative torque in the second quadrant, but since this is a passive control area, we choose to ignore it for the moment and call the regulator speed control a *single quadrant control.*

The bidirectional servo control system, on the other hand, can provide motor speed and torque in both negative and positive directions. Fig. 3.1.8 shows the areas within which servo control is possible, and explains why we call the bidirectional servo system a *four quadrant control.*

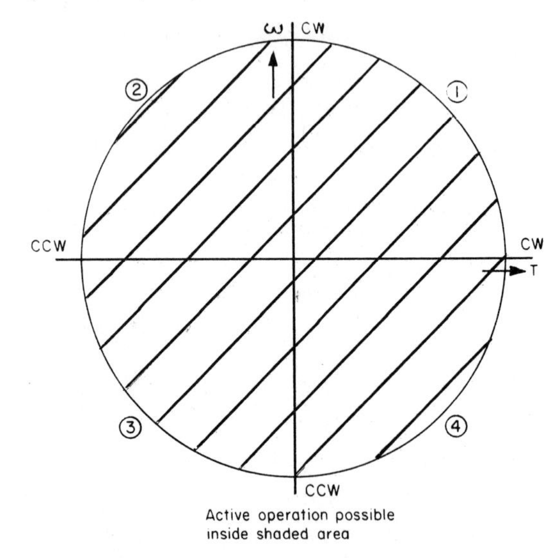

Fig. 3.1.8. A servo system operates in four quadrants.

3.2. VELOCITY CONTROL – SINGLE QUADRANT CONTROLLER

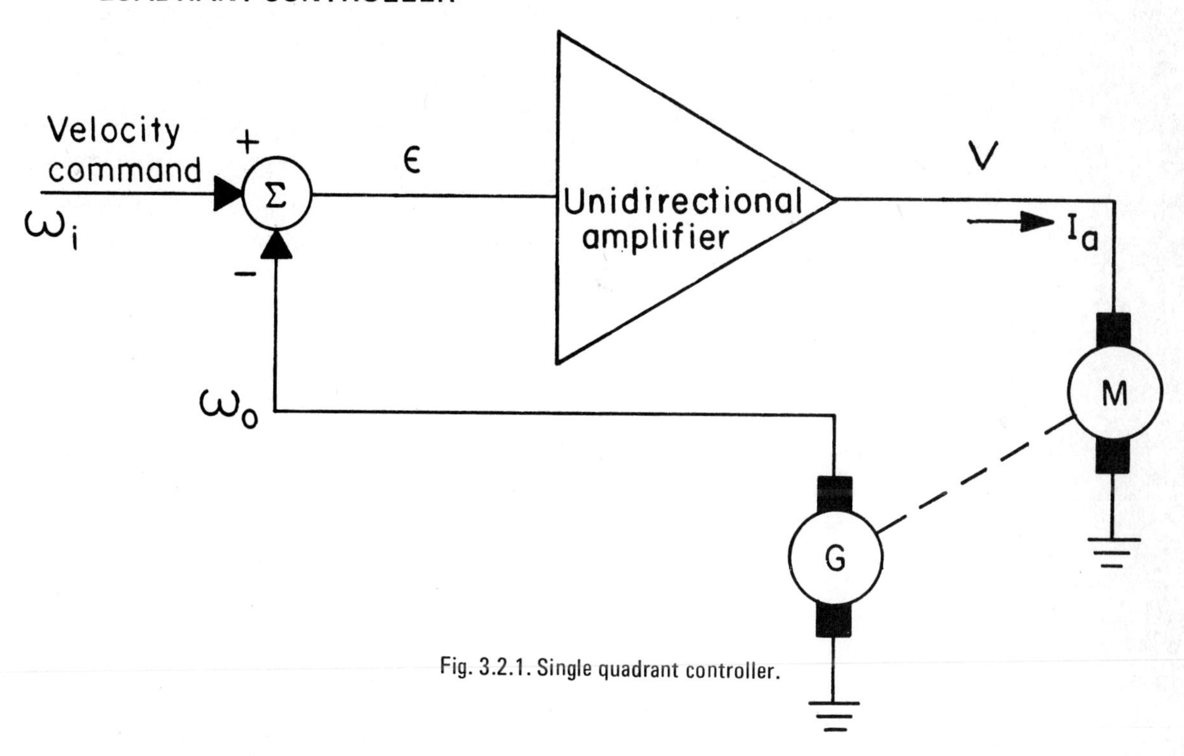

Fig. 3.2.1. Single quadrant controller.

VOLTAGE AND CURRENT OF A SINGLE QUADRANT CONTROLLER

A single quadrant controller is a control system which employs a unidirectional amplifier, as seen in Fig. 3.2.1.

As such, it can deliver to the motor only positive (or only negative) voltage and current, and therefore, control velocity only in one direction with the system load torque as an opposing torque. In Fig. 3.1.7, where the motor velocity ω corresponds to the vertical axis and the torque **T** to the horizontal axis, we find that the controllable velocity and torque range of a single quadrant controller is in the first quadrant of the plane.

The amplifier characteristic of the single quadrant controller will be described by:

$$\left.\begin{array}{lll} V = 0 & \text{if } \epsilon < 0 \\ V = A\epsilon & \text{if } 0 \leqslant \epsilon \leqslant \dfrac{V_{max}}{A} \\ V = V_{max} & \text{if } \epsilon > \dfrac{V_{max}}{A} \end{array}\right\} \quad (3.2.1)$$

where **A** is the amplifier gain, or graphically as shown in Fig. 3.2.2.

Current delivered from the amplifier to the motor can be only positive, therefore:

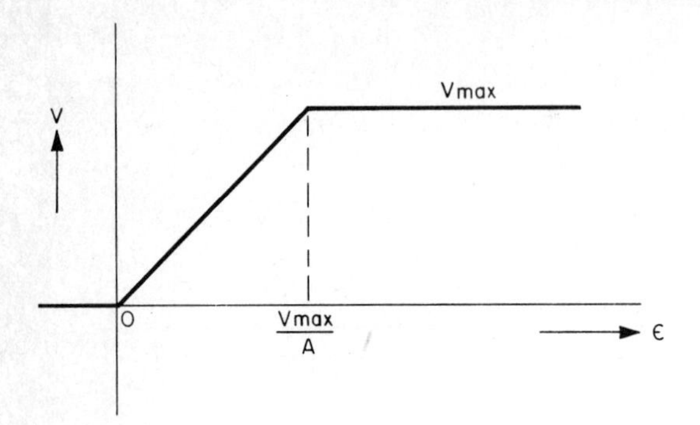

Fig. 3.2.2. Amplifier output voltage vs. error signal for a single quadrant controller.

$$\left. \begin{aligned} I_a &= \frac{V-E_g}{R} && \text{if } V > E_g \\ I_a &= 0 && \text{if } V \leqslant E_g \end{aligned} \right\} \qquad (3.2.2)$$

where

R is the motor resistance

E_g is internally generated voltage (counter emf) in the motor

SYSTEM OPERATION

The simplified dynamic equation of a motor is:

$$K_T I_a = J \frac{d\omega}{dt} + T_L \qquad (3.2.3)$$

When the motor is connected in the single quadrant controller system, the system will control the velocity and maintain it close to the desired value ω_i in the following way.

Suppose that due to disturbances or change of the command signal the actual velocity ω_o is larger than the desired one ω_i. In this case, the velocity error is negative:

$$\epsilon = k \; (\omega_i - \omega_o) < 0 \qquad (3.2.4)$$

This results in zero current and voltage as can be seen from (3.2.1) and (3.2.2); consequently, the velocity will be reduced as can be found from (3.2.3). When the current is zero, the rate of change of the velocity will be:

$$a = \frac{d\omega}{dt} = -\frac{T_L}{J} \qquad (3.2.5)$$

The deceleration rate equals the ratio between the opposing load torque (usually friction) and the combined moments of inertia of the motor and the load. The deceleration will continue as long as the velocity ω_o is too high and the velocity error ϵ is negative.

On the other hand, if the motor velocity is too low, causing an error ϵ which is larger than $\frac{V_{max}}{A}$, this results in the output voltage $V = V_{max}$ and current

$$I_a = \frac{V_{max} - E_g}{R} \qquad (3.2.6)$$

Consequently, the motor will accelerate at the rate

$$a = \frac{d\omega}{dt} = \frac{1}{J} \, (K_T I_a - T_L)$$

$$a = \frac{1}{J} \left(K_T \; \frac{V_{max} - E_g}{R} - T_L \right) \qquad (3.2.7)$$

The acceleration at this rate will continue until the velocity error ϵ becomes smaller than V_{max}/A and it moves to the linear region.

This discussion shows that in case of large deviations from the desired velocity the system tends to correct the error. It remains to analyze system performance under steady-state conditions and small deviations that keep the amplifier output voltage within the linear range, i.e. $0 \leqslant V < V_{max}$.

Note that in the steady state the velocity is constant. Therefore, the steady-state current, as found from (3.2.3) is

$$I_a = \frac{T_L}{K_T} \qquad (3.2.8)$$

The required voltage V and the velocity error ϵ can be found from (3.2.1), (3.2.2) and (3.2.8) as follows:

$$V = E_g + I_a R = E_g + \frac{T_L R}{K_T} \qquad (3.2.9)$$

and

$$\epsilon = \frac{V}{A} = \frac{1}{A}\left(E_g + \frac{T_L R}{K_T}\right) \qquad (3.2.10)$$

Thus, the steady-state velocity error is positive, which indicates that the motor velocity is lower than the set velocity. This result is identical to that obtained in any velocity control system.

When the performance of the system around the steady-state point is analyzed, it is found that the system is identical to that of velocity control. Thus, the system will be overdamped when the gain is low, and will tend to overshoot when the amplifier gain, A, is increased. And, considering the lag in the amplifier, the system can become unstable if the gain is increased beyond a certain limit.

To illustrate these results, suppose that the motor is running at a given velocity ω_1 and the command is increased to ω_2 for some time t_1, then decreased again to ω_1. The corresponding motor velocity profile is shown in Fig. 3.2.3.

It can be seen that when the command is increased, the motor will accelerate at the rate

$$a_1 = \frac{1}{J}\left(K_T \frac{V_{max} - E_g}{R} - T_L\right) \qquad (3.2.7)$$

The deceleration rate in $t > t_1$ is

$$a_2 = -\frac{T_L}{J} \qquad (3.2.5)$$

When the deceleration rate is too small, due to small friction or large moment of inertia, it is possible to increase the deceleration rate by the use of dynamic braking.

DYNAMIC BRAKING

The amplifier used in the single quadrant controller system has the feature that its output impedance is low as long as $V > E_g$ and the current I_a is positive. When this is

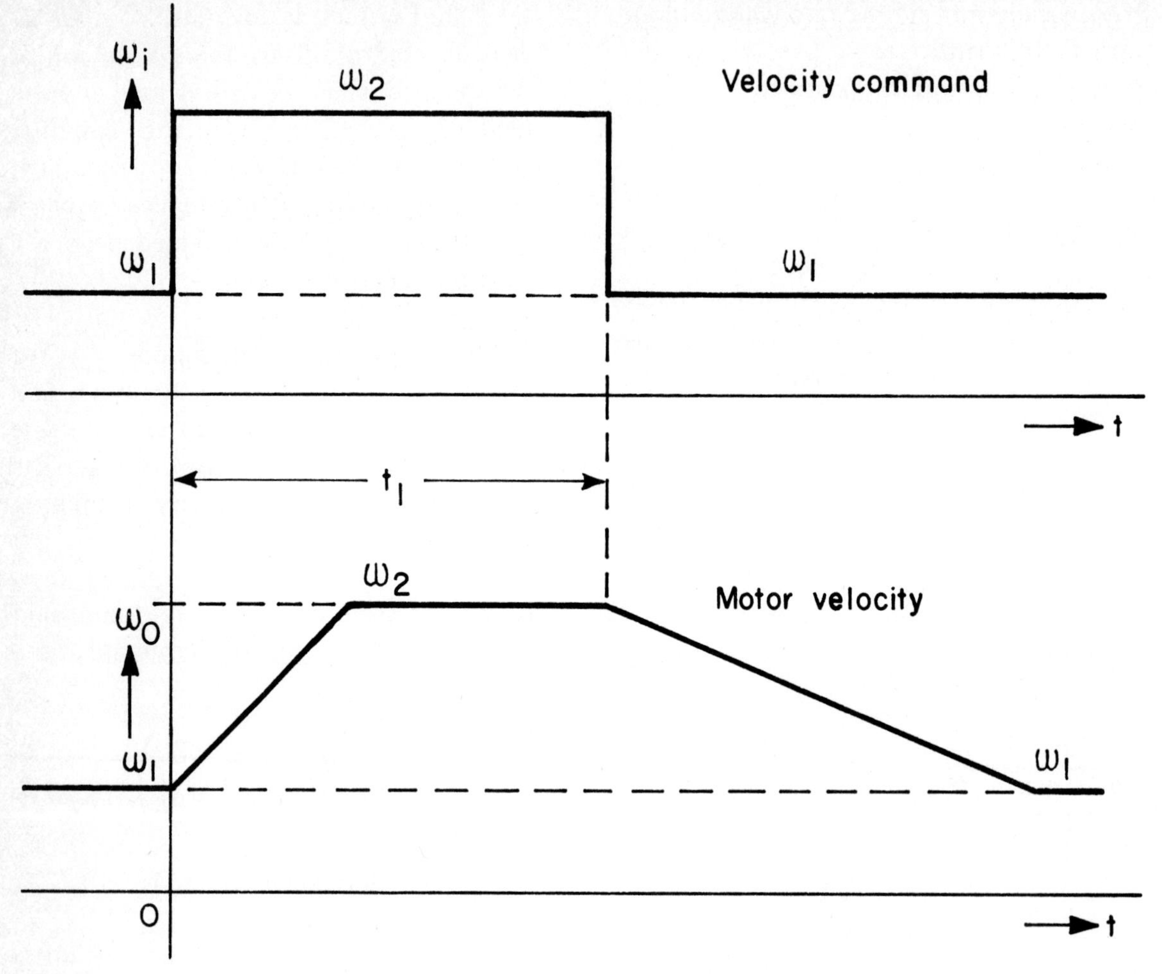

Fig. 3.2.3. Example of a stepwise changed command and motor velocity profile.

changed, output impedance becomes very large, and does not allow negative current. This can be modified by introducing a circuit which senses current I_a. As long as the current is positive, the circuit is inactive; but if I_a becomes zero, the circuit shorts the motor terminals, allowing a negative current to circulate in the armature, thus stopping the motor. This method is called *dynamic braking.*

The voltage equation (3.2.1) for this case is not changed, but the current equation (3.2.2) and the motor dynamic equation (3.2.5) will change.

The current equation becomes

$$\left.\begin{aligned} I_a &= \frac{V - E_g}{R} \qquad \text{if } V > E_g \\ I_a &= -\frac{E_g}{R} \qquad \text{if } V \leqslant E_g \end{aligned}\right\} \qquad (3.2.11)$$

If the velocity error ϵ is positive and large, the amplifier output voltage is $\mathbf{V} = \mathbf{V}_{\mathbf{max}}$, the current is given by (3.2.6) and the acceleration by (3.2.7), so that there is no change in system function.

However, if the velocity error ϵ becomes negative, which indicates that the motor runs too fast, the amplifier output voltage becomes $\mathbf{V} = \mathbf{0}$ and the current is negative:

$$I_a = -\frac{E_g}{R} \tag{3.2.12}$$

This results in the deceleration of:

$$a = \frac{d\omega}{dt} = -\frac{1}{J}\left(\frac{E_g}{R} K_T + T_L\right) \tag{3.2.13}$$

This is a larger rate than the one given by (3.2.5) and, hence, enables the system to respond faster. □

3.3. AMPLIFIERS

Transistor amplifiers used in speed control systems fall into two broad categories; *linear* amplifiers (also called *class A* amplifiers) and *switching* amplifiers. Linear amplifiers are almost exclusively transistor devices, whereas the switching amplifier can be designed using either transistors or SCR's (silicon controlled rectifiers, also known as thyristors).

THE LINEAR AMPLIFIER

The linear amplifier is very desirable from a control analysis standpoint, since it has linear control characteristics and no significant control lag within the operating bandwidth. It is easily compensated, and is capable of providing speed control over a wide range. It is the least troublesome in terms of transient problems in adjacent circuits.

The simplest closed-loop control system is typified by the single transistor amplifier (Fig. 3.3.1). It has been used widely in control applications where the loads are essentially constant and a limited speed range was acceptable in view of the extreme simplicity and consequent low cost. The circuit consists simply of a 5W potentiometer and a power transistor, mounted on a heat sink.

Circuit operation can be explained in the following way: the potentiometer is set at a point such that a voltage V_1 appears be-

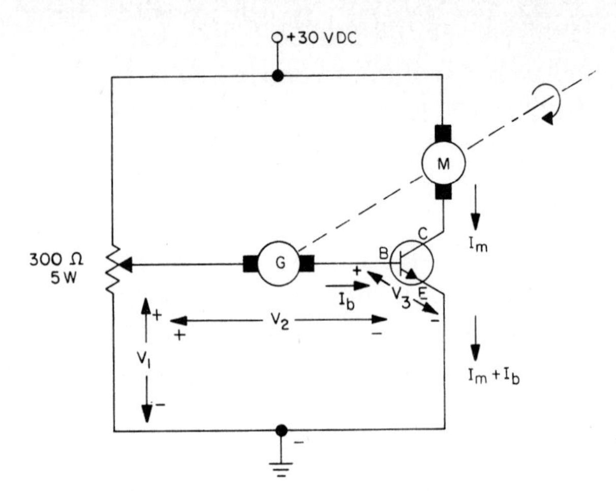

Fig. 3.3.1. A typical single-transistor speed control system.

tween the wiper and the negative terminal (ground) of the circuit. This voltage V_1 will cause a current I_b to flow through the base **(B)** and the emitter **(E)** of the transistor. The base current causes the transistor to conduct, and a current I_m will flow from the positive terminal through the motor winding **(M)** to the collector **(C)** and emitter **(E)** of the transistor to the negative terminal. The motor current I_m will cause the motor to rotate and, if the tachometer is properly connected with respect to its polarity, an opposite voltage V_2 will appear at the tachometer terminals. Since any transistor in conduction has a given associated base-to-emitter voltage for each conduction level, a voltage V_3 will appear across the transistor with the polarity shown in the figure.

Equilibrium (speed control) will appear when the voltage V_2 reaches its steady-state value. The motor cannot run faster than the speed at which it produces V_2, since if it did, I_b would cease flowing, thus making the transistor non-conductive, hence slowing the motor down to the "equilibrium" speed. Conversely, if a load change would cause the motor to tend to slow down, the voltage V_2 would be smaller, allowing an increase in I_b. This would tend to make the transistor to increase its conductivity, thus allowing an increase in I_m, which in turn would overcome the increased load. We can see that this circuit has a self-regulating quality, and therefore qualifies as a closed-loop regulator system. This explanation is somewhat simplified, but it brings out the general principles involved.

The single transistor amplifier has the basic limitation of low gain. In other words, it requires a significant error (ΔI_b) to correct a speed disturbance. Providing more gain is a simple task, however, and Fig. 3.3.2 shows a four-transistor amplifier, employing a two-stage voltage amplifier and a dual emitter-follower current amplifier.

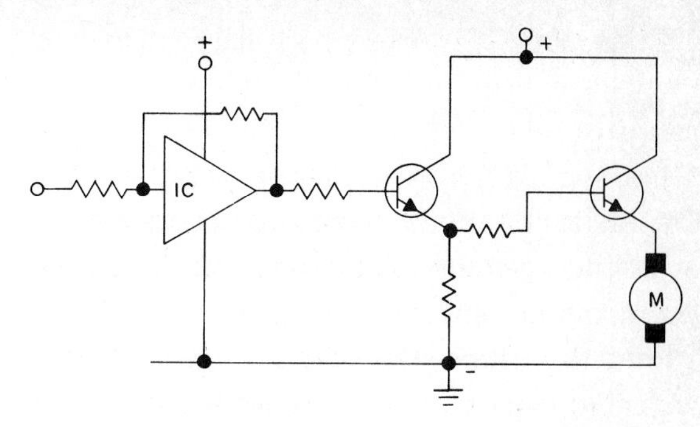

Fig. 3.3.3. Speed control system utilizing an integrated circuit.

Error signals on the order of a few millivolts are able to control the amplifier from "full off" to "full on" conditions, which makes the closed loop control more sensitive to disturbances. This yields a more precise speed control system. The availability of integrated circuits (**IC**) has simplified circuit design; in Fig. 3.3.3 the voltage amplifier of Fig. 3.3.2 was replaced by a single **IC**. This technology can be expanded to include the power stages (hybrid amplifiers).

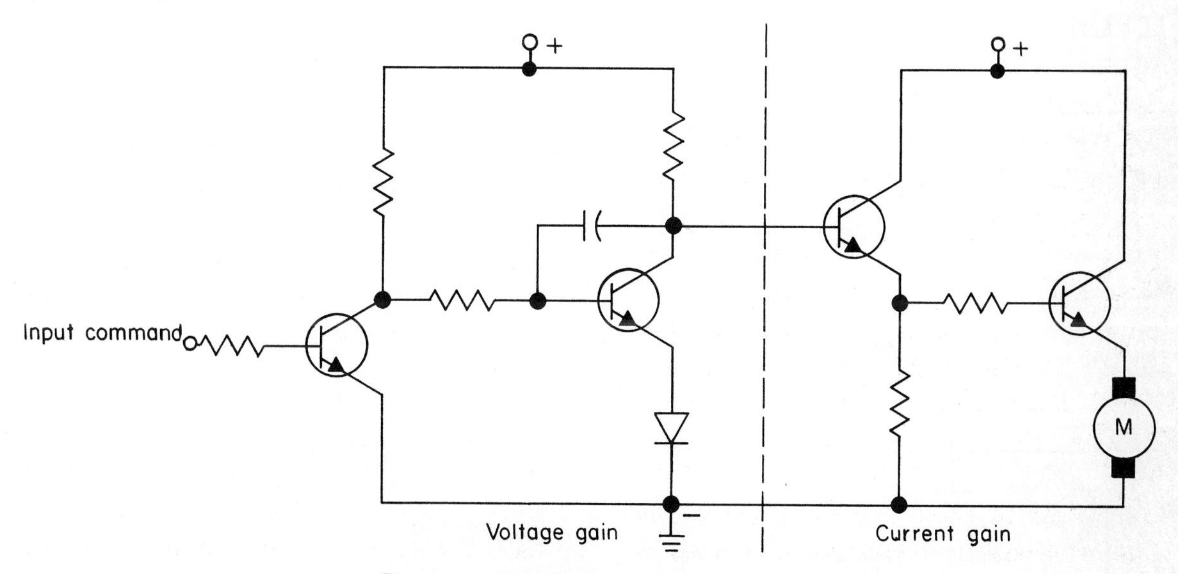

Fig. 3.3.2. Elementary multi-transistor amplifier.

If the amplifier offers high gain, then one would expect an almost error-free speed control.

In reviewing the possible sources of errors, we find that the command stage (in speed controls usually the speed setting potentiometer) must be stabilized so that the reference voltage does not vary with changes in load current, line voltage and ambient temperature. This is done either by the use of Zener diodes, or by recently available **IC** regulators. A simple example of a regulated reference supply employing the Zener diode is shown in Fig. 3.3.4.

The Zener diode stabilizes the voltage across the potentiometer R_2 against changes in supply voltage. This circuit is sufficient for many applications, but for a higher degree of reference voltage stability a two-stage Zener regulator is used. Sometimes series stacks of Zener diodes with low temperature coefficients are employed to minimize effects of changing ambient temperature.

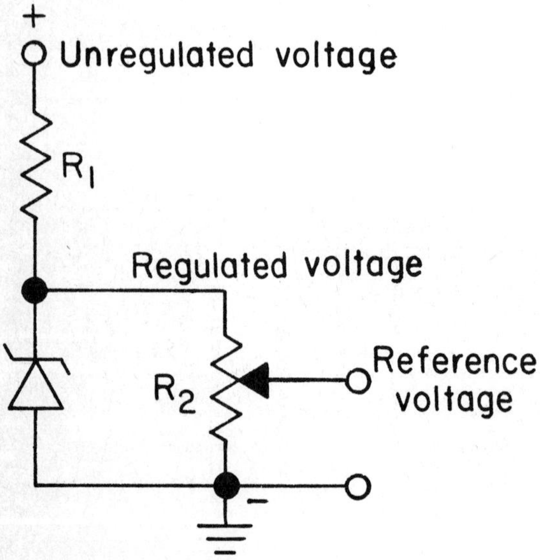

Fig. 3.3.4. Example of regulated reference voltage supply.

A high-performance transistor speed control system will basically consist of a regulated reference supply, an error comparator stage, a voltage amplifier, and a power amplifier.

An example of the steady-state operating conditions of such a system will demonstrate the performance advantage of a high gain system compared with the earlier mentioned single-transistor control.

The system has the following parameters:

Motor:

K_T = 8 oz-in/A
= 5.65×10^{-2} Nm/A

K_E = 5.916 V/krpm

R = 2Ω

Generator:

K_{Eg} = 5.916 V/krpm

Load:

T = $T_L + T_f$ = 6 oz-in
= 4.237×10^{-2} Nm

n = 2000 rpm

Amplifier:

DC gain A = 1000 V/V

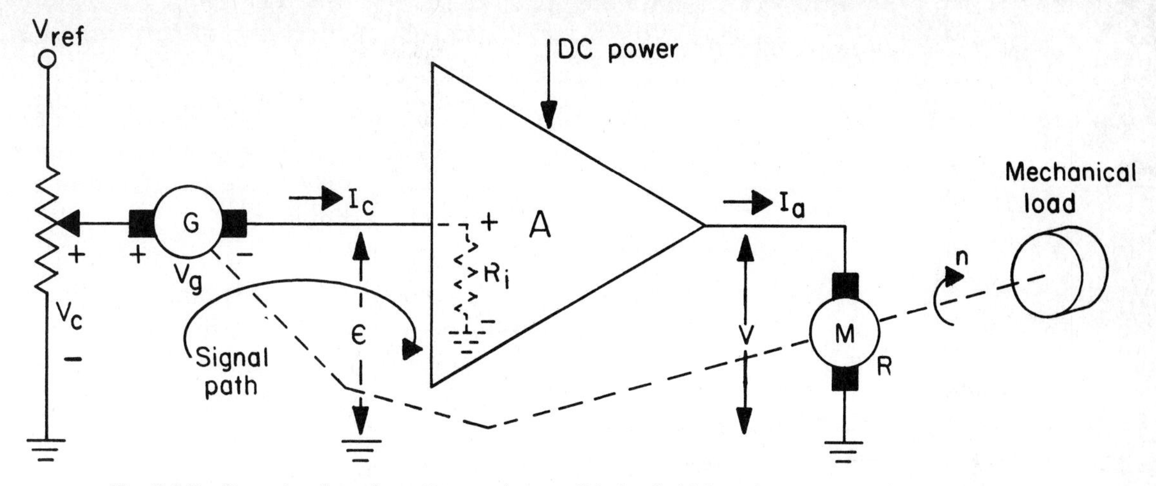

Fig. 3.3.5. Example of steady-state operating conditions of a high performance speed control system.

Fig. 3.3.5 shows the system connection and the input and output signal polarities.

Following the input signal path:

$$\epsilon = R_i I_c \quad (3.3.1)$$

$$-V_c + V_g + \epsilon = 0 \quad (3.3.2)$$

In a properly operating closed-loop speed control system the input error

$$\epsilon \ll V_c \quad (3.3.3)$$

meaning that the generator voltage V_g is always nearly equal to the command voltage V_c.

The steady-state equation of motor is:

$$V = RI_a + K_E n \quad (3.3.4)$$

Since

$$I_a = \frac{T}{K_T} = \frac{6 \text{ oz-in}}{8 \text{ oz-in/A}} = 0.75 \text{ A}$$

we have:

$$V = 2 \times 0.75 + 5.916 \times 2$$

$$= 13.332 \text{ V}$$

and since

$$V = A\epsilon \quad (3.3.5)$$

the required input error signal to produce this voltage is:

$$\epsilon = \frac{V}{A} = \frac{13.332}{1000} \cong 13.3 \text{ mV}$$

Assume that the load torque is now changed from 6 to 10 oz-in. We now have new equilibrium conditions and the following changes occur.

First, combine (3.3.4) and (3.3.5) we obtain:

$$A\epsilon = RI_a + K_E n \quad (3.3.6)$$

The armature current I_a must change from

$$0.75 \text{ A to } I_a = \frac{10 \text{ oz-in}}{8 \text{ oz-in/A}} = 1.25 \text{ A}$$

The armature voltage drop is now RI_a = **2.5 V,** which is **1.0 V** higher than at the 6 oz-in load condition. This must be made up for by the amplifier,thus by a change in error signal of

$$\Delta\epsilon = \frac{\Delta V}{A} = \frac{1}{1000} = 1 \text{ mV}$$

To accommodate the load change, the error signal had to change from 13.3 to 14.3 mV. From (3.3.2) we have:

$$\epsilon = V_c - V_g = V_c - K_E n \qquad (3.3.7)$$

$$V_c = \epsilon + K_E n \qquad (3.3.8)$$

Thus in the first case (6 oz-in load torque)

$$V_c = 13.3 \times 10^{-3} + 5.916 \times 2 = 11.8453 \text{ V}$$

Since V_c was constant during the event, we can find the change in the motor speed **n** after the load torque became 10 oz-in from (3.3.7):

$$n = \frac{V_c - \epsilon}{K_E} = \frac{11.8453 - 14.3 \times 10^{-3}}{5.916} \times 10^3$$

$$\cong 1999.8 \text{ rpm}$$

$$\Delta n \cong 0.2 \text{ rpm}$$

The resulting change in speed is so small (approx. 0.01%) that it is not worth pursuing the exact number of the final speed.

In actual practice, the effects from such sources as amplifier input summing errors, generator ripple voltage changes, and temperature coefficients of the generator voltage gradient may cause larger errors than this example shows.

A linear amplifier can minimize errors in velocity in inverse proportion to its gain due to external sources such as load changes, temperature changes of K_T, R_m, R, but it cannot correct feedback errors due to changes in generator voltage constant K_{Eg}, amplifier input drift errors and reference voltage errors. The three last mentioned error sources can, of course, be minimized to a point where they are not significant to the user.

One may also want to increase amplifier gain to lessen the effects of the errors correctable by the feedback loop, but for each system there is a point where an increase in gain will cause instability due to the mechanical or electrical characteristics of the feedback loop.

Any speed control system has performance characteristics which are results of a balance between speed range, gain-bandwidth product, long and short-term stability and selling price. The Motomatic® speed controls by Electro-Craft are used worldwide, and have a reputation for great versatility in a variety of applications. Fig. 3.3.6 shows a typical ECC master control and motor, model E-650.

One general limitation of the linear transistor amplifier is that the output stage must

Fig. 3.3.6. Electro-Craft E-650 master control and motor-generator.

dissipate heat energy not absorbed by the motor. Thus, the worst amplifier power dissipation condition occurs when the motor is run at very low speed at maximum torque. The output stage will then have to absorb the power proportional to the product of the full armature current and almost all available DC voltage. Thus, ample provisions for heat dissipation must be made. Fig. 3.3.7 shows a typical open chassis amplifier configuration, capable of dissipating 160 W at 40 °C environment temperature.

Fig. 3.3.7. Open chassis amplifier configuration.

The speed control systems described above are generally augmented by accessories which improve operation in a variety of applications.

TORQUE LIMITING

To prevent inadvertent overloading of the system, Electro-Craft supplies a current sensor which turns off the amplifier at adjustable current levels. Since torque is proportional to current in a permanent magnet DC motor, the current limiter becomes in effect a torque limiter. The torque-speed characteristics will then exhibit two distinct regions: the speed control region and the torque control region. Fig. 3.3.8 illustrates typical characteristics of the adjustable speed control system with current limiting. The system assumes an almost constant torque mode when the

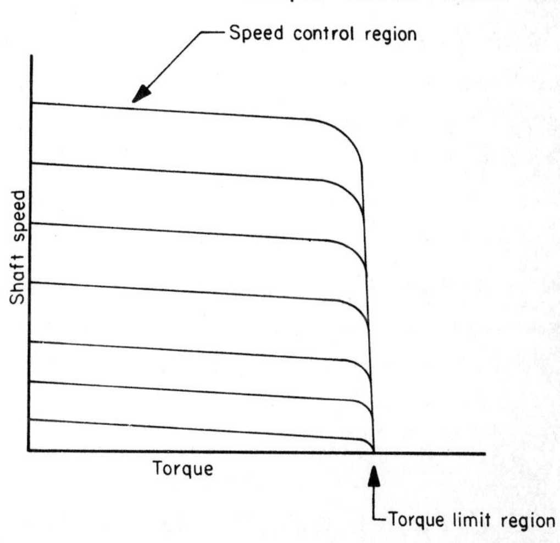

Fig. 3.3.8. Speed torque curves of a high gain speed control system with torque limiting.

preset current limit has been exceeded. As soon as the overload is removed, the system will resume its preset speed.

In some cases, it is important to provide an acceleration torque which is many times higher than the torque limit necessary to protect the system from overload. This requirement is met by providing a time delay in function of the torque limiter. The time delay is usually of the order of 150 ms for iron-core armature motors, and somewhat less for low inertia motors. This provides ample time for rapid acceleration with virtually unrestricted current, and a subsequent introduction of current limit after acceleration has been completed.

Some users have found that the current limit works to advantage when very high inertia loads are to be accelerated. The current limit circuit in such a case permits controlled acceleration to the required speed without damage to the control circuit or motor.

DYNAMIC BRAKING DEVICES

When the speed setting in a basic speed control circuit is rapidly reduced, the motor coasts to the lower speed setting. During this time the motor is basically "out of control" until the circuit is again in equilibrium.

The dynamic braking circuit shown in Fig. 3.3.9 minimizes this effect by providing a controlled "short circuit" around the motor whenever the control system is not

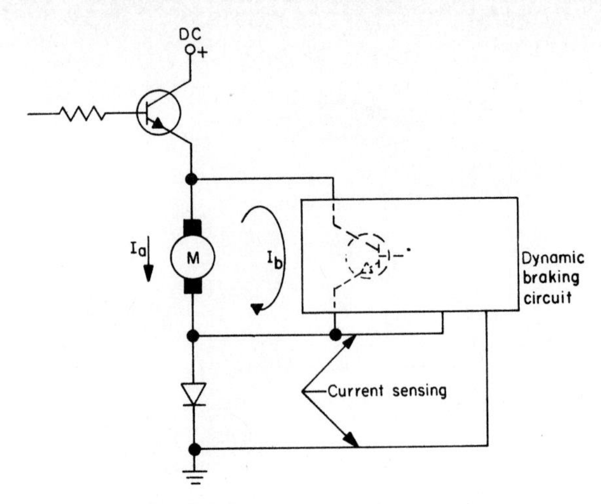

Fig. 3.3.9. Dynamic braking circuit.

providing a current through the motor — a condition which is prevalent during a "coasting" situation.

The dynamic braking depends on the armature *counter emf* to provide the needed braking current; the braking torque is the greatest at high speeds and less effective at lower speeds. For example, the dynamic braking brings an Electro-Craft E-650 control system to a stop in about 400 ms.

REVERSIBLE SPEED CONTROLS

Fig. 3.3.10 illustrates a switch arrangement which gives a single polarity speed control reversible features.

This reversing switch arrangement provides for a middle position (B) which gives a controlled dynamic braking condition, minimizing the possibility of quickly switching a reverse voltage to the motor still coasting in previous direction, which can cause transistor breakdown. The reversing circuit

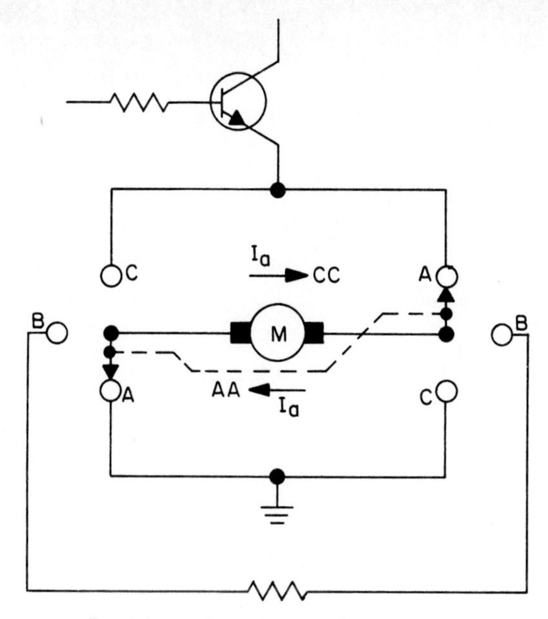

Fig. 3.3.10. Reversing switch arrangement.

shown is used on all Electro-Craft Motomatic standard and master controls.

The reversing circuit can be operated by relays instead of by a manually operated switch, giving the speed control system a bidirectional characteristic for remote control manual operations. It is not practical, in most cases, to attempt to make a closed-loop position or velocity servo by this method, however. A servo system is usually subject to endless reversals around the null point, and relay contacts would rapidly wear. Furthermore, such a system would be subject to time delays and nonlinearities due to relay switching action.

CONTROLLED ACCELERATION-DECELERATION SYSTEM

While torque limit circuits to some extent can control the acceleration and deceleration characteristics of a speed control sys-

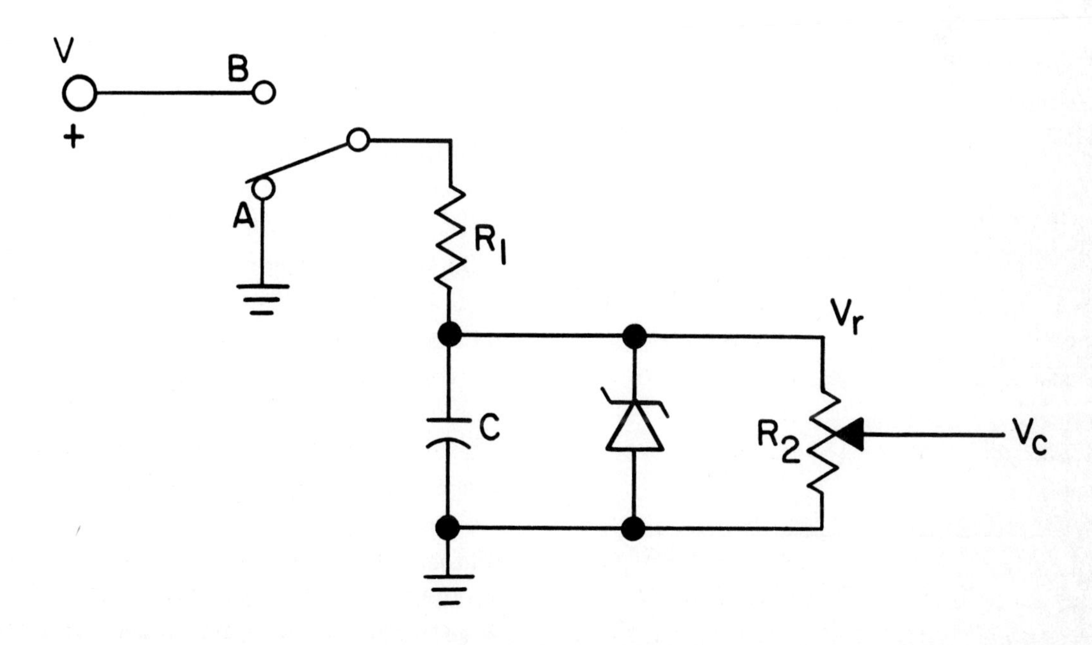

Fig. 3.3.11. Simple ramp command circuit.

tem, it is clear that they are sensitive to variations in both load and friction torque.

In cases where it is desirable to have the acceleration and/or deceleration under velocity control, several methods can achieve the desired results.

The simple RC circuit of Fig. 3.3.11 can be employed to control acceleration. First, assume that the switch is initially in the **A** position so that the command voltage $\mathbf{V_c}$ is zero. The speed control is, therefore, holding the motor at rest. With the switch set in the **B** position, capacitor **C** is charged through voltage divider $\mathbf{R_1}$, $\mathbf{R_2}$. The voltage $\mathbf{V_r}$ will follow an exponential rise (Fig. 3.3.12) until the Zener voltage $\mathbf{V_Z}$ is achieved, at which point the reference voltage $\mathbf{V_r}$ (as also the command voltage $\mathbf{V_c}$) is stabilized.

As the switch is returned to **A**, the voltage will decline in an inverse exponential fashion.

It should be noted that while the speed control system will follow the acceleration

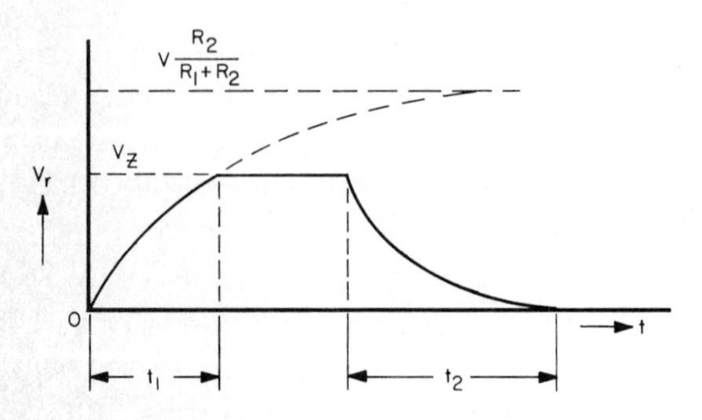

Fig. 3.3.12. Turn-on and turn-off characteristics of the RC circuit in Fig. 3.3.11.

curve (if acceleration rate is within the torque capacity of the system), it may not follow the deceleration curve unless the load conditions and the dynamic braking capacity allow. If this is not the case, then a bidirectional system (servo system) must be used where continuous velocity tracking is necessary.

Another circuit with linear features is shown in Fig. 3.3.13. This is a ramp generator which produces linear, independently adjustable slopes.

The circuits discussed above have been shown as manually operated devices. In many applications, they can be operated by various logic circuits, thereby taking their place in modern process controls.

SWITCHING AMPLIFIERS

While the linear amplifier discussed above performs excellently in high-performance speed controls, it has the problem of heat generation in the output stage, requiring forced-air cooling in amplifiers over 100 to 200 W (depending on ambient temperature and heat sink design). The switching amplifiers overcome this problem by letting its output stage rapidly switch from a nonconductive state to a fully conductive state, thereby minimizing operation of the output stage in the high dissipation region.

Three basic methods are used to control power in switching amplifiers: *pulse-width modulation* (PWM), *pulse-frequency modulation* (PFM), and *silicon controlled rectifier*

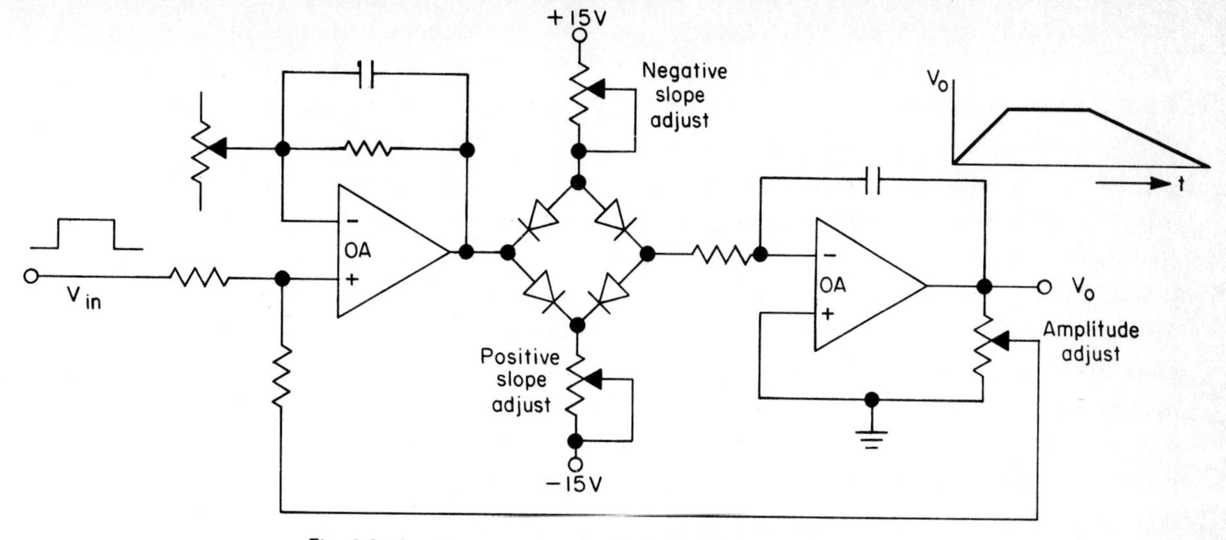

Fig. 3.3.13. Ramp generator with independently adjustable slopes.

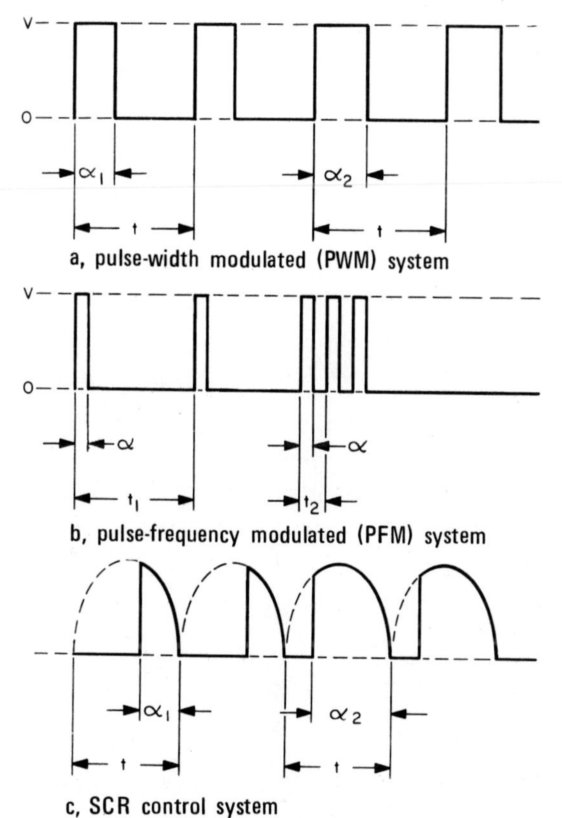

Fig. 3.3.14. Voltage waveforms in switching amplifiers.

(SCR) *controls.* Their principal differences are shown in Fig. 3.3.14.

The PWM system usually utilizes a DC supply, and the amplifier switches the supply voltage on and off at a fixed frequency and at a variable "firing angle" a (see Fig. 3.3.14a) so that an adjustable average voltage across the load is established. The amount of power transferred to the load (motor) will depend on switching rate and the load inductance. In many of the PWM circuits, the pulse frequency is allowed to shift over a given range – in some cases for a good purpose.

The PFM system has a fixed firing angle and a variable repetition rate (Fig. 3.3.14b), achieving essentially the same results as the PWM, but when used in motor control circuits, the widely variable pulse frequency required causes dissipation problems which makes the PWM more attractive. Incidentally, a PWM circuit with a variable pulse rate is really a hybrid between the two.

The SCR circuit for DC control is usually used with a rectified AC supply voltage, although a rectification circuit can be placed before or after the control section of the amplifier. Fig. 3.3.14c shows a full-wave rectified supply voltage of a fixed frequency. The firing angle can be varied to cover a portion from 0 to 180 ^{o}el of a half-wave. The average output voltage is not proportional to the firing angle, and this part requires special attention in the control design.

In the following discussion only PWM and SCR circuits will be covered, since the pure PFM is not particularly suited to motor control applications.

PULSE-WIDTH MODULATED AMPLIFIERS

As already mentioned, the PWM amplifier generally is powered by a DC supply. It can be designed using either transistors or SCR's as switches. Actually, the SCR amplifier is a PWM in spirit; but the functional differences, the output voltage frequency and wave shapes are different enough to warrant a separate discussion.

Then what about SCR's powering a PWM? Since PWM's are based on a continuously available DC power source, and since SCR's cannot "turn off" current conduction by themselves as transistors can, they must be provided with a separate "turn-off" circuit. The problem associated with such circuitry has limited the use of SCR's in PWM systems to high current, low switching rate applications, such as lift truck and vehicle traction controls. Due to the unique nature

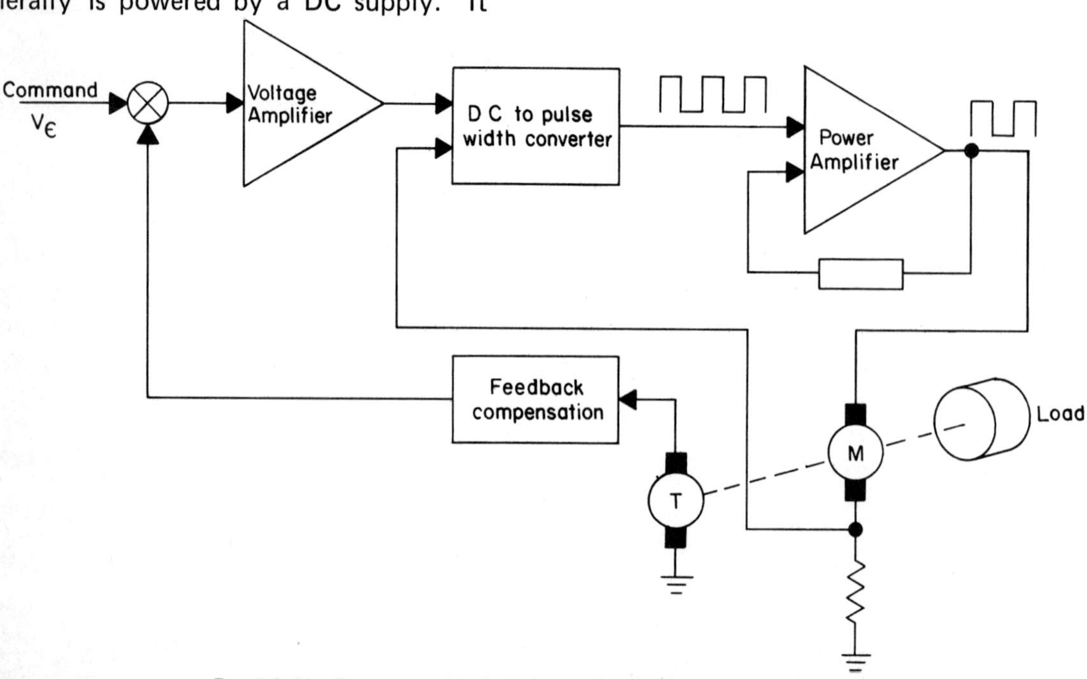

Fig. 3.3.15. Elementary block diagram of a PWM speed control system.

of these controls, they are not discussed here.

Today, then, transistor-operated PWM switching amplifiers are used in most high-performance, high power speed control systems and servo sytems.

A typical block diagram of a PWM speed control system is shown in Fig. 3.3.15.

The pulse repetition rate is usually above 1 kHz (often about 10 kHz), and this frequency is mainly dictated by the required system response bandwidth, motor inductance and motor high frequency loss characteristics. At times audio noise emissions through wiring, heat sinks and motor frame components can be loud enough to be disturbing, and in such applications one can raise the pulse frequency to a point where the noise is not audible.

In examining voltage and current characteristics of a PWM system, we can first look at the ideal motor and its behavior in a PWM system. This motor equivalent circuit is shown in Fig. 3.3.16.

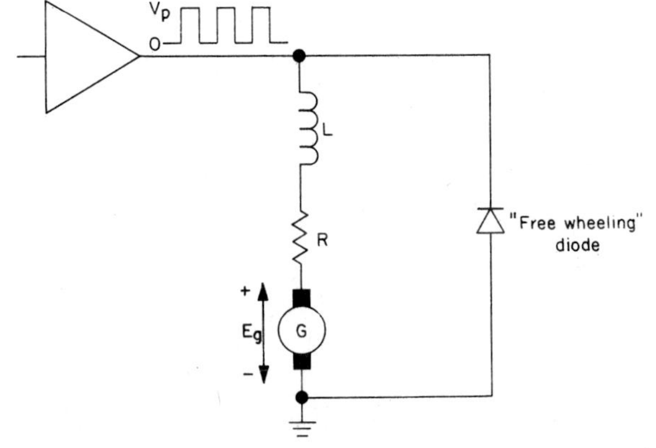

Fig. 3.3.16. Equivalent circuit of a DC motor in a PWM control system.

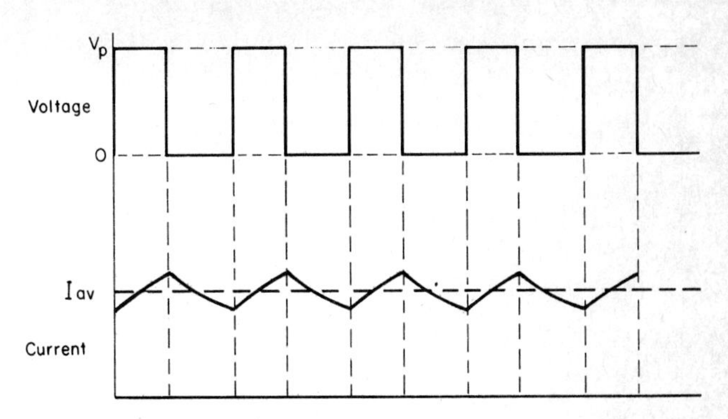

Fig. 3.3.17. Current and voltage relationship in a PWM control system.

The waveform of the motor current during the switching mode is dependent not only on the switching rate, but on the motor speed **n**, the total inductance **L**, motor resistance **R** and the current level in the last cycle. Fig. 3.3.17 shows a steady-state relationship of current and voltage.

Since the supply voltage is being switched on and off at a high frequency, we should evaluate the power losses in the motor. The power losses in such a system may be due to a host of factors, depending on the design of the motor: eddy current losses, hysteresis losses, armature commutation losses, viscous friction losses and armature resistance losses. But considering modern permanent magnet servomotors with small electrical time constants and minimal high frequency losses, a first approximation to evaluate motor power losses due to the armature resistance can be expressed by:

$$P_L = RI^2_{RMS} \tag{3.3.9}$$

In order to relate RMS current to average current in such cases, an expression relating the two has been established, called the *form factor* (k):

$$k = \frac{I_{RMS}}{I_{av}} \qquad (3.3.10)$$

The average motor current will determine the motor torque produced:

$$T_g = K_T\ I_{av} \qquad (3.3.11)$$

Substituting (3.3.10) into (3.3.9), the motor losses under PWM conditions are:

$$P_L = R\ k^2 I_{av}^2 \qquad (3.3.12)$$

and we see that the armature losses depend on average current, form factor, and armature resistance. Another way of looking at the sources of the losses is to substitute (3.3.11) into (3.3.12). Then

$$P_L = \frac{R}{K_T^2}\ k^2 T_g^2 \qquad (3.3.13)$$

Here it is clear that losses are due to several factors; the motor constants **R** and $\mathbf{K_T}$ inherent in a given design, the form factor, and the output torque of the motor. Thus, the form factor has a strong influence on motor heating. In the case of a **k = 1**, the heating effect of a linear amplifier from the previous section is:

$$P_L = RI_{av}^2 \qquad (3.3.14)$$

But for a form factor **k = 2**, we have an armature power loss of *four times the amount* shown above. Thus, we can construct a graphical solution of the increase in armature loss due to form factor. This is shown in Fig. 3.3.18.

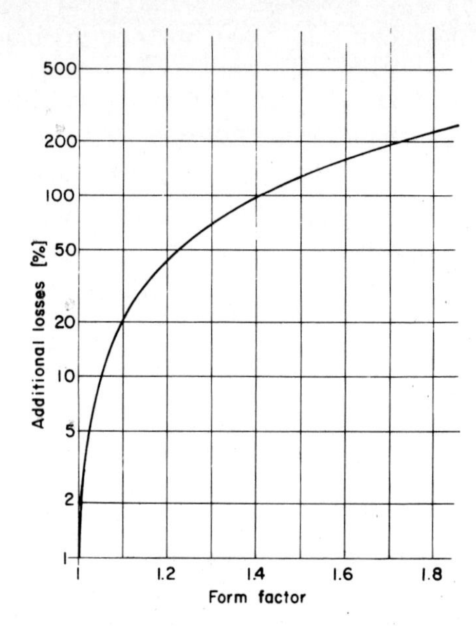

Fig. 3.3.18. Relationship of additional armature losses in a servomotor with form factor.

The form factor greatly influences armature losses, as for example a **k = 1.2** will increase losses by 44%.

Since servomotor performance in speed control systems is often limited mainly by power dissipation, it is important to investigate the form factor of a given amplifier source. In the case of a PWM amplifier the form factor will depend on the pulse frequency, on the electrical time constant of the motor and on any associated series inductances.

The net result of using a PWM circuit is usually that the power dissipation in the amplifier is vastly decreased, and the total system dissipation is improved; but in some cases the motor power dissipation may be higher than if a linear (class A) amplifier is used. One should also check the application to see if the audio noise (if any) will be of consequence. This may be particularly irritating if the pulse frequency is shifting greatly with load variations. Another factor to consider in using a PWM system is the electrical noise generation, which can be transmitted into low level circuitry if care is not taken to properly shield and ground the high current portion of the system.

SCR CONTROLS

SCR speed control systems are nearly always based on power line frequency (50-60 Hz). This limits the control bandwidth of the system from 2-3 Hz for a half-wave, single-phase system to about 25-30 Hz for a three-phase, full-wave system. This translates into a speed control range of 5:1 for a half-wave control to 20:1 for a full-wave control with tachometer feedback.

While the SCR control system is not capable of controlling as wide a speed range as is the linear transistor amplifier, it has a high power conversion efficiency due to its switching mode and the low forward conduction voltage drop. The relative simplicity of its circuitry further adds to the positive features of the SCR control. The popularity of the SCR control for limited speed range DC motor control in the fractional and integral horsepower sizes has been largely due to these features, and to the fact that SCR's can operate directly from the power line without AC-to-DC power conversion.

One serious problem of operating motors from SCR controls is the high form factor, which may cause serious derating of the motor. The form factor depends on the type of control used (half-wave, full-wave, three-phase), on the conduction angle and on motor inductance; therefore, it is difficult to obtain form factor values without a detailed knowledge of the application.

In the case of high-performance servomotors with low inductance, an appropriate determination of the form factor can be made on the assumption that the load is resistive. The graph in Fig. 3.3.19 shows the form factor versus conduction angle for resistive loads. The power dissipation in a motor is

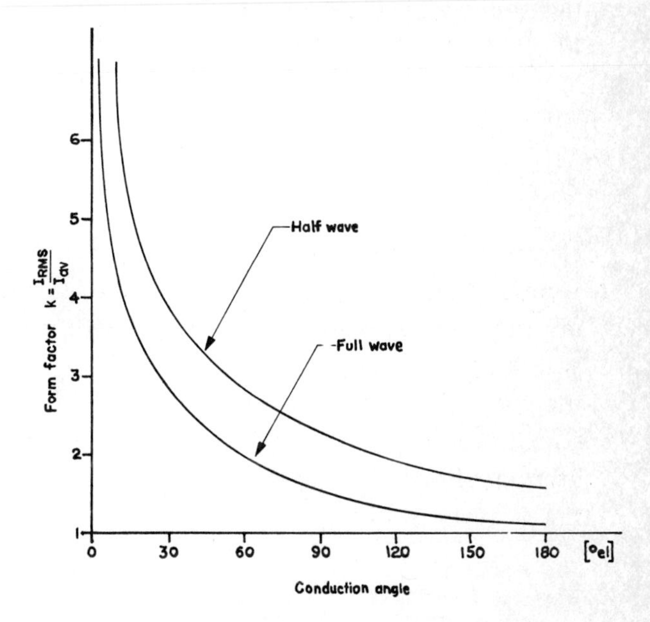

Fig. 3.3.19. Form factor vs. conduction angle for resistive loads in SCR control circuits.

the highest at very low speeds and high torque conditions.

The addition of an inductance in the motor circuit will minimize the RMS current, and therefore make the form factor smaller. However, this may have the effect of decreasing system bandwidth which could affect the speed range or system stability.

Another important problem in SCR speed control systems is the so-called transportation lag in the system, caused by the discrete firing angle conduction. In attempting to control fractional horsepower motors over a wide speed range with SCR controls, the result is often momentary instability, especially at low speeds. In such situations, the system is operating at very small "firing angles" (Fig. 3.3.20) and if a sudden torque disturbance occurs, the resulting change in speed causes the error signal to switch on the output stage, sometimes at an "early" firing angle, as shown in the diagram.

The result will then be a heavy surge of power for the remaining part of the half cycle (since the SCR cannot turn off until forward current goes to zero), even if the error signal in the meantime has returned to normal. The motor now has a large overspeed error, and will coast down to the predetermined controlled speed again, in which case the former firing pattern will resume. This behavior will be observed as a jerky, cogging action.

The most obvious remedy for this type of system behavior is to make the system transient response slower, using *lag networks* to overcome the effects of instability. This tends to make the system more sluggish in its response to sudden changes in load and speed command changes. This is not usually important in some applications with steady loads or loads with large moments of inertia, but for wide range, high-performance controls it is not desirable.

A properly compensated SCR control in a suitable application can be a fine speed con-

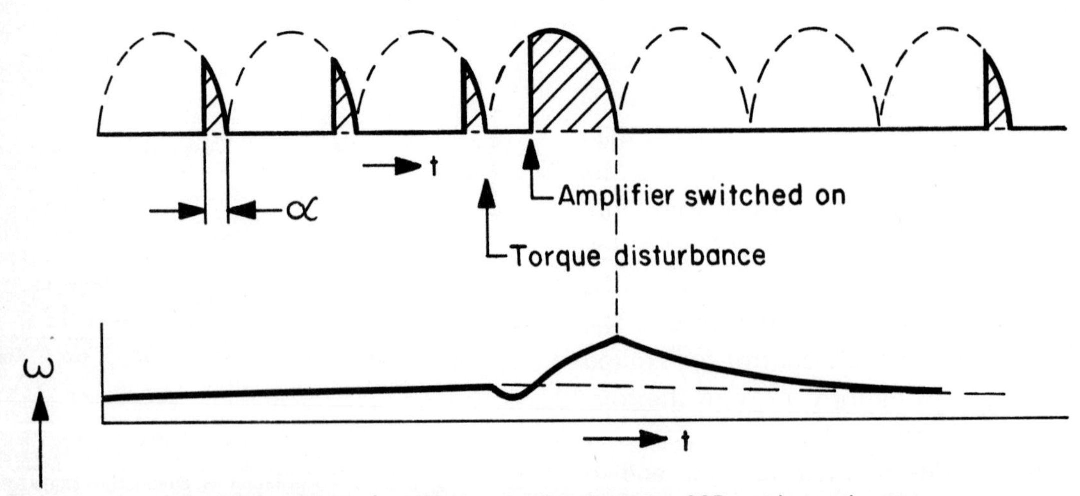

Fig. 3.3.20. Result of a sudden torque disturbance in a SCR speed control system.

trol system. Each unique application has an equally unique solution where a host of qualities have to be considered, such as performance, price, versatility and utility. The selection of the "best" control is, indeed, a complex responsibility. □

Chapter 4
Speed Controls and Servo Systems

A servo system in its most elementary form consists of an amplifier, actuator and feedback element. Paramount to the system function is stability, and a fair amount of a system designer's time is spent analyzing system parameters and arranging for conditions of servo stability and proper response time.

We will in this chapter analyze system fundamental equations, establish "block diagram" concepts and formulate transfer functions. We will cover Laplace transformation and use it to develop the input-output relationships (transfer functions) for a motor, amplifier and feedback elements. Also, in our discussion, we will cover system stability and describe the root locus method for determining stability.

The closed-loop servo system mentioned above is an example of just one of the many various types of control systems. We shall discuss the most common types, namely: velocity control, position control, torque control, and hybrid control systems. We will then describe the feedback components, and present their transfer functions. Various types of servo amplifiers will also be covered.

Besides describing the most common types of controls, we will also cover system characteristics, such as system step response and system bandwidth.

We will also discuss phase-locked servo systems, "how to make systems work", and how to optimize your system (velocity profile, coupling, and capstan size).

4.1. SERVO THEORY

LAPLACE TRANSFORMATION

In engineering analysis and design, a linear system is, first of all, described mathematically, then the system is studied to determine response to a variety of input excitation signals. The differential or integro-differential equation or equations which mathematically describe the system must be manipulated to obtain a desired equation characterizing the input-output relationships of the system.

Manipulation of these equations which contain exponential, transcendental or non-sinusoidal functions can be simplified when they are "transformed" from differential form to an easier-to-handle algebraic form by applying the *Laplace transformation.* The algebraic mathematics are equated into input-output relationships, then the result is transformed back into the original differential or integro-differential form by applying an *inverse Laplace transformation.*

In the first two sections, the Laplace transformation will be explained.

The defining equations and symbols of Laplace transformation are:

If

f(t) is any function of time such that f(t) = 0 for t < 0 and the integral (4.1.2) has a finite value, then

f(s) is the Laplace transform of f(t).

The Laplace operator **s** is defined as the complex variable:

$$s = \sigma + j\omega \tag{4.1.1}$$

and the Laplace transformation is defined as:

$$f(s) = \int_0^\infty f(t)\, e^{-st}\, dt \tag{4.1.2}$$

Thus, to find the Laplace transform of a function **f(t)**, the function **f(t)** is multiplied by e^{-st}, then integrated from **t=0** to **t=∞**.

Examples:

1. Let **f(t)** = A for t > **0** (a constant); then the Laplace transform will be:

$$f(s) = \int_0^\infty Ae^{-st}\, dt = A\frac{-1}{s}\, e^{-st}\Big|_0^\infty$$

$$= -\frac{A}{s}\left(e^{-\infty} - e^{0}\right) \tag{4.1.3}$$

$$f(s) = \frac{A}{s} \tag{4.1.4}$$

2. Let $f(t) = Ae^{-at}$. Then:

$$f(s) = \int_0^\infty Ae^{-at}\, e^{-st}\, dt$$

$$= A \int_0^\infty e^{-at}\, e^{-st}\, dt \tag{4.1.5}$$

$$f(s) = \frac{A}{a + s} \tag{4.1.6}$$

INVERSE LAPLACE TRANSFORMATION

Once an equation has been transformed, the solution for the unknown variable may be determined through algebraic manipulations. The solution expressed in terms of the complex variable, **s**, may be sufficient in some cases; if not, the solution as a function of time can be obtained by either taking the inverse Laplace transformation

$$f(t) = \frac{1}{2\pi j} \int_{\sigma - j\infty}^{\sigma + j\infty} f(s)e^{st}\, ds \tag{4.1.7}$$

or by using standard tables such as Tab. 4.1.1.

As an example, assume that through algebraic manipulation, the Laplace transform is:

$$f(s) = \frac{2}{s\,(s + 1)} \tag{4.1.8}$$

f(s)	f(t)
$\frac{1}{s}$	1
$\frac{1}{s^n}$	$\frac{1}{(n-1)!}t^{n-1}$ (n=1, 2, 3. . .)
$\frac{1}{s-a}$	e^{at}
$\frac{1}{(s-a)^n}$	$\frac{1}{(n-1)!}t^{n-1}e^{at}$ (n=1,2,3,. . .)
$\frac{1}{s^2-a^2}$	$\frac{1}{a}\sinh at$
$\frac{s}{s^2-a^2}$	$\cosh at$
$\frac{1}{(s-a)(s-b)}$ $(a \neq b)$	$\frac{1}{a-b}[e^{at}-e^{bt}]$
$\frac{s}{(s-a)(s-b)}$ $(a \neq b)$	$\frac{1}{a-b}[ae^{at}-be^{bt}]$
$\frac{1}{s^2+a^2}$	$\frac{1}{a}\sin at$
$\frac{s}{s^2+a^2}$	$\cos at$
$\frac{1}{(s-a)^2+b^2}$	$\frac{1}{b}e^{at}\sin bt$
$\frac{s-a}{(s-a)^2+b^2}$	$e^{at}\cos bt$
$\frac{s}{(s^2+a^2)^2}$	$\frac{1}{2a}t\sin at$
$\frac{s^2-a^2}{(s^2+a^2)^2}$	$t\cos at$
$\frac{1}{(s^2+a^2)^2}$	$\frac{1}{2a^3}[\sin at - at\cos at]$

Tab. 4.1.1. Laplace transforms

To determine the inverse transform, **f(t)**, the function **f(s)** will be factored into:

$$f(s) = \frac{2}{s} - \frac{2}{s+1} \tag{4.1.9}$$

From this section, or the table, we know that the inverse transform of $\frac{2}{s}$ is a constant, 2, and the inverse transform of **2/(s+1)** is $\mathbf{2e^{-t}}$. Therefore the inverse transformation of the above **f(s)** gives

$$f(t) = 2 - 2e^{-t} \tag{4.1.10}$$

TRANSFER FUNCTIONS

In control theory, the relationship between input driving signal and output response is expressed by means of a *transfer function.* The transfer function is defined as a *ratio of the Laplace transforms of the output and input signals, with the assumption that all initial conditions are zero.*

As an example, we will examine a circuit in Fig. 4.1.1., write the network equations and, utilizing the Laplace transformation, manipulate the equations so as to obtain a ratio of the output and input transforms.

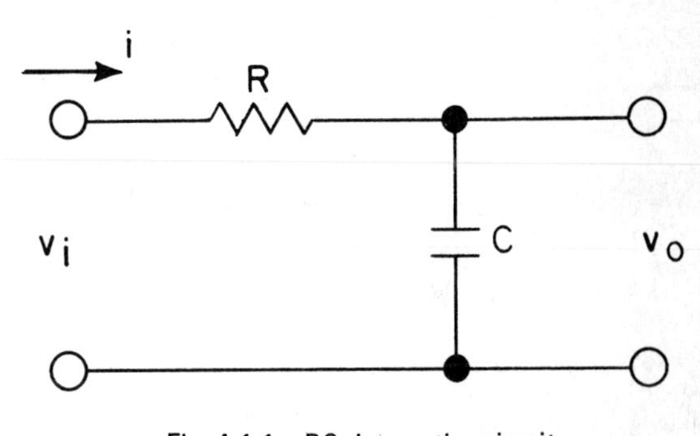

Fig. 4.1.1. RC integrating circuit.

In writing the network equations we obtain:

$$v_i = iR + \frac{1}{C}\int i\,dt$$

$$= R\frac{dq}{dt} + \frac{q}{C} \tag{4.1.11}$$

$$v_o = \frac{q}{C} \tag{4.1.12}$$

where $q = \int i\,dt$. These transform to:

$$v_i(s) = \left(sR + \frac{1}{C}\right)q(s) \qquad (4.1.13)$$

$$v_o(s) = \frac{1}{C}q(s) \qquad (4.1.14)$$

Dividing, we arrive at a ratio of input to output transforms:

$$\frac{v_o(s)}{v_i(s)} = \frac{\frac{1}{C}q(s)}{\left(sR + \frac{1}{C}\right)q(s)} = \frac{1}{sRC + 1} \qquad (4.1.15)$$

Since this expression is the ratio of the Laplace transforms of output and input voltages, it represents the transfer function of the RC network of Fig. 4.1.1.

BLOCK DIAGRAMS

To show the relationship between the various components and the flow of signals between them, a diagram is drawn in block form. Each block describes a component by means of its transfer function and is associated with arrows to illustrate direction of signal flow. Such a block is shown in Fig. 4.1.2.

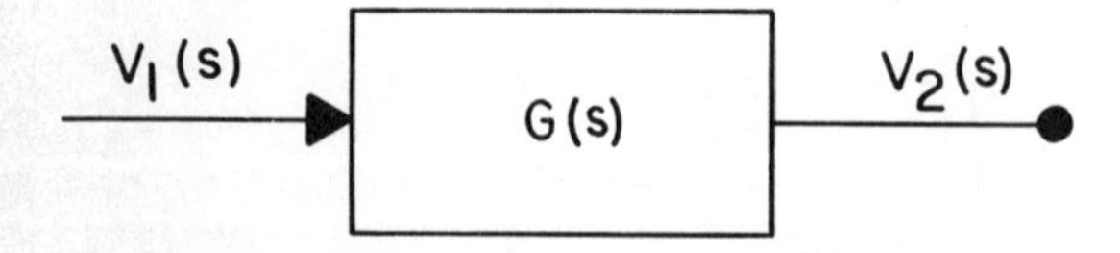

Fig. 4.1.2. A block diagram showing input signal, block transfer function, and output signal.

The arrow indicates that $V_1(s)$ is the input, $V_2(s)$ is the output, and $G(s)$ is the transfer function [the ratio of $V_2(s)$ to $V_1(s)$].

Several signals are joined by summing points such as shown in Fig. 4.1.3. A summing point has numerous inputs, but only one output which equals the sum of the inputs.

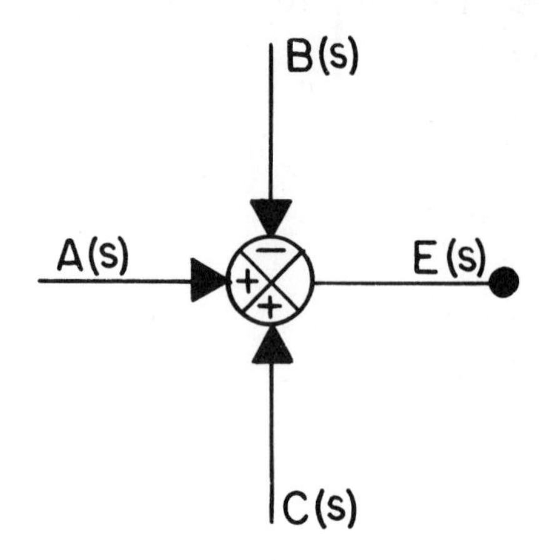

Fig. 4.1.3. A summing point may have several inputs. The output is the sum of the input signals.

The output of the summing point in Fig. 4.1.3 is:

$$E(s) = A(s) - B(s) + C(s) \qquad (4.1.16)$$

A block diagram representation of a closed-loop system, where the output signal is fed back and compared to the input, is illustrated in Fig. 4.1.4.

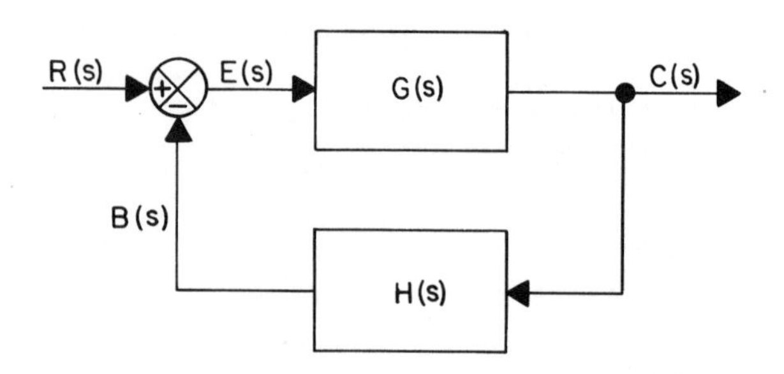

Fig. 4.1.4. Block diagram of a closed-loop system.

An input signal **R(s)** is compared to feedback signal **B(s)** at the summing point, the difference **E(s)** is input to the block **G(s)**, and an output **C(s)** is obtained. The relation between the variables are given by the equations:

$$\left.\begin{aligned} C(s) &= G(s)\,E(s) \\ B(S) &= H(s)\,C(s) \\ E(s) &= R(s) - B(s) \\ &= R(s) - H(s)\,C(s) \end{aligned}\right\} \quad (4.1.17)$$

By eliminating **E(s)** from the equations, we obtain:

$$C(s) = G(s)\,R(s) - H(s)\,G(s)\,C(s) \quad (4.1.18)$$

The closed-loop transfer function, or the relationship between output and input becomes:

$$f(s) = \frac{C(s)}{R(s)} = \frac{G(s)}{1 + G(s)\,H(s)} \quad (4.1.19)$$

TRANSFER FUNCTION OF A DC MOTOR

Now that we have reviewed the basic block diagram, what a transfer function is, and how to derive it, we will derive the transfer function of one specific block used in the closed-loop system - the block representing the motor.

A simplified model of a DC motor with permanent magnet field can be derived by assuming armature inductance to be zero and ignoring the resonance effect. A complex model of a DC motor was presented in section 2.3. and will be further discussed in section 4.2.1. With these stipulations, the DC motor equations are:

$$V = I_a R + K_E \omega \quad (4.1.20)$$

[V; A, Ω, V/rad s^{-1}, rad/s]

$$T_g = K_T I_a \quad (4.1.21)$$

[Nm; Nm/A, A]

$$T_g = J\alpha \quad (4.1.22)$$

[Nm; kg m^2, rad/s^2]

If we make a small closed-loop system, and assign each block a specific symbol, the block diagram would appear as shown in Fig. 4.1.5.

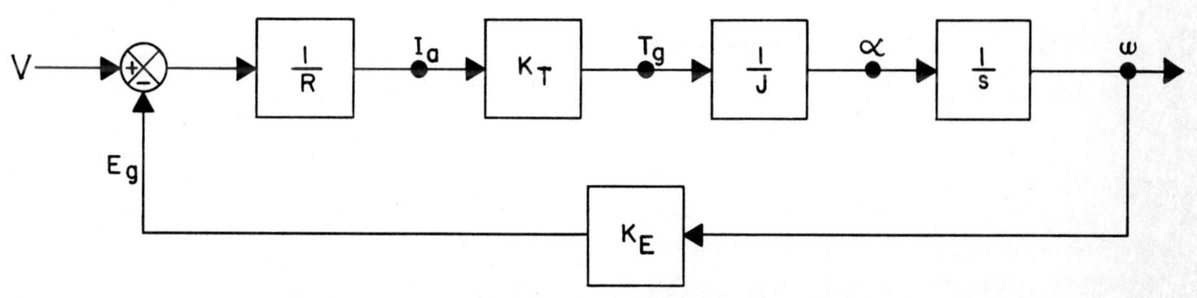

Fig. 4.1.5. Block diagram of a DC motor as the closed-loop system with each block representing a specific motor parameter.

Combining blocks to simplify the diagram, we arrive at the form shown in Fig. 4.1.6, from which the motor transfer function relating velocity ω to the input voltage **V** can be written:

$$G_m(s) = \frac{\omega(s)}{V(s)} = \frac{\dfrac{K_T}{R\,Js}}{1 + \dfrac{K_T K_E}{RJs}}$$

$$= \frac{1/K_E}{1 + s\dfrac{RJ}{K_T K_E}} \qquad (4.1.23)$$

or:

$$G_m(s) = \frac{1/K_E}{1 + s\tau_m} \qquad (4.1.24)$$

where τ_m is the mechanical time constant.

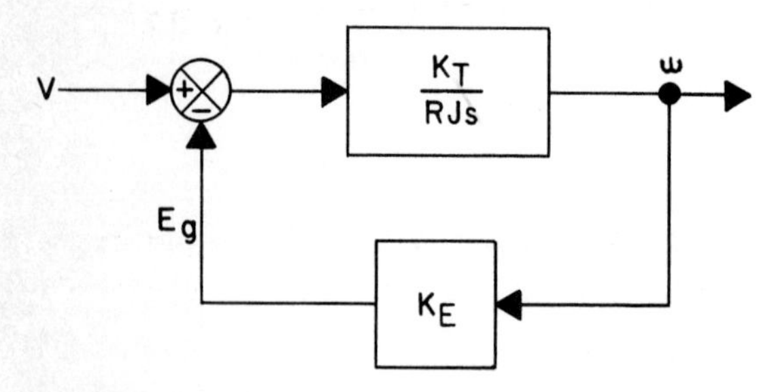

Fig. 4.1.6. By combining the blocks of Fig. 4.1.5, we will arrive at this simplified block diagram for a DC motor transfer function.

This transfer function contains terms that describe motor response to cyclic or variable frequency signals. We can plot the frequency response characteristic of the motor by setting **s** equal to **j**ω. Then:

$$s = j\omega = j2\pi f \qquad (4.1.25)$$

The transfer function is then evaluated for an entire range of frequencies and the results can be plotted. It is customary to plot the transfer function magnitude in decibels vs. frequency plotted in a logarithmic scale, and is termed the gain-frequency characteristic of the motor. Such a plot is called the *Bode plot.*

At a frequency where $2\pi f$ is equal to $1/\tau_m$ the slope of the curve shows a sharp change as can be seen in Fig. 4.1.7.

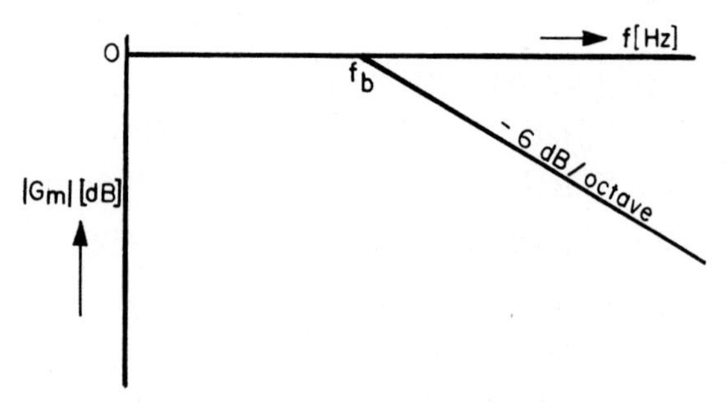

Fig. 4.1.7. Bode plot of a simplified motor model which ignores motor inductance.

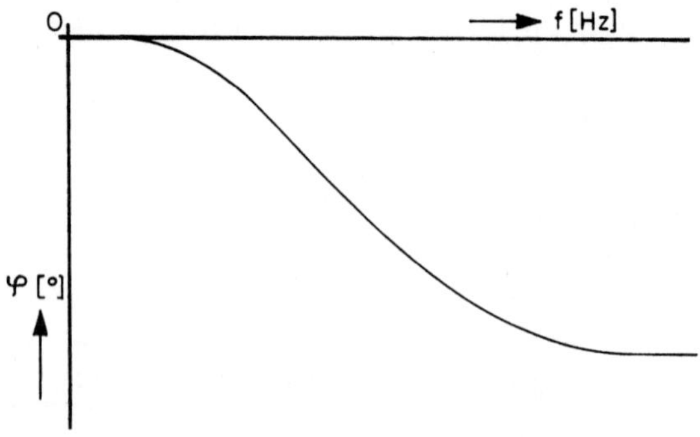

Fig. 4.1.8. Phase angle-frequency characteristic of a DC motor.

This frequency, which can be evaluated by the ratio $K_E K_T/R\,J$, is called the *break frequency* of the motor:

$$f_b = \frac{1}{2\pi \tau_m} = \frac{K_E K_T}{2\pi JR} \quad (4.1.26)$$

$[\text{Hz}; \text{V/rad s}^{-1}, \text{Nm/A}, \text{kg m}^2, \Omega]$

The phase angle-frequency characteristic of the motor can also be plotted as illustrated in Fig. 4.1.8.

The phase angle is determined by the denominator of the transfer function, so that:

$$\varphi = -\tan^{-1}\left(\frac{RJ}{K_E K_T}\right) \quad (4.1.27)$$

$[\text{rad}; \Omega, \text{kg m}^2, \text{V/rad s}^{-1}, \text{Nm/A}]$

EXAMPLE: Two motors are under consideration for an application where they are required to respond to control signals which recur at rates as high as 125 times per second. Without regard to the torque and heat dissipation requirements, which motor would be the first choice if the following data is available:

Motor "A"

$K_T = 9.5 \text{ oz-in/A} = 6.7 \times 10^{-2} \text{ Nm/A}$

$K_E = 7 \text{ V/krpm} = 6.7 \times 10^{-2} \text{ V/rad s}^{-1}$

$R = 0.5\ \Omega$

$J = 0.68 \times 10^{-3} \text{oz-in-s}^2$

$= 4.8 \times 10^{-6} \text{ kg m}^2$

Motor "B"

$K_T = 12 \text{ oz-in/A} = 8.48 \times 10^{-2} \text{Nm/A}$

$K_E = 8.9 \text{ V/krpm} = 8.48 \times 10^{-2} \text{V/rad s}^{-1}$

$R = 0.75\ \Omega$

$J = 2.4 \times 10^{-3} \text{ oz-in-s}^2$

$= 1.695 \times 10^{-5} \text{ kg m}^2$

The application requires the motor to accelerate the load with moment of inertia of 0.3×10^{-3} oz-in-s^2 = 2.12×10^{-6} kg m^2. Calculating the break frequency for motor "A", we obtain:

$$f_b = \frac{K_E K_T}{2\pi JR}$$

$$= \frac{6.7^2 \times 10^{-4}}{2\pi\ (4.8 + 2.12) \times 10^{-6} \times 0.5}$$

$$f_b = 206.5 \text{ Hz}$$

Then calculate break frequency for motor "B":

$$f_b = \frac{8.48^2 \times 10^{-4}}{2\pi\ (16.95+2.12) \times 10^{-6} \times 0.75}$$

$$f_b = 80 \text{ Hz}$$

It can be seen that insofar as dynamic characteristics are concerned, motor "A" provides significant advantage in its higher break frequency which would permit easier compensation of the servo system to achieve the desired bandwidth.

TRANSFER FUNCTION OF AN AMPLIFIER

An ideal DC amplifier would have a constant gain for all input frequencies. However, practical amplifiers are limited to a certain frequency range, and the corresponding transfer function in most cases is:

$$G_a(s) = \frac{V_o(s)}{V_i(s)} = \frac{A}{1 + s\tau_a} \tag{4.1.28}$$

where τ_a is the time constant of the amplifier and A is the DC gain.

STABILITY

We have just presented the transfer function for two blocks in the servo system, the motor and the amplifier. We shall now discuss the three possible types of system response to a step input.

Let a system be described by the transfer function:

$$G(s) = \frac{N(s)}{D(s)} \tag{4.1.29}$$

Roots of N(s) = 0 are zeros of the system, and roots of D(s) = 0 are poles of the system. Poles play a major role in the system's stability, and in order to analyze stability, these poles must be determined.

If a pole is given by:

$$p = \sigma + j\omega \tag{4.1.30}$$

then three different types of responses may be possible. These are (if all poles meet the following equations):

Case 1: $\sigma < 0$ and $|\omega| < |\sigma|$ (4.1.31)

Case 2: $\sigma \leqslant 0$ and $|\omega| > |\sigma|$ (4.1.32)

Case 3: $\sigma > 0$ (4.1.33)

Case 1: the system is overdamped and response to a step input will be as illustrated in Fig. 4.1.9.

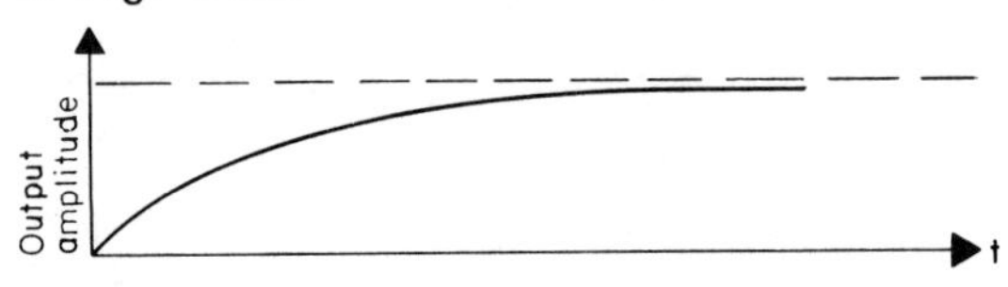

Fig. 4.1.9. Overdamped system response to step input.

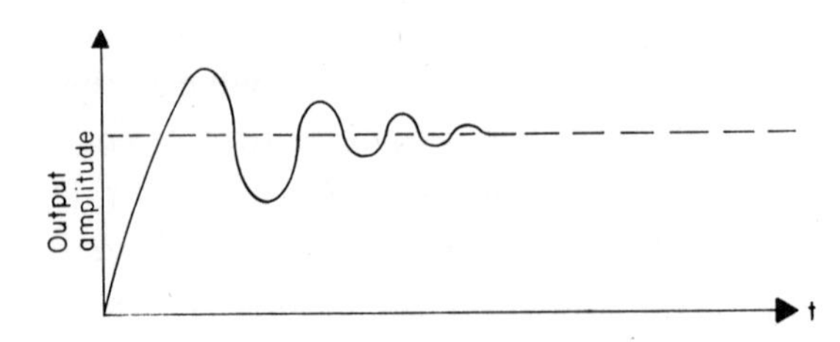

Fig. 4.1.10. Underdamped system response.

Case 2: the system is underdamped and a step input will cause overshoot in the response, as can be seen in Fig. 4.1.10.

Case 3: when one or more poles exhibit the property of $\sigma > 0$, the system will be unstable and response will increase with time.

ROOT LOCUS METHOD

There are several analytical and graphical methods to determine system stability, all of which depend on the fact that if $\sigma > 0$ the system will be unstable. We will describe the *root locus method* for determining servo system stability.

Consider the closed–loop system of Fig. 4.1.11, whose transfer function is:

$$f(s) = \frac{C(s)}{R(s)} = \frac{G(s)}{1 + G(s)\,H(s)} \qquad (4.1.34)$$

The entire expression is termed the *closed-loop transfer function;* **G(s) H(s)** is called the *open-loop transfer function.*

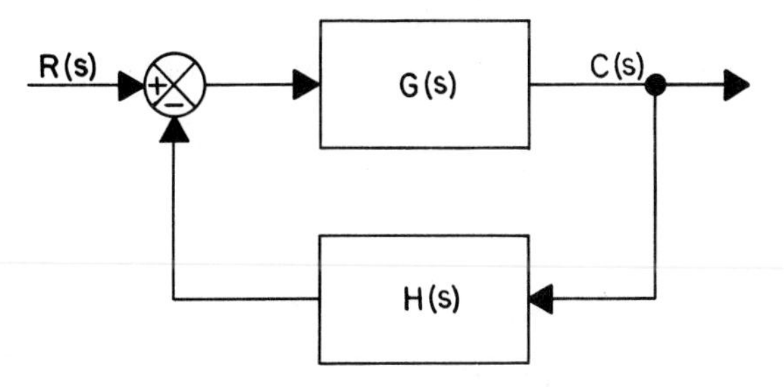

Fig. 4.1.11. Block diagram used for explanation of root locus method.

The objective is to determine poles of the closed–loop transfer function since they characterize the response of the system. Therefore, the equation to be solved is:

$$1 + G(s)\,H(s) = 0 \qquad (4.1.35)$$

which is called the *characteristic equation.*

The equation can be solved numerically to determine the poles. However, we want to find the dependence of the poles on some parameter such as the system gain. This is necessary since the parameter may not be known exactly, or because we may wish to change the parameter and will want to observe how this will affect location of poles.

A way to accomplish this is by utilizing the root locus method, which basically gives the closed-loop poles in relation to the open-loop poles and zeros, and a parameter **K**, which is the parameter of interest (usually gain). The method is described by the following set of rules, each of which is illustrated by an example:

RULE # 1:

In order to construct the root loci, obtain the characteristic equation and rearrange it to the format:

$$1 + K\frac{(s - z_1)(s - z_2)\ldots(s - z_m)}{(s - p_1)(s - p_2)\ldots(s - p_n)} = 0 \qquad (4.1.36)$$

where **K** is the parameter of interest (usually gain) which is assumed to be positive. The second term in the left side of (4.1.36) is the open-loop transfer function. Then locate the open-loop poles and zeros in the s-plane.

Example: consider a system with:

$$G(s) = \frac{K}{s\,(s + 1)} \qquad (4.1.37)$$

$$H(s) = \frac{(s + 2)}{(s + 3)\,(s + 4)} \qquad (4.1.38)$$

rearranging to the desired format:

$$1 + K \frac{(s+2)}{s(s+1)(s+3)(s+4)} = 0 \quad (4.1.39)$$

The open loop poles and zero are located in the s-plane as shown in Fig. 4.1.12.

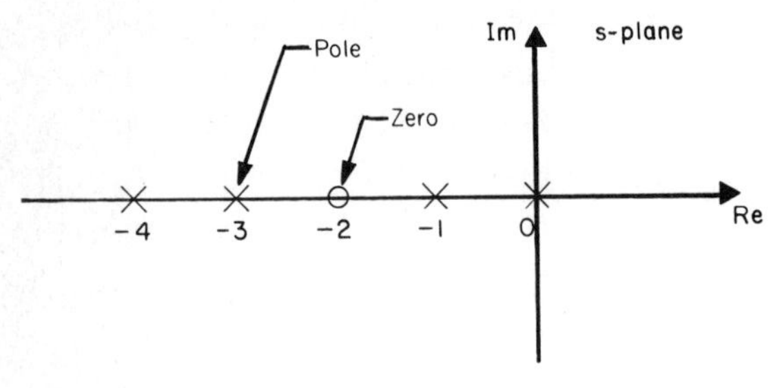

Fig. 4.1.12. Open-loop poles and zero plot of (4.1.39).

RULE # 2:

Find the starting and termination points of the root loci. Since in all real systems the number of open–loop poles is greater or equal to the number of zeros (**n ⩾ m**), the root loci start for **K** = **0** at the open-loop poles and terminate at either an open-loop zero or at infinity. There are **n** branches, **m** of which will terminate at a zero, and **(n – m)** branches will terminate at infinity along asymptotes.

Example: for the system of the previous example, the number of open-loop poles is **n** = **4,** and the number of open-loop zeros is **m** = **1**. So that one branch will terminate at the zero, and three will terminate at infinity.

RULE #3:

Determine the root loci on the real axis. A point on the real axis lies on a root locus *if* the total number of open-loop poles and zeros on the real axis to the right of the point is *odd.*

Example: for the given example, portions of the root loci on the real axis are marked by the heavy line illustrated in Fig. 4.1.13.

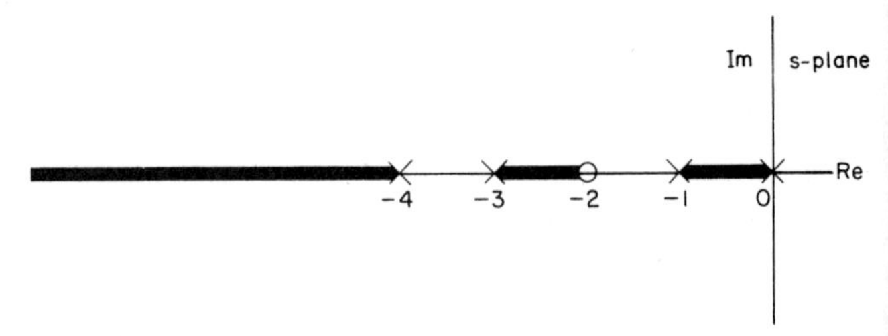

Fig. 4.1.13. Illustration of Rule #3 in which portions of the root loci on the real axis are marked.

Notice that all the points left of –4 up to – ∞ are part of the root loci. Therefore, this branch of the root loci approaches infinity along the negative axis.

RULE #4:

Determine the asymptotes of the root loci. **(n–m)** branches of the root loci terminate at infinity along asymptotes. The angles of the asymptotes equal:

$$\beta[^{o}] = \frac{\pm 180\,(2N+1)}{n-m} \qquad (4.1.40)$$

$$[N = 0, 1, 2, \ldots, (n-m-1)]$$

All the asymptotes intersect the real axis at the point σ which is given by:

$$\sigma = \frac{(p_1+p_2+\ldots+p_n)-(z_1+z_2+\ldots z_m)}{n-m} \qquad (4.1.41)$$

Example: for the given example we have **n − m = 3.** Therefore, the angles of asymptotes are

$$\beta = \frac{\pm 180\,(2N + 1)}{3} \qquad [N = 0,1,2] \qquad (4.1.42)$$

Thus, the angles will be + **180°**, + **60°**, and **-60°**, and they intersect at a point given by:

$$\sigma = \frac{(0 - 1 - 3 - 4) - (-2)}{3} = -2 \qquad (4.1.43)$$

The asymptotes are shown in Fig. 4.1.14.

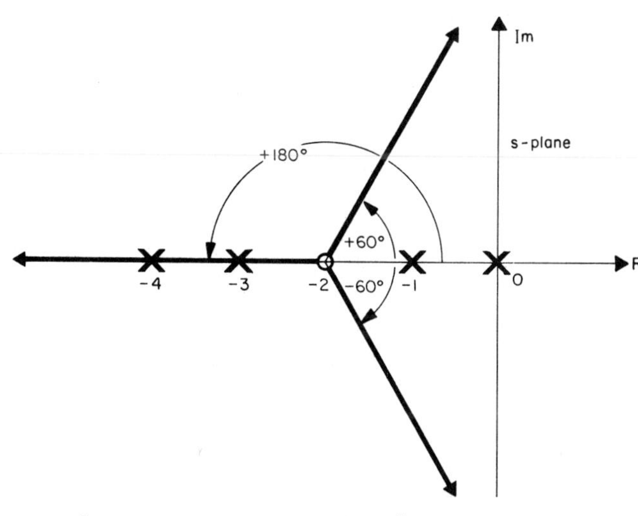

Fig. 4.1.14. Illustration of Rule #4 in which the angles of the asymptotes are plotted.

RULE #5:

Find the breakaway and break-in points. If the root locus lies between two *adjacent poles* on the real axis, there is at *least one breakaway point.* Similarly, if the root locus lies between two *adjacent zeros* on the real axis, there is at *least one break-in point.* However, if the root locus is *between a zero and a pole* on the real axis, there may exist *no breakaway or break-in points.*

If the characteristic equation is given by:

$$1 + \frac{K\,B(s)}{A(s)} = 0 \qquad (4.1.44)$$

then the location of breakaway or break-in points is given by:

$$A'(s)\,B(s) - A(s)\,B'(s) = 0 \qquad (4.1.45)$$

where the prime indicates differentiation with respect to **s**. The value of **K** at those points equals the product of distances to all poles divided by distances to all zeros.

Example: in the given example, the charteristic equation is:

$$1 + K\,\frac{(s+2)}{s(s+1)(s+3)(s+4)} = 0 \qquad (4.1.39)$$

Thus:

$$B(s) = s + 2 \qquad (4.1.46)$$

$$A(s) = s\,(s + 1)(s + 2)(s + 3)$$

$$= s^4 + 6s^3 + 11s^2 + 6s \qquad (4.1.47)$$

Then, differentiating with respect to s:

$$B'(s) = 1 \qquad (4.1.48)$$

$$A'(s) = 4s^3 + 18s^2 + 22s + 6 \qquad (4.1.49)$$

The breakaway point can be found from (4.1.45):

$$(4\,s^3 + 18\,s^2 + 22\,s + 6)\,(s + 2) -$$

$$- (s^4 + 6\,s^3 + 11\,s^2 + 6\,s) = 0$$

$$3\,s^4 + 20\,s^3 + 47\,s^2 + 44s + 12 = 0 \tag{4.1.50}$$

Eq. (4.1.50) will have to be solved numerically, and for this case it is found that the solution is:

$$s \approx -0.5 \tag{4.1.51}$$

RULE #6:

Find the points where the root loci cross the imaginary axis. The points where the root loci intersect the imaginary axis can be found by substituting $s = j\omega$ in the characteristic equation. Equating both the real and imaginary part to zero allow the solution for K and ω. The value of K at that point is important since it determines the value of the parameter which will cause the system to become unstable.

Example: using the characteristic equation from the above example, we set $s = j\omega$ and obtain:

$$1 + K\frac{(j\omega+2)}{j\omega\,(j\omega+1)(j\omega+3)(j\omega+4)} = 0 \tag{4.1.52}$$

$$j\omega\,(j\omega+1)(j\omega+3)(j\omega+4) + K\,(j\omega+2) = 0 \tag{4.1.53}$$

This becomes:

$$\omega^4 - 8j\omega^3 - 19\omega^2 + 12\,j\omega + Kj\omega + 2K = 0 \tag{4.1.54}$$

Separating this equation into real and imaginary parts:

$$\text{Real:}\quad \omega^4 - 19\omega^2 + 2K = 0 \tag{4.1.55}$$

$$\text{Imaginary:}\ -8\omega^3 + 12\omega + K\omega = 0 \tag{4.1.56}$$

Using (4.1.56), we obtain:

$$K = 8\omega^2 - 12 \tag{4.1.57}$$

This is substituted into (4.1.55):

$$\omega^4 - 19\,\omega^2 + 16\omega^2 - 24 = 0$$

$$\omega^4 - 3\omega^2 - 24 = 0 \tag{4.1.58}$$

If we substitute $X = \omega^2$ the equation becomes:

$$X^2 - 3X - 24 = 0 \tag{4.1.59}$$

which has a solution of:

$$X_{1,2} = \frac{3 \pm \sqrt{9 + 96}}{2}$$

$$X_1 = 6.62$$

$$X_2 = -3.62$$

Since only positive values are allowed for X, we obtain $X = 6.62$; so that $\omega^2 = 6.62$, and the root locus crosses the imaginary axis at ω = **2.57**. The value of the gain, **K**, at that point is:

$$K = 8\omega^2 - 12 = 41 \qquad (4.1.60)$$

The root loci for this example are shown in Fig. 4.1.15.

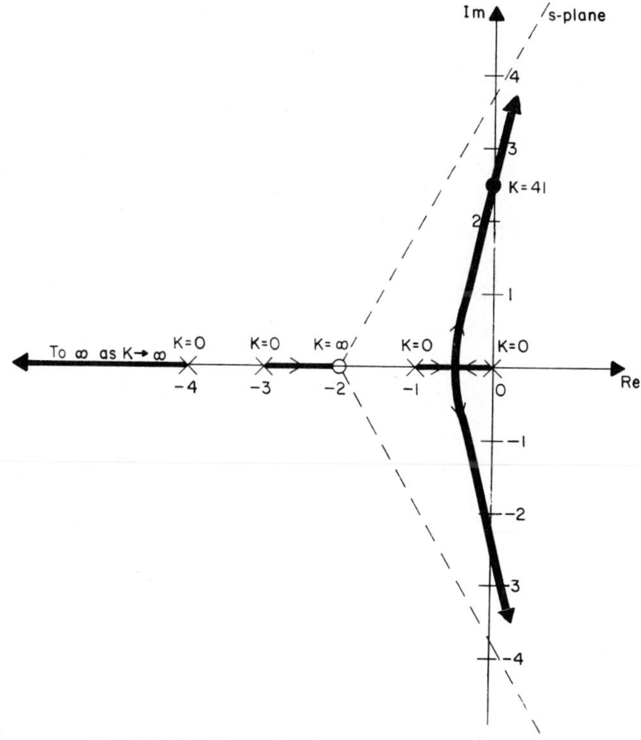

Fig. 4.1.15. Example of locating the point where root locus crosses the imaginary axis.

This concludes the presentation of the six rules for constructing of the root locus. To further illustrate construction of root loci, another problem is presented. The open-loop transfer function will be **G(s) H(s)**, where

$$G(s) = \frac{K}{s} \qquad (4.1.61)$$

$$H(s) = \frac{s+2}{s+1} \qquad (4.1.62)$$

Step 1: The characteristic equation is arranged to the prescribed format:

$$1 + K\frac{s+2}{s(s+1)} = 0 \qquad (4.1.63)$$

and plotted in Fig. 4.1.16.

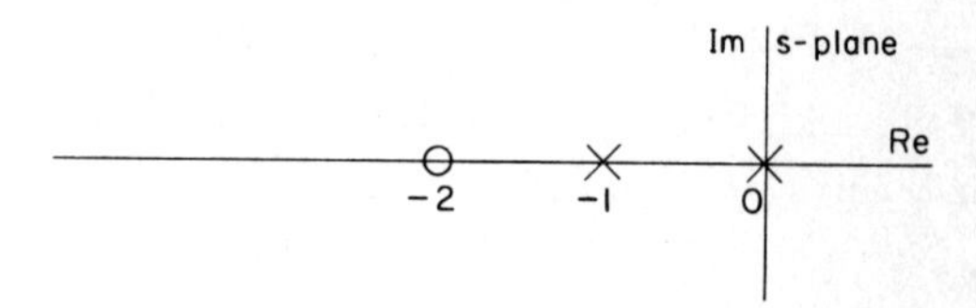

Fig. 4.1.16. An example of locating and plotting poles and zeros. This is the plot of Eq. (4.1.63).

Step 2: The starting points are **0** and **—1**; termination points are **—2** and since **n—m=1**, another termination point is at **infinity.**

Step 3: The root loci on the real axis are between **0** and **—1** and to the left of **—2.**

Step 4: There is one asymptote (since **n—m = 1**) and the angle of that asymptote is:

$$\beta = \pm\frac{180\,(2N+1)}{n-m} = \pm\,180^{o} \qquad (4.1.64)$$

$$(N = 0)$$

Step 5: To determine the breakaway and break-in points (note that there must be a breakaway point between **0** and **—1** and a break-in point to the left of **—2**) solve the equation:

$$A'B - AB' = 0 \qquad (4.1.45)$$

where:

$$A(s) = s(s+1) = s^2 + s \qquad (4.1.65)$$

$$B(s) = s + 2 \qquad (4.1.66)$$

Then, differentiating:

$$A' = 2s + 1 \qquad (4.1.67)$$

$$B' = 1 \qquad (4.1.68)$$

and substituting (4.1.65) thru (4.1.68) into (4.1.45) we obtain:

$$(2s + 1)(s + 2) - (s^2 + s) = 0$$

$$s^2 + 4s + 2 = 0 \qquad (4.1.69)$$

Then:

$$s_1 = -2 + 1.414 = -0.586$$

$$s_2 = -2 - 1.414 = -3.414$$

Thus, a breakaway point is at $s = -0.586$ and a break-in point is at $s = -3.414$. The gain at the breakaway point equals:

$$K_1 = \frac{0.586 \times 0.414}{1.414} = 0.17 \qquad (4.1.70)$$

and at the break-in point:

$$K_2 = \frac{3.414 \times 2.414}{1.414} = 5.8 \qquad (4.1.71)$$

Step 6: To see whether the root loci cross the imaginary axis substitute $s = j\omega$ in the characteristic equation:

$$1 + K \frac{j\omega + 2}{j\omega\,(j\omega + 1)} = 0 \qquad (4.1.72)$$

Then:

$$j\omega\,(j\omega + 1) + K\,(j\omega + 2) = 0$$

$$-\omega^2 + j\omega\,(1 + K) + 2K = 0$$

equating real and imaginary parts:

$$\omega^2 + 2K = 0 \qquad (4.1.73)$$

$$\omega\,(1 + K) = 0 \qquad (4.1.74)$$

The only point that satisfies these equations is when $\omega = 0$ and $K = 0$, which indicates the starting point. Thus, the root loci do not cross the imaginary axis. The construction of the root loci is shown in Fig. 4.1.17.

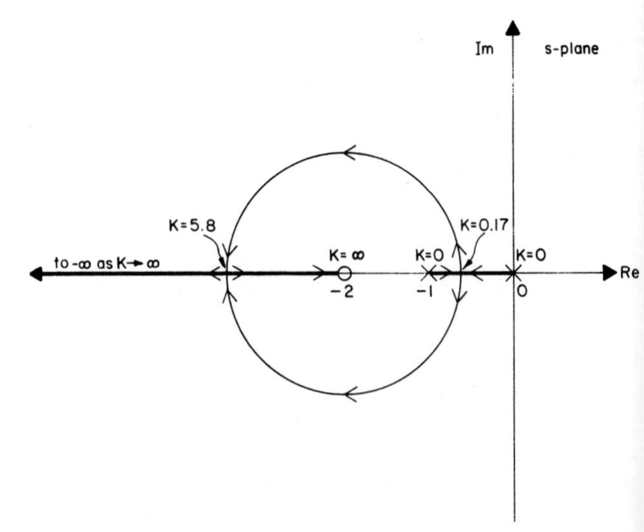

Fig. 4.1.17. Construction of root loci for Step 6 of the example. □

4.2. SERVO COMPONENTS

After reviewing servo theory, we will describe the common servo components by their transfer functions and block diagrams. This is necessary before we attempt to analyze the system as a whole.

4.2.1. DC MOTOR

A detailed model for a DC permanent magnet motor was derived in Chapter 2. The motor with the load and the tachometer was approximated as three bodies, coupled by inertialess shafts, and the torsional resonance was considered.

The model shows the relations between the following variables:

$V(s)$ = **motor voltage**

$I_a(s)$ = **armature current**

$\theta_m(s)$ = **angular position of armature**

$\theta_1(s)$ = **angular position of load**

$\theta_2(s)$ = **angular position of tachometer**

The relation among the variables was expressed (2.3.44) in a matrix form:

$$\begin{bmatrix} sL_a+R & sK_E & 0 & 0 \\ -K_T & s^2J_m+sD & s^2J_1 & s^2J_2 \\ 0 & sD_1+K_1 & -(s^2J_1+sD_1+K_1) & 0 \\ 0 & sD_2+K_2 & 0 & -(s^2J_2+sD_2+K_2) \end{bmatrix} \begin{bmatrix} I_a(s) \\ \theta_m(s) \\ \theta_1(s) \\ \theta_2(s) \end{bmatrix} = \begin{bmatrix} V(s) \\ 0 \\ 0 \\ 0 \end{bmatrix} \tag{4.2.1}$$

The above model may be too general and unnecessarily complicated in many cases. When the structural resonance is of no concern, one may assume

$$\theta_m(s) = \theta_1(s) = \theta_2(s) \tag{4.2.2}$$

and the motor equations may be simplified:

$$\begin{bmatrix} sL_a+R & sK_E \\ -K_T & s^2J+sD \end{bmatrix} \begin{bmatrix} I_a(s) \\ \theta_m(s) \end{bmatrix} = \begin{bmatrix} V(s) \\ 0 \end{bmatrix} \tag{4.2.3}$$

where **J** is the total moment of inertia:

$$J = J_m + J_1 + J_2 \tag{4.2.4}$$

Since, in many cases, the motor velocity, $\omega(s)$, is of interest, the motor equations may be written in terms of the voltage, current, and velocity.

$$\begin{bmatrix} sL_a+R & K_E \\ -K_T & sJ+D \end{bmatrix} \begin{bmatrix} I_a(s) \\ \omega(s) \end{bmatrix} = \begin{bmatrix} V(s) \\ 0 \end{bmatrix} \tag{4.2.5}$$

The transfer function between the input voltage and the velocity is then

$$G_m(s) = \frac{\omega(s)}{V(s)} = \frac{K_T}{(sJ+D)(sL_a+R)+K_EK_T} \tag{4.2.6}$$

The above models of (4.2.1), (4.2.5), and (4.2.6) assumed that the motor is driven by a voltage source. In case the amplifier is a current source, the model of (4.2.5) will be reduced to the form

$$(sJ+D)\ \omega(s) = K_T I_a(s) \tag{4.2.7}$$

and the motor transfer function is

$$G'_m(s) = \frac{\omega(s)}{I_a(s)} = \frac{K_T}{sJ + D} \tag{4.2.8}$$

Since, in most cases, the amplifiers are voltage sources and the resonance is not significant, (4.2.6) is the most common description of a DC motor. This equation may be further simplified when some of the parameters are small and can be ignored, e.g., $L_a \approx 0$ for moving coils motors, or $D \approx 0$ for some motors.

4.2.2. AMPLIFIER

Servo amplifiers may be divided into two main groups, according to the type of feedback they use: *voltage amplifiers* and *current amplifiers.* The two types will be studied separately, and the transfer function model for each one will be given. Further details on the amplifier design can be found later in this chapter.

Voltage Amplifier

Ideally, an amplifier should have constant gain for all frequencies. However, this is not the case in practice, and all amplifiers have limited bandwidth. If the amplifier transfer function has a single pole, the transfer function between the input command, $V_{in}(s)$, and the output, $V_{out}(s)$, is

$$\frac{V_{out}(s)}{V_{in}(s)} = \frac{A}{s\tau_a + 1} \tag{4.2.9}$$

where A is the DC gain of the amplifier and τ_a is the time constant.

If the amplifier has more than one pole, the transfer function should be constructed accordingly.

Current Amplifier

Current amplifiers differ from voltage amplifiers by the fact that the current is fed back and, therefore, the output current is proportional to the input voltage. Consequently, the transfer function in the case of a single pole is

$$\frac{I_{out}(s)}{V_{in}(s)} = \frac{A_I}{s\tau_a + 1} \tag{4.2.10}$$

Note that A_I has a unit of [A/V].

4.2.3. AMPLIFIER-MOTOR SYSTEM

In some cases, both current and voltage feedback are used to achieve a certain performance. The general arrangement in such a system is shown in Fig. 4.2.1. The block of $\frac{A}{s\tau_a + 1}$ represents an amplifier with a voltage feedback. We will show how a current feedback may be added and discuss its effect on the transfer function.

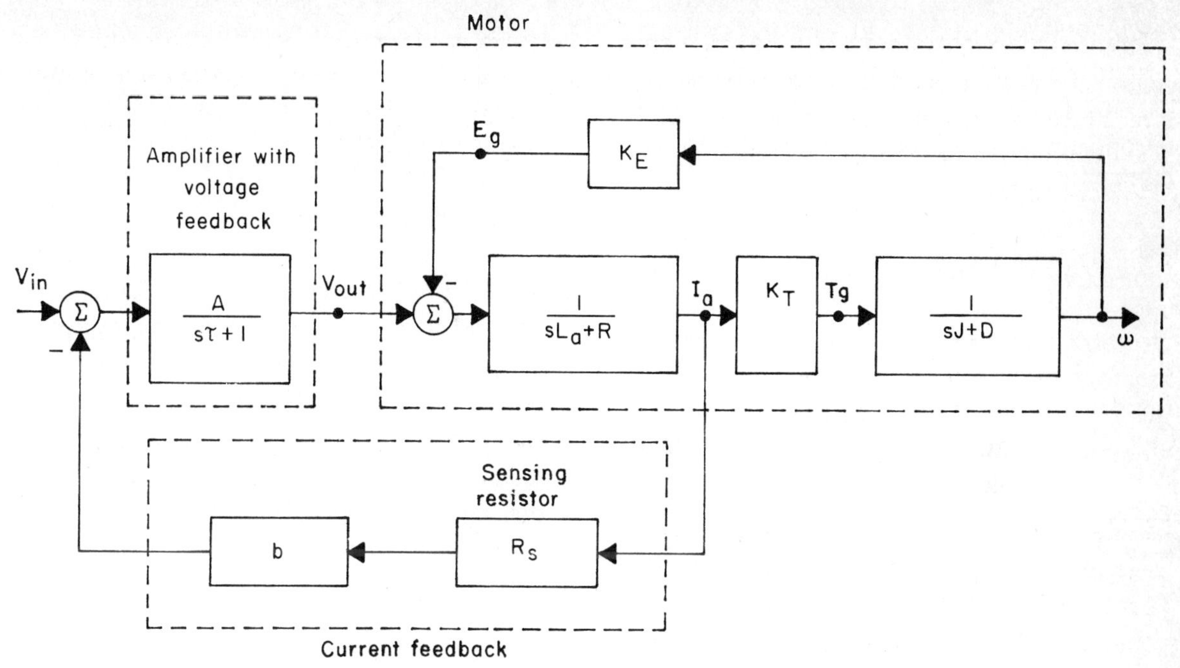

Fig. 4.2.1. A system of motor and amplifier with current and voltage feedbacks.

The transfer function between $\mathbf{V_{in}}$ and the velocity, ω, equals

$$\frac{\omega(s)}{V_{in}(s)} = \frac{\frac{A}{s\tau_a+1}\cdot\frac{1}{sL_a+R}\cdot\frac{K_T}{sJ+D}}{1+\frac{A}{s\tau_a+1}\cdot\frac{1}{sL_a+R}\cdot R_s\cdot b+\frac{1}{sL_a+R}\cdot\frac{K_T}{sJ+D}\cdot K_E}$$

$$= \frac{A\cdot K_T}{(s\tau_a+1)(sL_a+R)(sJ+D)+A\,R_s b(sJ+D)+K_E K_T(s\tau_a+1)} \tag{4.2.11}$$

Note that if $\mathbf{b} = \mathbf{0}$, then (4.2.11) is the product of (4.2.6) and (4.2.9), indicating that the system is a series combination of a voltage amplifier and a motor. If, on the other hand, **b** becomes large, the values of the system poles are altered considerably.

When a positive current feedback is used, then (4.2.11) holds under the condition $\mathbf{b} < \mathbf{0}$. Note that in this case large values of **b** will cause instability.

4.2.4. TACHOMETER

An ideal tachometer is a device whose output voltage, $\mathbf{V_g}$, is proportional to the shaft angular velocity ω. Thus:

$$\mathbf{V_g} = \mathbf{K_g}\omega \tag{4.2.12}$$

In the real case, however, the voltage will have other components, which appear as ripple in the voltage given by (4.2.12). Expanding (4.2.12) to include ripple frequency and amplitude terms, we arrive at:

$$\mathbf{V_g}=\mathbf{K_g}\omega\,[1+k_1\cos(u\omega t)+k_2\cos(2u\omega t)] \tag{4.2.13}$$

where **u** is the number of commutator segments.

It is desirable to have k_1 and k_2 as small as possible to reduce the ripple factor, ξ, so that tachometer output voltage would be proportional to ω only. The ripple factor is:

$$\xi = \sqrt{k_1^2 + k_2^2} \tag{4.2.14}$$

In most cases, the tachometers are selected so that the ripple is not significant. Then (4.2.12) holds, and the tachometer transfer function becomes:

$$\frac{V_g(s)}{\omega(s)} = K_g \tag{4.2.15}$$

The block diagram is shown in Fig. 4.2.2.

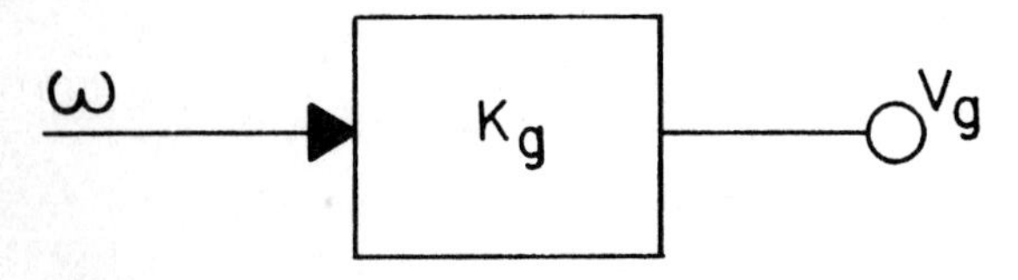

Fig. 4.2.2. Block diagram of tachometer showing velocity input and voltage output.

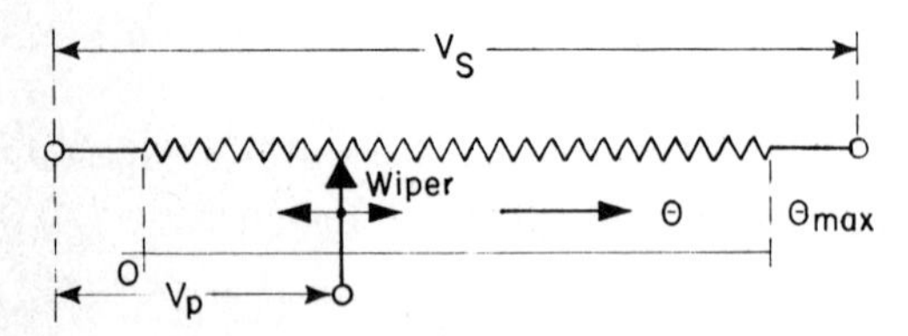

Fig. 4.2.3. A linear potentiometer.

4.2.5. POTENTIOMETER

A linear potentiometer provides an output voltage, V_p, proportional to the angular position, θ, of its shaft (Fig. 4.2.3). It operates on the principle of a voltage divider, where rotation moves the wiper across a fixed resistor, resulting in an output voltage proportional to the angular position. The output voltage of an ideal potentiometer is given by:

$$V_p = K_p\theta \tag{4.2.16}$$

In real potentiometers, the relationship of V_p and θ is not truly linear, as indicated by (4.2.16), due to two main factors: first, potentiometer resistance is not uniform, which results in non-linear response; and second, in connecting the wiper, we load the circuit slightly, also affecting potentiometer linearity.

The transfer function of an ideal potentiometer is:

$$\frac{V_p(s)}{\theta(s)} = K_p \tag{4.2.17}$$

and since the shaft position is an integral of the shaft angular velocity, we may write:

$$\frac{\theta(s)}{\omega(s)} = \frac{1}{s} \tag{4.2.18}$$

Combining (4.2.17) with (4.2.18) yields:

$$\frac{V_p(s)}{\omega(s)} = \frac{K_p}{s} \tag{4.2.19}$$

This is represented by the block diagram in Fig. 4.2.4.

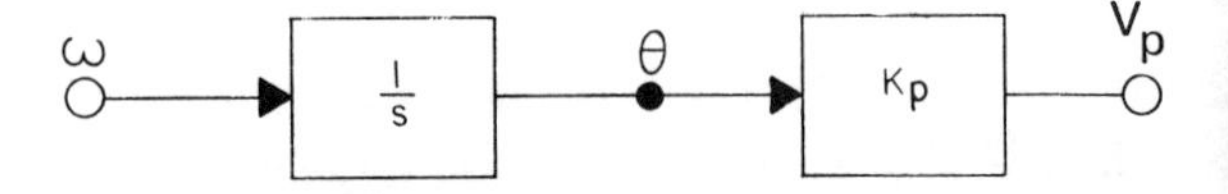

Fig. 4.2.4. Block diagram of potentiometer.

In some systems linear motion is required. In that case, this motion can be converted into rotation and a potentiometer can be used. The other alternative is to use a linear variation differential transformer, which is described in the following paragraph.

4.2.6. LINEAR VARIATION DIFFERENTIAL TRANSFORMER (LVDT)

The LVDT is the linear equivalent of the potentiometer, and produces a voltage V_L proportional to the linear position **x** of its movable part so that:

$$V_L = K_L x \qquad (4.2.20)$$

Since **x** is the integral of the linear velocity **v**, we can write:

$$\frac{x(s)}{v(s)} = \frac{1}{s} \qquad (4.2.21)$$

We can equate linear velocity **v** to angular velocity ω by:

$$v = r\omega \qquad (4.2.22)$$

Thus, (4.2.20), (4.2.21) and (4.2.22) can be combined to:

$$\frac{V_L(s)}{\omega(s)} = \frac{K_L r}{s} \qquad (4.2.23)$$

Eq. (4.2.23) is represented by the block diagram in Fig. 4.2.5.

4.2.7. ENCODERS

Incremental Encoder

An incremental encoder is a device which generates a pulse for a given angular increment of shaft rotation. The pulse may be square, trapezoidal, or sinusoidal. For example, assume that an unidirectional encoder generates **N** sinusoidal waves per revolution. The output voltage will be given by:

$$V_e = A \sin (N\theta) \qquad (4.2.24)$$

where θ is the angular position of the shaft in radians.

Note that the relationship of θ and V_e is non-linear and, therefore, cannot be described by a transfer function. However, in the case where the angle θ is kept within some small variations, (4.2.24) can be linearized, resulting in the equivalent equations of a potentiometer.

Bidirectional encoders, on the other hand, have two output signals shifted by some phase angle, φ, thus:

$$V_{e1} = A \sin (N\theta) \qquad (4.2.25)$$

$$V_{e2} = A \sin (N\theta + \varphi) \qquad (4.2.26)$$

The relative timing of V_{e1} and V_{e2} yields direction of the rotation and rotation angle.

A common use of the incremental encoder is to input encoder signal into a digital

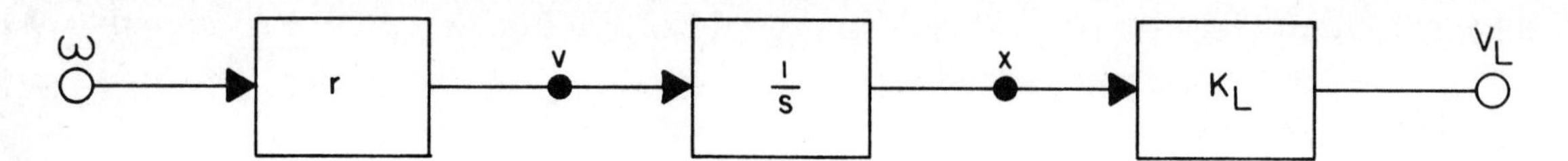

Fig. 4.2.5. Block diagram of a linear variation differential transformer.

counter, whose output is proportional to the angular position, θ. Other applications use an encoder in a hybrid control, in which a motor is driven in a velocity control mode for some angle and then switched to position control mode which stops the motor in the next neighboring pulse position.

Absolute Encoder

An absolute encoder is equivalent to an incremental encoder, connected to a digital counter with channels whose outputs form a number proportional to the shaft position. Output data is periodic for every revolution, as shown in Tab. 4.2.1 and in Fig. 4.2.6.

θ [rad]	Output binary numbers		Equivalent decadic number
	Channel #1	Channel #2	
$0 - \pi/2$	0	0	0
$\pi/2 - \pi$	1	0	1
$\pi - 3\pi/2$	0	1	2
$3\pi/2 - 2\pi$	1	1	3

Tab. 4.2.1. An absolute encoder output data.

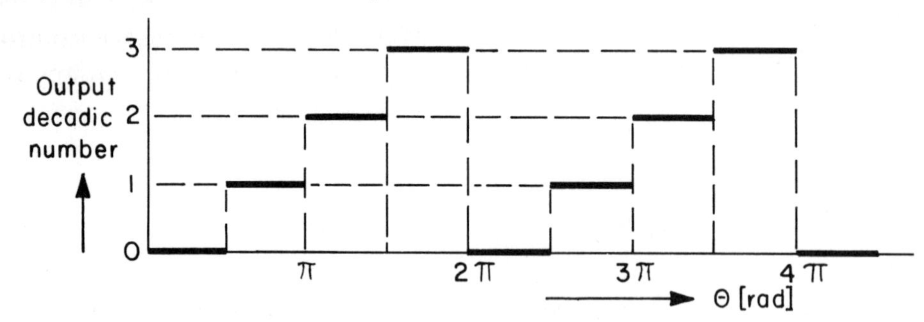

Fig. 4.2.6. Plot of an absolute encoder output. □

4.3. SERVO SYSTEMS

4.3.1. INTRODUCTION

One means of controlling the velocity or angular position of a motor is by an open-loop control system as shown in Fig. 4.3.1.

In the open–loop system, the output will follow the desired function as long as all the system variables are constant. Any change in load, amplifier gain, or any other system variable will, however, cause a deviation from the desired value. In order for the motor to conform to a desired function independently of changes in these variables, a closed-loop servo system, such as the one illustrated in Fig. 4.3.2, must be constructed.

In the closed-loop system, the output variable is measured, fed back and compared to the desired input function. Any difference between the two is a deviation from the desired result; the deviation is amplified and used to correct the error. In this manner, the closed-loop system is essentially insensitive to variations in parameters, and therefore performs correctly despite changes in load condition and other system parameters. However, now the response of the system depends on the closed–loop configuration and as such it may be overdamped, underdamped, or even unstable. Special care must be given to the design of the closed loop in order to obtain the desired response.

Servo systems are divided according to the variable being controlled. The most common systems are:

1. *Velocity control systems* - motor velocity is to follow a given velocity profile;

2. *Position control systems* - motor angular position is to be controlled;

3. *Torque control systems* - motor torque is to be controlled;

4. *Hybrid control systems* - the system switches from one control mode to the other. For example, we may want to control the velocity for some time and then switch to position control mode.

Since these modes of system control are the most common, the following paragraphs describe them in detail.

4.3.2. VELOCITY CONTROL SYSTEMS

In a velocity control system, motor velocity is controlled to follow a signal describing the desired velocity profile. In the following analysis, simplified transfer functions are used to describe the motor and the amplifier in order to describe the method of velocity control. The block diagram for such a system is shown in Fig. 4.3.3, using the following to describe the system:

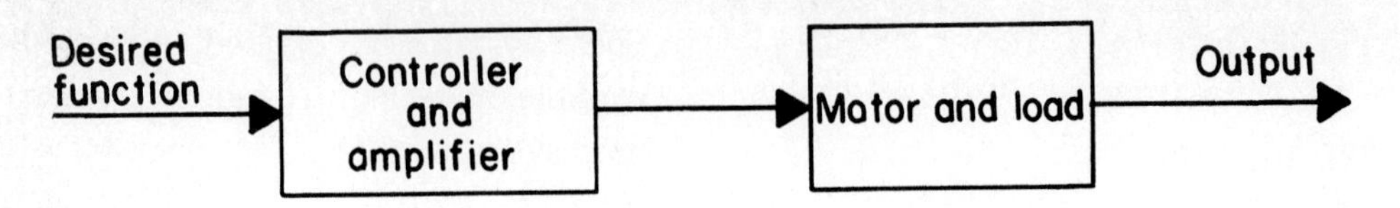

Fig. 4.3.1. A typical open-loop control system.

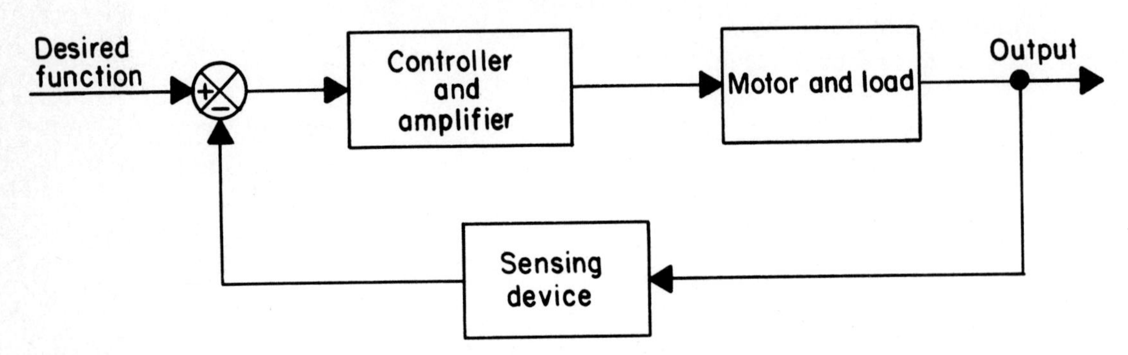

Fig. 4.3.2. A typical closed-loop control system.

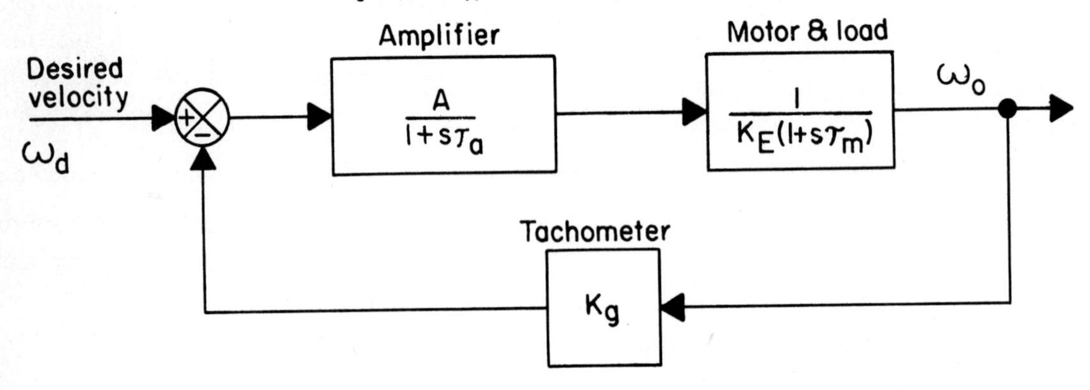

Fig. 4.3.3. Velocity control system.

A DC gain of the amplifier

τ_a time constant of the amplifier

τ_m mechanical time constant of loaded motor

K_E motor voltage constant

K_g tachometer voltage constant

The closed–loop transfer function for the velocity control system, which is the ratio of the Laplace transforms of the output velocity ω_o and the desired velocity ω_d, is:

$$G_v(s) = \frac{\omega_o(s)}{\omega_d(s)} = \frac{\dfrac{A}{1+s\tau_a}\,\dfrac{1}{K_E(1+s\tau_m)}}{1+\dfrac{A}{1+s\tau_a}\,\dfrac{1}{K_E(1+s\tau_m)}\,K_g} \tag{4.3.1}$$

In order to determine the system response, the root locus method is used to evaluate the effect of the amplifier gain **A** on the system response. The root locus for the velocity control system is shown in Fig. 4.3.4. Note that for the small values of **A** the roots are negative real, which indicates that the system is overdamped. As the gain is increased, the system becomes less and less damped, and will tend to overshoot in response to a step input. Because of the simplification used on the transfer functions, the system remains stable for any value of **A**. This is not the case in reality, where the system has some more poles and, therefore, will become unstable for a certain value of **A**.

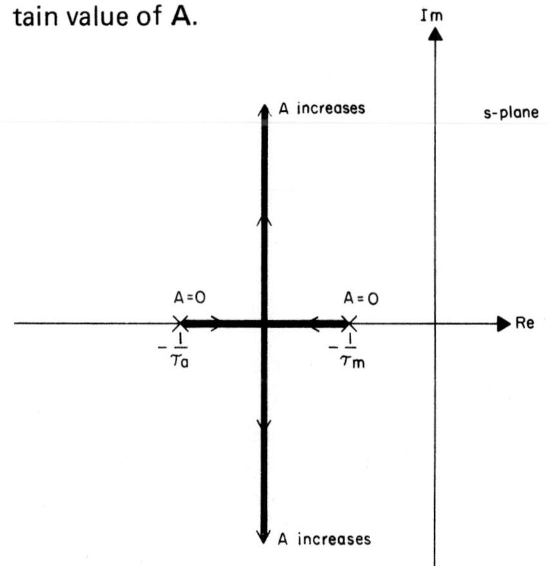

Fig. 4.3.4. Root locus for a velocity control system.

4.3.3. POSITION CONTROL SYSTEMS

The objective in position control systems is to control the angular position of the motor shaft, either locking the shaft at a desired position, or rotating it at a given rate. The block diagram for such a system, with simplified transfer functions for motor and amplifier, is given in Fig. 4.3.5.

The system parameters are explained as follows:

Motor velocity is denoted by ω_o and the angular position is θ_o. Since θ_o is the integral of ω_o, this corresponds to a transfer function **1/s** as shown in the diagram. K_g represents the tachometer voltage constant [**V/rad** s^{-1}], and K_p is the gain of the position sensor [**V/rad**]. The sensor can be a potentiometer or equivalent device. Note that the combination $K_p\theta_o + K_g\omega_o$ is fed back and subtracted from θ_d. This is done to achieve stability, as will be shown later.

The transfer function of the system is:

$$G_p(s) = \frac{\theta_o(s)}{\theta_d(s)}$$

$$= \frac{\dfrac{1}{s}\,\dfrac{A}{1+s\tau_a}\,\dfrac{1}{K_E(1+s\tau_m)}}{1+\dfrac{A}{1+s\tau_a}\,\dfrac{1}{K_E(1+s\tau_m)}\left(K_g+\dfrac{K_p}{s}\right)} \tag{4.3.2}$$

In order to analyze the system response, we look at the roots of the second term of the characteristic equation:

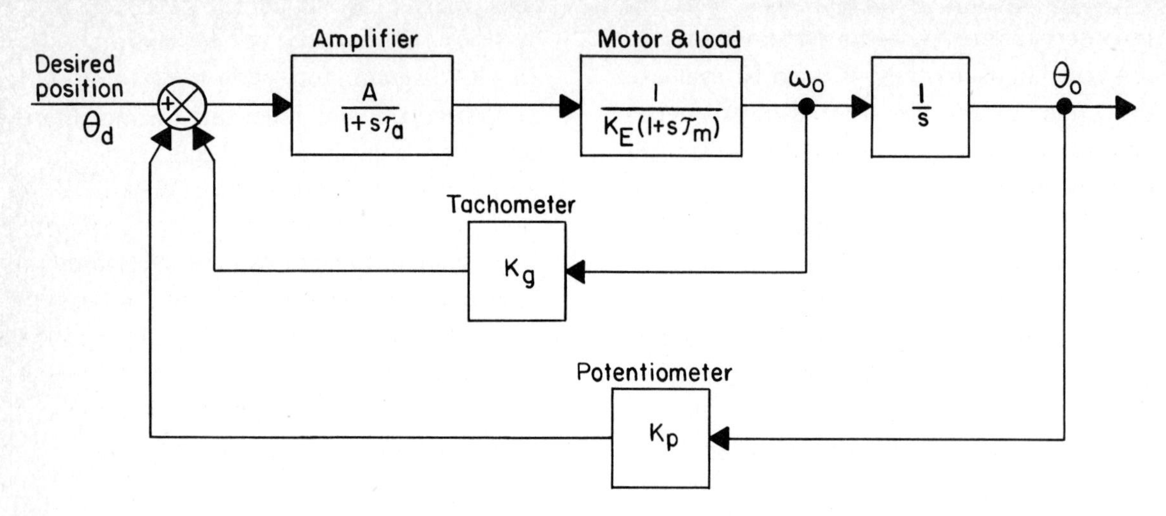

Fig. 4.3.5. Position control system.

$$1 + \frac{\frac{A}{K_E}}{1+s\tau_a} \; \frac{K_g s+K_p}{s(1+s\tau_m)} = 0 \qquad (4.3.3)$$

The root locus for this equation is shown in Figs. 4.3.6 and 4.3.7. Note that if no tachometer feedback is used, $K_g = 0$ and the characteristic equation will have three poles and no zeros. The root loci for this case are shown in Fig. 4.3.6, where it can be seen that the system becomes unstable for large values of gain A.

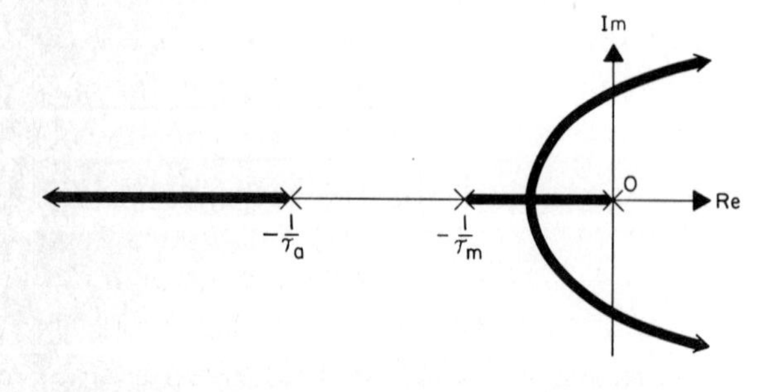

Fig. 4.3.6. Root locus plot of position control system with no tachometer feedback.

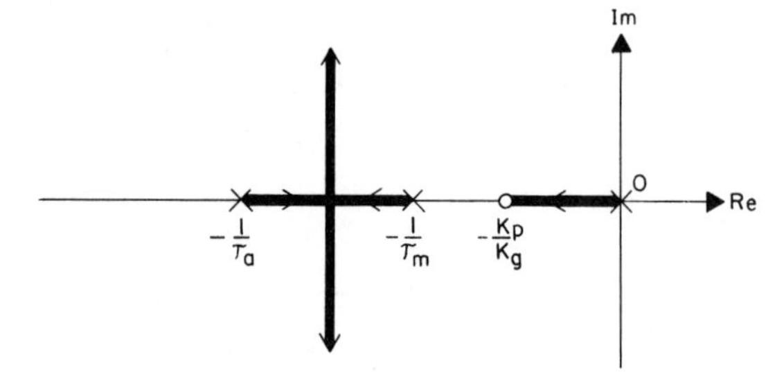

Fig. 4.3.7. Root locus plot of position control system with tachometer feedback. Note the improvement in stability.

The introduction of K_g adds a zero to the open loop transfer function and alters the root loci as shown in Fig. 4.3.7. The new configuration is more stable and can accommodate higher gain.

Note again that the above analysis is based on *simplified* transfer functions of the motor and amplifier. In the real case, the transfer functions will contain more poles and the system will tend to become unstable more rapidly. However, the effect of the velocity feedback is to increase the stability of the system as shown here. Moreover, one can control the system response by changing the value of K_g. When K_g is small, the system may be unstable or underdamped, and as K_g increases, the system becomes more damped.

4.3.4. TORQUE CONTROL SYSTEMS

In some cases it is desired to keep the torque of the motor constant. Since torque is proportional to the motor current, it means that a constant current must be provided to the motor. The method for achieving this is illustrated in Fig. 4.3.8.

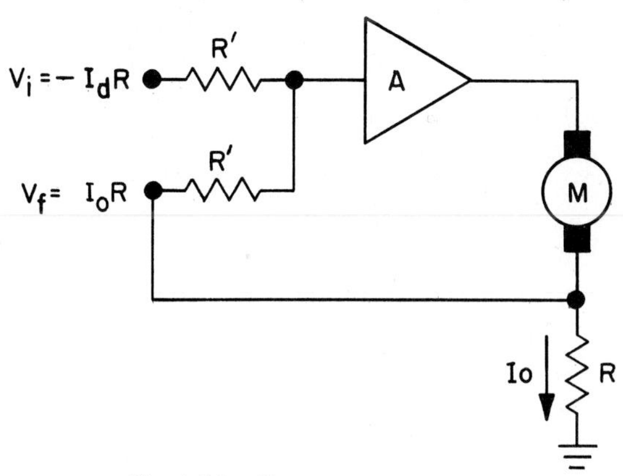

Fig. 4.3.8. Torque control system.

The input voltage $V_i = -I_dR$ is compared with the feedback voltage $V_f = I_oR$ and input to a high gain amplifier. If the output current I_o is different from the desired current I_d, the difference is amplified and used to correct the situation.

4.3.5. HYBRID CONTROL SYSTEMS

In some applications, the advantages of various control modes can be combined to establish the desired performance by switching the system from one control mode to another.

One example of a hybrid system is one which alternates between velocity and position control modes. Such a system can be used for incremental motion, where the motion is performed under velocity control in obedience to a desired velocity profile, whereas stopping is done by position control mode to achieve greater accuracy.

Another example would be when forward-backward motion is to be performed, and different velocity profiles are required for each direction. If, for instance, the forward motion is to be done at a constant velocity and the backward, or return, motion is to be in minimum time, it could be done by a hybrid control in which velocity control is used in the forward mode and position control is used for the return or "homing" mode.

A velocity/position hybrid control system is shown in Fig. 4.3.9. When the mode selector switch is *open*, the position feedback is disconnected, and the system operates as a velocity control servo. *Closing* the switch adds the circuitry required for position control. This hybrid system provides the advantages of both velocity and position controls, since it allows to follow a desired velocity profile in velocity control mode, while stopping with position accuracy. The shaft will be locked in the desired position after stopping. Achieving these advantages results in additional costs. First, a switch-

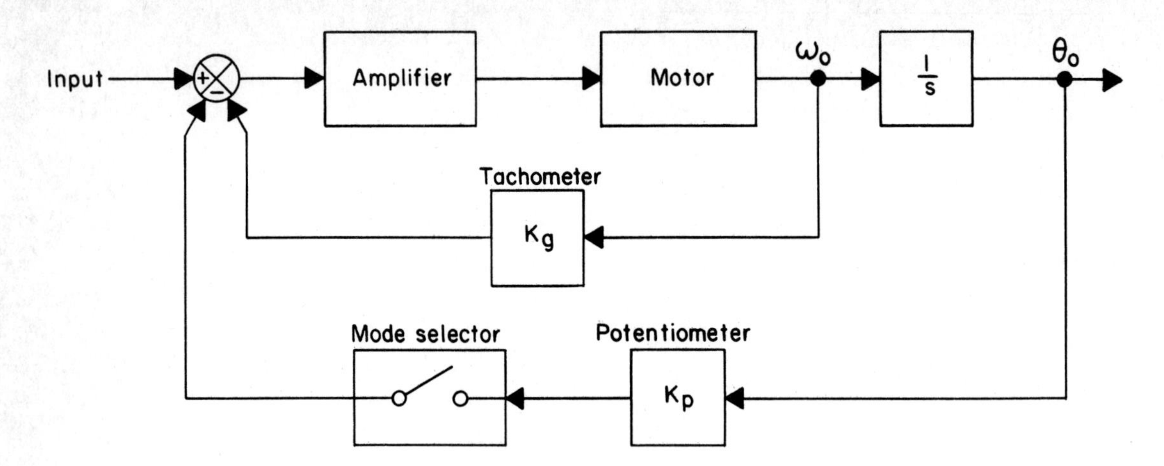

Fig. 4.3.9. Velocity/position hybrid control system.

ing circuit is required to select system modes at the right time. Secondly, the system has to be designed so that it responds to a desired manner in both modes, which makes the design more difficult. However, the additional design effort is justified if alternate control modes are dictated by specific applications. □

4.4. SYSTEM CHARACTERISTICS

The performance of a servo system is completely determined by its closed-loop transfer function. However, there are certain system characteristics which are especially significant, and even though they can be derived from the transfer function, it is worthwhile to describe them in detail. These characteristics are discussed below.

4.4.1. RESPONSE OF THE SYSTEM TO A STEP COMMAND

System response to a step input command depends on all the poles and zeros of the closed-loop transfer function. However, in most cases, one can obtain a good approximation of this response by considering only the dominant pair of poles - which may be real or complex. The system response may be any one of the following:

Overdamped Response

The two dominant poles are either negative real or complex, with the real part negative and with the imaginary part smaller (in absolute value) than the real part. Thus, if the poles are:

$$\left.\begin{aligned} p_1 &= \sigma_1 < 0 \\ p_2 &= \sigma_2 < 0 \end{aligned}\right\} \quad (4.4.1)$$

or

$$\left.\begin{aligned} &p_{1,2} = \sigma \pm j\omega \\ &\sigma < 0 \\ &|\sigma| > \omega \end{aligned}\right\} \quad (4.4.2)$$

then the system will be overdamped, as shown by curve (1) of Fig. 4.4.1.

Critically Damped Response

The dominant poles are complex with equal negative real and imaginary parts. Thus:

$$|\sigma| = \omega \quad (4.4.3)$$

The response in this case is fast and without an overshoot. It is shown by curve (2) of Fig. 4.4.1.

Underdamped Response

The imaginary part of the poles is larger than the absolute value of the real part; that is:

$$\omega > |\sigma| \quad (4.4.4)$$

The system response overshoots, but finally settles at the steady state value. The settling time is reciprocal of $|\sigma|$. Therefore, as $|\sigma|$ increases, the settling time becomes shorter. Curve (3) in Fig. 4.4.1 illustrates an underdamped response.

Unstable Response

When any of the system poles has a positive real part, the response is unstable. In

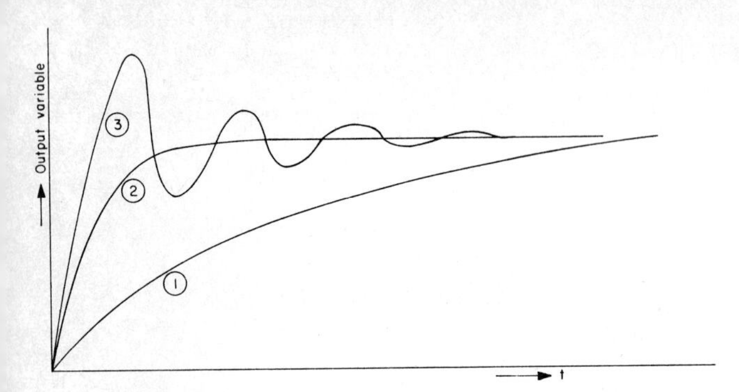

Fig. 4.4.1. Response of a system to a step input command: Curve 1 - overdamped response; Curve 2- critically damped response; Curve 3 - underdamped response.

practical systems instability leads to oscillations or saturation, and they are both undesired results.

The above discussion was based on the approximation of the system transfer function by its dominant pair of poles. The contribution of the remaining poles and zeros may vary, depending on the case. However, there are two common features caused by the additional poles:

1) Additional poles tend to introduce delay in the system response.

2) Complex poles which are close to the imaginary axis (e.g., poles due to structural resonance) will produce some "ringing" that rides on the response, and may have a long settling time before it decays. Fig. 4.4.2 shows a curve of the approximate response and a curve of the exact response with "ringing".

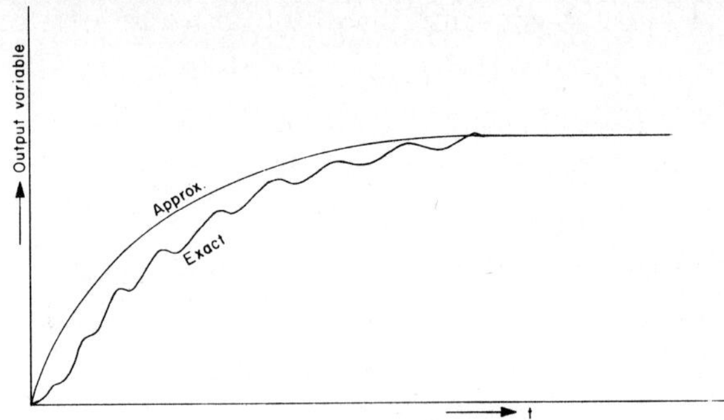

Fig. 4.4.2. Complex poles close to imaginary axis will cause "ringing" which will ride on the response.

Another question concerning the response of a system is its ability to follow input signals of different types. In analyzing a system, we observe that the system's following ability depends on the number of integrators in the open loop, which is defined as the *type of the system.* For this system, the open-loop transfer function has generally the form

$$G(s) = K \frac{(s-z_1)(s-z_2)(s-z_3)....(s-z_m)}{s^N(s-p_1)(s-p_2)(s-p_3)...(s-p_n)} \tag{4.4.5}$$

The integer **N** indicates the number of integrators in the loop and it determines the type of the system. Thus, a system is called Type 0, Type 1, Type 2. . . if **N= 0, 1, 2 . . .** respectively.

As mentioned above, the type of the system determines its following ability. The higher the type, the better the following. A Type 0 system will follow a step input signal with a constant steady-state error. If the input signal is a ramp, the error in following will increase with the time.

A Type 1 system, on the other hand, follows a step input signal with no steady-state error and follows a ramp input signal with a constant error. Clearly, as the type of a system is increased, its following ability is improved; a Type 2 system can follow either step or ramp input signals with no steady-state error.

4.4.2. SYSTEM BANDWIDTH

If the input signal to the system is sinusoidal with a low frequency, the system can respond and follow the input signal. Beyond a certain point when the input frequency becomes too high, the system cannot follow the command. *The range of frequencies that the system can follow is called the bandwidth of the system.*

For the system with open–loop transfer function **G(s)**, the closed-loop transfer function **F(s)** is given by:

$$F(s) = \frac{G(s)}{1 + G(s)} \tag{4.4.6}$$

Now, if we substitute $s = j\omega$ and plot the absolute value of $F(j\omega)$ in **dB** versus the frequency $f = \omega/2\pi$ in logarithmic scale, we find that $|F(j\omega)|_{dB} \cong 0$ for a certain range of frequencies before dropping down, as shown in Fig. 4.4.3.

The bandwidth (or the break frequency) of a system, f_b, is defined as the frequency where $|F(j\omega)| = 1/\sqrt{2} = 0.707$ or, equivalently, where:

$$|F(j\omega)|_{dB} = -3 \text{ dB} \tag{4.4.7}$$

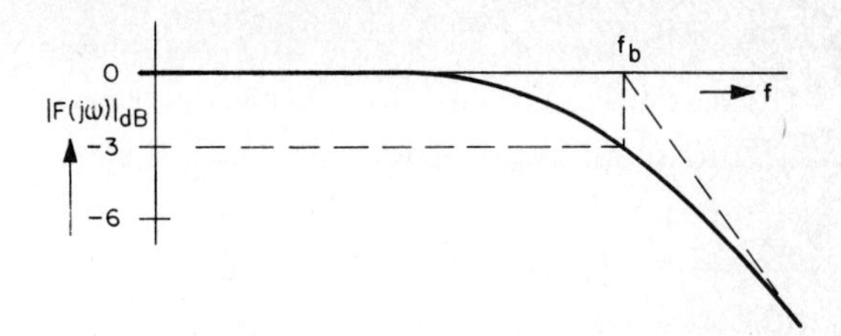

Fig. 4.4.3. Bode plot illustrating the system bandwidth.

The bandwidth of a servo system varies according to the components and the need. In general, velocity control systems using moving coil motors can reach a bandwidth of 600 Hz and more. Position control systems are more limited in general and could reach a bandwidth of about 200 Hz.

4.4.3. EFFECT OF TORSIONAL RESONANCE

In most servo systems, the open loop transfer function includes complex poles. As long as these poles are far enough from the imaginary axis, they do not pose a special problem. However, it becomes more and more difficult to compensate the system for complex poles as they come closer to the imaginary axis. A typical source for those poles is the effect of torsional resonance of the load-motor-tachometer combination (see section 2.3.4), and it is highly desirable to reduce the effect of those poles.

Since it is impossible to eliminate the effect of resonance completely, it is desirable to be able to measure or express quantitatively the effect of resonance, or the sensitivity of the system to sinusoidal inputs at the resonance frequency.

A simple way to describe the resonance effect is by observing the system response to sinusoidal input signals at various frequencies. The ratio of the response amplitude to the amplitude of the input signal indicates the closed-loop gain, or sensitivity of the system at various frequencies.

A typical frequency response is shown in Fig. 4.4.4, where it can be seen that the effect of resonance is represented by the peak at the frequency f_r.

It is desirable that the amplitude of the peak at f_r is kept as low as possible to reduce the effect of resonance. A good indication of the system sensitivity at resonance is the height of this peak, which could be described as the absolute value of system gain or in decibels.

□

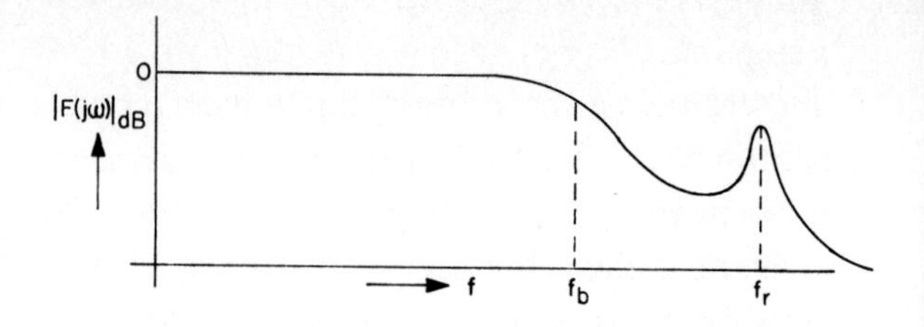

Fig. 4.4.4. Bode plot showing resonance point.

4.5. SERVO AMPLIFIERS

The discussion of amplifiers in Chapter 3 centered around the speed control systems operating in the "first quadrant". We will now describe the true servo amplifiers, capable of providing positive and negative output voltages or currents, and thus being able to operate in four quadrants.

Again, we can classify the amplifiers into three categories: the linear transistor amplifiers, the SCR amplifiers, and the switching amplifiers. The following paragraphs describe them in detail.

4.5.1. LINEAR AMPLIFIERS

Transistor servo amplifiers are characterized by two principal output stage designs, the "H" and "T" type basic configurations, as shown in Figs. 4.5.1 and 4.5.2. The "H" or bridge output stage consists of four transistors using a single DC power supply. This output stage has the advantage of a simple, unipolar power supply and some sharing of voltage protection between the transistors. However, it is not easy to drive in linear cases, and current and voltage feedback is not easy to achieve, since the motor is "floating".

The "T" stage needs two power supplies and complementary transistors, but is easy to drive and can supply voltage and current feedback signals in a simple fashion. For this reason the "T" is used most often in linear amplifiers. The biasing of the output transistors requires careful attention,

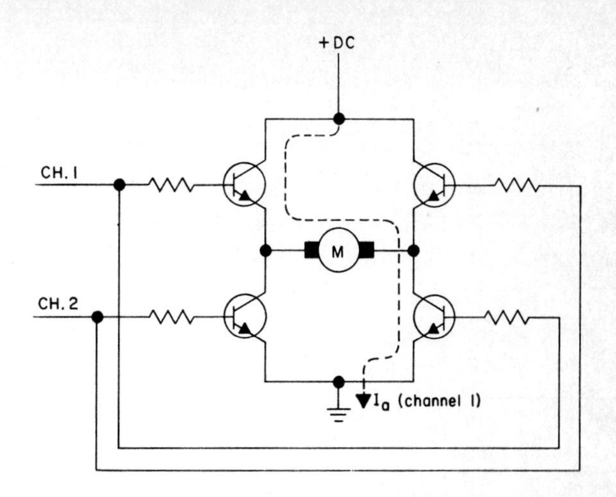

Fig. 4.5.1. "H" type output stage (one power supply required).

since a simultaneous conduction of both transistors would result in a short circuit between the two power supplies.

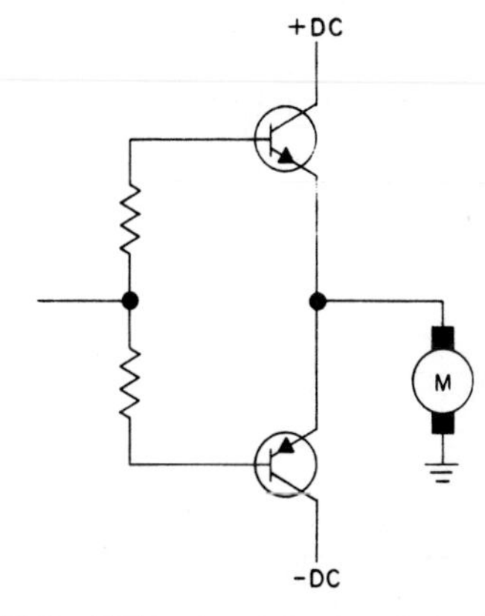

Fig. 4.5.2. "T" type output stage (two power supplies required).

The input-output characteristics of the output stage will generally have some dead zone - an undesirable feature from linear feedback point of view. Negative feedback around the amplifier will generally reduce such nonlinearity to a negligible amount. In cases where the linear amplifier will be used in a "bang bang" mode, the dead zone is of no primary consequence, since the amplifier will be used only in its conducting or nonconducting state.

The primary limitations on the output power handling capacity of the linear amplifier are the thermal characteristics of its output state. Since power dissipation of the output stage is the product of current and voltage across the transistors, the transistor and its associated heat sink assembly must be capable of dissipating this heat. In addition, the designer has to consider the secondary breakdown characteristics of the transistor, which will limit peak current time duration at given voltage levels, in order to ensure that such limits are not exceeded. Since amplifiers discussed in this book are designed to drive servomotors, which have a small electrical impedance, it is easy to envision situations in which an amplifier could be easily overloaded under conditions of stall or excessive torque. For this reason, current limiting circuits are generally found useful in servo amplifiers to prevent amplifier breakdown or blown fuses.

In some instances where high peak acceleration currents are needed with protection for prolonged current overload conditions, two-stage current limiters are used where the first current limit may be set at a high level with a time limit of a fraction of a second, after which a lower sustaining current limit will be brought into action. Such a circuit will not only protect the amplifier output stage, but also may prevent the motor from accidental destructive overload.

4.5.2. SCR AMPLIFIERS

In the last decade, the bidirectional SCR amplifier has found increasing use for servo circuits handling bandwidths up to 30 Hz. To achieve these bandwidths, a two-phase or three-phase power supply is generally used. Such systems are used to drive from one to five horsepower servo motors for the machine tool industry.

Just as in the case of the single quadrant amplifiers described in Chapter 3, close attention must be given to the form factor of the amplifier and its influence on the servo motor heating.

4.5.3. SWITCHING AMPLIFIERS

Introduction

A switching amplifier is perhaps the most versatile and popular servo amplifier today. With power transistors available which can switch in the MHz region, it is now easy to design power switching amplifiers capable of handling square wave switching rates of 50 kHz. These high switching rates are obtainable through the use of positive current feedback techniques; this enables the servo designer to achieve bandwidths on the order of several kHz.

Power dissipation in such an amplifier is caused by two elements: the output transistor forward voltage drop, which may be on the order of 1 or 2 V; and the finite transition time from one polarity to the other. The latter effect is especially prominent in amplifiers with high switching rates, where the transition time may be a significant portion of total switching time. Therefore, while a high switching rate may seem desirable from the point of view of optimum bandwidth, the switching rate may be tempered by the practical effect of output stage power dissipation.

In general, both the frequency and the duty cycle of a switching amplifier will vary with the load. While the variation in duty cycle is required, the frequency change is undesirable since it may excite resonance modes in the system. Another disadvantage of frequency variation is that it produces an audible noise, which in most cases is undesired.

In order to prevent frequency variations the amplifier can be designed to have a constant switching frequency. Such a switching amplifier is called "pulse-width-modulated" (PWM) amplifier, since only the width of the pulses will vary with the load.

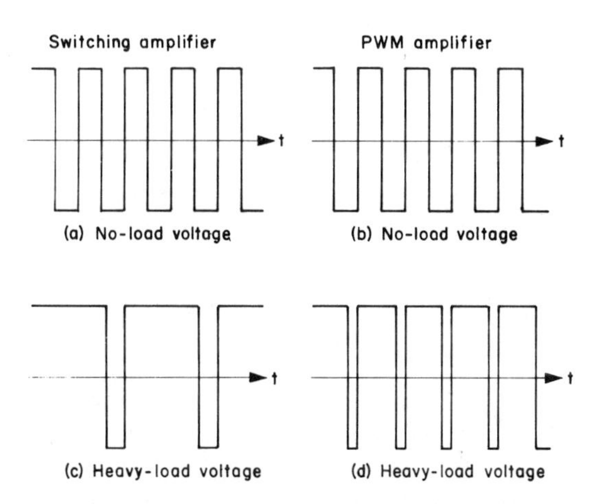

Fig. 4.5.3. Voltage of switching and PWM amplifiers. Both frequency and duty cycle of switching amplifiers vary with load (a) and (c), whereas, only the duty cycle varies in a PWM amplifier (b) and (d).

Figure 4.5.3 illustrates the difference between a general switching amplifier and a PWM amplifier. It can be seen that as the load is increased, the frequency of the switching amplifier decreases, while the PWM amplifier retains a fixed switching frequency.

A switching amplifier may be of an "H" or "T" type, as is the case for linear amplifiers. The main advantage of the "H" type is that only an unipolar power supply is needed. The main disadvantage is that the motor is floating and it is more difficult to construct a current feedback circuit. It appears that the advantages outweigh the disadvantages, and the "H" type is the more common form, especially when high voltage is required.

In the following sections we discuss the structure of a switching amplifier and analyze its performance.

Block Diagram and System Components

The block diagram of a switching amplifier and a motor is shown in Fig. 4.5.4. The system includes the following components:

1. *Comparator* — high–gain amplifier with positive feedback. This results in a hysteresis as shown in Fig. 4.5.5, so that the comparator output voltage will be either V_{max} or $-V_{max}$.

2. *Driver* – power amplifier whose output is equal to that of the supply, V_S, when the comparator voltage is positive, and $-V_S$ when the comparator voltage is negative. Therefore, it can be represented as a constant gain block.

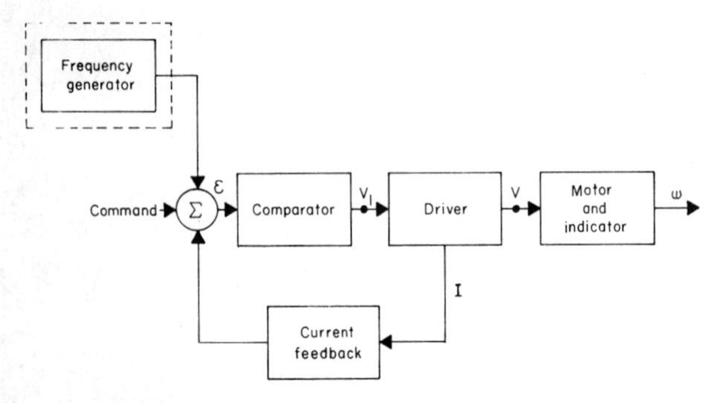

Fig. 4.5.4. Block diagram of a switching amplifier. The frequency generator is used only in PWM amplifiers.

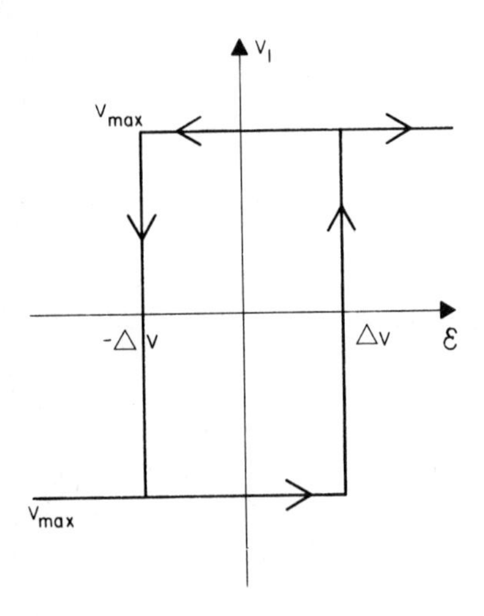

Fig. 4.5.5. Hysteresis effect in a high-gain amplifier.

In order to prevent a short circuit across the power supply, the switching of the amplifier voltage is done in two steps. First, the conducting transistors are turned off, and after a time delay, δ, the other pair is turned on. This is illustrated in Fig. 4.5.6. Since the interval δ is much shorter than the switching period, it will be assumed in the following analysis that $\delta = 0$.

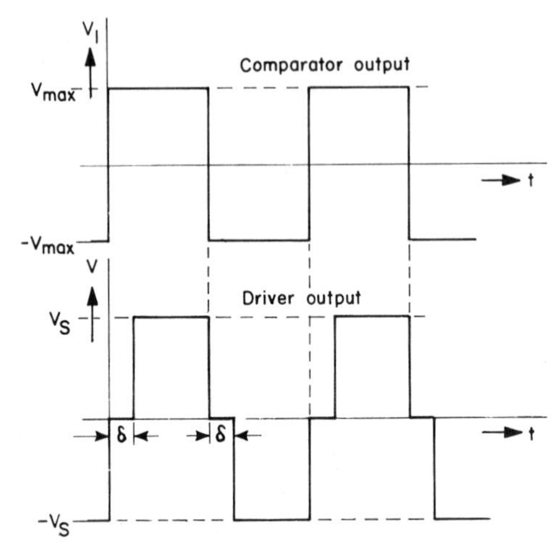

Fig. 4.5.6. The comparator and driver outputs. Note that the comparator switching is instant, whereas, the driver has an intermediate "off" interval, δ.

3. *Current feedback* – in a "T" type amplifier, the current is sensed on a series resistor and is fed back. When the amplifier is of "H" type, the currents are sensed on the two legs of the H, and their difference is fed back. The current feedback network may have a constant gain or dynamic components according to the required characteristics.

4. *Motor and inductor* – the inductor in series with a motor is used to limit the current variation over a

switching cycle. In some cases, the motor inductance may be sufficient and no additional inductor is needed. Since the additional inductor is in series with the motor, one can consider it as the motor inductance for the purpose of analysis.

5. *Frequency generator* – a frequency generator is used in PWM amplifiers to produce a constant switching frequency. The most simple generator signal is a triangular signal of fixed frequency, fed into the comparator.

System Transfer Function

The next step in understanding the operation of the switching amplifier is to find a mathematical model for it – preferably in the form of a transfer function.

The system components may be represented by their transfer functions, as shown in Fig. 4.5.7. The comparator and the driver are nonlinear elements and their output depends on the input signal ϵ. It is assumed here that $N(\epsilon)$ is a describing function which represents their behavior.

The motor impedance is considered to be a resistance R and an inductance L_t. The inductance L_t includes the internal and the external parts.

The transfer function of the system is given by

$$\frac{\omega(s)}{V_{in}(s)} = \frac{N(\epsilon) \cdot \frac{1}{sL_t+R} \cdot \frac{K_T}{sJ}}{1+N(\epsilon) \cdot \frac{1}{sL_t+R} \cdot a + \frac{1}{sL_t+R} \cdot \frac{K_T}{sJ} \cdot K_E} \tag{4.5.1}$$

or

$$\frac{\omega(s)}{V_{in}(s)} = \frac{N(\epsilon)\,K_T}{(sL_t+R)\,sJ+N\,(\epsilon)a sJ+K_E K_T} \tag{4.5.2}$$

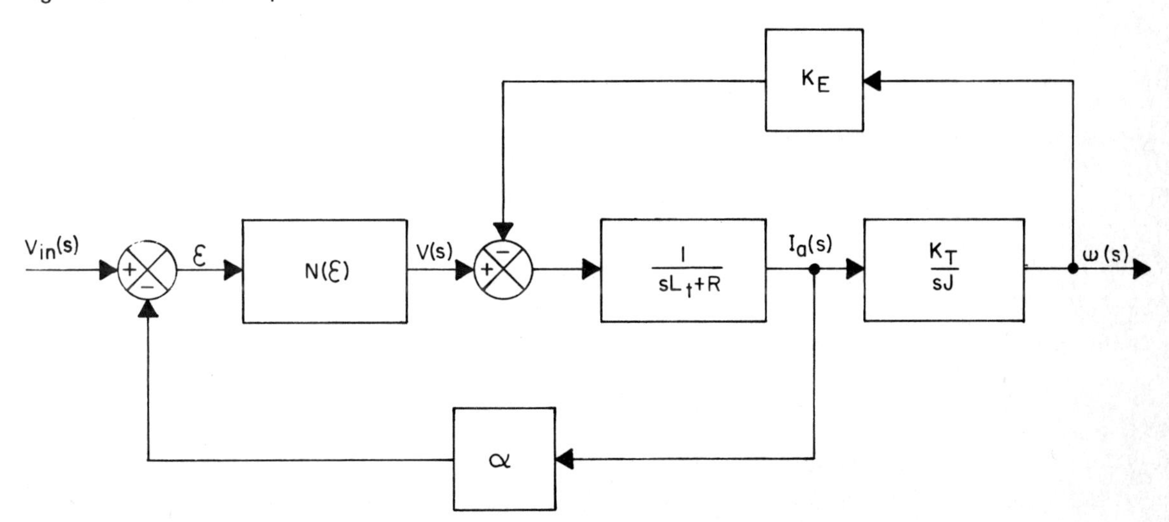

Fig. 4.5.7. Block diagram of a switching amplifier and a motor.

In most cases, the amplifier gain is very high and $N(\epsilon)$ is very large. Under those conditions, the transfer function may be approximated by

$$\lim_{N(\epsilon)\to\infty} \frac{\omega}{V_{in}}(s) = \frac{K_T}{a\, sJ} \qquad (4.5.3)$$

The simplified transfer function is represented by a block diagram in Fig. 4.5.8. The interpretation of this simplified model is that due to high forward gain in the switching amplifier, its response is determined by the feedback. In this case, the amplifier becomes a current amplifier, and for such a source, the transfer function of the motor is simplified considerably [see (4.2.8)].

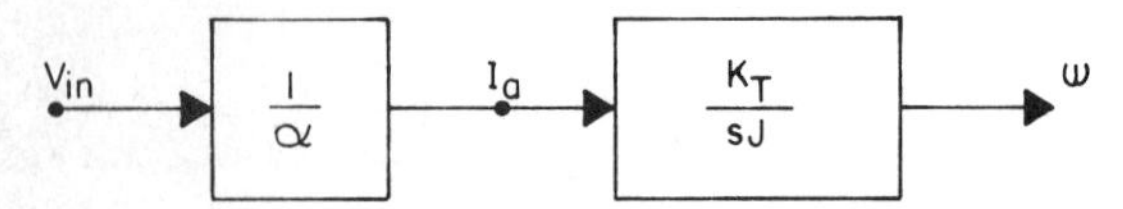

Fig. 4.5.8. Block diagram of simplified transfer function model for switching amplifier and motor.

It can be seen how one can alter this transfer function by changing the current feedback. If a dynamic element with a transfer function $a(s)$ is used, the transfer function becomes

$$\frac{\omega(s)}{V_{in}(s)} = \frac{1}{a(s)} \frac{K_T}{sJ} \qquad (4.5.4)$$

Thus, all the poles of $a(s)$ become zeros of the transfer function and vice versa.

The model of (4.5.4) may be used to analyze the behavior of the amplifier and the motor as parts of a larger system. It is useful in analyzing system stability, bandwidth, and other general characteristics. However, it cannot describe the unique characteristics of a switching amplifier, and for those we have a special discussion below.

Switching Frequency and Current Variations

The switching frequency of the amplifier can be determined as follows.

Suppose that the switching frequency is high, so that the motor velocity is nearly constant over a switching period. Let the average motor current be I_{av}, and define the average motor voltage as

$$V_{av} = RI_{av} + K_E\omega \qquad (4.5.5)$$

Now suppose that the amplifier voltage is $\pm V_S$ and that the current variation may be approximated by straight lines. The amplifier voltage and the motor current are shown in Fig. 4.5.9. The total current variation, ΔI, is such that when it is multiplied by a, it produces a voltage change of $2\Delta V$ to switch the comparator [see Fig. (4.5.5)].

Therefore,

$$\Delta I = \frac{2\Delta V}{a} \qquad (4.5.6)$$

During the positive interval, t_1, the amplifier voltage is V_S and the average motor voltage is V_{av}. Consequently, the current slope is $\dfrac{V_S - V_{av}}{L_t}$, and we can write the

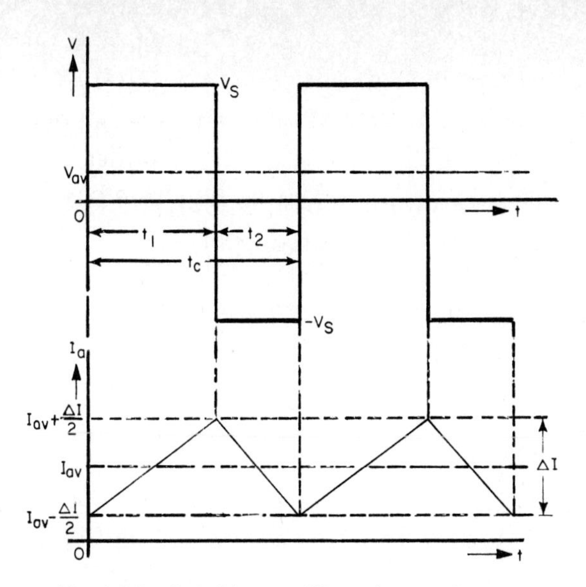

Fig. 4.5.9. Switching amplifier voltage and motor current.

expression

$$t_1 \frac{V_S - V_{av}}{L_t} = \Delta I \quad (4.5.7)$$

or

$$t_1 = \frac{\Delta I\, L_t}{V_S - V_{av}} \quad (4.5.8)$$

Similarly, the negative voltage interval, t_2, is

$$t_2 = \frac{\Delta I\, L_t}{V_S + V_{av}} \quad (4.5.9)$$

The total switching period, t_c, is given by:

$$t_c = t_1 + t_2 = \frac{2\Delta I\; L_t\, V_S}{V_S^2 - V_{av}^2} \quad (4.5.10)$$

or, substituting (4.5.6)

$$t_c = \frac{4\,\Delta V\, L_t\, V_S}{a\left(V_S^2 - V_{av}^2\right)} \quad (4.5.11)$$

and the switching frequency is

$$f_s = \frac{a\left(V_S^2 - V_{av}^2\right)}{4\,\Delta V\, L_t\, V_S} \quad (4.5.12)$$

Note that the switching frequency is a parabolic function of the average motor voltage, V_{av}. It attains its maximum when $V_{av} = 0$

$$f_{s_{max}} = \frac{a\, V_S}{4\,\Delta V\, L_t} \quad (4.5.13)$$

The frequency variation is illustrated by Fig. 4.5.10.

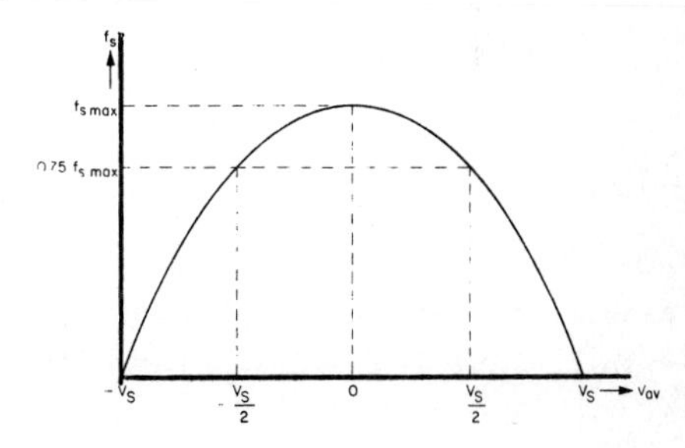

Fig. 4.5.10. Parabolic variation of switching frequency versus average motor voltage.

Power Dissipation in the Motor

A switching amplifier reduces considerably the dissipation in the amplifier by switching the transistors either completely "on" or completely "off". However, how this affects the power dissipation in the motor is not clear, and it is discussed below.

Consider a given motor and load which have to be driven following some velocity profile. Suppose that the motor is driven by both a linear amplifier and a switching amplifier, and consider the difference in power dissipation between the two cases. The output power of the motor, P_o, is independent of the amplifier. Therefore, any difference in power dissipation is equal to the difference in input power.

Let the motor current and voltage, when driven by the linear amplifier, be I_a and V, respectively. Then the input power in that case is

$$(P_i)_{linear} = I_a V \qquad (4.5.14)$$

To determine the input power from a switching amplifier, the current variations are approximated by straight lines. The current and voltage are assumed to be as in Fig. 4.5.9, with the average values

$$\left.\begin{aligned} I_{av} &= I_a \\ V_{av} &= V \end{aligned}\right\} \qquad (4.5.15)$$

The input power in this case equals

$$P_i = f_s \int_0^{t_1+t_2} V_{out}(t)\ I(t)\,dt \qquad (4.5.16)$$

where f_s is the switching frequency, as given by (4.5.12). This can be written as

$$P_i = f_s \left[V_S \int_0^{t_1} I(t)\,dt - V_S \int_{t_1}^{t_1+t_2} I(t)dt \right]$$

$$= f_s \left[V_S I_a t_1 - V_S I_a t_2 \right] \qquad (4.5.17)$$

Now substitute (4.5.8), (4.5.9), and (4.5.12) into (4.5.17). This results in

$$P_i = \frac{I_a\, a\, \Delta I\, V}{2 \Delta V} \qquad (4.5.18)$$

This can be further simplified by substituting (4.5.6) for ΔI:

$$\left(P_i\right)_{switching} = I_a V \qquad (4.5.19)$$

When this is compared to (4.5.14), it is seen that the input power is the same in both cases, and so is the motor power dissipation.

Note, however, that this result is based on the assumption that the current variations are linear in time. Since this is not exact, there may be some variations in power dissipation, but they are small and insignificant.

Inductance Selection

The total inductance of the motor and the external inductor determines the magnitude of the current variations. The relation between these variables can be found from (4.5.8) and (4.5.9) under symmetric operation of $V_{av} = 0$.

Under that condition, the switching intervals are

$$t_1 = t_2 = \frac{1}{2f_s} \qquad (4.5.20)$$

and the inductance can be found from

$$L_t = \frac{V_S}{2f_s \Delta I} \qquad (4.5.21)$$

Since L_t is usually larger than the motor inductance L_a, an additional inductance of $(L_t - L_a)$ has to be added in series with the motor. □

4.6. PHASE-LOCKED SERVO SYSTEMS

4.6.1. INTRODUCTION

As the needs for precise velocity control systems increase, and with the availability of suitable integrated circuits, *phase-locked servo loops* are becoming increasingly popular because of their excellent speed regulation and the inherent insensitivity to parameter changes.

The basic phase-locked loop and its components are shown in Fig. 4.6.1. When the system is "phase-locked", the frequencies of the command and feedback signals become identical, and if the frequency of the command signal is constant, it produces a constant motor velocity.

As long as the system is "phase-locked", the motor velocity follows the frequency of the command independently of motor and amplifier parameter variations. This guarantees a motor speed regulation which is as good as the stability of the command frequency.

It should be noted that although phase-locked servo systems are used to control the motor velocity, they really are *position control systems* where the continuous position control results in a velocity control.

4.6.2. SYSTEM COMPONENTS

The phase-locked servo loop includes the following components:

1. Phase comparator
2. Low-pass filter
3. Amplifier
4. Motor
5. Encoder (optical tachometer)

The operation of these components is discussed below.

Phase Comparator

The purpose of the phase comparator is to detect differences between the phases of the command signal and the feedback signal, and produce an error signal to indicate them.

Some comparators are designed to detect frequency differences as well as phase differences between the command and feedback signals, and produce an output voltage which is a function of the phase error and the frequency error. Such detectors are called phase-frequency comparators.

In the following analysis, we derive the mathematical equations for the most simple phase comparator, the multiplier. For the simplicity of derivation, it is assumed that both the command signal and the feedback signal are sinusoidal. This is only a simplifying assumption and the results are valid for square-wave signals as well. Let the command signal and the feedback signal be $S_1(t)$ and $S_2(t)$, respectively, and denote the multiplier gain by K_m. The multiplier output is

$$V_c = K_m S_1 S_2 \qquad (4.6.1)$$

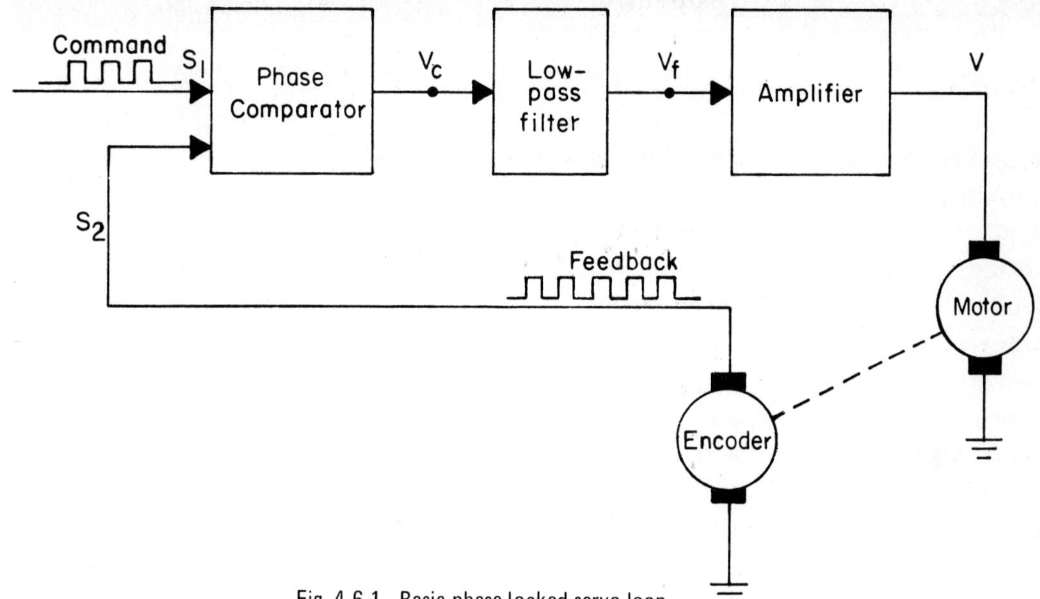

Fig. 4.6.1. Basic phase-locked servo loop.

Since the command signal is sinusoidal, it can be written as

$$S_1 = -V_s \cos \theta_i \tag{4.6.2}$$

Let θ_m be the angular position of the motor shaft and define the electric angle of the encoder, θ_o, as

$$\theta_o = n\theta_m \tag{4.6.3}$$

where **n** is the number of pulses generated by the encoder each revolution. The feedback signal, S_2, may be written as

$$S_2 = V_o \sin \theta_o \tag{4.6.4}$$

The comparator output voltage, V_c, becomes

$$V_c = -K_m V_s V_o \cos \theta_i \sin \theta_o \tag{4.6.5}$$

This can be written as

$$V_c = \frac{K_m V_s V_o}{2} \left[\sin (\theta_i - \theta_o) - \sin (\theta_i + \theta_o) \right] \tag{4.6.6}$$

The voltage component proportional to $\sin (\theta_i + \theta_o)$ is sinusoidal with high frequency and will be blocked by the low-pass filter. Therefore, we may ignore it and consider only the voltage component proportional to $\sin (\theta_i - \theta_o)$. Also define the comparator gain, K_p, as

$$K_p = \frac{K_m V_s V_o}{2} \tag{4.6.7}$$

Then the effective comparator output voltage is

$$V_c = K_p \sin (\theta_i - \theta_o) \tag{4.6.8}$$

Define the phase error, θ_e, as

$$\theta_e = \theta_i - \theta_o \tag{4.6.9}$$

and note that when θ_e is small, (4.6.8) may be approximated by

$$V_c = K_p (\theta_i - \theta_o) = K_p\theta_e \tag{4.6.10}$$

Equation 4.6.10 indicates that for small phase errors, the comparator output is proportional to the phase difference. This is illustrated in Fig. 4.6.2.

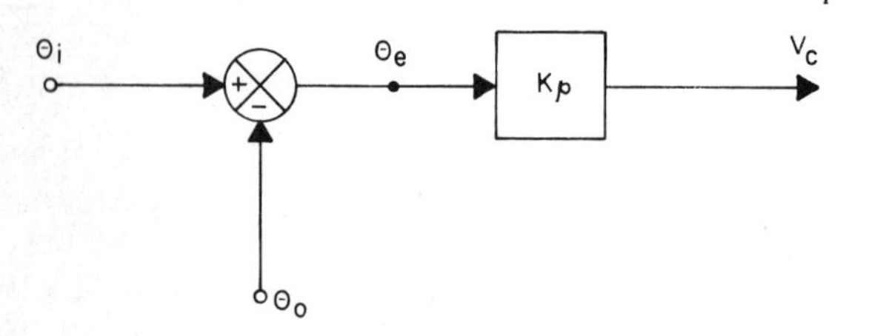

Fig. 4.6.2. Linearized model for phase comparator.

The model (4.6.10) was developed for a multiplier, but it is valid for all the phase comparators, as long as the phase error, θ_e, is small. In cases where phase-frequency comparators are used the output voltage may be approximated as the sum of two components, the phase error voltage and the frequency error voltage:

$$V_c = K_p\theta_e + K_v \frac{d\theta_e}{dt} \tag{4.6.11}$$

When Laplace transformation is used, this becomes

$$V_c(s) = \left[sK_v + K_p\right] \theta_e(s) \tag{4.6.12}$$

A block diagram model for the phase-frequency comparator is illustrated in Fig. 4.6.3.

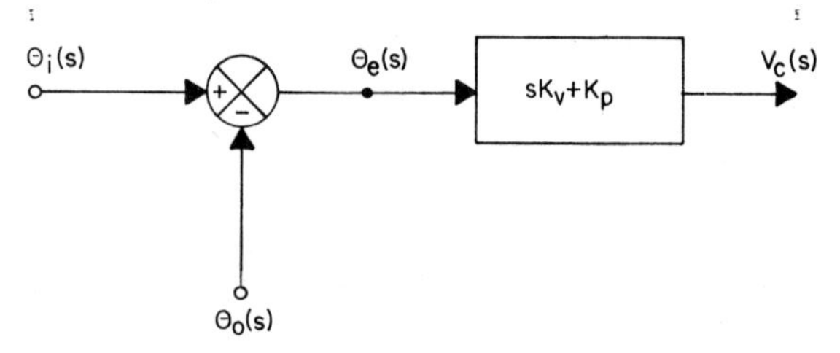

Fig. 4.6.3. Linearized model for phase-frequency comparator. Note that K_p gives the phase difference, whereas K_v indicates frequency difference.

Low-Pass Filter

The low-pass filter is needed to block the high frequency components of the comparator output. A single-pole filter may be sufficient for attenuation, but the final selection is made on the basis of loop design. The transfer function of the filter is denoted by **F(s)**.

Amplifier

The amplifier is needed to raise the power level of the error signal in order to drive the motor. The amplifier transfer function, **H(s)**, may be selected to provide additional compensation if needed.

Motor

The transfer function of the motor was discussed in detail in section 4.2. In the following analysis, it is denoted by **G(s)**.

Encoder

The encoder, or optical tachometer, is a nonlinear device which generates n sinusoidal pulses for each revolution of the motor shaft. Thus the encoder output, $S_2(t)$, is

$$S_2(t) = V_o \sin n\theta_m = V_o \sin \theta_o \quad (4.6.13)$$

However, since the comparator model is linearized, the encoder may be represented as a block with constant gain of **n** (Fig. 4.6.4).

Now we are ready to construct a detailed block diagram for the phase-locked servo system. This is shown in Fig. 4.6.5.

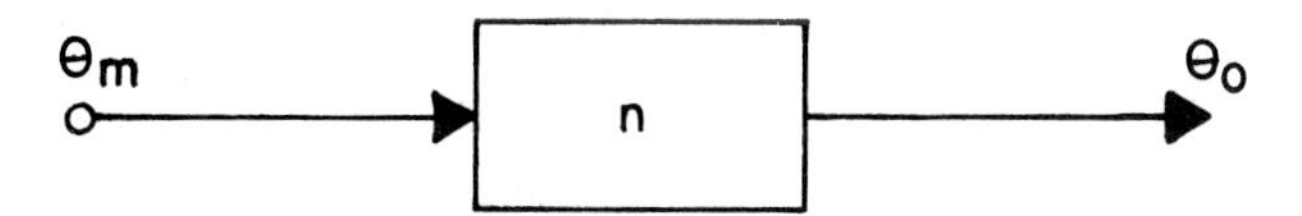

Fig. 4.6.4. Block diagram model for the encoder, relating the mechanical shaft position θ_m to the encoder electrical angle θ_o.

4.6.3. SYSTEM DESIGN AND STABILITY

When we examine the open–loop transfer function of the system in Fig. 4.6.5, we find it to be

$$L(s) = n(sK_v+K_p)\ F(s)\ G(s)\ H(s)\ \frac{1}{s} \quad (4.6.14)$$

It can be seen that the term **1/s** introduces 90° phase shift to the loop. Additional lag is produced by motor and the filter. If we couple this with the fact that the open–loop gain is very high (**n** is between 500 and 2000), it is apparent that the system will have stability problems, unless carefully designed.

The design is made more difficult by the fact that the filter **F(s)** is needed, and any lead network on the comparator will cancel the filtering effect. Consequently, all efforts are directed at determining compensations which will reduce the lag and loop phase shift at frequencies about the break frequency. This can be achieved by lead-lag, or similar type compensations.

Next, we want to examine the effect of a lead-lag compensation network on the loop stability.

Consider a phase-locked loop where the phase detector has a gain K_p. The low-pass filter has a single pole at −1000 rad/s, and a transfer function

$$F(s) = \frac{1000}{s + 1000} \quad (4.6.15)$$

The amplifier is assumed to have a constant gain

$$H(s) = 5 \quad (4.6.16)$$

The motor transfer function is

$$G(s) = \frac{1000}{s + 100} \quad (4.6.17)$$

and the encoder line density is

$$n = 500 \quad (4.6.18)$$

The corresponding open-loop transfer function is

$$L(s) = \frac{2.5 \times 10^9\ K_p}{s(s + 100)(s + 1000)} \quad (4.6.19)$$

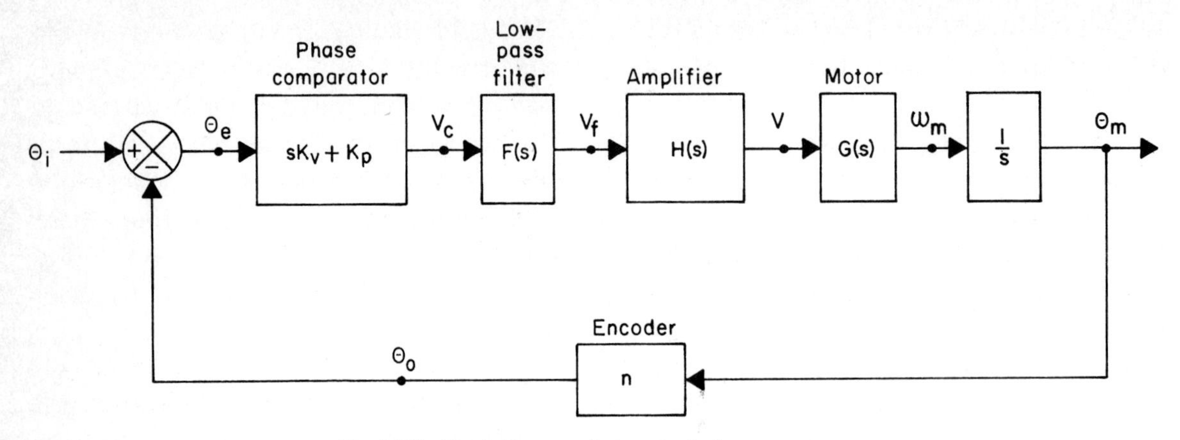

Fig. 4.6.5. Block diagram of phase-locked servo system.

and the corresponding root-locus diagram is shown in Fig. 4.6.6.

In order to determine the limit on the comparator gain, solve the equation

$$1 + L(j\omega) = 0 \tag{4.6.20}$$

or

$$1 + \frac{2.5 \times 10^9 \ K_p}{j\omega(j\omega + 100)(j\omega + 1000)} = 0 \tag{4.6.21}$$

$$j\omega(j\omega+100)(j\omega+1000) + 2.5\text{x}10^9 \ K_p = 0 \tag{4.6.22}$$

Separating the real and imaginary parts of (4.6.22) and solving for ω^2 and K_p yields

$$\left.\begin{aligned} \omega^2 &= 10^5 \\ K_p &= 0.044 \end{aligned}\right\} \tag{4.6.23}$$

Thus, the comparator gain has to be limited below 0.044 V/rad to guarantee stability. Next, suppose that we add a lead-lag compensation to the amplifier by introducing a zero at -100 rad/s and a pole at -5000 rad/s. The modified amplifier transfer function becomes

$$H_m(s) = \frac{250\ (s + 100)}{(s + 5000)} \tag{4.6.24}$$

Note that at low frequencies, the amplifier gain remains unchanged.

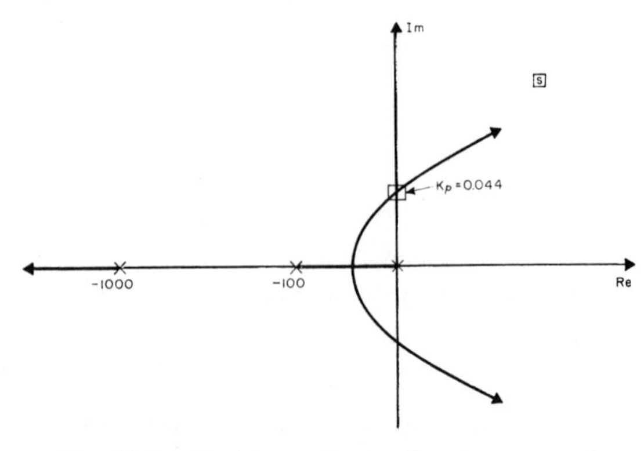

Fig. 4.6.6. Root locus diagram for the system of (4.6.19).

The modified open-loop transfer function becomes

$$L_m(s) = \frac{1.25 \times 10^{11} K_p}{s(s+1000)(s+5000)} \qquad (4.6.25)$$

The maximum gain for the modified system can be determined by the same procedure as before. When this is done, we find that the maximum comparator gain and the corresponding frequency are

$$\left.\begin{aligned} \omega^2 &= 5 \times 10^6 \\ K_p &= 0.24 \text{ V/rad} \end{aligned}\right\} \qquad (4.6.26)$$

Thus, by introducing the compensation, one can increase the gain by a factor of 5.5.

4.6.4. SPECIAL CHARACTERISTICS

In the previous analysis the model of the phase-locked servo system was linearized and the loop was treated as a regular position control system. While this analysis is valid, there are some special characteristics that result from the nonlinear nature of phase-locked loops, and these are described below.

Velocity Acquisition and Locking Range

In order for the system to remain locked, it has to satisfy the following conditions:

a) The system must be stable.
b) The frequency of the effective error signal, V_c, should be within the bandwidth of the system.

As long as the two conditions are satisfied, the system can correct phase errors and remain locked. On the other hand, if the frequency of the error signal is higher than the system bandwidth, the loop cannot respond fast enough and the system will "lose lock".

The frequency of the error signal is proportional to the difference between the motor velocity and the desired velocity. Large differences between the two may occur during start-up condition when the motor starts from rest, or due to large speed disturbances.

To illustrate this point consider a stable phase-locked servo system where the encoder line density is $n = 1000$. Let the frequency of the input signal be $f_i = 20$ kHz, and suppose that the system bandwidth is 500 Hz. The system will remain locked as long as the frequency of the feedback signal is between 19 500 and 20 500 Hz. With the given encoder, this corresponds to a motor velocity in the range of 19.5–20.5 r/s, or 1170–1230 rpm.

In other words, as long as the motor velocity is within the range of 1200 ± 30 rpm, locking is guaranteed. Any change in velocity beyond those limits will result in loss of lock.

In order to guarantee velocity acquisition and prevent loss of lock, one can use a phase-frequency comparator, where the velocity error signal may be used to accelerate the motor until lock is achieved. This enables the system to acquire the desired velocity; however, the acquisition range is limited, as explained by the following example.

A motor with an encoder of line density $n = 1000$ is required to rotate at a velocity of 2400 rpm in CW direction. A stable phase-locked system is used where a positive error signal accelerates the motor in the CW direction. The frequency of the input signal is $f_i = 40\,000$ Hz, and the feedback frequency, f_o, is proportional to the velocity. Furthermore, note that the velocity error of the frequency comparator is proportional to $(f_i - f_o)$. Thus,

$$V_c = K_c (f_i - f_o) \tag{4.6.27}$$

Next, consider various initial velocities, and the system response.

a) Initial velocity above 2400 rpm in CW direction. Here the error signal is negative and the motor will decelerate until the required velocity is achieved.

b) Initial velocity below 2400 rpm in either direction. Here the error signal is positive and the motor is accelerated in CW direction to the required velocity.

c) Initial velocity above 2400 rpm in CCW direction. Now the error signal is negative, which accelerates the motor further in CCW direction until the amplifier is saturated. Under this condition, velocity acquisition is not possible.

In conclusion, phase-frequency comparator will accelerate the motor to a desired velocity, ω_d, if the initial velocity, ω_m, is

$$\omega_m > -\omega_d \tag{4.6.28}$$

Otherwise, the motor will run away in the opposite direction.

There are several methods to solve this problem. The most simple one is to use a unidirectional amplifier when possible.

Speed Variations

Phase-locked systems are used primarily for velocity control because of the high accuracy and regulation they can provide. It is therefore important to investigate the sources of the velocity errors and the methods for reducing them. The most common sources of velocity disturbances are listed below.

1. *Variations in input frequency* – As the frequency of the command signal varies with time or temperature, the system will follow the command frequency and produce a velocity error. Clearly, the speed regulation of the system cannot be better than that of the command signal and a stable oscillator is needed for good regulation.

2. *Load variations* — Variation in load torque, T_L, will produce speed disturbances. These variations will be minimized by increasing the loop gain and bandwidth. An alternative method for reducing velocity disturbances is to add a balanced inertial load to the system which attenuates the effect of the load changes.

3. *Motor torque disturbances* — The torque produced in the motor during rotation is not constant, but it varies with the angular position. The torque ripple is equivalent to load disturbance and results in velocity variations. In order to minimize those, one can select a motor with low torque ripple, in addition to increasing the loop gain or using an additional inertial load.

4. *Encoder-generated errors* — In the previous analysis it was assumed that the encoder is an ideal device with an output

$$S_2 = V_o \sin \theta_o \qquad (4.6.29)$$

In reality, the encoder has some error and its output may be written as

$$S_{2a} = V_o \sin \theta_a \qquad (4.6.30)$$

where θ_a, the actual electrical angle, is given by

$$\theta_a = \theta_o + \theta_L + \theta_h \qquad (4.6.31)$$

θ_L represents the low-frequency phase errors which result from mounting errors. This can be approximated by

$$\theta_L = K \sin \theta_m \qquad (4.6.32)$$

θ_h, on the other hand, results from errors in encoder mask, and therefore will have high-frequency elements.

When the actual feedback signal, S_{2a}, is fed back to the phase comparator, the output voltage becomes

$$\begin{aligned} V_c &= K_p [\theta_i - \theta_a] \\ &= K_p [\theta_i - \theta_o - \theta_L - \theta_h] \end{aligned} \qquad (4.6.33)$$

θ_h will be blocked by the filter and hence may be ignored. θ_L, which may be within the system bandwidth, will pass through the filter as an error, and will try to correct the system phase accordingly. Thus, the phase-locked loop will attempt to correct errors which are produced in the encoder, and when the frequency of those errors is low, they result in speed disturbances. □

4.7. HOW TO MAKE SYSTEMS WORK

Some system requirements may be contradictory to the basic requirement of stability, and compensation must be added. There is no simple rule for compensation, but some worthwhile ideas will be discussed.

The designer of a servo system usually desires high gain in the open-loop transfer function, as this reduces the sensitivity of the system to disturbances and parameter variations, and widens the system bandwidth. For example, a position control system with high open-loop gain will appear as a "stiff" system that can follow a position command despite disturbing torques.

On the other hand, as gain is increased, the system tends to become unstable. In fact, all practical servo systems become unstable if their open-loop gain is increased beyond a certain limit. This indicates that the plurality of the number of poles over the number of zeros of the characteristic equation is at least three. Then, as gain is increased, at least one of the poles moves to the right half-plane, as follows from the asymptotes rule of the root loci.

Thus, the requirements for high gain and wide bandwidth are contradictory to the basic requirement for stability. To overcome this difficulty, compensation networks are used to change the open-loop transfer function so that higher gain can be achieved under stable conditions.

There is no simple selection rule for compensation networks. Instead, a network has to be tailored for a given case after careful analysis. There are, however, some general ideas which are worth discussing here.

In general, servo system stability is enhanced if the phase shift, or the lag, in the open-loop transfer function is reduced. Thus, phase shift should be reduced whenever possible. The first major source of phase shift is the motor; it is possible to reduce system phase shift *by selecting a motor with a short mechanical time constant.* This possibility should be considered, however, only if gain or bandwidth requirements are extremely high.

The other major source of lag is the amplifier. It is easy to stabilize an amplifier by using a large capacitor in the feedback, but such an amplifier will cause stability problems when it is used in a closed loop. Therefore, as a general rule, *capacitive feedback in the amplifier should be avoided as much as possible.* Actually, an ideal amplifier should have a constant gain and no phase shift over a wide bandwidth. Electro-Craft's Servo Control Amplifiers (SCA) can maintain a constant gain up to a frequency of 20 kHz, so for the purpose of analysis they can be considered as constant gain devices.

Thus, in general, by reducing the phase shift, a system can stand higher gain without disturbing its stability. In cases where further reduction in phase shift is needed, a lead-lag compensation network can be used with the transfer function:

$$H(s) = K\frac{s-a}{s-b} \quad (4.7.1)$$

where both **a** and **b** are positive and:

$$b > a \quad (4.7.2)$$

Such a network has a zero at the frequency ω = **a** and a pole at ω = **b**. This combination results in some phase lead, whose amplitude depends on the parameters **a** and **b**. Such a compensation network can be constructed using either active or passive components. The two possible circuits are shown in Figs. 4.7.1 and 4.7.2. The two transfer functions are of the lead-lag type, and the circuit components are selected to obtain the right values of **a** and **b**.

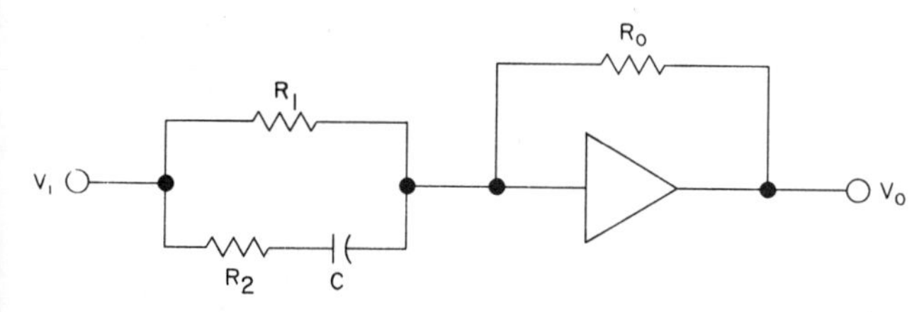

Fig. 4.7.1. Compensation network using an active component (amplifier).

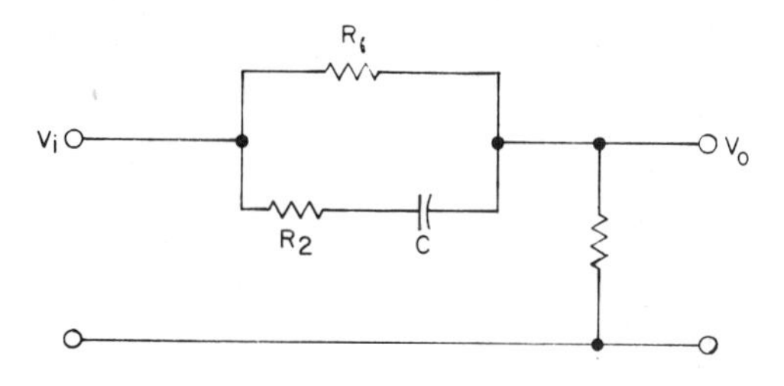

Fig. 4.7.2. Compensation network using passive components.

In some cases, the desired velocity profile is to have a constant acceleration and deceleration. While this is extremely difficult to achieve with linear compensation networks, it can be obtained easily using some nonlinear techniques. The simplest device to be used here is a current limiter. Thus, amplifier output current is limited to a certain adjustable value. During acceleration and deceleration the amplifier current saturates, resulting in constant velocity change. Because of the simplicity and usefulness of the current limiter, all Electro-Craft SCA amplifiers are equipped with this feature, which can be used or disconnected according to individual application requirements.

In some applications, it is desired to follow the input command with zero error so that the system output is less sensitive to parameter variations. This requires the use of an additional integrator in the loop, which increases the DC gain to infinity. However, an additional integrator adds to the phase shift and, hence, causes stability problems. To design an integrator without the phase shift, a lag-lead compensation network can be used. The transfer function of the network is:

$$H(s) = K\frac{s-a}{s} \quad (4.7.3)$$

Thus, the network has a pole at **s = 0** and a zero at the frequency **s = a**. The compensation network causes a phase shift which is noticed at the low frequencies. However, beyond the frequency of **s = 3a** the effect of the network is minor, and can be ignored. A possible network is shown in Fig. 4.7.3, having a transfer function of:

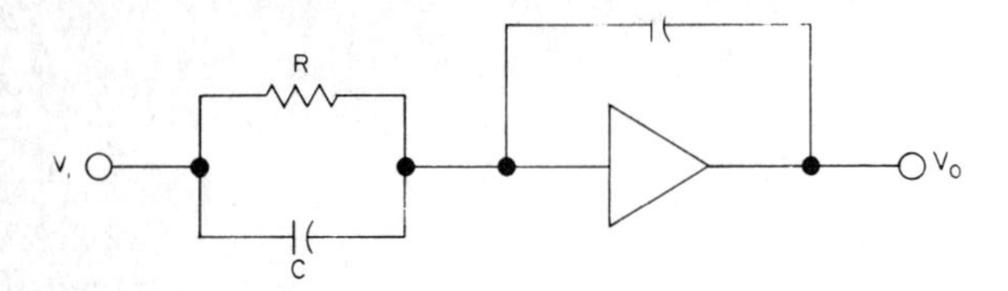

Fig. 4.7.3. Example of a compensation network.

$$H(s) = \frac{V_o(s)}{V_i(s)} = -\frac{1 + sRC_i}{sRC_f} \qquad (4.7.4)$$

When a system has to operate in several control modes, it is inefficient to design one compensation network for all the modes. Instead, it will be more desirable to design the best compensation for each mode and switch between them at the appropriate time.

For example, a system that has to run at a specified velocity and then stop exactly at a certain position is a two control mode system. It operates on either velocity or position control; the system, as shown in Fig. 4.7.4, is suggested for variable compensation.

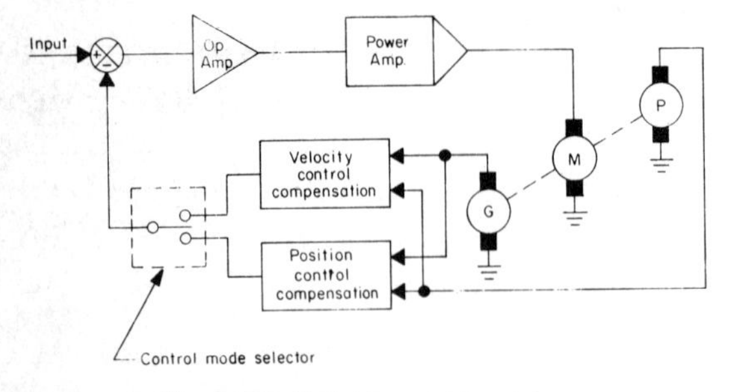

Fig. 4.7.4. Hybrid control system.

The position sensor, **P**, may be a potentiometer or a digital encoder, and the compensation network may use analog circuits, digital circuits, or both. The mode selector switch will be in the desired position according to whether the velocity or the position are to be controlled. For example, suppose that the desired velocity profile is trapezoidal, as shown in Fig. 4.7.5, and it is required that the motor will stop after rotating through a given angle θ.

During this time interval between **0** and t_2, the system will be in velocity control mode, and between t_2 and t_3 it operates in position control mode.

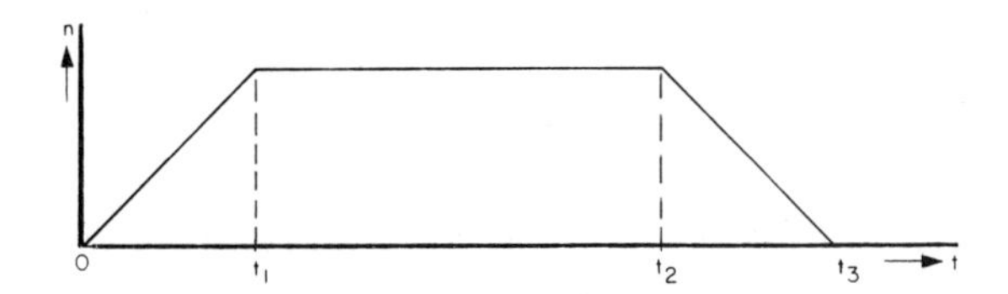

Fig. 4.7.5. A trapezoidal velocity profile.

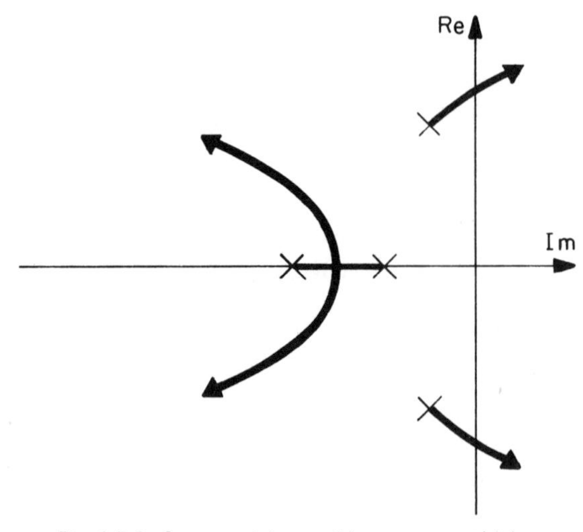

Fig. 4.7.6. System with unstable response at higher values of gain.

A system which is difficult to compensate is one where the open-loop transfer function has a pair of complex poles which are close to the imaginary axis. A common source for such a pair is the torsional resonance between the motor and the load or in the load. The difficulty can be seen easily if we observe the root locus curves for this system in Fig. 4.7.6. Assuming that the system has at least two more open-loop poles (from motor and amplifier), the root loci branches will tend to move into the right half-plane, causing underdamped response when the gain is low and unstable response for a higher gain.

Such a response is not acceptable, and the system has to be compensated to divert the root locus branches which originate at the complex poles and move them to the left, away from the imaginary axis. Since open-loop zeros have the tendency to attract root locus branches, a good approach will be to use some zeros in the compensation network and locate them in the appropriate place for best results. As the detailed design of the compensation network depends on the location of all poles and zeros of the open-loop transfer function, an analysis should be done to locate them. This can be done by measuring exactly the frequency response of the system by recording amplitude of gain and phase shift.

On the basis of this information, the structure of the compensation network is determined, and a realizing circuit can be synthesized. □

4.8. OPTIMIZATION

4.8.1. INTRODUCTION

In recent years the automation industries have pressed for increased performance (and at the same time higher reliability) in automatically controlled processes and equipment. As a result, in many instances mechanical systems are being replaced by all-electronic servo controls driving a highly responsive electromechanical transducer. The field of incremental motion control by this method is relatively new; and, therefore, many potential pitfalls await the system designer. This discussion is intended to provide insights into the selection of velocity profiles and coupling ratios so that these system considerations may aid in realizing an optimum system design.

The incremental motion system to be analyzed consists of a load with moment of inertia J_L which is to be accelerated, rotated through an angle θ, decelerated and brought to rest, all in time t_c. A retarding load torque, T_L, will be considered constant. The driving torque will be provided by a DC servomotor having a torque constant K_T, an armature resistance R, and a moment of inertia J_m.

4.8.2. VELOCITY PROFILE OPTIMIZATION

The velocity profile (i.e., the motor and load angular velocity as a function of time) could be optimized around minimum peak speed or minimum peak current or any of several parameters. In general, however, the limiting factor in the performance of the type of incremental motion control system we are analyzing is motor armature heat dissipation. Therefore, it is the parameter we will minimize.

The basic equations of this incremental motion are:

$$J \frac{d\omega}{dt} + T_L = K_T I_a \tag{4.8.1}$$

where $J = J_m + J_L$, and

$$\int_0^{t_c} \omega(t)\, dt = \theta \tag{4.8.2}$$

The resulting energy dissipation in the motor armature per step, W_c, is given by:

$$W_c = \int_0^{t_c} I_a^2(t)\, dt \tag{4.8.3}$$

The problem now reduces to the determination of the optimum velocity profile $\omega(t)$, which minimizes W_c and still satisfies (4.8.1) and (4.8.2). We shall discuss three types of velocity profiles (parabolic, triangular, and trapezoidal) and determine energy dissipation for each.

Parabolic Velocity Profile

The solution to this problem results in two conclusions:

1. The constant opposing torque T_L adds a fixed energy dissipation:

$$\frac{R}{K_T^2} T_L^2 t_c$$

which is independent of velocity profile.

2. The optimal velocity profile is parabolic and is given by the equation:

$$\omega_o(t) = 6\theta \frac{t_c - t}{t_c^3} t \qquad (4.8.4)$$

A graph of this profile is shown in Fig. 4.8.1. The resulting energy dissipation per step becomes:

$$W_{co} = \frac{R}{K_T^2} \left[12 \frac{J^2 \theta^2}{t_c^3} + T_L^2 t_c \right] \qquad (4.8.5)$$

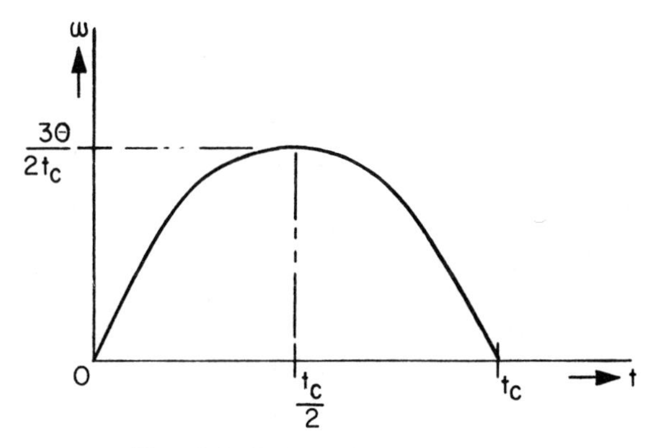

Fig. 4.8.1. Optimized velocity profile.

Triangular Velocity Profile

As a result of limitations in practical control system complexity, velocity profiles other then optimum are frequently used. In the triangular velocity profile the load is accelerated at a fixed rate and then decelerated

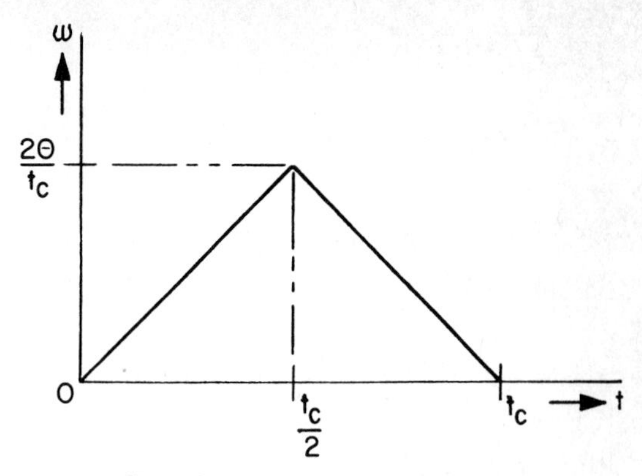

Fig. 4.8.2. Triangular velocity profile.

at the same rate, as shown in Fig. 4.8.2. The energy dissipation per step is:

$$W_c = \frac{R}{K_T^2} \left[16 \frac{J^2 \theta^2}{t_c^3} + T_L^2 t_c \right] \qquad (4.8.6)$$

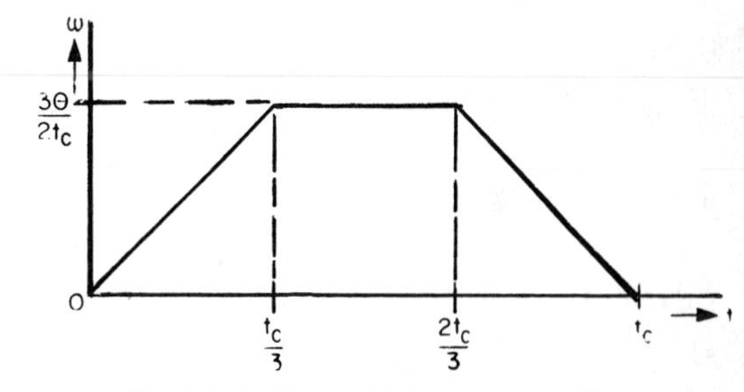

Fig. 4.8.3. Trapezoidal velocity profile.

Trapezoidal Velocity Profile

The trapezoidal profile is divided into three parts: acceleration, run, and deceleration as shown in Fig. 4.8.3. Frequently there is a feeling among engineers that the triangular velocity profile is more efficient than the trapezoidal, as the latter's run time seems wasted. The results show, however, that for a trapezoidal velocity profile in which all three parts are of equal time interval, the energy dissipation per step is:

$$W_c = \frac{R}{K_T^2}\left[13.5\,\frac{J^2\,\theta^2}{t_c^3} + T_L^2\,t_c\right] \quad (4.8.7)$$

Again, if we assume $T_L = 0$, the energy dissipation in the motor armature is 12.5% higher than in the case of a parabolic velocity profile, but 20.83% less than in the case of a triangular velocity profile. For convenience at a later point, let us now define η as the velocity profile efficiency when $T_L = 0$:

$$\eta = \frac{W_{co}}{W_c} \quad @\ T_L = 0 \quad (4.8.8)$$

Then the energy dissipation per step equals:

$$W_c = \frac{R}{K_T^2}\left[\frac{12}{\eta}\,\frac{J^2\,\theta^2}{t_c^3} + T_L^2\,t_c\right] \quad (4.8.9)$$

Thus, η = **1** for the parabolic velocity profile; η = **0.75** for the triangular velocity profile; and η = **0.889** for the trapezoidal velocity profile.

4.8.3. COUPLING RATIO OPTIMIZATION

A second important system parameter affecting motor armature power dissipation is the *coupling ratio* between the motor and the load. The term coupling ratio is loosely defined as the angular movement of the motor compared to the movement of the load. For a system in which the load is rotated this would be more commonly called the gear ratio. For a system in which the load is linearly translated, the coupling ratio is the motor rotation for one unit of linear motion (i.e., it is expressed in units of [rad/m] or [rad/in]). Three methods of coupling will be considered; (1) the gear transmission, (2) the belt and pulley, and (3) the lead screw. The objective will be to determine the coupling ratio in which the motor armature energy dissipation is minimized. For the sake of simplicity it will be assumed that total load moment of inertia is fixed and does not change with coupling ratio and that all mechanical losses are contained in the T_L term.

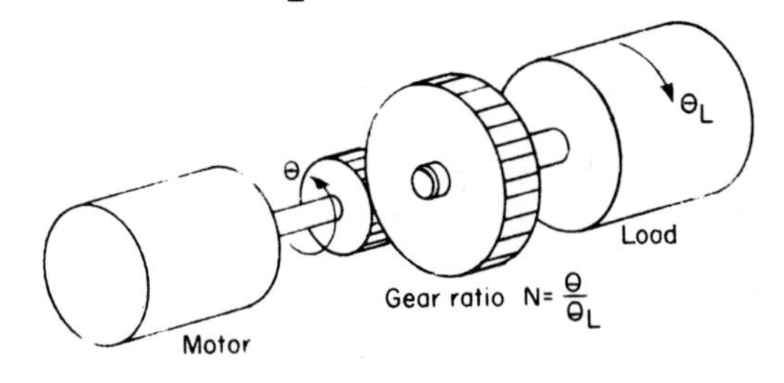

Fig. 4.8.4. Gear coupled system.

Gear Ratio Selection

In a coupling system such as the one shown in Fig. 4.8.4 using a gear ratio of **N:1**, the following relationships reflect all parameters to the motor shaft:

$$\theta = N\theta_L \quad (4.8.10)$$

$$J = J_m + \frac{1}{N^2}\,J_L \quad (4.8.11)$$

$$T_L' = \frac{1}{N}\,T_L \quad (4.8.12)$$

The equation (4.8.9) for energy dissipation, derived from (4.8.5) and (4.8.8) therefore becomes:

$$W_c = \frac{R}{K_T^2}\left[\frac{12}{\eta}\,\frac{\left(J_m + \frac{1}{N^2}J_L\right)^2 N^2\theta_L^2}{t_c^3} + \frac{T_L^2}{N^2}t_c\right] \quad (4.8.13)$$

For convenience now define:

$$\gamma = \frac{\eta}{12}\left[\frac{T_L\, t_c^2}{\theta_L\, J_L}\right]^2 \quad (4.8.14)$$

Substituting into (4.8.13) and rearranging gives:

$$W_c = \frac{R}{K_T^2}\,\frac{12}{\eta}\,\frac{\theta_L^2\, J_L^2}{t_c^3}\left[N^2\left(\frac{J_m}{J_L} + \frac{1}{N^2}\right)^2 + \frac{\gamma}{N^2}\right] \quad (4.8.15)$$

Performing the differentiation of the part of (4.8.15) in brackets, i.e.

$$\frac{d}{d(N^2)}\left[\cdots\cdots\right] = 0$$

we obtain the gear ratio N_o which minimizes W_c:

$$N_o^2 = \frac{J_L}{J_m}\sqrt{1+\gamma} \quad (4.8.16)$$

If there were no load opposing torque, $T_L = 0$, then $\gamma = 0$ and (4.8.16) becomes:

$$N_o = \sqrt{\frac{J_L}{J_m}} \quad (4.8.17)$$

This result is known as an *inertia match* since the motor moment of inertia equals the reflected load moment of inertia, $J_m = \frac{J_L}{N^2}$

In cases where for practical reasons a non-optimum gear ratio must be used it would be of interest to examine the resultant increase in armature energy dissipation. Therefore, let

$$\rho = \frac{W_c(N)}{W_c(N_o)} = \frac{N^2\left(J_m + \frac{1}{N^2}J_L\right)^2}{N_o^2\left(J_m + \frac{1}{N_o^2}J_L\right)^2} = \frac{1}{4}\left(\frac{N}{N_o} + \frac{N_o}{N}\right)^2 \quad (4.8.18)$$

The energy ratio ρ increases as shown in Fig. 4.8.5 if the gear ratio changes from optimum value. It can be seen that a small deviation from optimum is not critical, however, as the deviation increases the penalty becomes increasingly severe. If we deviate by ±10% the energy increases by less than 1%, but if we deviate by a factor of 2 the energy increases by 56%.

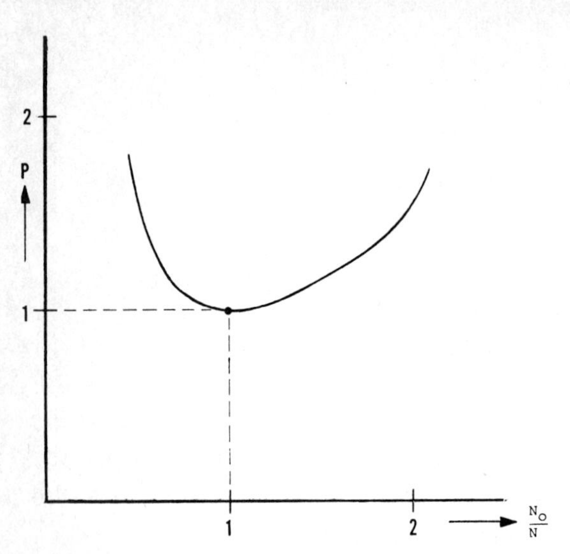

Fig. 4.8.5. Energy dissipation increase due to non-optimum gear ratio.

Belt-Pulley Drive

The belt-pulley drive is shown in Fig. 4.8.6, and consists of a motor turning a pulley which pulls a belt to which the load is attached; thus converting rotary motion into translation. The coupling ratio in this case is the reciprocal of the radius **r** of the pulley on the motor. The load mass, **m**, is to be moved a distance **x** in one step, i.e. in time t_c. The constant opposing force is **F**. Thus:

$$\theta = \frac{x}{r} \tag{4.8.19}$$

$$J = J_m + mr^2 \tag{4.8.20}$$

$$T_L = Fr \tag{4.8.21}$$

According to the concepts previously developed [using (4.8.9)], the energy dissipated in the motor armature per step is:

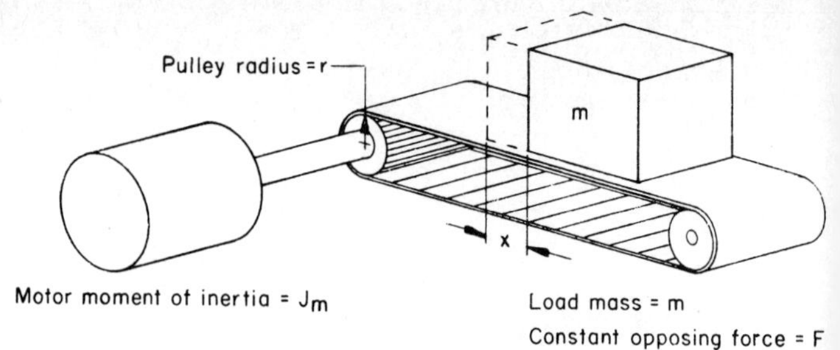

Fig. 4.8.6. Belt-pulley drive system.

$$W_c = \frac{R}{K_T^2}\left[\frac{12}{\eta}\,\frac{\left(\frac{x}{r}\right)^2}{t_c^3}\left(J_m + mr^2\right)^2 + (Fr)^2\,t_c\right] \tag{4.8.22}$$

Now define the coupling ratio

$$G = \frac{1}{r} \tag{4.8.23}$$

and:

$$\beta = \frac{\eta}{12}\left[\frac{F\,t_c^2}{m\,x}\right]^2 \tag{4.8.24}$$

Then the energy equation (4.8.22) becomes:

$$W_c = \frac{R}{K_T^2}\,\frac{12}{\eta}\,\frac{m^2\,x^2}{t_c^3}\left[G^2\left(\frac{J_m}{m} + \frac{1}{G^2}\right)^2 + \frac{\beta}{G^2}\right] \tag{4.8.25}$$

Note that (4.8.25) is similar to (4.8.15) where **G** is equivalent to **N**, **m** is equivalent to J_L and β instead of γ. Clearly, the optimum coupling ratio, according to (4.8.16), will be:

$$G_o^2 = \frac{m}{J_m} \sqrt{1 + \beta} \qquad (4.8.26)$$

The optimum pulley radius is then:

$$r_o = \left(\frac{J_m}{m \sqrt{1 + \beta}} \right)^{\frac{1}{2}} \qquad (4.8.27)$$

If the constant opposing force **F** = **0**, then β = **0**, and we again have an exact inertia match:

$$r_o = \sqrt{\frac{J_m}{m}} \qquad (4.8.28)$$

Lead Screw Drive

The lead screw drive is shown in Fig. 4.8.7, and consists of a motor turning the lead screw which moves a mass **m**, much in the same way as the belt-pulley drive. Assuming the motion problem variables **x**, θ, T_L, **F**, and **m** remain the same as in the belt-pulley system, the problem becomes to determine the optimum pitch, P_o, of the lead screw. Pitch here is taken as number of revolutions per unit length and therefore:

$$P = \frac{G}{2\pi} \qquad (4.8.29)$$

Since **G** is the motor rotation in radians per unit length, the optimum pitch is therefore:

$$P_o = \frac{1}{2\pi} \left(\frac{m \sqrt{1 + \beta}}{J_m} \right)^{\frac{1}{2}} \qquad (4.8.30)$$

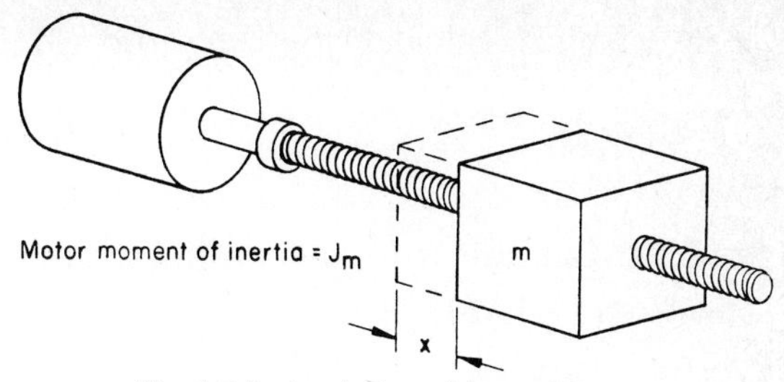

Fig. 4.8.7. Lead screw drive system.

EXAMPLES

For clarity, example solutions to coupling optimization problems are given below. Assume in all cases the motor has the following parameters:

K_T = 6 oz-in/A = 0.04237 Nm/A

R = 1 Ω

J_m = 5 x 10^{-4} oz-in-s^2 = 35.3 g cm^2

Assume also that a trapezoidal velocity profile is used. Therefore

η = 0.889

Example 1: The motor has to drive a load by means of a gear transmission. The load has the following parameters:

J_L = 8 x 10^{-2} oz-in-s^2

= 5.65 x 10^{-4} kg m^2

T_L = 12 oz-in = 0.0847 Nm

θ_L = 0.2 rad

t_c = 0.05 s

First we must calculate γ from (4.8.14):

$$\gamma = 0.260$$

Now we can calculate the optimum gear ratio from (4.8.16):

$$N_o^2 = 179.6$$

$$N_o = 13.4$$

Example 2: The second example involves a mass to be moved linearly by means of a belt-pulley drive. The following load and performance parameters apply:

$$m = 0.01 \text{ kg}$$
$$F = 0.003 \text{ N}$$
$$x = 0.02 \text{ m}$$
$$t_c = 0.01 \text{ s}$$

From (4.8.24) we have:

$$\beta = 1.667 \times 10^{-7}$$

By observing (4.8.27) it can be seen that the factor β is insignificant and that the inertia match formula

$$r_o = \sqrt{\frac{J_m}{m}} \qquad (4.8.28)$$

can be used. The optimum pulley radius is thus

$$r_o = 18.8 \text{ mm} = 0.74 \text{ in}$$

Example 3: In the final example, a mass is to be driven by a lead screw in the same manner as in Example 2. The problem is to find the optimum lead screw pitch. The following load and performance parameters apply:

$$m = 1 \text{ kg}$$
$$F = 0.02 \text{ N}$$
$$x = 0.02 \text{ m}$$
$$t_c = 0.01 \text{ s}$$

Let us also consider the moment of inertia of the lead screw to be

$$J_{LS} = 0.4 \times 10^{-3} \text{ oz-in-s}^2$$
$$= 28.25 \text{ g cm}^2$$

This moment of inertia is to be added to the motor moment of inertia, so that:

$$J'_m = J_m + J_{LS} = 63.55 \text{ g cm}^2$$

We can now calculate the opposing force factor β from (4.8.24):

$$\beta = 7.4 \times 10^{-10}$$

This factor can be ignored. The optimum pitch is thus from (4.8.30):

$$P_o = \frac{1}{2\pi}\left(\frac{m}{J'_m}\right)^{\frac{1}{2}}$$

$$= 63.134 \text{ turns/m} \cong 1.6 \text{ turns/in}$$

4.8.4. CAPSTAN OPTIMIZATION

In the following let us briefly examine some of the practical aspects of a typical servo problem in computer peripheral equipment.

Introduction

Fast-moving technological changes in incremental tape transports of the past decade have made 200–250 in/s (5 – 6.4 m/s) machines commercially available. Single capstan drive systems have emerged as virtually sole survivors in the drive methods contest. As the quest for increasingly higher tape speeds and repetition rates continues, tape transport servo and capstan designers as well as users must face the increasing demands on motors, feedback elements, capstan design and power supplies.

Thus, in the decade we can expect demands on servomotors with higher K_T/R ratio, lower moment of inertia, and lower thermal resistances (and, lest we forget, increasing downward price pressure). We believe that capstan motor power dissipation problems and torsional resonance problems are the major technical hurdles we have to face in the future.

This discussion covers one aspect of this problem: how to pick the right motor and design the optimum capstan radius to minimize power dissipation. It also discusses what added power dissipation the application would need under non-optimum conditions.

Optimum Capstan Selection

If we define a tape drive system with a given linear velocity profile **v(t)**, the linear acceleration of the tape is:

$$a = \frac{dv}{dt} \tag{4.8.31}$$

A direct-coupled-to-the-motor capstan of radius **r** drives the tape. The motor angular velocity ω is therefore:

$$\omega = \frac{v}{r} \tag{4.8.32}$$

Differentiating (4.8.32) gives the angular acceleration:

$$\alpha = \frac{d\omega}{dt} = \frac{1}{r}\frac{dv}{dt} \tag{4.8.33}$$

Combining (4.8.31) and (4.8.33) gives:

$$\alpha = \frac{a}{r} \tag{4.8.34}$$

The moment of inertia of the capstan, J_c, depends on the radius **r** and on the capstan design. Suppose that the capstan structure is approximated by a solid disc and a cylinder (Fig. 4.8.8, A and B).

If the dominant portion of capstan moment of inertia would be represented by a disc of length w_1, radius **r**, density ρ and mass m_1, then:

$$J_c = \frac{1}{2} m_1 r^2 = \frac{1}{2} \rho \pi w_1 r^4 \tag{4.8.35}$$

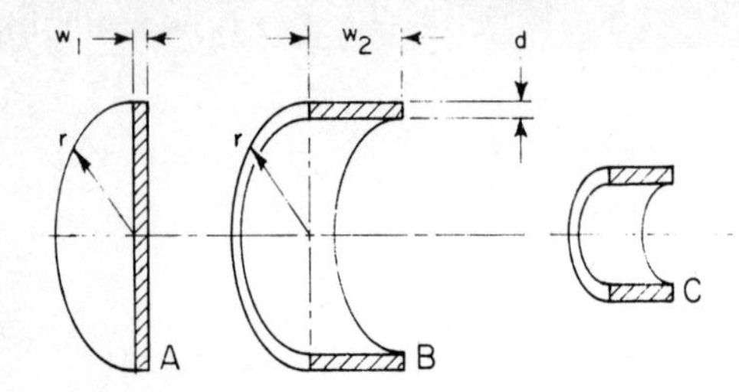

Fig. 4.8.8. Typical capstan construction can take the form of (A) a disc, and (B) a cylinder. (C) indicates the hub and its small mass. Hence, it contributes a negligible amount to the capstan moment of inertia.

On the other hand, if the dominant portion of J_c lies in the outside cylinder of thickness **d**, length w_2 and mass m_2, then:

$$J_c = m_2 r^2 = 2\pi\rho w_2 d\, r^3 \tag{4.8.36}$$

From these extreme conditions, we can conclude that a capstan made from elements A and B (Fig. 4.8.8) provides a moment of inertia which is proportional to the power of **r** between 3 and 4. We can usually ignore the inertia contribution of the hub C. We can therefore define the inertia equation as:

$$J_c = k\, r^p \tag{4.8.37}$$

where the exponent **p**:

$$3 \leqslant p \leqslant 4 \tag{4.8.38}$$

A subsequent example shows how to arrive at a value for **p** in a specific case. The total system moment of inertia is the sum of motor moment of inertia J_m and the capstan moment of inertia J_c.

$$J = J_m + J_c \tag{4.8.39}$$

Combining (4.8.37) with (4.8.39) gives:

$$J = J_m + k\, r^p \tag{4.8.40}$$

The torque **T** needed to run the motor at acceleration a is:

$$T = Ja \tag{4.8.41}$$

If we combine (4.8.34), (4.8.40) and (4.8.41), we get:

$$T = \frac{a}{r}\left(J_m + k\, r^p\right) \tag{4.8.42}$$

From the chosen motor specification we can define the armature current during acceleration as:

$$I_a = \frac{T}{K_T} \tag{4.8.43}$$

Hence the instantaneous power dissipation in the motor (neglecting friction - a realistic assumption in most cases of incremental motion) is:

$$P(t) = R\, I_a^2 \tag{4.8.44}$$

or, using (4.8.43):

$$P(t) = R\frac{T^2}{K_T^2} \tag{4.8.45}$$

Clearly, minimizing **P(t)** minimizes **T**.

Performing the differentiation

$$\frac{dT}{dr} = 0$$

of (4.8.42) gives:

$$-a J_m r^{-2} + (p-1) a k r^{p-2} = 0$$

or:

$$J_m = (p-1) k r^p \qquad (4.8.46)$$

Substituting (4.8.37) into (4.8.46) gives the relationship:

$$J_m = (p-1) J_c \qquad (4.8.47)$$

Equation (4.8.47) shows how the motor moment of inertia relates to the capstan's. For the given range of **p**, minimum power dissipation requires a capstan moment of inertia between 1/2 and 1/3 of the motor moment of inertia. This requirement differs from the usual rule-of-thumb concept in analog servo design that calls for motor moment of inertia equal to load moment of inertia for optimum performance conditions.

Non-optimum Choices

Frequently, additional constraints or available equipment prevent the selection of optimal capstans. Since in this case we must choose a non-optimal capstan, how does power dissipation increase?

Recalling that the capstan moment of inertia is $J_c = k r^p$ (4.8.37), and if we call the optimal capstan radius r_o and the realizable (non-optimum) radius **r**, and denote the power dissipation corresponding to r_o and **r** by P_o and **P**, respectively, then from (4.8.45) and (4.8.42) we obtain the power dissipation.

$$P = \frac{R a^2}{K_T^2 r^2} \left(J_m + k r^p\right)^2 \qquad (4.8.48)$$

and the minimum power dissipation:

$$P_o = \frac{R a^2}{K_T^2 r_o^2} \left(J_m + k r_o^p\right)^2 \qquad (4.8.49)$$

A good indicator of the "efficiency" is η, the ratio of **P** and P_o:

$$\eta = \frac{P}{P_o} = \left[\frac{r_o (J_m + k r^p)}{r (J_m + k r_o^p)}\right]^2 \qquad (4.8.50)$$

Now recall the r_o satisfied (4.8.46), therefore:

$$k r_o^p = \frac{J_m}{p-1} \qquad (4.8.51)$$

If we denote

$$\gamma = \frac{r}{r_o} \qquad (4.8.52)$$

and substitute into (4.8.50) yields:

$$\eta = \left[\frac{r_o (J_m + k \gamma^p r_o^p)}{\gamma r_o (J_m + k r_o^p)}\right]^2 \qquad (4.8.53)$$

This can be further simplified by substituting (4.8.51):

$$\eta = \left[\frac{p - 1 + \gamma^p}{\gamma p}\right]^2 \qquad (4.8.54)$$

Equation (4.8.54) indicates the increase in power dissipation when a non-optimal radius is used. Note that for **p = 3** and **p = 4** it becomes:

$$\left.\begin{aligned} \eta &= \left[\frac{2+\gamma^3}{3\gamma}\right]^2 \quad @\ p = 3 \\ \eta &= \left[\frac{3+\gamma^4}{4\gamma}\right]^2 \quad @\ p = 4 \end{aligned}\right\} \quad (4.8.55)$$

Since for most cases $3 \leqslant p \leqslant 4$, the power dissipation increase is within the limits given by (4.8.55) and presented in Fig. 4.8.9. Note that a small deviation from the optimal radius does not result in a large power dissipation increase. However, as the deviation increases, the power dissipation increases considerably. For example, if the capstan radius is half the size of the optimal, power dissipation increases between 100% and 134%.

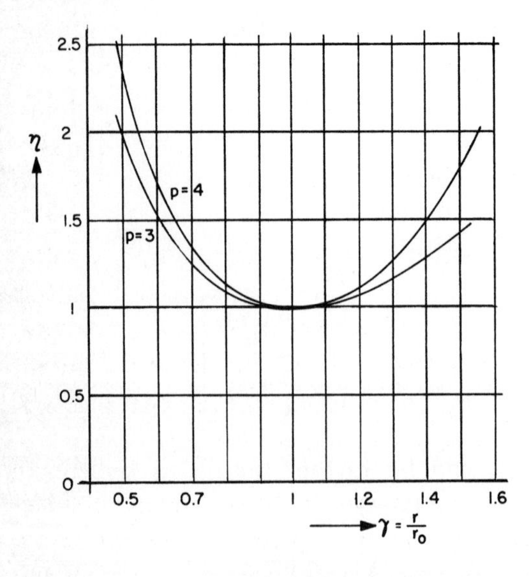

Fig. 4.8.9. Motor power dissipation increase as function of capstan radius.

State-of-the-art systems not uncommonly work the servo system close to the dissipation limits of the motor. Therefore, you should know the penalty for non-optimum capstan choice.

4.8.5. OPTIMUM MOTOR SELECTION FOR INCREMENTAL APPLICATION

So far, we learned how to optimize capstan design for a given motor. Now we can start over again and assume we have not chosen a specific motor. Let's select the best motor for a general capstan shape.

If we combine (4.8.44) and (4.8.45), we define power dissipation for any motor as:

$$P = R\, I_a^2 = R\, \frac{T^2}{K_T^2} \qquad (4.8.56)$$

Inserting (4.8.34) and (4.8.41) into (4.8.56) produces:

$$P = \frac{R\, J^2}{K_T^2} \left(\frac{a}{r}\right)^2 \qquad (4.8.57)$$

Substitution from (4.8.39) and (4.8.47) gives:

$$P = \frac{R}{K_T^2} \left[J_m \frac{p}{p-1} \frac{a}{r}\right]^2 \qquad (4.8.58)$$

Rearranging (4.8.46) gives us the optimal radius:

$$r_o = \left[\frac{J_m}{(p-1)\,k}\right]^{\frac{1}{p}} \qquad (4.8.59)$$

and substituting into (4.8.58) gives:

$$P = \frac{R}{K_T^2}\left[J_m \frac{ap}{p-1}\left(\frac{[p-1]\ k}{J_m}\right)^{\frac{1}{p}}\right]^2 \tag{4.8.60}$$

If we isolate motor dependent parameters in (4.8.60), we see that:

$$P = \frac{R\ J_m^{(2-2/p)}}{K_T^2}\left[\frac{ap}{p-1}\left([p-1]\ k\right)^{\frac{1}{p}}\right]^2 \tag{4.8.61}$$

and the term to be minimized is:

$$\frac{R\ J_m^{\beta}}{K_T^2} \tag{4.8.62}$$

where

$$\beta = 2 - 2/p$$

For the given range of **p**, it becomes:

$$\frac{4}{3} \leqslant \beta \leqslant \frac{3}{2} \tag{4.8.63}$$

Thus, we now observe that for optimum power dissipation we must minimize the following term:

$$\frac{R\ J_m^{(1.33\ \text{to}\ 1.5)}}{K_T^2} \tag{4.8.64}$$

In some cases, we may find that the design objective is to minimize the motor temperature rise, Θ_r. We find that Θ_r is related to the power dissipation by:

$$\Theta_r = PR_{th} \tag{4.8.65}$$

where R_{th} is the thermal resistance of the motor. Then the motor dependent parameters to be minimized are:

$$\frac{R\ J_m^{\beta}\ R_{th}}{K_T^2} \tag{4.8.66}$$

Thus, we note that (4.8.66) matches the formula of (4.8.62) with the addition of R_{th}.

We can establish the value of β from the design philosophy of the capstan design, which gives a value of **p** (see subsequent example). To select the optimum motor, we can now take specification data for various motors and plug it into (4.8.64). The lowest value indicates the optimum motor from the power dissipation point of view. Of course, we must consider other reasons for selecting a motor, such as compatibility with power supply voltage and current limits, resonant frequencies, tachometer accessories, physical size, available shaft dimensions and any other parameters specific to the user's need.

We have now acquired an insight into factors which affect power dissipation in an incremental motion system. In the following example we compare two motors of rather diverse performance characteristics to show the drastic differences in power dissipation.

EXAMPLE

When designing a capstan drive, we can use either of two available typical capstans with the following parameters:

$$r_1 = 0.75 \text{ in}, \; J_{c1} = 0.3 \times 10^{-3} \text{ oz-in-s}^2$$

$$r_2 = 1.0 \text{ in}, \quad J_{c2} = 0.85 \times 10^{-3} \text{ oz-in-s}^2$$

We can therefore assume that we can design the optimum capstan of a radius that ranges between 0.55 in to 1.0 in. Which of the following two available motors should we select?

Motor	K_T [oz-in/A]	R[Ω]	J_m [oz-in-s^2]
M-1030	5.8	0.85	0.45×10^{-3}
M-1438	9.5	0.55	0.65×10^{-3}

To determine the equation of capstan moment of inertia, we want to arrive at the form:

$$J_c = k\, r^p \tag{4.8.37}$$

Then:

$$J_{c1} = k\, r_1^p \text{ and } J_{c2} = k\, r_2^p \tag{4.8.67}$$

Dividing the two equations yields:

$$\frac{J_{c2}}{J_{c1}} = \left(\frac{r_2}{r_1}\right)^p \tag{4.8.68}$$

Now take the logarithm of the two sides:

$$p = \frac{\log\left(\frac{J_{c2}}{J_{c1}}\right)}{\log\left(\frac{r_2}{r_1}\right)} \tag{4.8.69}$$

Here:

$$p = \frac{\log\left(\frac{8.5}{3}\right)}{\log\left(\frac{1.0}{0.75}\right)} = 3.6$$

To determine **k**:

$$k = \frac{J_{c2}}{r_2^p} = \frac{0.85 \times 10^{-3}}{1^{3.6}} = 0.85 \times 10^{-3}$$

Since **p** = **3.6**, then:

$$\beta = 2 - \frac{2}{p} = 2 - \frac{2}{3.6} = 1.45$$

To select the motor, evaluate (4.8.64) the term:

$$\frac{R\, J_m^{1.45}}{K_T^2}$$

for the two motors.

In evaluating the previous term, use any unit system, as long as you use it for all the motors. The ratio of power dissipation of the two motors will then be:

$$\frac{(P)_{M\text{-}1030}}{(P)_{M\text{-}1438}} = \frac{\dfrac{0.85 \times 4.5^{1.45}}{5.8^2}}{\dfrac{0.55 \times 6.5^{1.45}}{9.6^2}}$$

$$= \frac{0.223}{0.072} \cong 3.1$$

The second motor, M-1438, will dissipate power at a much lower rate than M-1030, by a factor of **3.1**!

If M-1438 motor fulfills other requirements, such as price, size, power supply needs, we will select it as the capstan driver.

Since the optimal capstan moment of inertia J_c (4.8.47) for M-1438 motor is:

$$J_c = \frac{J_m}{p - 1} = \frac{0.65 \times 10^{-3}}{2.6}$$

$$= 0.25 \times 10^{-3} \text{ oz-in-s}^2$$

The capstan radius should be (4.8.37):

$$0.85 \times 10^{-3}\, r^{3.6} = 0.25 \times 10^{-3}$$

$$r^{3.6} = \frac{0.25}{0.85} = 0.294$$

$$3.6 \log r = \log 0.294$$

From this,

$$r_o = 0.71 \text{ in}$$

The optimum capstan radius is 0.71 in and the optimum capstan moment of inertia is **0.25×10^{-3} oz-in-s^2**. These values come close to those of one of the sample capstans listed earlier. If we use the 0.75 in radius capstan rather than the 0.71 in optimum, then

$$\gamma = \frac{r}{r_o} = \frac{0.75}{0.71} = 1.056$$

and the power dissipation increase due to non-optimal radius approximates 1% — a small penalty. □

4.9. PERMANENT MAGNET MOTORS FOR SERVO APPLICATIONS

In this section, we will discuss a few reasons why permanent magnet motors are useful for servo applications.

The easily controlable speed feature of the DC motor, coupled with its extremely linear torque-speed control characteristic makes the PM motor an ideal servo component. An expression relating motor speed to applied voltage is:

$$n = \frac{V}{K_E} \tag{4.9.1}$$

This relationship allows motor speed to be controlled by proper adjustment of the applied terminal voltage. If the terms of the above expression are rearranged we obtain:

$$V = K_E n \tag{4.9.2}$$

In this form it is easy to see that if driven at the shaft and used in a generator mode, a PM device will provide an output voltage signal that is proportional to the shaft speed. This feature makes the PM generator an ideal and compatible feedback device for use in combination with a PM motor in a DC servo system. The basic elements of a simple speed control system are shown in Fig. 4.9.1. The armature current is regulated by the transistor Q which in turn is controlled to be more or less conductive by the generator connected between its base and the wiper arm of a potentiometer. Each setting of the potentiometer corresponds to a given shaft speed. If a given setting is supplied by the potentiometer, the circuit will then function automatically, holding the motor shaft at the desired speed, independent of variation in the load torque.

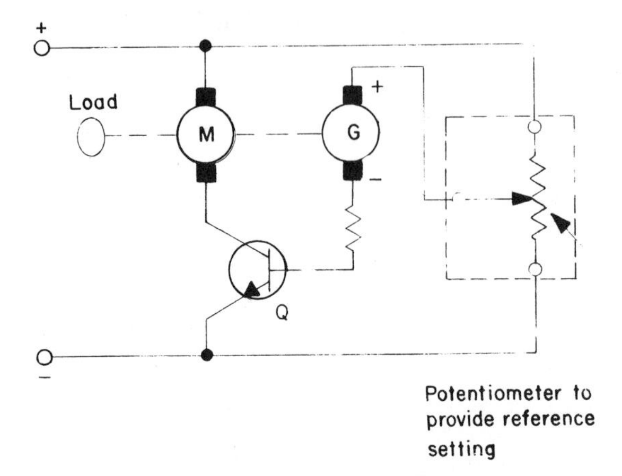

Fig. 4.9.1. Elementary speed control system.

In order to exploit the adaptability to servo systems, PM machine design has assumed some rather unique forms not normally seen on general purpose devices. In the speed control scheme shown in Fig. 4.9.1, it is necessary for the motor shaft and the generator shaft to be rigidly coupled together. To accomplish this, many servo systems are built with both motor and generator on one shaft. Such a device is called a motor-tachometer-generator (tachometer means speed counter) or simply a *motor-tach.*

While motor or generator action is reversible, depending on whether the shaft is driven or the armature excited, the electrical characteristics of the devices are often quite different. The motor is designed to convert electrical power to a mechanical form. Because of this power handling role it must have large wire size and is usually characterized by low armature resistance. On the other hand, the generator is not required to convert significant amounts of power, but rather to provide a voltage signal. The generator can function well using relatively fine wire sizes. These different demands of motor and generator function have resulted in a unique motor-tach design patented and marketed by Electro-Craft under the trade name "Motomatic". The concept of the Motomatic design is that the fine wire for generator armature winding could easily occupy the interstices of the regular large motor armature winding, and thus occupy the same armature core without actually losing space for the motor winding. The widespread acceptance of the Motomatic components has proved the design concept to be correct. Motomatic armatures are shown in Figs. 4.9.2 and 4.9.3. The motor and generator windings may be wound simultaneously or consecutively. In either case, the windings occupy the same slots and are electrically isolated from one another. Electrical power flows in through a set of brushes at the motor end, develops torque and mechanical power to a load, and at the same time

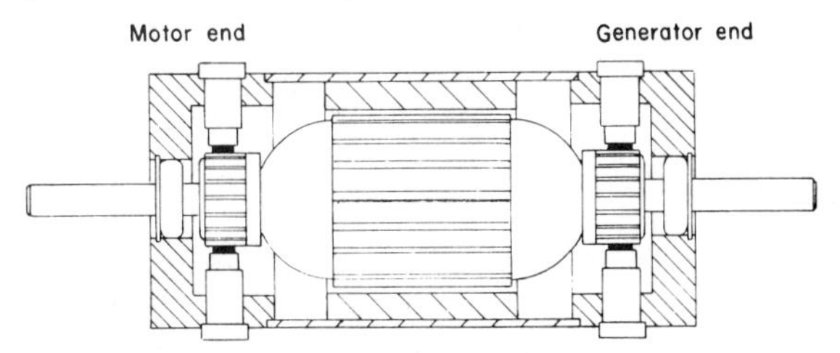

Fig. 4.9.2. Motomatic design concept – combining motor and generator on one armature core.

the voltage proportional to shaft speed is provided off the set of brushes at the opposite (generator) end. The concept allows very significant economies in manufacturing costs.

The Motomatic style (integral motor-tach), of construction provides acceptable characteristics for most applications. There are two reasons, however, why it might be desirable in some instances to place the motor and generator windings on separate cores (Fig. 4.9.3) but still on the same shaft. One is that in the case of an SCR type controller the pulsating DC current in the motor produces a large voltage ripple in the generator voltage because of the transformer action of the common iron core.

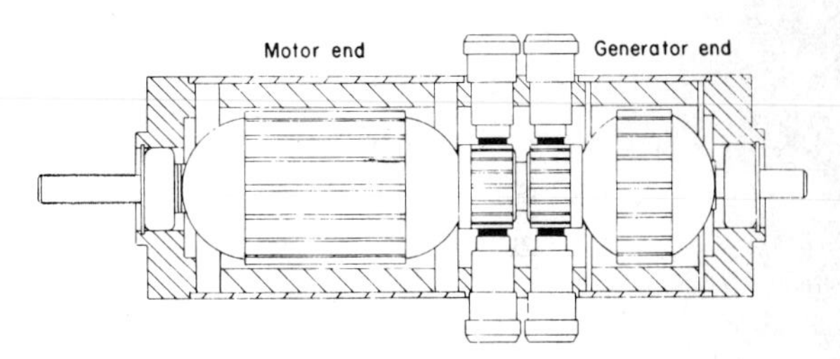

Fig. 4.9.3. Motor-tach design for magnetic isolation of motor and generator windings.

The other – and entirely different – reason is that if either a very wide controlled speed range is desired or a high degree of instantaneous or long–term speed stability is required at the very low speed, then these objectives might be better achieved with a higher voltage gradient tachometer on a separate core.

In general applications where the motor operates for long periods at very low speeds and at the same time extremely close speed control (better than 0.2% of the set speed) is required and any drift in speed over a long time cannot be tolerated, then it may be necessary to use a silver commutator instead of the conventional copper one in the generator. Under the above conditions, the film on a copper commutator tends to change, introducing a drift in generator voltage – hence in motor speed. This is not the case when silver is used as a commutator surface. □

4.10. SYSTEMS AND CONTROLS TROUBLESHOOTING

Prior to undertaking the troubleshooting of a system, a basic comprehension of each part of the system is essential, as well as a knowledge of operating and calibration controls. Cognizance of a normally operating system, such as voltage and current waveforms at key points, will facilitate isolation of the defective components.

To aid troubleshooting, we can divide failure modes into three classes:

1) Total

2) Partial

3) Intermittent

In the case of a *total* failure, start troubleshooting at the incoming power lines. If power is present and no fuses or circuit breakers are defective, check the system power supply. Then follow normal power flow through the system toward the load until the defective stage is located. Substitution of a function generator signal for normal command will facilitate signal tracing.

When troubleshooting a *partially* defective system, a more unique approach is required. By substituting a signal midstream in the system, say at the driver amplifier, it is possible to localize the fault to a before or after stage. Removing the load from the amplifier, opening the velocity or position feedback loop, the loading of an amplifier with a resistive load will allow conventional amplifier signal tracing.

The *intermittent* fault can best be isolated by long-term monitoring of key points with an events' recorder or similar instrument. Knowing the rate of the intermittent fault may lead to causes. Also the application of heat, vibration, cold, moisture, stress or other variables to areas of the system may cause the intermittent fault to appear.

When the failure is isolated, the symptom which causes the failure should be noted and questioned as to why the symptom occurred. Refer to the theoretical failure symptom chart, and as an example take an inoperative motor, noting that the symptom was a blown fuse or tripped circuit breaker. Before replacing the fuse or resetting the breaker, question why the fuse blew. Consider the following:

1) Did the fuse blow because of age or fatigue, or was it overloaded?

2) Did the motor fail, causing the fuse to blow or was it a case of overload?

3) Did the amplifier go into an oscillatory mode, producing high ripple currents which blew the fuse?

4) Did the motor make any unusual sounds or emit smoke prior to failure of the system?

THEORETICAL FAILURE-SYMPTOM CHART

Nature of Failure	Symptom Noted
Motor will not run	1. Motor or amplifier over-heated 2. Fuses blown 3. Fuses or pilot light off
Uncontrollable operation at full speed	1. Immediately when power is applied 2. When load is applied
Poor load regulation	1. At high speed only 2. At low speed only 3. Through entire speed range 4. In one direction only
Erratic performance	1. At high speed 2. At low speed 3. At all speeds 4. When motor is hot 5. In both directions of operation 6. In one direction only 7. Under extreme load conditions
Speed variation with time	1. Increases or decreases 2. Starts or stops
Speed variation with temperature	1. Increases or decreases 2. Stops or starts

Repairing and replacing components must be done with the same techniques and care that the original part was installed. This is to say all heat sink mounted semiconductors must be installed with the proper insulators, hardware and thermal mounting components as originally used. Low-power devices should be installed carefully, noting the amount of heat used for soldering, lead strain, and cleaning. Coated printed circuit boards should be thoroughly cleaned, then recoated.

When testing a unit which has been repaired, it is good practice to run the supply voltage up from zero with a variable transformer. As line voltage is slowly applied any serious defects which remain may be noticed soon enough to prevent further damage by monitoring the input power on a wattmeter. The repaired unit should be tested to its original specifications. Calibration to original specifications will disclose any additional shortcomings, or verify that the proper repair has been made. □

Chapter 5

Applications

5.1. INTRODUCTION

The following section has two aims: one is to provide information and guidance to the engineer who plans to design his own servo system; and the second is to illustrate the considerations which relate to the application of servo systems, so that the engineer who plans to buy a complete system can have a better understanding of the problems involved.

It is not possible to overstress the importance of complete cooperation between customer and supplier when designing a special motor or system. In many cases both the customer's engineers and the servo manufacturer's engineers may find themselves facing problems which are foreign to their previous experiences. For example, if the application is a knitting machine, the servo engineer may be confronted with problems involving stitch formation and thread handling; while the knitting machine designer may be facing problems associated with supplying electronic interfacing with the servo system. This kind of situation cannot be avoided where a new technology is applied in a traditional industry; however, the difficulty need not be an obstruction. Electro-Craft's experience as a principal manufacturer of servomotors and components for more than ten years has shown that when a high level of cooperation exists the chances of an application project being mutually successful are high.

5.2. SYSTEM CLASSIFICATION AND SPECIFICATION

More and more industries remote from the technology of electronics and servo systems are making use of the many advantages that DC servo systems can offer. In many cases, difficulty occurs in communication between a design engineer in such an industry and the servo systems designer. This section is intended to show how a potential user can examine his application, pick out the important performance criteria and describe them in terms meaningful to the servo designer.

5.2.1. CLASSIFICATION

Controlled Variable

An important step in examination of a system is *identification of the controlled variable.* By definition, a servo system monitors the condition of an output variable, comparing it to an input command and making adjustments to maintain an equality.

The most common output quantities used as controlled variables are:

1. Position
2. Speed
3. Torque (or force)

It is possible for a system to have more than one controlled variable, or for a system to switch from one variable to another during operation. Identification of the controlled variable is not always as straight forward as it might appear. For example, consider a drive system used to synchronize an audio tape recorder with a movie camera. At first, it would appear the purpose of the drive is to keep the speed of the tape recorder exactly equal to the speed of the camera. Closer study will show that the correct approach is to control the relative position of the two drives to ensure that the appropriate sounds occur while the correct picture is being projected when the film is played back. If this is done, the speeds will automatically be equalized.

The following are examples of controlled variables in typical applications:

Position	Velocity	Torque
line printer	metering pump	reel drive (constant tension)
matrix printer	matrix printer	
teleprinter	crystal grower	muscle exercizer
machine tool (x–y)	blueprint machine	weighing machine

Operating Mode

A servo system operates in one of two modes: incremental or continuous.

The *incremental mode* of operation is defined as the mode where the input command causes the controlled variable to change its value in discrete steps, with a brief static condition occurring between each step.

The *continuous mode* is one where the input command signal causes the controlled variable to remain constant, or to change in a linear (rather than step) mode.

Examples

Incremental	Continuous
line printer	audio tape recorder
photo typesetter	rewinding drive
digital tape transport	digital tape transport
disc head drive	pumps
	energy chopper
	laboratory stirrer

The digital tape transport is an example of a *dual mode* system. During normal reading the capstan drive motor moves the tape in discrete variable length steps, but during rewinding – or when scanning through a length of tape – the speed is held constant. The system, therefore, can operate in both incremental and continuous modes.

Load Classification

Loads can be classified under two headings. The first describes the kind of the motion, either rotational or linear. The second describes the means of coupling the load to the motor (capstan, lead screw, belt, gears, etc.).

5.2.2. SPECIFICATION

When the above classifications are established for a particular application, the next is the specification of performance. The way in which this is done will vary with the system classification.

Incremental Mode Systems

The most common incremental mode systems are position systems.

Their function is usually to move a load from one position to another, by steps. The steps may be of varying length, or may occur at different repetition rates. A *velocity profile* must be drawn to describe the motion accurately (Fig. 5.2.1).

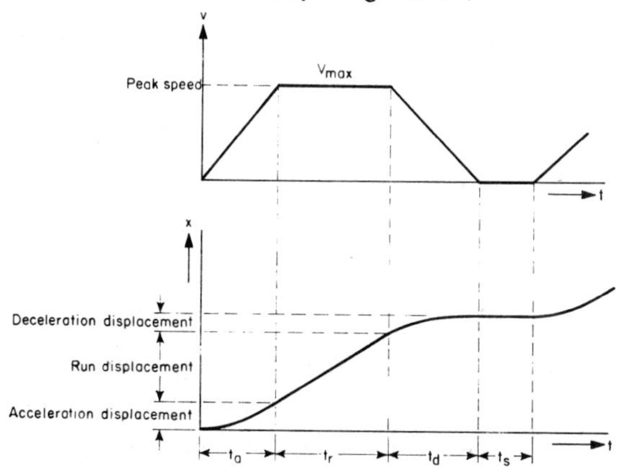

Fig. 5.2.1. Velocity profile diagram.

The velocity profile diagram represents the motion of the load during one increment. Where the peak velocity $\mathbf{v}_{\mathbf{max}}$ is not fixed by other requirements, it should be selected to make $\mathbf{t_a} = \mathbf{t_r} = \mathbf{t_d}$, which minimizes motor heating. If the system is an extremely high performance drive, a time interval of approximately 1 or 2 ms should be subtracted from the motion time and added to the dwell time to allow the system to settle after each step.

The units used for speed and displacement description will depend on the load classification.

If there are several different types of steps which the system is required to execute, a velocity profile must be drawn for both the average and the worst cases. An explanation of how the step varies is also required.

Next, motor velocity profiles should be drawn. This will use the units [rad/s] for speed and [rad] for angular displacement. These represent the motion which the motor has to execute to drive the load through the previously established profiles. To draw these profiles, one simply applies the various gear and coupling ratios used to the original load velocity profiles. A section covering selection of optimum coupling is presented earlier in this book (section 4.8.).

NOTE: Where the coupling type or ratio is not fixed at the time of writing the specification, it is good policy to leave them open and supply the servo designer with load velocity profile diagrams. This gives the designer maximum flexibility in selecting a motor, and prevents the possibility of the coupling ratio forcing the designer to use a motor of higher cost than necessary.

Load Description

All loads will possess the following properties: inertia, friction and viscous friction. Values for each of these should be given. Also, if the load is coupled to the motor by a method other than direct mounting on the motor shaft, each component of the coupling system will possess these three properties, and their values must be given. All of these values will be used to calculate the moment of inertia, friction torque and viscous damping factor as reflected to the motor shaft by the coupling system.

Accuracy

Accuracy is defined as the *deviation of the controlled variable from the desired level.* Accuracy specifications divide into two categories: short-term and long-term.

Short-term accuracies are normally defined by the system design, and include such factors as gain, line densities on encoders, bandwidth and motor mechanical features.

Long-term accuracies are normally controlled by the system *component* design, temperature coefficients of tachometers and operational amplifiers, etc. To meet long-term accuracy specifications, the system designer employs component selection and compensation techniques. Values of required accuracies are needed to determine the type of feedback transducer and other system components.

Continuous Mode Systems

The function of a continuous mode servo system is to maintain the value of the controlled variable constant against externally and internally introduced disturbances, and to execute changes in the value of the controlled variable when so commanded. Generally, the execution of intentional changes in the value of the controlled variable in a continuous mode system will either be made at a relatively low response rate, or the speed of response will not be a critical parameter. In either case, the torque required to accelerate or decelerate the load will not be a significant factor.

However, to ensure that an exception is not overlooked, it is wise to specify worst case and typical acceleration and deceleration rates; and, where applicable, repetition rates.

For a typical continuous mode servo system, the following data will be required.

Load Description

Inertia, friction and viscous friction data for the load and coupling mechanism will all be required exactly as for the incremental system. Special attention is required to evaluate all impulse type loadings which may be generated in gears and bearings.

Disturbances

Any impulse loading, whether generated internally or externally, should be included in this category. The important factor is the maximum impulse loading that the system will see. This may be one force or the sum of several forces. The customer needs to specify the peak value of the force, the duration and the average repetition rate. If the impulses are periodic, a graph similar to Fig. 5.2.2. showing a typical cycle, would be helpful.

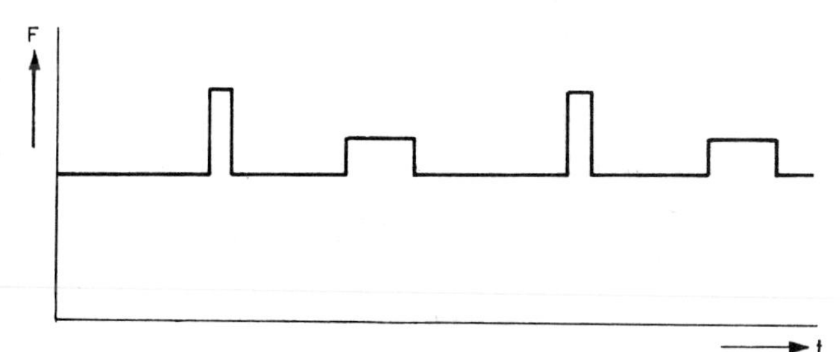

Fig. 5.2.2. Periodic load disturbance.

Although all the above information may be essential for complex high accuracy systems, the majority of systems will only require a statement describing the maximum value of impulse torque.

Error Size and Recovery Time

The maximum *error size* is the maximum value of instantaneous speed (or position) deviation due to the occurrence of a load or system disturbance. It should be expressed as a percentage of set speed (or position).

The *recovery time* describes the time the system takes to recover to the set speed or position after the disturbance ceases. A useful way to express this is the time the system should take to correct 63.2% of the error after the impulse force is removed. This is illustrated in Fig. 5.2.3.

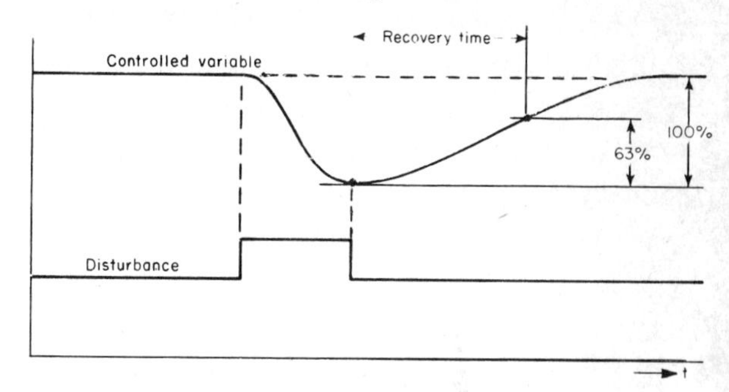

Fig. 5.2.3. System recovery.

Drift

Drift is a long term error due to changes in the temperature and other conditions of the system components. It should be expressed as the maximum percentage variation from the set level per unit of change in the temperature or alternative condition.

General Specification for Incremental and Continuous Servo Systems

In general, for both incremental and continuous mode systems, the following information should be provided:

1. Operating ambient temperature range
2. Operating ambient humidity range
3. Environment, i.e. laboratory, industrial, etc.
4. Available power supplies
5. Physical size and weight limitations
6. Life requirements

In addition, as much information as possible should be given regarding the operation and use of the final machine. Although the relevance of such information may not be apparent at first, it serves to make the servo designer as familiar as possible with the application. A complete understanding of the customer's machine is an invaluable advantage.

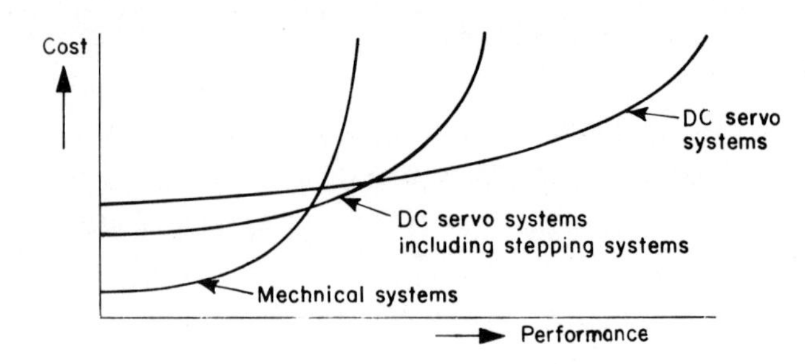

Fig. 5.2.4. Relationship between cost and performance of servo systems.

Cost

In general, the advantage of using a DC servo system is that higher performance, better reliability and improved marketability of the final product can be obtained at a lower cost than by upgrading more traditional devices. A typical relationship between cost and performance is shown in Fig. 5.2.4.

Performance can be speed, accuracy, reliability or any other part of a specification which presents a significant design problem.

The exact points where the alternative methods become uneconomical will vary, depending on the application and what the user defines as performance. It is true that in any application where the aim of the designer is to increase the marketability of a product by increasing its capabilities beyond present standards, then utilization of a DC servo system should always be investigated.

5.3. MOTOR SELECTION CRITERIA

5.3.1. INTRODUCTION

There is no known formal procedure for the optimum selection of a PM servomotor. One reason is that optimum is difficult to define. Here, *optimum* will be defined as *the most economical choice which satisfies specifications.* No matter what choice is made, it is important to understand how it affects the selection of adjacent system components, and to determine if the selected motor is adequate.

The following is a discussion of PM servomotor selection factors and analytical methods for analysis of the selected motor.

5.3.2. WHEN TO SPECIFY

The motor, drive train, and power supply should be given simultaneous consideration as early as possible in the system design process. It is a mistake to delay motor selection. Severe compromises may be required if the choice is limited by the remaining available space, or if the drive train and amplifier are fixed without having considered their effect on motor selection. For example, the moments of inertia of solid cylinders of homogenous materials (shafts, gears, pulleys) of equal length vary as the fourth power of diameter. Therefore, early consideration should be given to the materials and the geometry of these elements, because the motor and amplifier size will depend on it.

5.3.3. WHAT TO CONSIDER

A thorough description of the application should be given. This information is required to select the most suitable motor type with the proper ratings and design features. The information which is required to make this decision is given below. Later, there appears a detailed discussion of how the information is used.

1. Load Analysis

A thorough analysis of the machine loading is required. The analysis will not only help in the selection of the motor, but may also suggest design options which could lead to a more economical design approach.

Friction loads should be described from actual measurements, if possible. A plot of load torque vs. speed over the entire operating range is useful to evaluate viscous loading. Inherent design imbalances (i.e., eccentrics) should be described completely. The cyclical actuation of brakes, cams, clutches, or random disturbances on the system should be described as accurately as possible.

Inertial loads should be described. It is instructive to separate load moment of inertia into its component parts to identify the most significant sources.

A *velocity profile* should be described for each operational mode. The object of examining each operational mode is to determine what constitutes "worst case" operation.

2. Environmental Conditions

The following should be specified:

- Temperature range
- Altitude
- Humidity range
- Presence of airborne contaminants, chemical fumes, etc.
- Shock and vibration

3. Mechanical

The selection of all interfacing mechanical components should be given consideration to assure compatibility with the motor:

Heat sinking can have a dramatic effect not only on the motor, but also on the driven load. Good design practice will allow for proper heat conduction away from temperature sensitive machine members, as well as away from the motor.

A *coupling method* should be selected to prevent damage to the motor due to axial or radial forces. Conversely, motor shaft end play, radial play, and concentricity should be considered to assure proper mating to the load.

Motor mounting position should be specified.

Shaft *radial and axial loads* should be described, including both magnitude and direction.

Resonance problems can be avoided by closely coupling all elements of the drive train. Direct coupling is normally more economical, quieter and less prone to resonance problems.

4. Life

An indication of the desired life should be given. There is no analytical method for calculating motor life. But we do know that motor life is a complex function of: environment (temperature, altitude, humidity, air quality, shock, vibration, etc.), design (commutation design, bearing design, etc.), application variables (shaft loads, velocity profile, current form factor, etc.). The only conclusive method of determining motor life for a specific application is to perform life tests on an *actual* machine under *real* operating conditions. It is always good policy to perform life tests, providing that the test conditions are realistic. Attempts to simulate machine loads or alter the velocity profile for accelerated life tests can yield misleading results.

5. Efficiency

The PM motor is the most efficient type of electrical motor. However, in applications where the motor is operated about stall, or in an incremental mode, output/input power ratios are not primary considerations. In applications in which the motor is operated in the torque mode, output speed is zero; therefore, efficiency is zero by definition. For incrementers, a large portion of the output power is consumed changing the system energy during acceleration and deceleration.

6. Motor Type

PM servo motors are available in five basic configurations: torque, conventional iron core, surface wound, "printed" armature and moving coil. An attempt to make a relative comparison of them is given in Tab. 5.3.1. For certain, contradictions to the table exist because of the variation of performance available from each type. There perhaps should be an additional comparison made which indicates the design flexibility of each type: armature winding, commutation system, magnetic circuit, etc. Some types can be modified more easily and extensively than others to cause significant changes in all of the performance factors listed in the table.

For example, the low speed performance of an iron core motor can be altered significantly by: adding skewed pole shoes, skewing the lamination stack, increasing the number of commutator bars and/or changing the lamination design, etc. In fact, iron core type motors may be the most flexible in terms of design features and the most versatile in terms of being capable of operating satisfactorily in a variety of modes. Each of the other types could be considered as having evolved from it, with features designed to satisfy the more special demands of a specific class of applications.

5.3.4. SELECTION ANALYSIS

The initial selection of a motor for a specific application is usually based on an educated guess, considering the selection factors discussed in the preceding paragraphs. The type is determined after weighing the relative importance of the characteristics of each type. The size is estimated after making a quick calculation of the required output. The choice can be analyzed by considering the following:

1. Thermal Effects

The *ultimate winding temperature* is a primary consideration for every motor application. The calculation of armature temperature is derived below. The derivation is based on the assumption that temperature rise is proportional to the product of power dissipation and armature-to-ambient thermal resistance.

Input power

$$P_i = V\, I_a \qquad [W; V, A] \qquad (5.3.1)$$

Motor voltage

$$V = R_h I_a + 0.7395\, K_T n \qquad (5.3.2)$$

[V; Ω, A, oz-in/A, krpm]

R_h, the motor terminal resistance at the operating temperature is given by

$$R_h = R\,(1 + \psi\Theta_r) \qquad (5.3.3)$$

where Θ_r is the temperature rise.

Characteristic \ Motor Type		Torque	Iron core	Surface wound	Printed	Moving coil
Moment of inertia		large	medium	small	small	very small
Electrical time constant		long	average	short	very short	very short
Internal damping losses		medium	medium	small	large	small
High speed performance		poor	good	very good	poor	average
Low speed performance		very good	average	good	very good	good
Armature thermal time constant		long	long	average	short	very short
Torque/inertia ratio		low	medium	high	high	very high
Relative cost		medium	low	high	medium	high
Construction (typical)	Brush springs type	leaf	coil	coil	coil	coil
	Brush orientation	radial	radial	radial	axial	radial
	Commutator radius	large	medium	medium	large	medium
	Magnets	Alnico	Alnico/ceramic	Alnico/ceramic	ferrite/Alnico	Alnico

Tab. 5.3.1. Motor type comparison.

Combining (5.3.2) and (5.3.3) with (5.3.1) gives:

$$P_i = R\,I_a^2 + R I_a^2\,\psi\,\Theta_r + 0.7395\,I_a K_T\,n \tag{5.3.4}$$

The output power P_o is given by

$$P_o = 0.7395\,T_o n \tag{5.3.5}$$

[W; oz-in, krpm]

where T_o is the motor output torque. The power dissipation equals

$$P_L = P_i - P_o = R\,I_a^2 + R\,I_a^2\,\psi\Theta_r + 0.7395\,n\,(K_T I_a - T_o) \tag{5.3.6}$$

The temperature rise Θ_r is then given by

$$\Theta_r = P_L R_{th} = \frac{\left[RI_a^2 + 0.7395\,n\,(K_T I_a - T_o)\right] R_{th}}{1 - RI_a^2\,\psi\,R_{th}} \tag{5.3.7}$$

The term $(K_T I_a - T_o)$ may be recognized as the difference between internally generated torque and output torque. This difference is due to internal friction losses (brush, bearing, viscous, etc.). For many applications this term can be ignored without significantly affecting accuracy. The term I_a in (5.3.7) is the RMS value of armature current, calculated from the given velocity profile and other system parameters.

It is not always obvious what constitutes "worst case" operating conditions. If the machine is cycled in a variety of operational modes, temperature rise should be calculated for each mode to establish what constitutes worst case.

If the motor selected on the first try is limited because of temperature rise, the following options may be available: forced-air cooling, increased heat sinking, relaxed performance specifications, reduced loading (moment of inertia, friction, viscosity, etc.) Each of these options has drawbacks. Forced–air cooling may be provided by static type blowers which direct air axially or radially over the armature; a fan type which directs air over the external housing; or an impeller attached either internally or externally to the housing.

High pressure static blowers are in some instances more costly than the motor. They require added space, and they produce noise. Forced-air cooling is generally limited to cases where an air supply is already available. Directing a small amount of air over the housing by a fan is reasonably effective for constant power output applications; but, since the most motors armature-to-ambient thermal time constants are relatively long, there will not be good short-term overload protection. Impellers are not generally attached to servomotors because they introduce additional inertial and viscous loading — which defeats their purpose. Proper attention should always be given to the natural heat sinking available by mounting the motor to a good thermal conductor. *Low inertia motors* are especially susceptible to armature failures which are caused by short-term high loads. If an overload is imposed for more than several thermal time constants (armature-to-housing), trouble can be expected. If it is anticipated that overloading will occur, the best method for protecting the armature is to provide forced-air cooling.

2. Acceleration Effects

Motor *demagnetization* can occur if the required acceleration current exceeds the motor pulse current rating. Caution should be observed when the calculated value of acceleration current approaches the rated limit. Acceleration current is:

$$I_{max} = \frac{T_{g\ max}}{K_{T\ min}} \tag{5.3.8}$$

where $K_{T\ min}$ is the minimum value of K_T, specified at 25 ^oC and derated by 0.2%/^oC or 0.05%/^oC of *magnet* temperature rise for ceramic and Alnico magnets, respectively. Demagnetization can be avoided by: limiting amplifier current, relaxing performance specifications, reducing the shaft load (moment of inertia, friction, viscosity), or selecting another motor.

If the motor selected has a *long electrical time constant* that is long compared to the specified response times, the motor will be unsuitable for the application. A general rule of thumb is that the motor electrical time constant should be not greater than one-fifth of the minimum allowable time required for speed changes. If the motor electrical time constant is a limitation, a different motor will have to be considered.

3. Power Supply Effects

If the power supply has already been specified, calculations should be made to determine if sufficient current and voltage are available. If the power supply has not been designed, the voltage and current requirement must be defined.

Maximum required current is defined by (5.3.8). Maximum torque $T_{g\ max}$ must be determined by examining the load and velocity profile to determine when the maximum combination of acceleration torque, friction torque, and viscous torque occur.

If insufficient current is available, the following options may be considered:

- Select a winding with a higher K_T;
- Couple the load through a gearhead to increase the torque delivered to the load;
- Select a larger motor with a higher K_T;
- Reduce the load by minimizing moment of inertia, friction, and viscosity.

The *maximum required terminal voltage* is given by:

$$V_{max} = \frac{T_{g\ max}}{K'_T} R_h + 0.7395\ K'_T\ n_{max}$$

[V; oz-in, Ω, oz-in/A, krpm]

(5.3.9)

It may not be obvious if the maximum or minimum value of K_T should be used to evaluate V_{max}. Therefore, K'_T is intended to indicate whichever value of K_T that results in V_{max}. Both maximum and minimum values of K_T should be considered and the one selected which results in the maximum voltage. □

5.4. APPLICATION OF MOTOMATIC SYSTEMS

The first problem the designer faces is the selection of a basic motor and control system. Next, he must choose a controller with the needed features. The system selection will principally be based on required speed and torque.

Applications in this category usually have three characteristics: (1) system acceleration and deceleration characteristics are not critical, i.e., they can be measured in time intervals of tenths of a second rather than in milliseconds; (2) speed should be held constant at the set value over a wide range of load torques; and (3) the speed should be controllable over a relatively broad range, say 10:1 to 1000:1. Applications fitting these categories are numerous; and while the transistor speed control solution may not always be as apparent to potential users as other drive methods, in sections 5.5. and 5.6. we will present a variety of successful applications which demonstrate the versatility of the transistorized control system. While a speed controller for a silicon crystal puller may not seem likely to be similar to one for a blood cell separator, from the motor viewpoint they are just the same. Once your machine requirement is analyzed from the control system viewpoint, motor and control selection is considerably simplified.

To select the optimum motor and speed controller for your apparatus, you need to know the following:

1. Speed range required
2. Output torque required
3. Maximum ambient temperature
4. Other features needed (speed meter, torque meter, reversing switch, dynamic braking, 1—turn or 10—turn speed control potentiometer, etc.)

5.4.1. SPEED RANGE

The speed range must be exactly determined. It is necessary to determine whether it is preferable to employ a direct-drive motor or to use a gearhead. For example, if the load speed range is 0.1 rpm to 10 rpm, it is apparent that a 100:1 ratio gearhead would change the motor speed range to 10 rpm to 1000 rpm. Better still, a 200:1 gearhead would place the motor speed range at 20 rpm to 2000 rpm — a region of better speed stability.

A brief explanation at this point will clarify how to select the operational speed range that will result in optimum speed stability and regulation. The speed stability and regulation properties of a speed control system depend primarily on the "signal-to-noise" level of the feedback voltage. In the upper portion of the speed range, the feedback signal is much larger than the noise voltages generated in the amplifier, wiring and tachometer. However, as the speed is decreased toward the lower end of the available speed range, noise signals tend to become increasingly prominent. This is a physical fact which limits *any* analog feedback system.

Many times engineers consider using Motomatic systems to operate continuously in the lowest portion of the available speed range, simply because this would permit use of the output shaft in a direct-drive fashion without any gearing. While this may seem at first to be practical, the result would be the minimizing of available speed stability. If high speed stability and regulation are important, the proper solution is to insert a speed reduction between the motor and the load — thereby using a motor speed range which will yield best performance. Fig. 5.4.1 presents the speed-torque characteristics of a Motomatic E-650 speed control with regulation contours, showing that the optimum performance range is between 50 and 2000 rpm. This is the range in which one should strive to keep control applications, especially when high performance is a factor.

There are, of course, special applications where a primary machine operation must be performed in the very low end of the speed range, with occasional excursions into the high speed range. An example of this performance requirement is microfilm scanning, where a high speed search for the general area is followed by visual scanning in a very low speed mode — where any speed instability would be optically magnified, making visual scanning difficult. In such instances, design features are incorporated in the tachometer and amplifier to minimize instabilities due to ultra low speed operation.

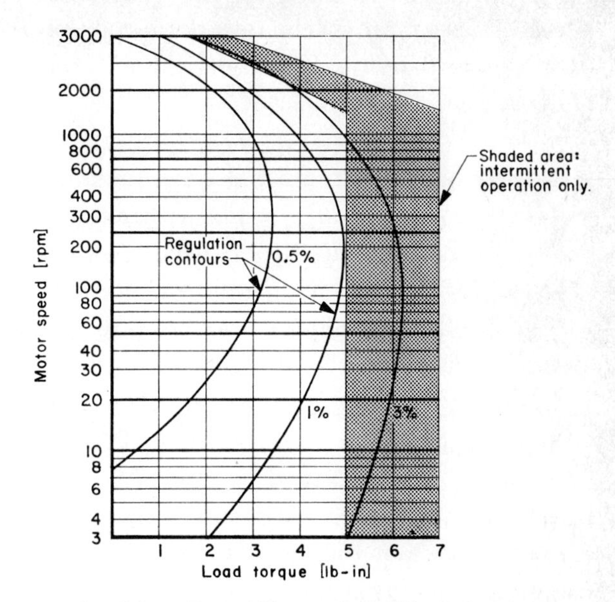

Fig. 5.4.1. Typical Motomatic E-650 speed-torque characteristics with regulation contours.

5.4.2. OUTPUT TORQUE AND AMBIENT TEMPERATURE

For determination of torque requirements of the load, two questions must be asked:

1. Is the precise time which the motor takes to accelerate or decelerate the load a critical factor?

2. Does stopping and starting occur frequently?

If the answer to either of these questions is *yes,* then the torque required to accelerate the load must be calculated, and the system selected to provide this torque. If both answers are *no*, then the maximum running torque is the rated torque required from the system.

Gear ratio selection must also be based on the *required output torque.* After the motor output torque figures have been adjusted for ambient temperature conditions (Fig. 5.4.2), the adjusted maximum torque can be multiplied by the gear ratio to get the theoretical output torque. The efficiency of gearheads varies between 60–70% for worm gearheads and between 75–85% for spur gearheads. We can arrive at a practical output torque figure by multiplying the theoretical torque by the efficiency figure. We must also consider the *maximum torque capacity* of the output gear. This is usually the limiting case in applications involving very high gear ratios.

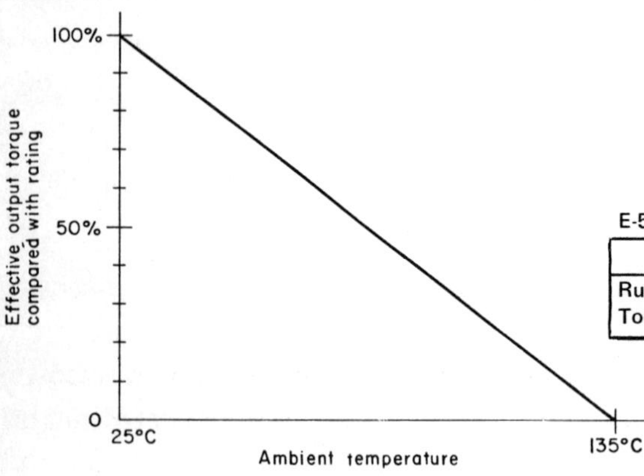

Fig. 5.4.2. Temperature derating curve for Motomatic motor-generators.

For example, consider the case of an Electro-Craft E-650 motor rated at 5 lb-in with a 1000:1 gear ratio, and an efficiency figure of 80%. The theoretical possible output torque would be calculated as follows:

$$T_p = 0.80 \times 10^3 \times 5 = 4000 \text{ lb-in}$$

but the maximum allowed output torque shown in the selection guide for this gearhead in Tab. 5.4.1 shows a *maximum allowed continuous torque* of only 80 lb–in; i.e., 1/50 of the possible torque at the output shaft.

If such a hypothetical gearhead-motor combination were stalled on the output side, the continuous torque rating could be exceeded by a factor of 20. In such a case, the gearhead would probably break, unless the torque limiter had been adjusted to hold the maximum input torque to a protective value.

E-550 MGHP (Precision Gearhead)

Ratio	10:1	30:1	50:1	100:1	500:1
Maximum Output Torque [lb-in]	6	20	25	45	50

E-550 MGHD (Heavy-Duty Gearhead)

Ratio	5:1	10:1	50:1	100:1	200:1	500:1	1000:1	5000:1
Running Torque [lb-in]	3	6	25	50	80	80	80	80

E-650 MGHR (Right Angle Gearhead)

Ratio	5:1	10:1	20:1	30:1	40:1
Maximum Output Torque [lb-in]	7.5	30	40	40	40

E-650 MGHS (Standard Spur Gearhead)

Ratio	50:1	100:1	200:1	500:1	1000:1
Running Torque [lb-in]	27	80	80	80	80

Tab. 5.4.1. Motomatic gearheads data.

Another factor which has to be considered is *the effect of inerital loads on the gearhead.* Because of the high acceleration and deceleration potential of a Motomatic motor-tachometer, it is relatively easy to induce high peak forces in the gearhead – at times high enough to break a gear tooth. If the application indicates that high inertia loads, input forces or overhung loads are apt to cause problems, contact Electro-Craft's Customer Engineering Department.

Figs. 5.4.3 and 5.4.4 illustrate motor speed ranges vs. torque for Motomatic systems with example reduction gear ratios.

Having determined the speed range and maximum torque requirements, the load characteristics can be superimposed on the system performance graphs (see Fig. 5.4.1) and the correct system selected.

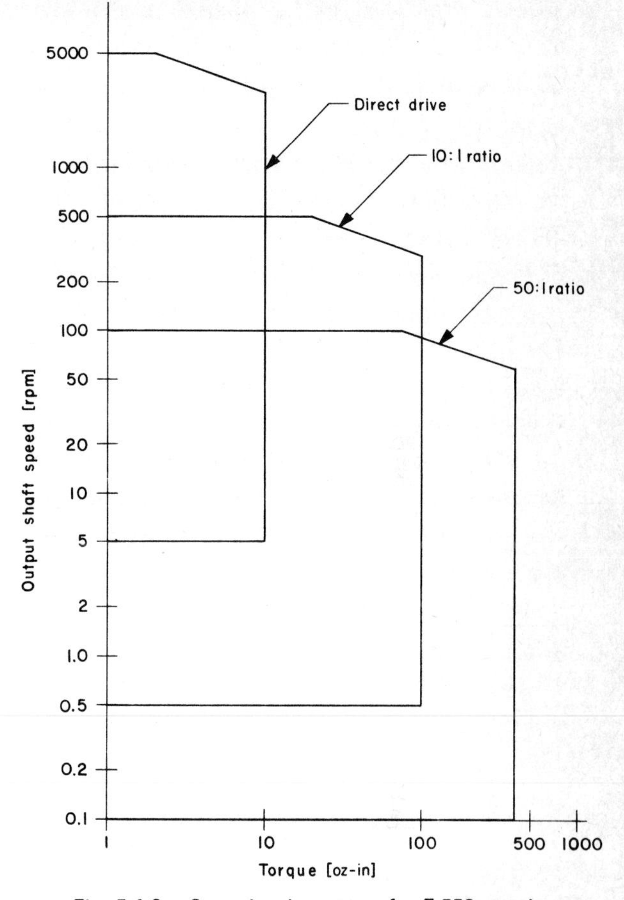

Fig. 5.4.3. Operational contours for E-550 speed control system with various gear ratios.

5.4.3. SPEED REGULATION AND STABILITY

The terms "speed regulation" and "speed stability" can sometimes cause misunderstandings between motor manufacturers and system designers.

Speed regulation is defined as the change of motor speed due to a change in output torque. Thus, at Electro-Craft, a change of motor speed from 1000 rpm at no-load to 990 rpm at full torque is considered 1% speed regulation. This is a most important distinction. At Electro-Craft, we rate speed regulation as a percentage of "set speed". Most controller specifications define it as a percentage of "top speed". This distinction becomes most important at the low end of the speed range. Thus, a 10 to 3000 rpm, 1% speed regulation Electro-Craft Motomatic controller at, say, 100 rpm may slow down to 99 rpm at full load torque. A conventional controller with a 1% speed regulation may slow down to 70 rpm (100 rpm – 1% of 3000 rpm top speed). Almost the same specification; quite a difference in actual performance!

As shown in Fig. 5.4.1, these regulation figures vary with the operational speed range. The regulation values for speed control systems should not be compared with open-loop motor speed regulation constant (R_m) discussed in Chapter 2. The reason why the *system* speed regulation is so much better than the *motor* speed regulation is that the tachometer and the amplifier minimize the regulation effect by the feedback connections.

Speed stability is a term consisting of two basic parts: short-term stability, covering such disturbances as oscillation, "jitter" and electrical noise "pickup" induced velocity errors; and long-term drift velocity errors, usually due to amplifier input stage drift and temperature-induced tachometer gradient changes. The latter effect is discussed in section 2.11.

The short-term effects, such as oscillations and jitter, can either be caused by the control system itself, or be induced by the load or the coupling between load and motor. Motomatic control systems are compensated to exhibit good speed stability over a wide range of speeds, and over a wide range of load conditions. It sometimes happens that a large inertia load is coupled to a motor-tachometer through a compliant (springy) coupling or through gears with backlash, and the result is that some instability will be noticed at certain speeds. This problem can usually be corrected by minimizing the offending condition.

Systems requiring a high degree of speed stability over a very wide speed range may sometimes experience unacceptable instan-

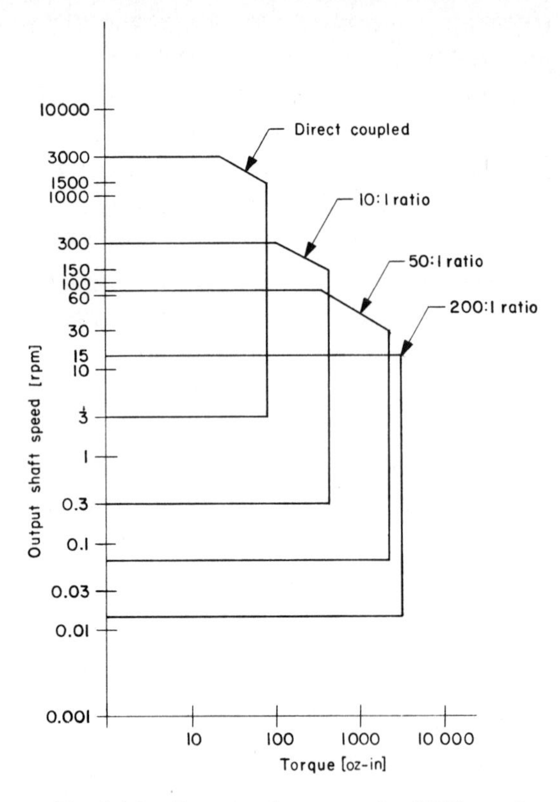

Fig. 5.4.4. Operational contours for E-650 speed control system with various gear ratios.

taneous speed variations (ISV) due to the magnetic coupling between the motor and tachometer windings. These ISV's occur because the bifilar winding of the Motomatic motor-generator acts as an electrical transformer, superimposing disturbances originating in the motor winding on the tachometer output. These disturbances normally occur outside of the frequency response of the system and have little effect. However, there are critical speeds at which the disturbances can cause measureable ISV's. The remedy for this situation is to use a motor with a tachometer generator mounted on the same shaft, but isolated electromagnetically from the motor winding. Such motors would be the E-576 as a substitute for the E-550 MG, and the E-676 as an alternative to the E-650 MG. This phenomenon is not usually significant, and can be ignored by 95% of users; but if a user believes that his system would be sensitive to such effects, he should contact the factory for advice.

The long-term drift is usually very small, since the Motomatic amplifier input stage has either temperature-matched differential input transistors or integrated circuits which serve to minimize drift.

The tachometer temperature coefficient errors are predictable, since for a given tachometer the maximum temperature coefficient is defined. When temperature conditions are known, it is a simple matter to predict the resulting errors. In cases where long-term errors cause problems, it is possible to provide air cooling of the tachometer, a heat radiator, or other means to minimize temperature effects.

5.4.4. MOTOMATIC SPEED CONTROL FEATURES

The outstanding standard features of Motomatic motor speed control systems are:

1. Wide speed range, 1000:1;

2. High speed regulation accuracy, up to ±1% of set speed depending on model;

3. Smooth, low cogging output at all speeds;

4. Small physical size;

5. High reliability.

In order to provide the greatest flexibility in adapting Motomatic systems to individual requirements, a number of features are offered, some of which are standard and others optional. Product description found in sections 5.4.5., 5.4.6. and in Chapter 6. lists which models contain particular features. A general description of what these features provide is given below.

Speed and Torque Readout

This is a meter which is calibrated to display either motor speed or output torque, as selected by a switch. This feature is supplied as standard on master controls and as an option for panel mounting of "open" controls.

Dynamic Braking

This technique is used in cases where a motor must be decelerated at a faster rate than load friction can provide. The dynamic braking feature consists of a load resistor which is automatically switched across the motor armature by a transistor circuit anytime the motor speed exceeds the control command. For example, if the load moment of inertia in an application is high and the friction very low, deceleration time may prove excessive when the speed control command is reduced. With dynamic braking, the deceleration time will be vastly improved; the system will follow commands both in acceleration as well as in deceleration.

Reversing Option

This feature consists of a switch which permits the selection of clockwise, stop, or counterclockwise operation. In the stop mode, the motor is dynamically braked by insertion of a resistor across the armature.

Adjustable Torque Limiting

This feature consists of a potentiometer which can be adjusted to provide a limit of the maximum output torque to any desired value. For example, if the load were such that it could be damaged by excessive motor torque, an adjustment of the torque limiter to a value below that in which damage could occur would provide the needed protection. The torque limiting circuit is completely automatic, and has a response time of a few milliseconds.

Remote Control Input

The speed setting or command may, in some applications, need to be controlled automatically from an external source rather than by the operator. Most Motomatic systems have a remote control input which can serve either of two functions: (1) speed can be controlled from any sensor whose output is a DC voltage sufficient to cover the needed input range, and (2) the speed adjustment potentiometer or selector switch can be panel-mounted at a separate location from the controller. Accompanying the remote input is a regulated voltage supply so that the remote input sensor can be as simple as a variable resistor, thermistor, photocell, linear potentiometer, etc.

5.4.5. SERIES E-550 MOTOMATIC CONTROLS

Fig. 5.4.5 shows an Electro-Craft E-550 control. The E-550 series controls have the following features:

Basic specifications:

Torque 10 oz-in continuous, speed range 5—5000 rpm.

Master controls only:

Electronic tachometer: Indicates motor speed in revolutions per minute.

Electronic torquemeter: Indicates motor load torque in ounce-inches.

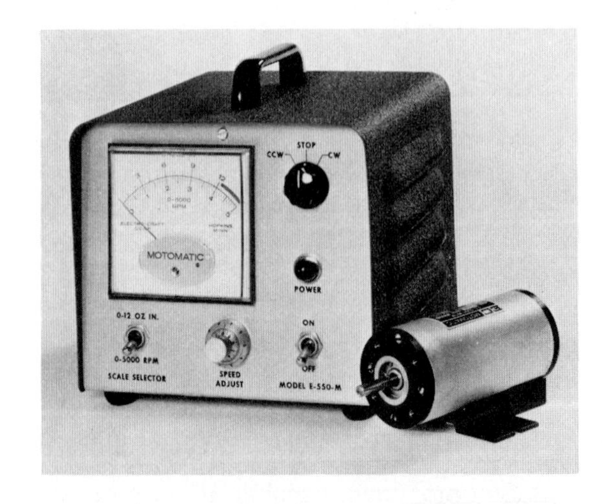

Fig. 5.4.5. Electro-Craft E-550 Master Control

Remote input: A provision for controlling motor speed by an external potentiometer or by a DC voltage derived from any source.

Master and Standard controls:

Reverse-brake switch: Reverses motor rotation through a center stop-brake position.

E-550 Front Control Panel

Power switch – this switch connects or disconnects the 115 VAC line for the control.

Power light – when this light is on, it indicates the power switch is in the ON position, and that the fuse and circuit breaker are normal.

Reverse-stop switch – this switch permits reversing of the motor through a center brake position. In the brake position, both the motor and generator leads are shorted through a current limiting resistor, bringing the motor to a stop in less than 0.5 s from full speed under no-load conditions.

Speed adjust control – this is a linear single turn potentiometer used to set speed between 0 and 5000 rpm.

Scale selector switch and meter – with the scale selector switch in the 0–5000 rpm position, the meter will indicate within 5% the speed of the motor in revolutions per minute. With the scale selector switch in the 0–12 oz-in position, the meter indicates the motor output torque within 10% of reading.

E-550 Rear Control Panel

Motor-generator connector – located in the lower left hand corner of the chassis.

Meter calibration control – this potentiometer is used to calibrate the tachometer to correctly indicate motor speed. To calibrate, use a strobe light locked to the line frequency at 3600 rpm and set motor speed at 3600 rpm. Loosen locknut on potentiometer, adjust meter to read 3600 rpm. Spot check at other speeds, and divide error if necessary.

Fuse – this is a 0.75 A slow blow fuse in series with the AC line.

Circuit breaker – this is a 2 A circuit breaker in series with the motor designed to prevent prolonged overloads of the motor and control.

Remote input switch and jacks – with the command switch in the remote position, the remote control jacks are switched into the circuit and the front panel control is switched out. Speed can be controlled by: (1) remotely placed potentiometer 5 to 10 kΩ; (2) a 0–20.5 VDC signal with a 5 to 10 kΩ source impedance; (3) selector switch.

Remote control connections are shown in Fig. 5.4.6. Examples of control voltage/speed relationships:

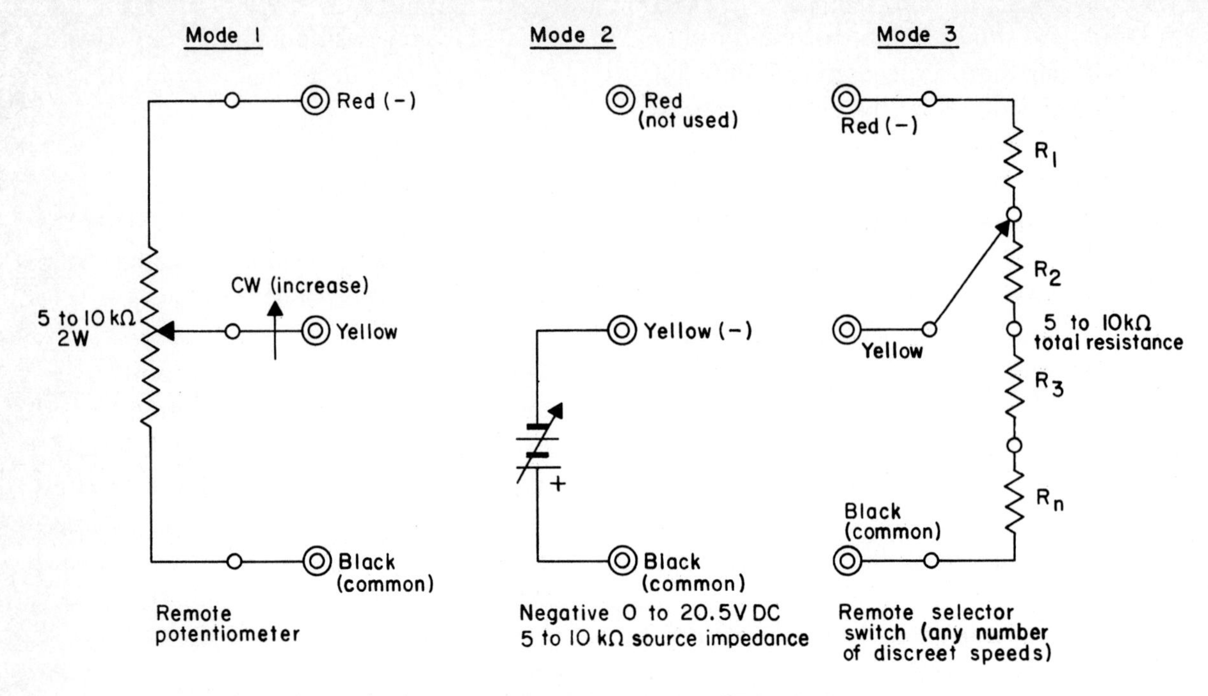

Fig. 5.4.6. Remote control mode connections for E-550 controls.

20.5 VDC input produces a motor speed of approximately 5000 rpm.

10 VDC input produces a motor speed of approximately 2500 rpm.

E-550 Operating Instructions

1. Make sure *POWER SWITCH* is in the *OFF* position.

2. Insert the motor-generator plug into the jack on the rear panel.

3. Connect the control *LINE CORD* to any approved 115 V, 50/60 Hz, grounded 3-wire receptacle.

4. Turn the *SPEED ADJUST CONTROL* to the counterclockwise stop.

5. Place the *COMMAND SWITCH* (on the rear panel) in the *NORMAL COMMAND* position.

6. Place the *SCALE SELECTOR* in the 0–5000 rpm position.

7. Place the *REVERSE-STOP SWITCH* in the *STOP* position.

8. Place the *POWER SWITCH* in the *ON* position. The red light just above the switch should light. If not, check the fuse and circuit breaker.

9. Place the *REVERSE-STOP SWITCH* in the *CW* position and slowly rotate the *SPEED ADJUST CONTROL* clockwise. The motor shaft should rotate smoothly in a clockwise direction looking at the shaft end of the motor. The tachometer will show an increasing reading as the *SPEED ADJUST CONTROL* is rotated clockwise.

10. Place the *REVERSE-STOP SWITCH* in the *CCW* position. The motor shaft should turn in a counterclockwise direction, looking at the motor at the shaft end.

11. Place the *SCALE SELECTOR* in the 0–12 oz-in position. As the load on the motor is increased, the torque meter will indicate an increasing reading.

Figs. 5.4.7 and 5.4.8 are outline drawings of the E-550 Master and E-550 Open Controls.

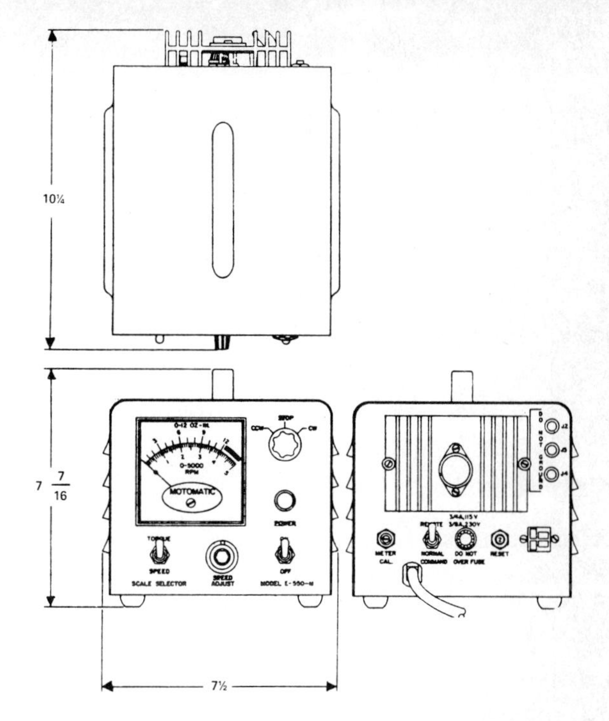

Fig. 5.4.7. Outline drawing, Electro-Craft E-550 Master Control.

Installation

When installing Electro-Craft motor-generators on any application, the following procedure should be carefully followed.

1. Mount securely on a rigid foundation.

2. Align all shafts accurately. The motor-generator shaft and the driven machine shaft should be in line as near perfectly as possible. Failure to maintain proper alignment can result in excessive noise, overheating and part wear. Flexible couplings can be used to compensate for slight misalignment. Care must be used in selection

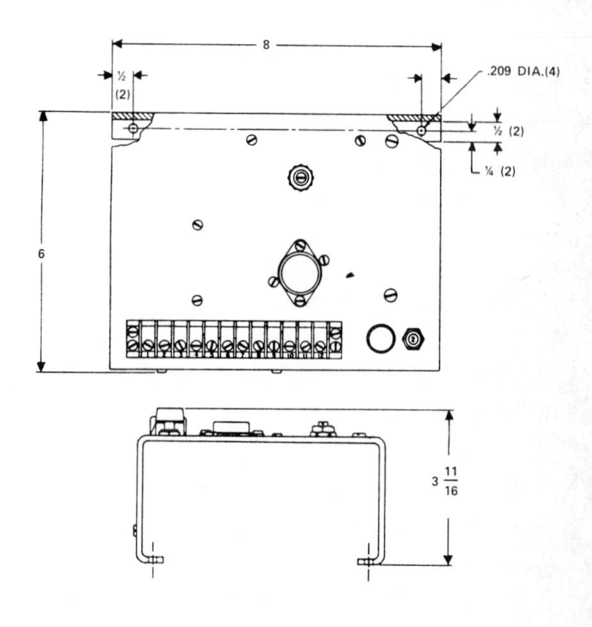

Fig. 5.4.8. Outline drawing, Electro-Craft E-550 Open Control.

of couplings, as any compliance in the coupling material may cause the control to sense a false change in load requirements. This can cause the motor speed to oscillate about its setting.

3. Observe all limits listed in the specifications concerning maximum overhung loads, side thrust and end thrust.

4. For best operation, the motor-generator should be mounted in an area where air circulation is available.

Maintenance

No lubrication is required, or should be used. If the motor-generator is equipped with a gearhead, lubrication instructions provided by Electro-Craft and/or the gearhead manufacturer should be followed carefully.

Under normal operating conditions, a 3000 hour brush life can be expected. If brushes are removed, they must be replaced in the same brushholder with the same orientation. If brushes are replaced, a 24 hour run-in period is recommended. Use only Electro-Craft brushes.

Disassembly of the motor-generator is *not* recommended. If disassembly is absolutely required, please contact Electro-Craft before proceeding.

5.4.6. SERIES E-650 MOTOMATIC CONTROLS

Shown in Fig. 5.4.9, Electro-Craft E-650 controls have the following features:

Basic specifications:

Torque 5 lb-in continuous; speed range 3–3000 rpm.

Master controls only:

Dynamic brake – this circuit allows more rapid deceleration of motor speed. It essentially places a short circuit across the motor winding during a down speed command. It is less effective at speeds under 100 rpm.

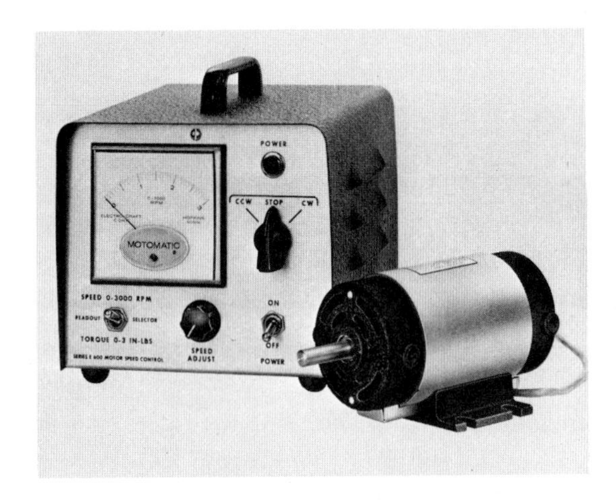

Fig. 5.4.9. Electro-Craft E-650 Master Control.

Torque limiting circuit – this circuit is, in effect, an electronic clutch. By limiting the current to the motor at any preselected level, the maximum torque can be adjusted to prevent damage to the motor, control and driven equipment in the event of a jam or stall.

Electronic tachometer – indicates motor speed in revolutions per minute.

Electronic torquemeter — indicates motor load torque in pound-inches.

Remote input — a provision for controlling motor speed by an external potentiometer or by a DC voltage derived from any source.

Master and Standard controls

Reverse-brake switch — reverses motor rotation through a center stop-brake position.

E-650 Front Control Panel

Power switch — this switch connects or disconnects the 115 VAC line for the control.

Power light — when this light is on it indicates the power switch is in the *ON* position, and that the fuse and circuit breaker are normal.

Reverse-stop switch — this switch permits reversing of the motor through a center brake position. In the brake position, both the motor and generator leads are shorted through a current limiting resistor bringing the motor to a stop in less than 0.5 s from full speed under no-load conditions.

Speed adjust control — this is a 10-turn potentiometer, and is calibrated to indicate the percentage of top speed (3000 rpm) within 3%.

Scale selector switch and meter — with the scale selector switch in the 0–3000 rpm position, the meter will indicate the speed of the motor in revolutions per minute. With the scale selector switch in the 0–5 lb-in position, the meter indicates the motor output torque within 10% of reading.

E-650 Rear Control Panel

Motor-generator connector — located in the rear lower left hand corner of the chassis. When the motor-generator is plugged into this connector, the lock should be turned so the connector cannot become disengaged through vibration or accident.

Fuse — this is a 5 A, 250 V, 3 AG fuse in series with the circuit breaker and the AC line.

Meter calibration control — this potentiometer is used to calibrate the tachometer to correctly indicate motor speed. To calibrate, use a strobe light locked to line frequency at 1800 rpm and set motor speed at 1800 rpm. Loosen locknut on the potentiometer and adjust meter to read 1800 rpm. Spot check at other speeds, and divide errors if necessary.

Torque control — this is the control for the electronic torque limiting circuit. If a load greater than the pre-selected level is applied, the motor and control cannot deliver additional torque, thus preventing damage to the driven equipment. Torque limiting can also be used to control acceleration. With the control in the minimum position,

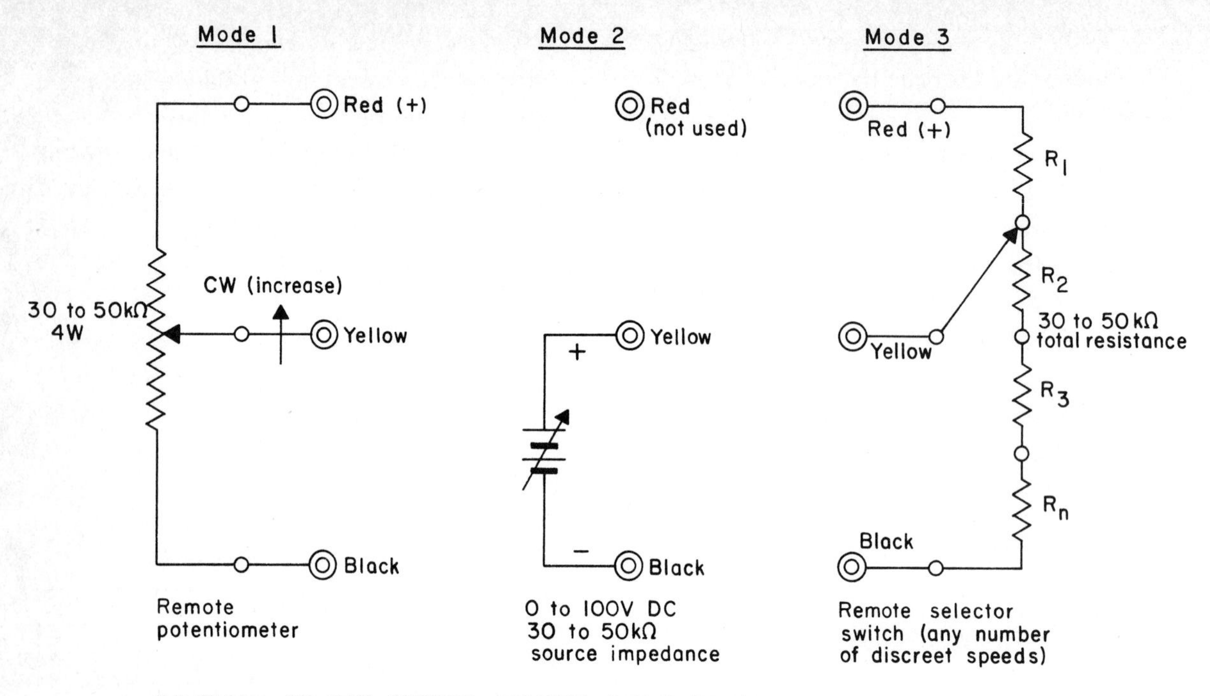

Fig. 5.4.10. Remote control mode connections for E-650 controls.

the maximum torque that can be delivered is approximately 1.5 lb-in. With the control in the maximum position, the control and motor-generator can deliver torque in excess of 5 lb-in. However, a load greater than 5 lb-in (red area) for a sustained period will result in damage to the control or motor-generator, or both.

Remote input switch and jacks – with the command switch in the remote position, the remote control jacks are switched into the circuit and the front panel control is switched out. Speed can be controlled by: (1) a remotely placed potentiometer of 30 to 50 kΩ; (2) a 0–100 VDC signal with a 30 to 50 kΩ source impedance; (3) the selector switch. Connections are shown in Fig. 5.4.10.

NOTE: In mode 2, speed stability is dependent on the stability of the remote control voltage. Maximum speed of the motor is also proportional to remote input voltage; e.g., a 100 VDC input voltage produces approximately 3000 rpm motor speed, and a 50 VDC input voltage produces approximately 1500 rpm. This can be changed by factory adjustment of the controller.

E-650 Operating Instructions

1. Make sure the *POWER SWITCH* is in the *OFF* position.

2. Connect the motor-generator cable to the connector on the rear of the chassis.

3. Connect the control *LINE CORD* to any approved 115 V, 50/60 Hz, grounded, 3-wire receptacle.

4. Turn the *SPEED ADJUST CONTROL* fully counterclockwise.

5. Place the *COMMAND SWITCH* on rear panel in the *NORMAL COMMAND* position.

6. Place the *SCALE SELECTOR* in the 0—3000 rpm position.

7. Place the *REVERSE STOP SWITCH* in the *STOP* position.

8. Place the *POWER SWITCH* in the *ON* position. The red light just above the power switch should light. If not, check the fuse located on the back of the chassis and reset circuit breaker.

9. Place the *REVERSE-STOP SWITCH* in the *CW* position and slowly rotate the *SPEED ADJUST CONTROL* clockwise. The motor shaft should rotate smoothly in a clockwise direction looking at the shaft end of the motor. The tachometer will show an increasing reading as the *SPEED ADJUST CONTROL* is rotated clockwise.

10. Place the *REVERSE STOP SWITCH* in the *CCW* position. The motor shaft should turn in a counterclockwise direction, viewing the shaft end of the motor.

11. Place the *SCALE SELECTOR* in the 0—5 lb-in position. As the load on the motor is increased, the torque meter will indicate an increasing reading.

Figs. 5.4.11 and 5.4.12 are outline drawings of the E-650 Master and Open Controls.

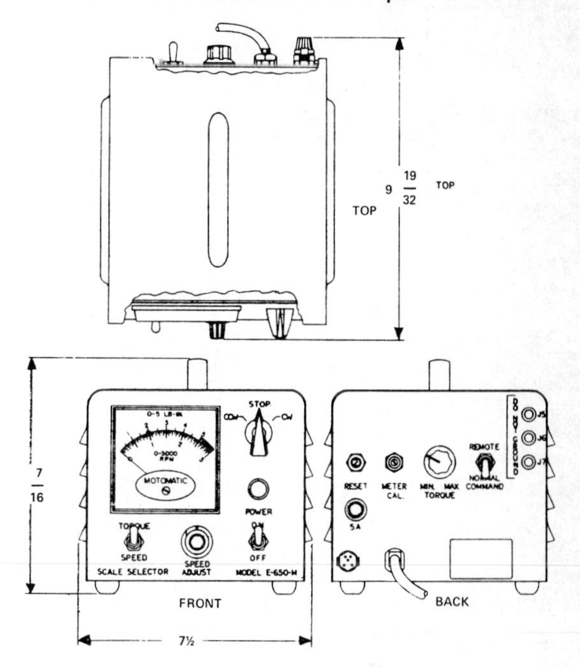

Fig. 5.4.11. Outline drawing, Electro-Craft E-650 Master Control.

Installation

When installing Electro-Craft motor-generators on any application, the following procedure should be carefully followed.

1. Mount securely on a rigid foundation.

2. Align all shafts accurately. The motor-generator shaft and the driven machine shaft should be in line as near perfectly as possible. Failure to maintain proper alignment can result in excessive noise, over-

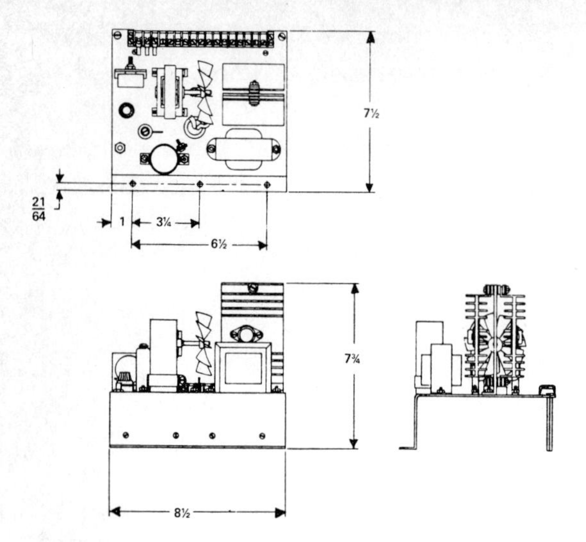

Fig. 5.4.12. Outline drawing, Electro-Craft E-650 Open Control.

heating and part wear. Flexible couplings can be used to compensate for slight misalignment. Care must be used in selection of couplings, as any compliance in the coupling material may cause the control to sense a false change in load requirements. This can cause the motor speed to oscillate about its setting.

3. Observe all limits listed in the specifications concerning maximum overhung loads, side thrust and end thrust.

4. For best operation, the motor-generator should be mounted in an area where air circulation is available.

Maintenance

No lubrication is required or should be used. If the motor-generator is equipped with a gearhead, lubrication instructions provided by Electro-Craft and/or the gearhead manufacturer should be followed carefully.

Under normal operating conditions a 3000 hour brush life can be expected. If brushes are removed they must be replaced in the same brushholder with the same orientation. If brushes are replaced, a 24 hour run-in period is recommended. Use only Electro-Craft brushes.

Disassembly of the motor-generator is *not* recommended. If disassembly is absolutely required, please contact Electro-Craft before proceeding. □

5.5. APPLICATION EXAMPLES

In this section we will discuss a variety of applications of Electro-Craft Motomatic and other controls taken from our customer files. Some figures and specific details have been altered to protect proprietary information. These examples are set forth to illustrate the versatility of the controls, and to show how they have met requirements in applications varying from office machine drives to medical instrumentation controls.

We hope these "case histories" will help readers realize that applications of packaged speed control systems are limited only by the imagination of the potential users.

OFFICE COPYING MACHINE

An office copying machine had a manually-operated mechanical speed control device, which provided exposure control for an infrared sensitive copy medium. In the initial design, the operator had to watch a temperature gage and continuously adjust the speed to match the varying temperature of the infrared source. The mechanical speed control had a reliability problem; the manual adjustment resulted in a high waste factor due to operator inattention.

The Electro-Craft E-500 motor-generator, and a modified version of our standard speed control system were selected for this application. This E-500 MG and control, as are any of the other Motomatic systems, is ideally suited for automatic response to changes in not only temperature, but light, pressure, or any other parameter which can be converted to a resistance of voltage change.

Fig. 5.5.1 shows the modifications to the standard control which were necessary to meet this application's requirements. The thermistor is exposed to the same infrared source as the original and sensitized paper. The potentiometer R_2 is a manual speed adjust, and is set to achieve desired copy contrast. The resistor R_3 sets a minimum speed limit to prevent heat damage from overexposure. The design of the combination of thermistor and resistors in the sensor-command chain had to be coordinated with the desired response of the control.

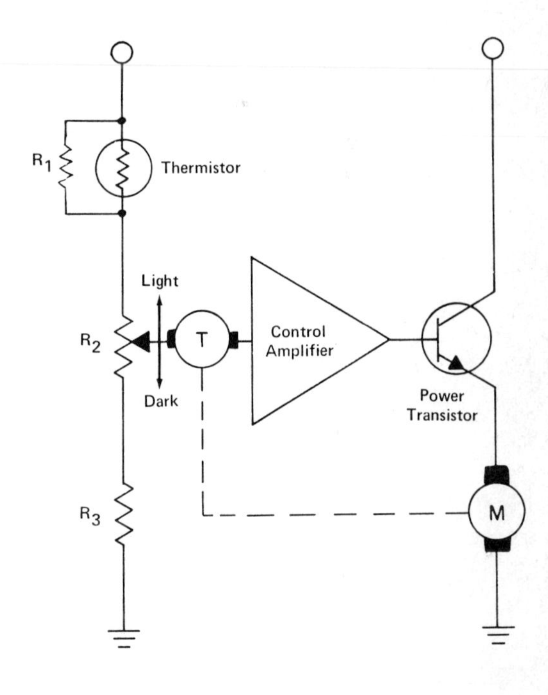

Fig. 5.5.1. Example of a temperature-sensitive speed control input stage.

During the copying process, the increase in temperature in the exposure area is sensed by the thermistor, causing a decrease in its resistance. The resistance decrease causes the command voltage to rise, increasing copying speed. As the machine cools, the thermistor's resistance increases and the drive speed decreases. These interactions maintain the selected heat/speed ratio.

The use of the E-500 system in the redesign improved the office copying machine in the following respects:

1. automatic response to temperature and speed command changes;
2. reliable exposure control, and increased customer satisfaction because of lower waste;
3. performance reliability was greatly increased;
4. reduced noise level;
5. moderate additional system costs.

HYDRAULIC MOTOR DRIVES

A hydraulic motor utilized a system of pressurized hydraulic vanes to perform torque amplification, thereby providing a very powerful rotating shaft output. The output shaft was to follow very closely the speed and position of an input shaft requiring very low input torque. The customer wanted a suitable servomotor and control which could command the hydraulic motor, giving it very broad speed range and optimum speed response.

The Motomatic E-550 system was selected for this application because its speed range more than met the manufacturer's needs, and the torque/moment of inertia ratio of the system permitted the responsive acceleration and deceleration needed by machine tool users. The E-550 MG and hydraulic motor are shown in Fig. 5.5.2.

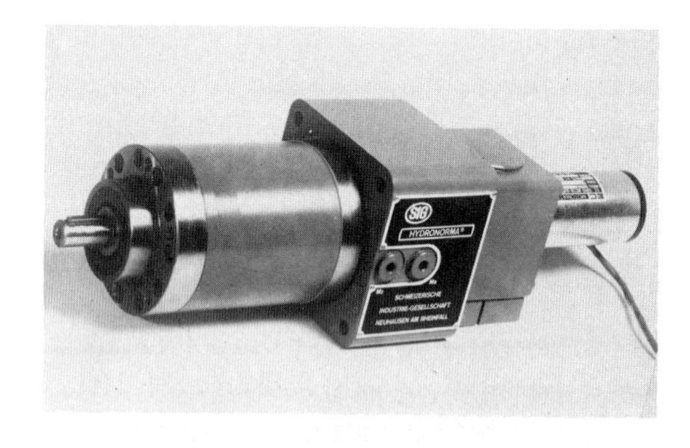

Fig. 5.5.2. The E-550 MG utilized as an input drive for hydraulic motor.

In much the same way as the rotary hydraulic motor is a *torque* amplifier, the linear hydraulic motor is a *force* amplifier. The input motor shaft (in this case an E-650 MG) controls speed and displacement of the actuating arm. The unit shown in Fig. 5.5.3 is capable of applying a force of several hundred tons through a displacement of several meters. Quite clearly, reliability and broad speed range are the two key advantages of a closed–loop control system in this application.

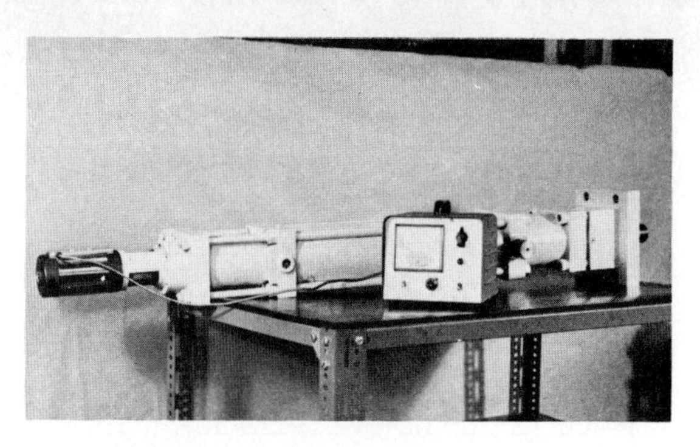

Fig. 5.5.3. Linear hydraulic motor utilizing an E-650 MG as input drive. (Photo courtesy of Précision Mécanique Labinal, France)

AUTOMATIC RETRIEVAL SYSTEM

A manufacturer of automated storage systems needed a high-response drive system to power the carriage in their **x—y—z** motion control arrangement. The customer selected the E-678 motor on the basis of its conformance to performance requirement.

Fig. 5.5.4. The CONSERV-A-TRIEVE automated storage system. (Photo courtesy of Supreme Equipment and Systems Corp., Brooklyn, N.Y.)

Since this customer operates a circuit design facility, he chose to design a special servo circuit to fit the unique requirements for a wholly automated retrieval system. A photograph of the system is shown in Fig. 5.5.4.

SPEED-TORQUE ANALYZER

A producer of custom-made gearing desired to measure accurately the friction torque of his gears. He provided his customers with data showing friction torque vs. speed, and on request produced friction data of the gearing at the customer's specified operating speed. This would allow his customers to take necessary measures to handle friction in their system designs. This application is typical of many situations where friction torque measurements are necessary. It can also be applied to the elimination of design errors due to inadequately stated data by providing static torque measurements and by monitoring shaft speed.

Using a Motomatic E-550 speed control system, his procedure for measuring any unknown torque is as follows:

1) with the motor unloaded, adjust motor speed to the desired setting by reading speed on the Motomatic speed meter.

2) with the motor operating at no load, record the torque reading on the Motomatic electronic torque meter. This reading will be the sum of internal motor friction torque and damping torque at the selected speed.

3) connect the unknown friction load and, again operating at the desired speed, record the reading on the torque meter.

4) the actual friction torque is the difference between the readings obtained in steps 2 and 3 above.

The E-550 speed control system gives both readout of speed and readout of the developed motor torque.

ENERGY CHOPPER

The measurement of radiant energy requires the energy input to the sensor be chopped at a given frequency to separate the true signal from undesirable background and noise. A reference synchronization pulse is generated synchronously with the resultant chopped energy signal as shown in Fig. 5.5.5. In laboratory testing, where it is desirable to provide optimum chopping for a detector, the blade must be changed in order to cover a total frequency range of 0.65 Hz to 1 kHz. To fulfill this requirement with a minimum of operator effort, a multi-speed motor control was required. The standard E-550 motor-generator and open control was selected for this application. The wide speed range (1000:1) of this control system made feasible the concept of a one plug-in selector card for the detector. With this selector card and its related chopper blade, the appropriate sensor detector instrument turning and mechanical chopping rate were automatically provided.

Utilizing the E-550 control system, the laboratory energy chopper was greatly improved over previous designs. Besides the feasibility of the one plug-in card for all adjustments, the system was improved in the following respects:

1) the wide speed range of the E-550 control system allowed a larger total frequency range to be scanned;

2) the accurate speed regulation of the control system (better than 0.5%) improved accuracy of readings;

3) laboratory tests were repeatable with the E-550;

4) the long term speed stability of the E-550 provided accurate reading whether the laboratory instruments were operating for just minutes or for 18 or more hours a day;

5) the constant speed provided by the E-550 system, which is independent of load torque variations and line variations, maintained a constant chopper frequency.

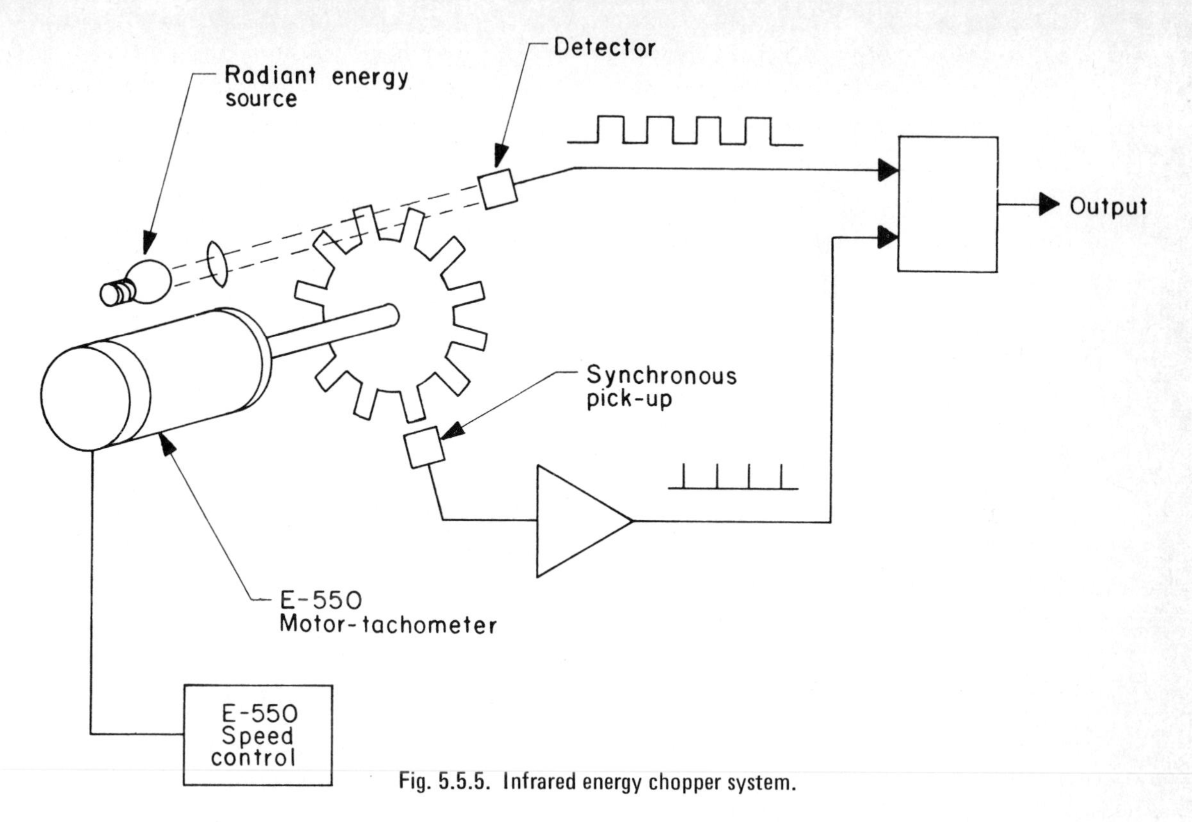

Fig. 5.5.5. Infrared energy chopper system.

This system, in effect a multispeed chopper, provides optimum chopping frequencies for virtually any type of detector.

Motomatic controls can be utilized in many other types of laboratory equipment, and will provide improved system performance as noted above. Applications quite similar to the above are: laboratory stirrers, industrial drilling machines, centrifuges and viscosimeters. Any application requiring high speed stability and regulation, and constant speed irrespective of load and line variations, can utilize Motomatic systems.

LABORATORY STIRRER

A laboratory instrument manufacturer needed a wide range speed control system to power a laboratory stirrer. The system should have high torque capacity and good regulation to accommodate a variety of viscosity levels and stirring speed requirements.

A Motomatic E-650 system was designed to fit the customer's requirements; the resulting design (Fig. 5.5.6) is a versatile, dependable laboratory stirrer which has found wide-spread use in laboratories.

In another case, a special automatic laboratory stirrer was designed to supply oxygen release at a controlled, adjustable rate for tissue-growing experiments.

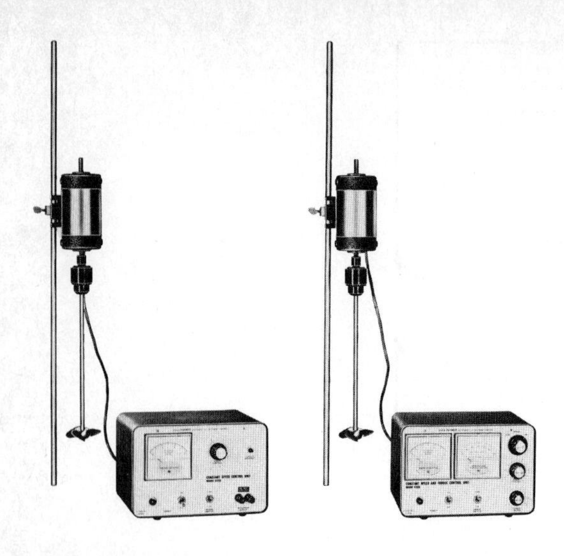

Fig. 5.5.6. Application of E-650 standard and master controls for Servodyne Laboratory Stirrer. (Photo courtesy Cole-Parmer Instrument Co., Chicago, Ill.)

The relationship between oxygen release and stirring speed is nearly exponential, as shown in Fig. 5.5.7. The "oxygen feedback gain" would vary depending on which portion of the curve the system would operate. However, since the working range was limited, the gain change was only 4:1, which was within the stability margin of the system.

Since the oxygen release time constant was on the order of 100s, it necessitated a special operational amplifier to provide feedback compensation in the "outer loop" to assure stability (Fig. 5.5.8). The "inner" velocity loop consisted of an E-550 Motomatic control.

FILAMENT WINDING MACHINE

A customer wanted a versatile filament winding machine which would feature variable mandrel speed, and also have a traverse mechanism with an adjustable traverse rate which would follow the varying mandrel speed in a proportional fashion.

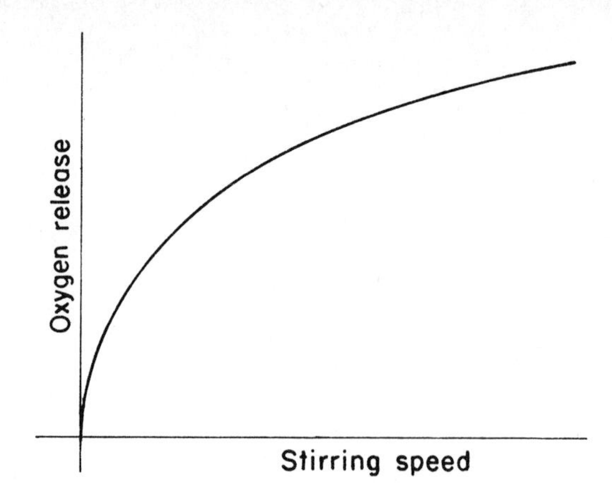

Fig. 5.5.7. Relationship between oxygen release and stirring speed in a tissue culture.

The solution was a dual speed control arrangement shown in Figs. 5.5.9 and 5.5.10. The E-650 gearhead motor-generator drives the winding mandrel through a coupling. The E-650 control system supplies generator feedback voltage to the E-550 control as a reference voltage; thus the traverse drive is always "slaved" to the speed of the winding drive. The ratio between the two speeds is adjustable over a wide range, however, and the filament winder can accommodate a range of filament widths and spacing requirements.

This control method is adaptable to a host of other operations, requiring close ratio control of two variables.

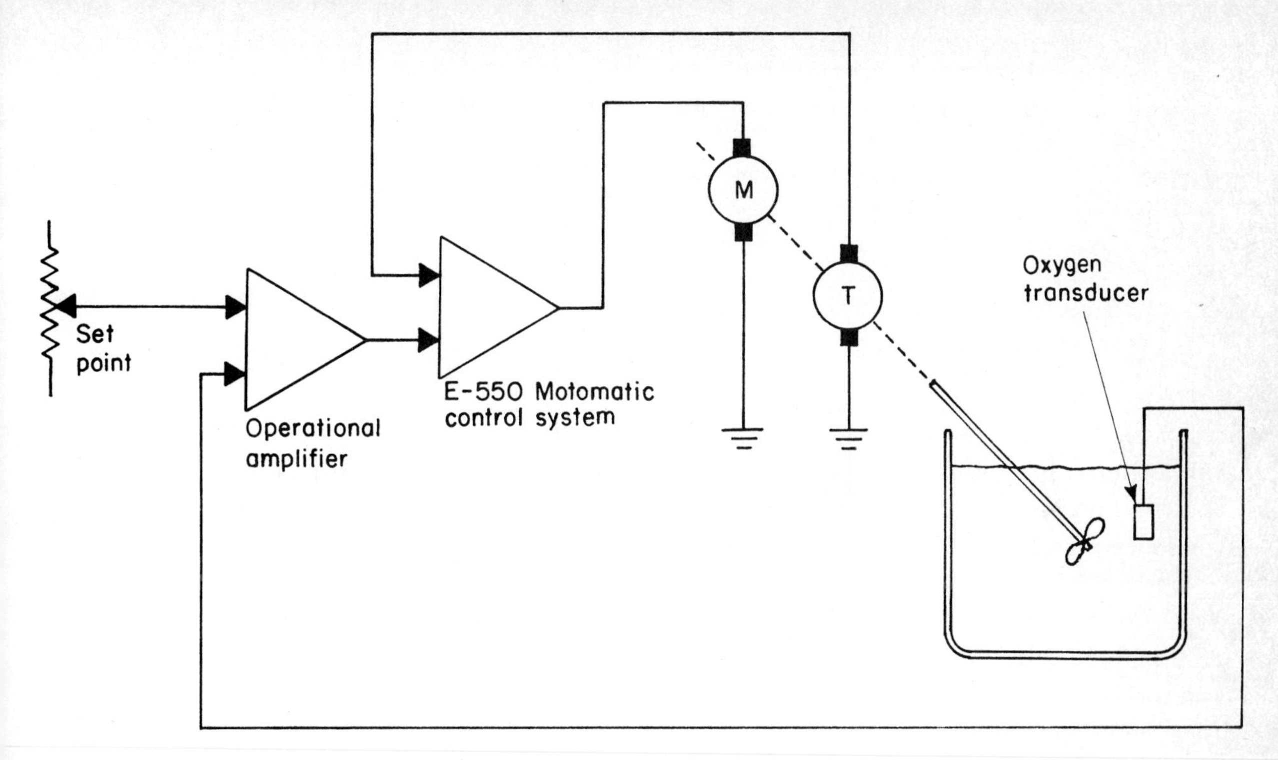

Fig. 5.5.8. Setup for oxygen release control.

REWINDER SPEED CONTROL

The application illustrated in Fig. 5.5.11 is a system used in textile fabrication for an operation called "center winding", but the method is usable for wire, fiber, paper, film, and any other material which must be rewound at a constant linear velocity.

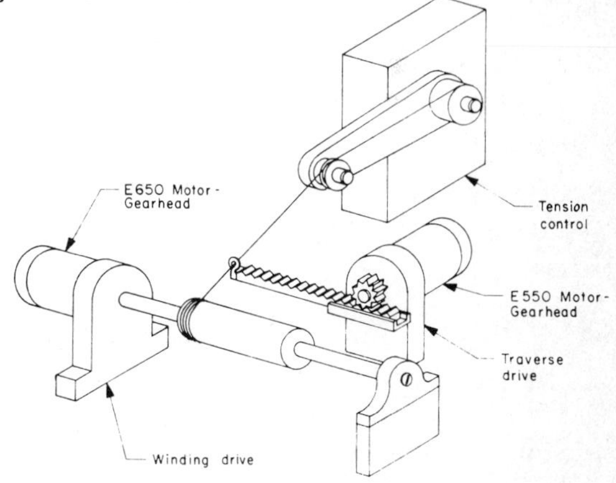

Fig. 5.5.9. Dual motor filament winding machine arrangement.

Since the center winding process requires constant yarn speed, the shaft speed of the drive motor must be inversely proportional to the material roll diameter.

An E-650 control system connected as shown provides the needed performance. A gear and a magnetic pickup provide indication of yarn velocity. The gear is mounted on a small roller which rides on the surface of the material roll. Pulses from the pickup are converted to a voltage

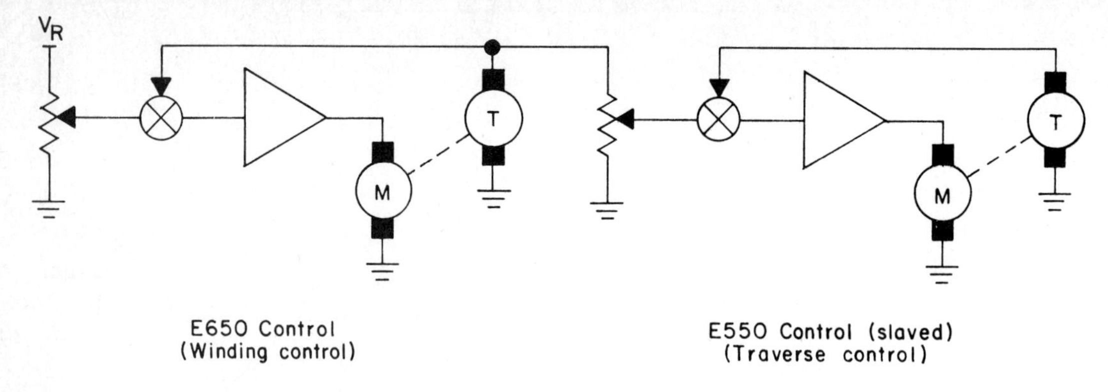

Fig. 5.5.10. Electrical diagram for the dual motor system of Fig. 5.5.9.

proportional to pulse rate (hence, to yarn speed). Comparing this voltage with the reference preset yarn speed, an error signal is obtained and fed to the DC control. The motor speed gradually decreases to maintain equality between the reference and the feedback voltage, thus holding yarn speed essentially constant during the roll buildup.

MULTI-MOTOR CONTROL FOR INDUSTRIAL KNITTING MACHINES

A typical multi-motor control system shown in Fig. 5.5.12 includes power supplies, control circuits, safety circuits and motor-generators to provide variable feed rates and sequences for feeding the material into knitting machinery.

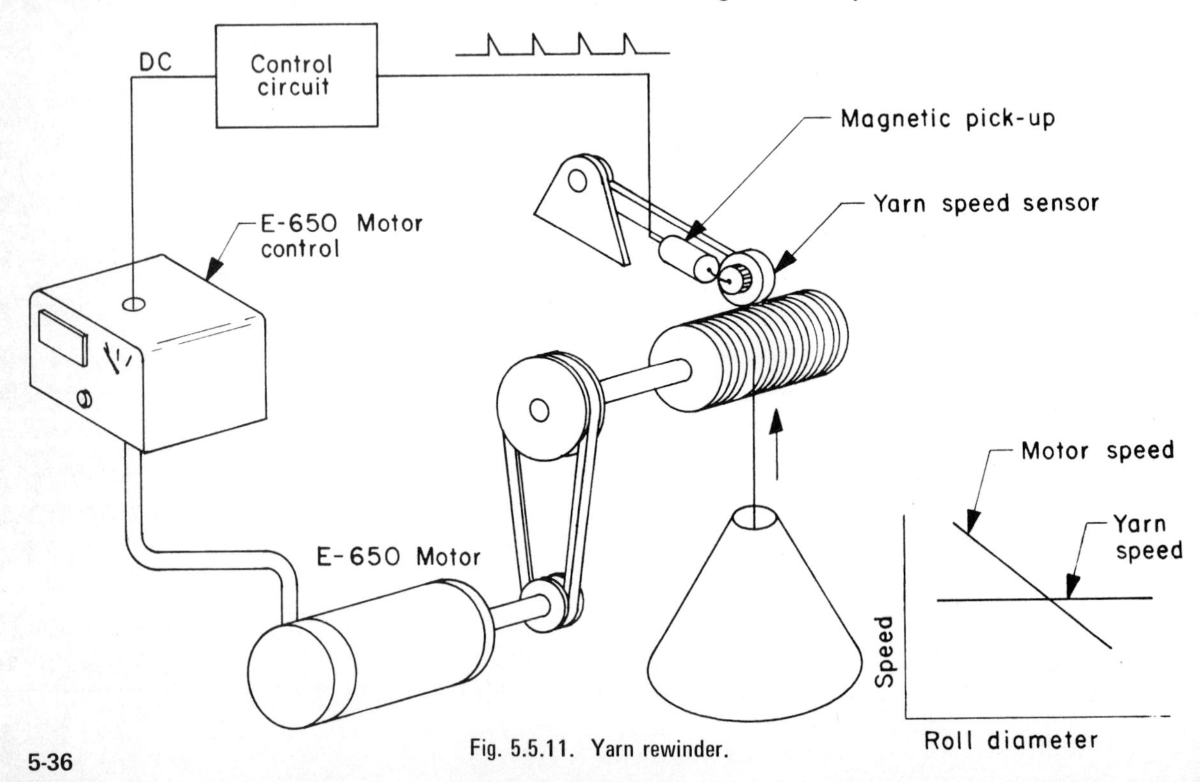

Fig. 5.5.11. Yarn rewinder.

The desired system was to provide adjustable output speeds in both incremental and continuous modes. High reliability and minimum downtime were also required.

Speed range and output torque requirements dictated that an Electro-Craft E-650 system be utilized. Off-the-shelf E-650 Motor-Generators with right angle gearheads (Fig. 5.5.13) were used. The control circuits are modified versions of off-the-shelf items.

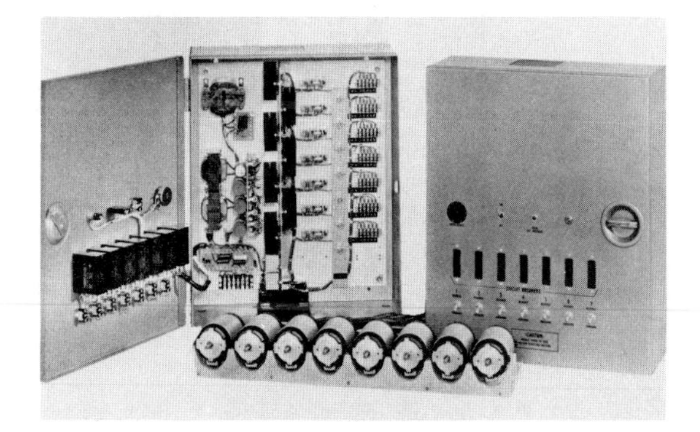

Fig. 5.5.12. Eight-motor knitting machine control.

Fig. 5.5.14 shows the circuit diagram of a typical multi-motor control system. The potentiometer **R1** is the master speed adjust, and sets the base speed for all systems. Differences between voltages on the wiper of **R1** and the output of **G1** produce the error signal for speed alterations. In a multi-motor system all generators have a common connection to **R1**. A speed input adjustment causes speed changes in all systems. Resistors **R2** and **R3** from a voltage divider for the generator output. **R2** is a gradient adjustment, setting a ratio between speed command signals from **R1** and speed changes at the motor-generator shaft. The gate switch , **S1**, opens and closes the control amplifier input, starting and stopping the motor-generator. In addition, a junction point is provided in the speed adjust line for common sequences of all systems. The transistorized control amplifier provides the drive for the power output stage which controls the motor and a circuit breaker. The contacts of the circuit breaker are in series with a main breaker and a safety line which monitors the roving. If overloading occurs in the motor-generators or the power supply, or a break in roving occurs, the entire feed system will shut down.

Fig. 5.5.13. Electro-Craft motor with right angle gearhead.

Reliability in environmental extremes and minimum downtime are designed into the system by selection of high grade industrial components, assembled in easily removable modules. Modifications or additions to this scheme include meters to monitor speed and/or torque for each motor-generator, remote speed control, variable torque limiting, dynamic braking and reversing.

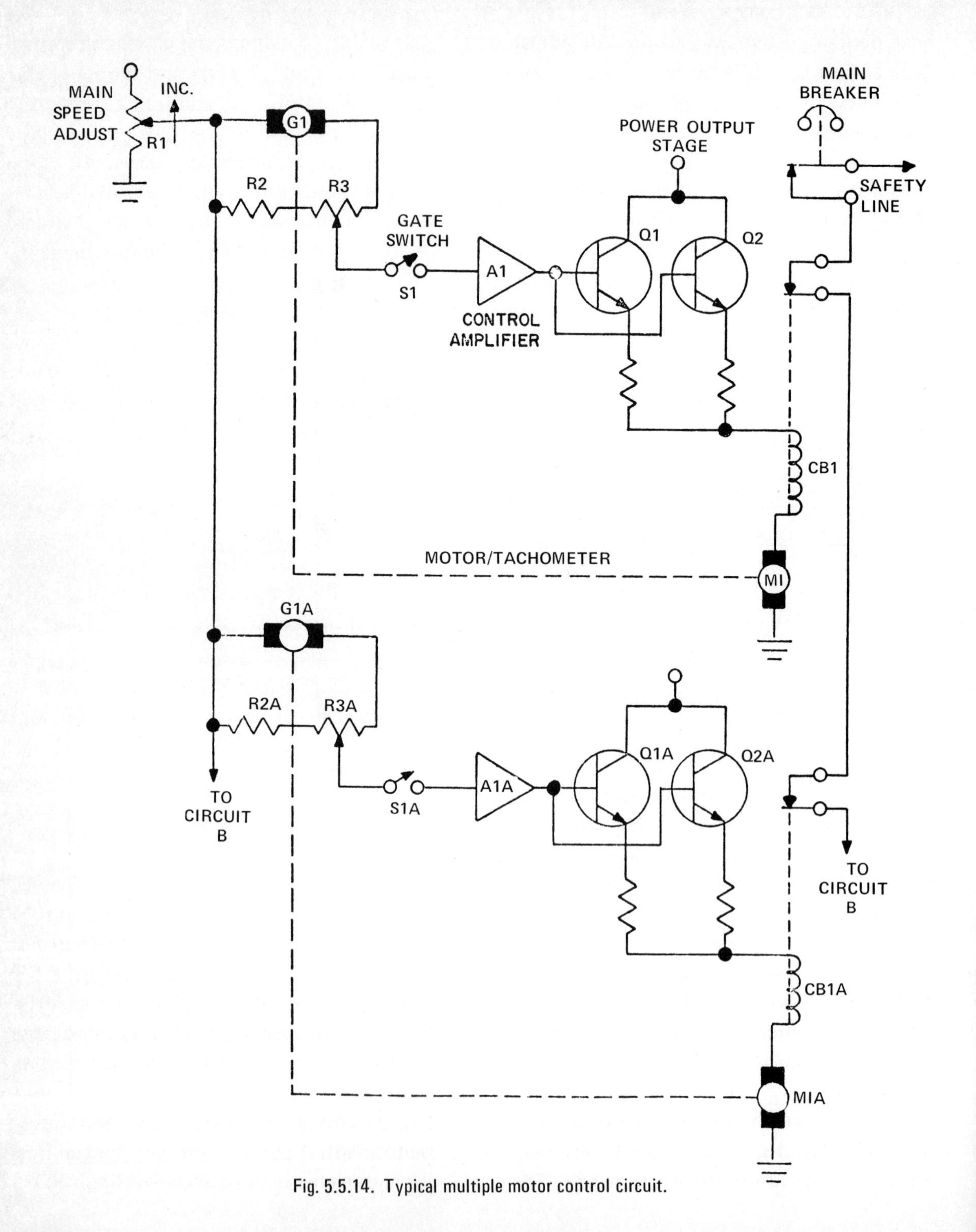

Fig. 5.5.14. Typical multiple motor control circuit.

Knitting, tension sensing winding, spooling, and pattern generating, are some of the many related applications which can be performed with Motomatic control systems.

SEMICONDUCTOR MANUFACTURING-PHOTORESIST SPINNER

A critical phase in the manufacturing of semiconductors is the deposition of photoresist on a silicone wafer. Photoresist is a liquid which provides a photosensitive surface on the wafer. To assure proper density, the resist is deposited while the wafer is rotated at a high speed. Speed regulation is important, since the deposition thickness of photoresist is dependent not only on viscosity, but also on rotational speed.

Utilization of a Motomatic control system with active dynamic braking is especially helpful in applications requiring shaft velocity to closely follow speed command variations.

The transistorized dynamic brake requires no contacts or excitation. The counter emf of the motor is used to provide braking torque. This is accomplished by inserting a transistor circuit, which in effect short-circuits the motor armature when the motor is "coasting". It is not active during normal speed control operation, nor when a speed increase command is entered. During these times it is isolated by a diode. When the speed command is decreased, the Motomatic control, for all practical purposes, turns off. In effect, the motor becomes a generator with opposite polarity current flowing in the external circuit, allowing the brake transistors to operate and short-circuit the motor. The effect of the dynamic brake is proportional to speed; it functions best at the higher speeds, where the photoresist spinner is required to operate.

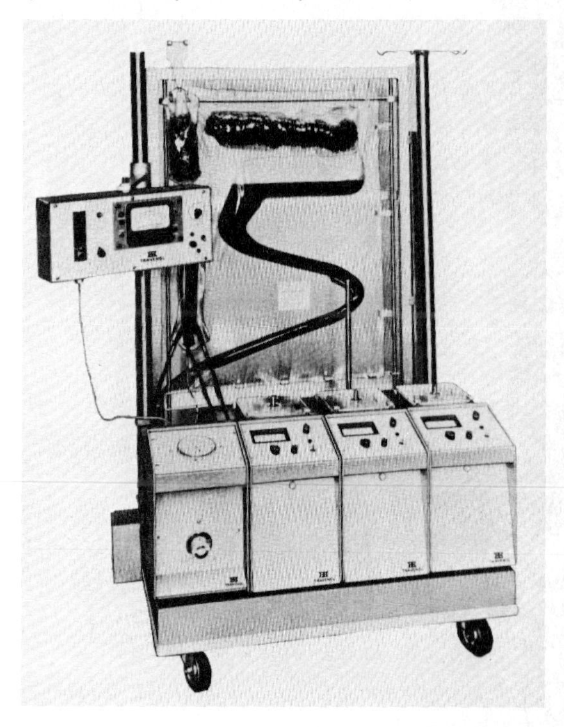

Fig. 5.5.15. Portable heart-lung machine.

BLOOD OXYGENATOR

A manufacturer of medical instrumentation needed a precise speed control system to drive blood pumps for an oxygenator. The system was designed to operate from 115 VAC line power, and is also capable of using 24 VDC battery power in emergencies.

Electro-Craft designed a special control system, adapting the Motomatic concept to this application. The control system can be switched from AC line power to DC battery power in a fraction of a second, insuring uninterrupted service in case of power interruptions. The assembled heart-lung machine is shown in Fig. 5.5.15.

The reliability and versatility of Electro-Craft drive systems make them eminently suitable for numerous critical medical applications, such as blood and kidney pumps, respirators, infusion pumps, and many other demanding applications in the rapidly expanding medical electronics field.

INFUSION PUMP

Infusion pumps are frequently used in medical laboratory experiments where accurate amounts of fluids are continuously added to a system under study.

Greater ease and speed of operation are some of the improvements attained by the laboratory equipment which incorporated a Motomatic motor, as shown in Fig. 5.5.16. The pump can control infusion fluid flows as low as 0.1 cm^3/h, or pump as fast as 10^5 cm^3/h with accuracies of 1%. Other models can pump 12 different fluids simultaneously without any of the fluids coming into direct contact with any other fluid being pumped.

Exceptionally wide flow range, safety and reliability, combined with economy and flexibility, provided improvements over previous techniques. Improved precision and reliable, repeatable performance were desirable characteristics of this design using the Motomatic speed control system.

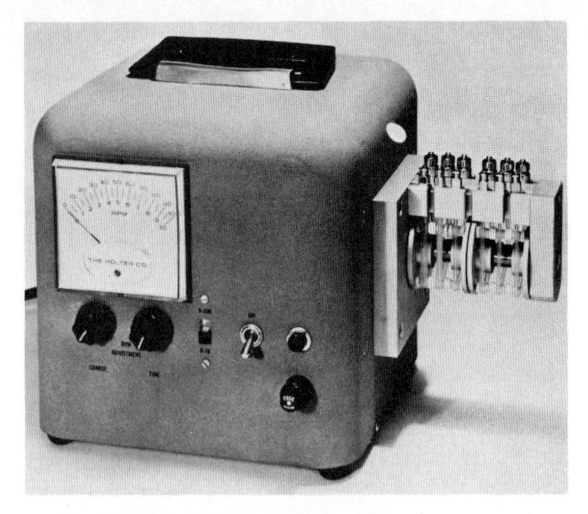

Fig. 5.5.16. Infusion pump. (Photo courtesy of Holter Company, a division of Extracorporeal Medical Specialties, Inc., King of Prussia, Pennsylvania)

CRYSTAL PULLING MACHINE

In the silicon growing process, precise speed control for withdrawal rates of the crystal growth form the melt is required. Silicon ingots are first broken up and melted down in a crucible through high frequency induction heating. Then a seed of perfect crystalline structure silicon is touched to the surface of the molten silicon and the seed is slowly withdrawn (see Fig. 5.5.17).

Three motions are involved during this withdrawal time:

1. the crucible is rotated;
2. the seed is rotated in either the same or the opposite direction;
3. the seed is withdrawn at a very precisely controlled rate.

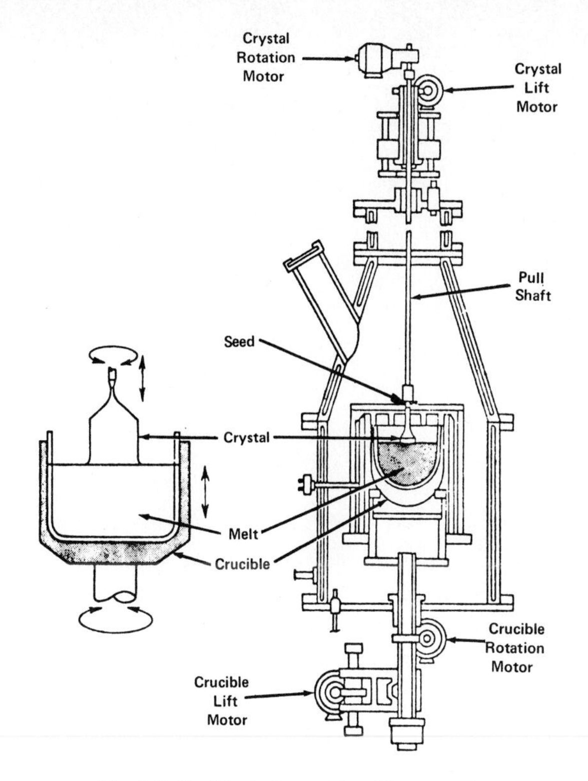

Fig. 5.5.17. Typical crystal pulling machine.

The Motomatic E-650 speed control system was selected for this application (Fig. 5.5.18) because of the following characteristics:

1. broad speed range;

2. fast system response;

3. the ability to provide constant withdrawal rates independent of the varying load.

The seed withdrawal rate determines crystal diameter; consequently, motor speed must be closely controlled. Indication of the melt junction temperature is obtained from a photocell detector positioned on the projected image of the melt surface. Radiation is sensed from the melt surface at the line of intersection between the melt and the crystal. Because of the fast response of the E-650 system, it is capable of using this information for controlling crystal growth rate and diameter. Another beneficial feature of this system is that the speed rate maintained is independent of external influences such as line voltage variations and crystal weight.

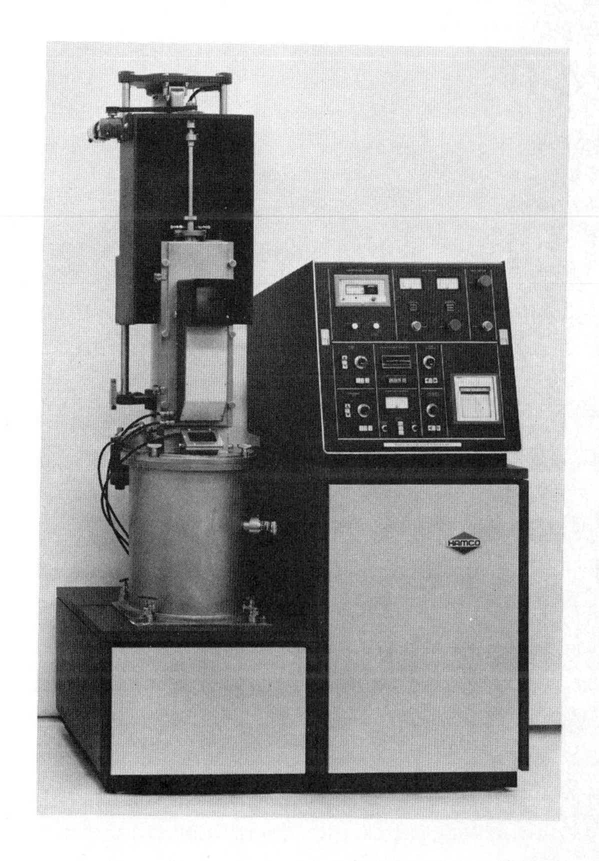

Fig. 5.5.18. Crystal growing furnace. (Photo courtesy of Hamco, Rochester, N.Y.)

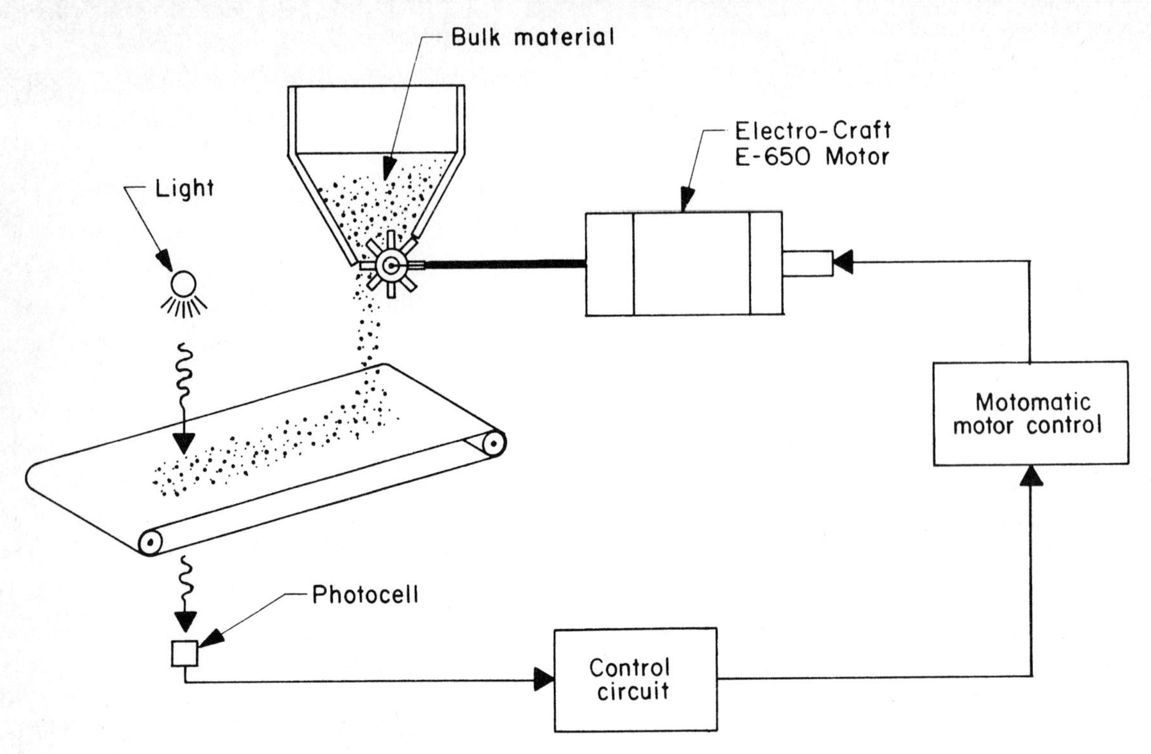

Fig. 5.5.19. An example of a feed rate controller.

CONVEYOR FEED RATE CONTROL

To maintain constant density of bulk material carried on a constant speed conveyor, the feed rate of the material to the conveyor must be adjusted. Such controls may be utilized in the processing of natural and synthetic fibers, or in many related applications which require constant material density.

Fig. 5.5.19 illustrates a method to control the material density. An Electro-Craft E-650 Motomatic speed control system drives feed rolls which feed fiber stock from a storage bin to a conveyor. An array of photocells across the width of the conveyor senses the intensity of light transmitted through the moving material from a source above the conveyor. The photocell output is converted to a voltage proportional to average light intensity. This voltage indicates material density, and is compared with a preset density reference voltage; the difference between the two voltages is the error signal which goes to the motor control. The error signal directs the motor to increase or decrease speed to maintain the requried density.

REFLOW SOLDERING

In an industrial process the manual procedure required numerous soldering operations on a cylindrical object. The solder joints were not always reliable, and the process was time-consuming. The manufacturer desired to eliminate these problems in order to increase production rates and improve reliability. A system was required which could position the work piece, solder, index to the next angular position, and repeat the process.

A Motomatic E-500 system was selected to drive both the solder positioning arm and the indexing axis. The arrangement is shown in Fig. 5.5.20.

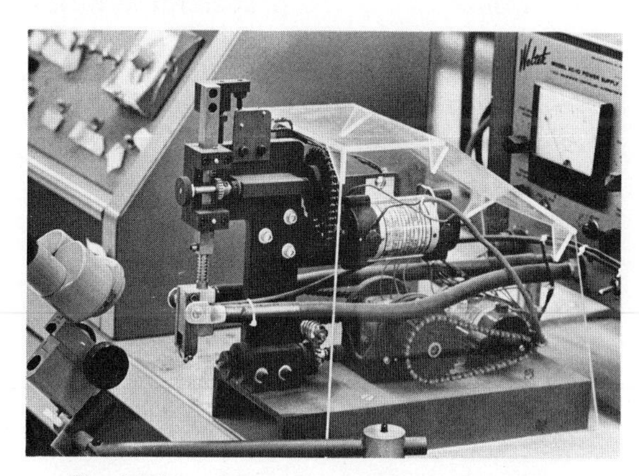

Fig. 5.5.20. Motomatic E-500 dual control for automatic positioning, soldering and indexing.

Linear positioning through gearing is accomplished by the E-500 positioning the solder arm. An input "go" command is received by the motor control, which provides a position setting proportional to a reference signal. The motor position feedback comes from the angular position of a potentiometer. Once the object is in position, soldering is performed.

The next step is indexing, which is also performed by an E-500 system driving a worm gear. The index position is sensed by a microswitch. This arrangement achieves automatic operation, and has both increased production and improved reliability.

DRIVE FOR DISC MEMORY TESTING

In the testing of standard 14-in computer memory discs, a disc is rotated at high speeds and checked for surface wobble.

The computer memory disc is pneumatically locked onto the air bearing spindle shown in Fig. 5.5.21. This unit is normally

Fig. 5.5.21. Universal air bearing spindle for holding 14-in memory discs.

mounted to the top of a specially designed soft spring mounting (provided with suitable damping) in which the drive motor is suspended, as shown in Fig. 5.5.22. The drive motor is connected to the air bearing spindle by precision couplings to assure the spindle's accuracy — better than 2 millionths

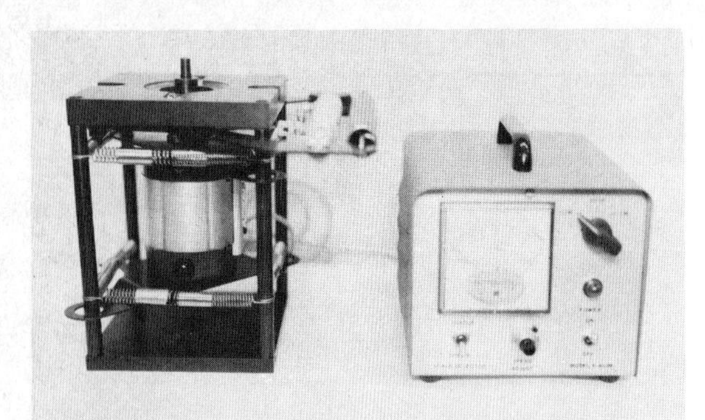

Fig. 5.5.22. The Motomatic variable speed spindle drive system, featuring the E-650 motor-generator and master control.

of an inch radially and axially. The surfaces of the chuck are ground to less than 10 millionths TIR. Testing requires high shaft speeds which are highly repeatable and closely regulated.

The E-650 speed control system provides a complete closed-loop system featuring a fast response E-650 motor-generator and the E-650 master controller, a transistorized velocity feedback amplifier incorporating a special low drift amplifier.

Under operating conditions, the E-650 provides a velocity feedback signal proportional to actual test speed, which is compared with the reference pre-set signal proportional to desired test speed. An error signal is generated if the two voltages differ.

By this means, the pre-set desired testing speed is accurately maintained. Overall speed regulation of the E-650 system is better than 1% over the greater part of the system dynamic range, thus providing very close regulation and test repeatability for this application.

VARIABLE SPEED TAPE RECORDER

An automobile manufacturer wished to test vehicle suspension components by simulating a variety of road conditions at various speeds. The solution to this problem was to tape-record road surface information and play back the recording at the desired test speed. The output of this special tape playback unit (Fig. 5.5.23) would then drive a high-powered, electromechanical transducer, which would apply the proper force function to the parts under test.

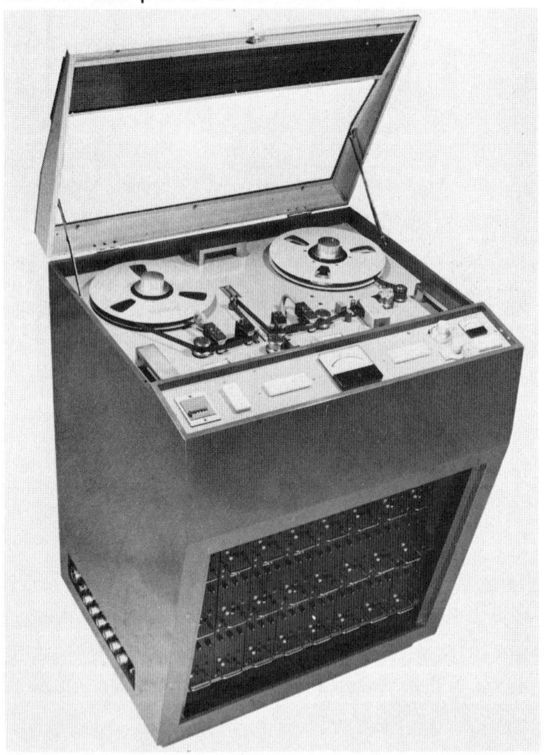

Fig. 5.5.23. Variable speed tape recorder. (Photo courtesy of Précision Mécanique Labinal, France)

The capstan speed control chosen for this play-back was an E-550 system, which gives an effective test speed range of from 0.3 to

300 km/h (0.186 to 186 mph). In addition, the responsiveness of the E-550 system was adequate to simulate rapid acceleration and braking over the test road conditions.

MOTOR TESTING

Modern production testing of permanent magnet DC motors requires two basic tests to determine armature resistance **R** and voltage constant K_E. The voltage constant K_E can best be obtained by driving the motor under test with another motor (in effect, the motor under test is being run as a generator). The generated voltage and shaft velocity are simultaneously read and recorded.

In order to perform such testing, a constant speed drive system is required. The Motomatic E-650 system is employed for this application, as shown in Fig. 5.5.24. This system provides a direct readout of shaft speed (and torque, if required), while accurately regulating the speed. These systems maintain an overall speed accuracy of better than 0.5% over the greater part of the system dynamic range. The long term speed stability of the system leads to test repeatability, which is important for production testing. An additional benefit of this system is constant speed, independent of torque variations and line voltage fluctuations.

While the motor under test is being driven at a constant speed, the speed and generated voltage are recorded. The voltage constant is then obtained by the following relationship:

$$K_E = \frac{E_g}{n} \qquad \text{[V/krpm; V, krpm]}$$

The motor torque constant K_T can be determined by the measurement of the motor voltage constant which was just described. The torque constant is always related to the voltage constant, as shown in section 2.1.

The ease with which this measurement can be made, makes it the most suitable production acceptance test for both motor voltage constant and motor torque constant.

Fig. 5.5.24. The E-650 utilized for final motor testing at Electro-Craft Corp.

TACHOMETER RIPPLE VOLTAGE TESTING

One of the important steps in testing precision tachometers is the ripple test. In the case of high performance tachometers having ripple voltages on the order of 1% peak-to-peak, the test setup must have constant speed drive arrangements with instantaneous velocity diviations of less than

0.1% peak-to-peak. Such an arrangement used by Electro-Craft Corp. is shown in Fig. 5.5.25.

In this setup, a Motomatic E-650 master control provides power for the E-650 motor-tachometer which controls the speed of an inertia wheel driving the motor-tachometer combination under test. A rubber belt couples the drive motor to the flywheel. By a direct reading of the speed indicated by the electronic speed meter, and knowing the belt pulley ratio, the test speed can be calculated. Due to the variable speed feature of the Motomatic system, data can be gathered for numerous speed points.

Fig. 5.5.25. Precision tachometer test station at Electro-Craft Corp.

TORQUE LIMITER

A manufacturer required a system which will accelerate a large inertial load without causing damage to the motor or control system.

Many of the Motomatic speed control systems feature transistorized, variable torque limiting. The torque limiter can be described as an "electronic clutch"; the motor can be stalled at a predetermined load torque and remain in this state indefinitely without damage to any component.

Acceleration of large inertial loads impose a heavy power demand on a motor. Once the load is rotating, power requirements fall off sharply. Utilizing a Motomatic system with variable torque limiting, a load of this nature can be accelerated using a much smaller motor (and control system) than would normally be required. The only sacrifice in performance would be an extension in acceleration time.

PAPER WEB WIND-UP CONTROL

A system for the paper industry was required to control the rewind of paper from a large roll onto smaller rolls. The web must be continuously wound on the smaller rolls (see Fig. 5.5.26). A series of six spindles are evenly spaced on a turntable. When one spindle is fully wound, the turntable is indexed, bringing the next spindle into the winding position. The fully wound roll is removed, and a new core is loaded.

As the spindle is moved into the winding position, it must be accelerated up to desired initial speed. The web is then placed onto the core in a flying splice manner. During the winding cycle, the spindle speed must be controlled in a programmed manner

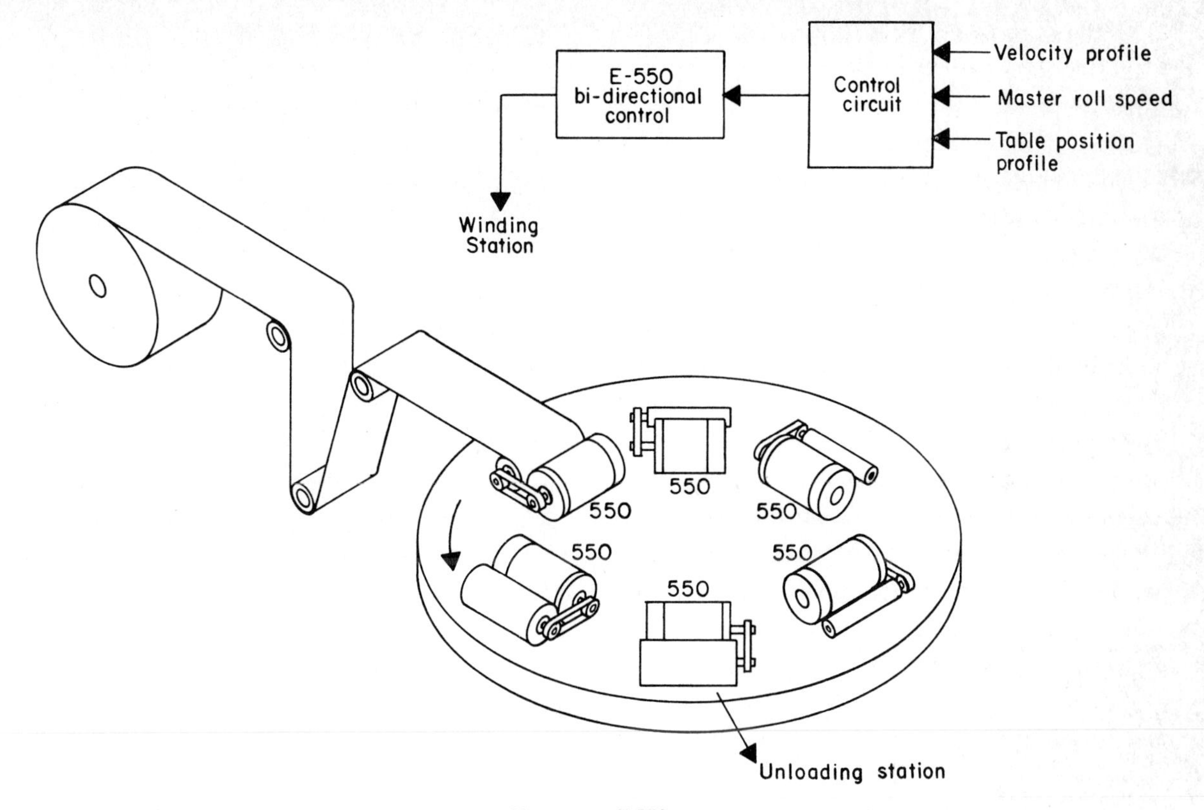

Fig. 5.5.26. Paper rewinding system, featuring Motomatic E-550 motor-generators and open chassis controllers.

to accommodate the roll buildup while maintaining peripheral speed equal to web speed.

The supply roll must be accelerated up to desired speed relatively slowly because of the large moment of inertia involved. Since it is desired to wind the smaller rolls during startup and slowdown times, the spindles of the smaller rolls must be programmed as a function of the master roll speed. The supply roll has a "dancer" arm storage to hold paper during spindle cycling.

In order to provide high dynamic response to match the desired acceleration and slow-down characteristics of the velocity profile, a bidirectional Motomatic E-550 control was used. The open chassis controls were mounted on the turntable to avoid ship rings carrying low level signals for all motors.

WELDING MACHINES

In a welding process the quality and uniformity of the weld is directly affected by the proper control of the welding wire feed rates, and by the relative motion between the work pieces and the welding joint.

Before the process was automated, a highly skilled operator would normally position the welds and feed the welding wire. By providing an automatic, accurately controlled feed rate for weld wire, a proper quantity of material is always supplied to the welding joint. Once a setup was completed, the operator had only to load and unload the work pieces, as the control systems assumed a very high degree of uniformity, thus minimizing operator skill requirements.

Various Motomatic systems are used in these applications, depending on the torque and speed requirements of the welder. In Fig. 5.5.27, an end-to-end pipe welder employs a pair of E-550 systems. The first system governs wire feed, and the second system rotates the pipes with respect to the welder. A very high degree of repeatability from piece to piece is the result of utilizing this closed-loop control system.

The Motomatic closed–loop system automatically adjusts motor input voltage and current, so that constant speed is maintained even though the load may be varying. The integral tachometer senses the effect of load variations and controls the transistorized servo amplifier, which maintains constant speed. Thus, despite load variation from work piece to work piece, the Motomatic system will maintain the set speed within a close range, usually 1%. Additionally, by providing a very smooth, stable motion for the work pieces, a highly uniform and repeatable welding operation was achieved.

Fig. 5.5.27. End-to-end automatic pipe welder. (Photo courtesy of Précision Mécanique Labinal, France)

Fig. 5.5.28 shows a larger welding machine, which requires the extra torque of an E-650 system.

Motomatic series E-650 speed controls were selected to drive both the carriage drive and the wire feed machanism for this application. The operator sets the feed rate and controls the welding function.

Fig. 5.5.29 shows an interesting extension of the welder control concept. An electron beam welder is operated along a linear direction, utilizing a "suction cup" technique to maintain the required vaccum. The exceptional requirements for broad speed range and speed stability resulted in the selection of Electro-Craft's DC servomotor/tachometers.

Fig. 5.5.28. Automatic welding machine utilizing Motomatic controls. (Photo courtesy of Guild Metal Joining Co., Bedford, Ohio)

BIDIRECTIONAL POSITION CONTROL

In a data retrieval system, a DC drive system is required to move a search mechanism and position it accurately for data recovery. The system must be bidirectional, and may be required to operate on a 50% accelerate/decelerate duty cycle.

Positional accuracy of the drive system is limited by two variables: 1) the drift characteristics of the reference power supplies, and 2) the accuracy of the command and feedback signals. Control system selection had to be made on the basis of the ability to cope with these variables. Using the potentiometer supplied with the standard Motomatic E-550 BPC (Bidirectional Position Control), system accuracy is approximately 0.1% per revolution or 0.36°.

Fig. 5.5.29. An electron beam welder. (Photo courtesy of Précision Mécanique Labinal, France)

The E-550 BPC system is a position control system employing a high gain, closed-loop electronic control and an E-550 motor-generator. As shown in Fig. 5.5.30, the input voltage to this electronic control commands the motor to move the search mechanism to a specific position. Motor angular position is proportional to the command setting, with the position feedback coming from the angular position of a potentiometer. Actually, a potentiometer, or any other position feedback voltage source may be employed for similar applications. As the search mechanism is "homing in" on the desired position, velocity feedback is employed to control the travel rate to the commanded location.

Optional features of the E-550 bidirectional controller allow adjustment of the velocity profile parameters. For example, the re-

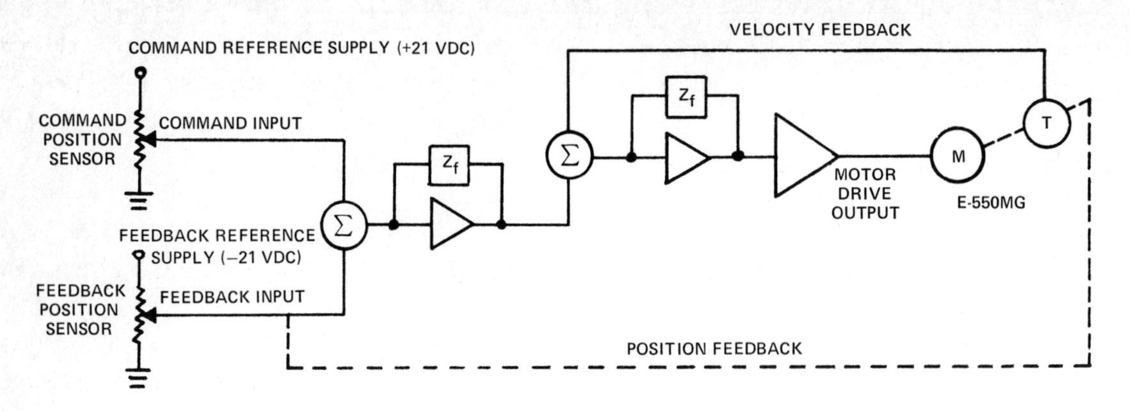

Fig. 5.5.30. Motomatic E-550 BPC block diagram.

sponse time can be adjusted to permit "soft" starting, and the travel rate can be set at any desired motor speed between 5 and 5000 rpm.

MUSCLE EXERCISER–MEDICAL REHABILITATION DEVICE

An exerciser used in muscle rehabilitation requires an accurately calibrated velocity control with torque readout in order to monitor the patient's recovery. Muscle strength recovery can be precisely monitored and the patient's progress accurately measured through this novel device, shown In Fig. 5.5.31.

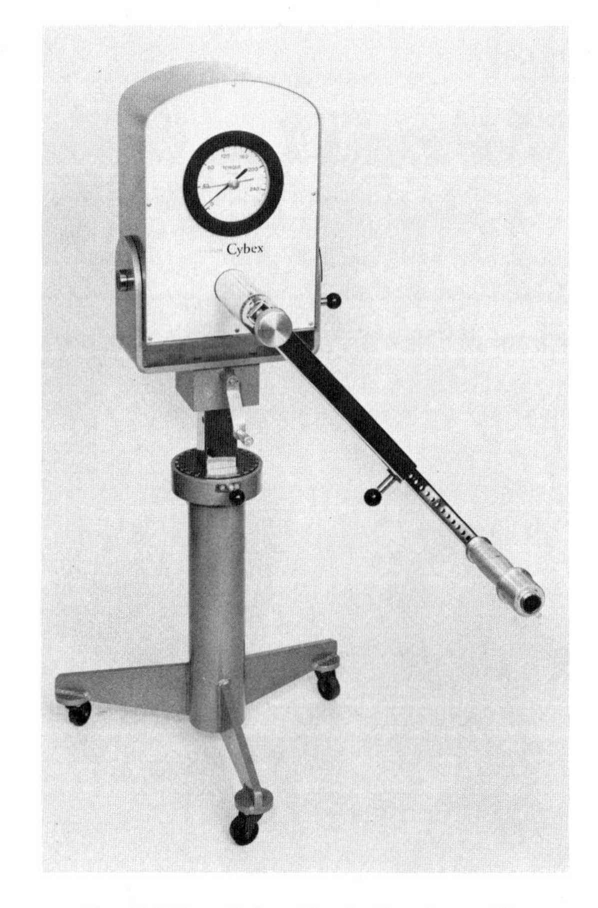

Fig. 5.5.31. Cybex Muscle Exerciser. (Photo courtesy of Technicon-Cybex, Inc., Tarrytown, N.Y.)

The rate of motion is set by the therapist. In the initial stage of rehabilitation, the patient merely follows the crank handle motion with his hands or feet. As he progresses, he may exert force to either retard or speed up the handle motion. The degree of muscle exertion can be read on a calibrated torque dial, and recorded on a separate device. The special features of this device have made this machine very useful in scientifically controlled hospital rehabilitation programs.

The system utilizes an E-650 motor-generator as a prime mover with a special bidirectional amplifier controlling speed and providing positive or negative torque, as the patient requires.

PUNCH PRESS FEED CONTROL

This application consisted of a clutch-brake operated material feed control for an automatic punch press. Due to the intermittent demand and the incremental feed technique, instability of the feed system would occur. The clutch-brake system would wear, and require frequent adjustment.

A new system was designed, using an E-650 speed control, commanded by a sensor arm to which a potentiometer was attached (Fig. 5.5.32). The redesigned system operates very smoothly, seeking an average velocity matching that of the punch press. This eliminates the jerky operation previously experienced, and prevents stretch damage to the material.

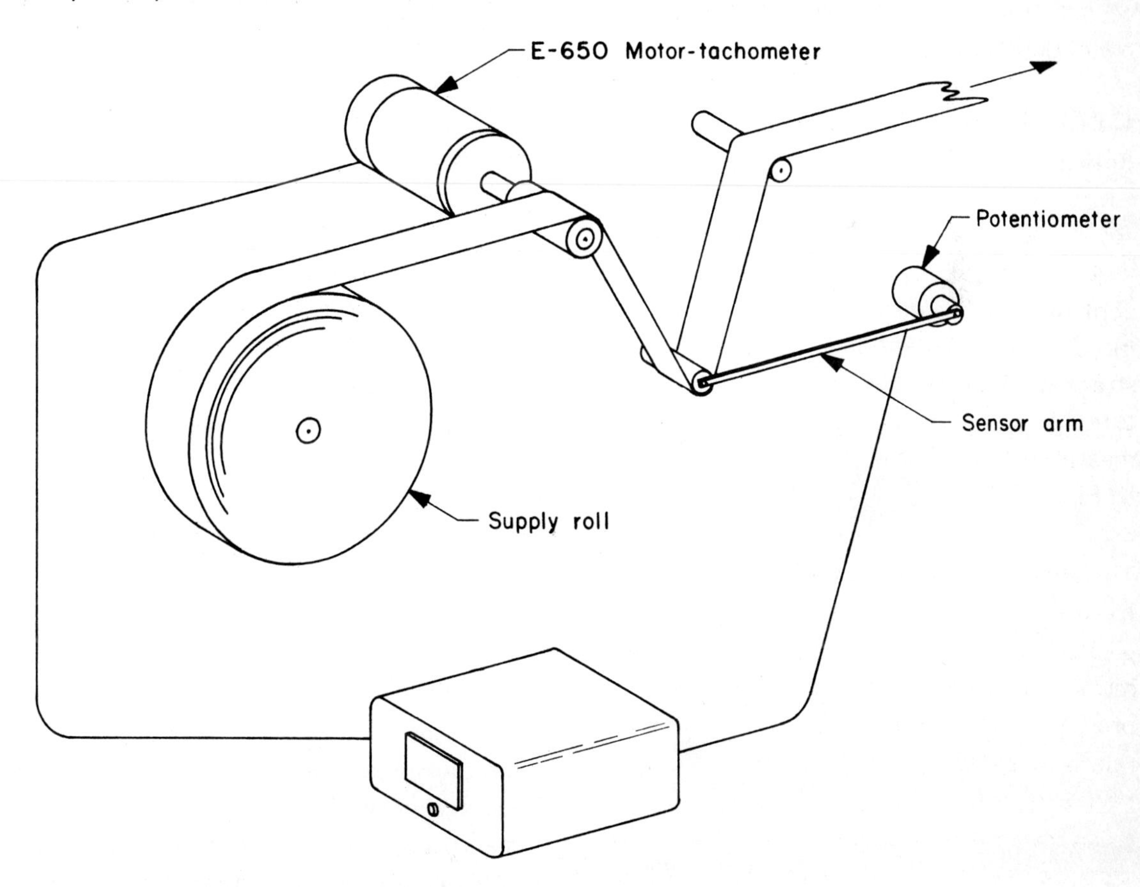

Fig. 5.5.32. Punch press feed control, utilizing a Motomatic E-650 speed control.

CONNECTOR TEST STATION

A major manufacturer of electrical connectors wished to build a connector test station, which would perform tests on the insertion force of connectors over a repeated number of cycles. Since the speed at which the mating halves were joined had a large bearing on the force measured, it was found necessary to automate the contacting speed in order to obtain testing repeatability. At the same time it was recognized that the additional load of pin engagement would potentially cause a speed regulation problem; i.e., the larger the force required to mate the connector, the slower the drive system would tend to operate, thus complicating the problem.

Fig. 5.5.33. Connector test station. (Photo courtesy of AMP de France)

An E-550 system was selected which, coupled through a vertical ball screw (Fig. 5.5.33), gave the high degree of speed regulation needed to achieve repeatable results.

BLUEPRINT MACHINE DRIVE

Diazo print machines ("blueprint" machines) traditionally have a manual speed adjustment which requires the operator to set the exposure time according to the transparency of the original material. The success of this process depends on the operator's judgment and skill, and the process often results in waste of both time and material.

A basic E-550 system, with added control circuitry, sets the machine speed accurately and automatically for consistently good quality reproductions. The purpose of the added circuitry is to generate a signal which is dependent on the original's transparency by generating a speed reference voltage, which in turn controls exposure time.

A photocell senses light passing from a mercury lamp through the original material to provide a measure of its transparency. The photocell provides a DC voltage proportional to the product of light intensity and transparency. This voltage is the speed reference to the motor drive. The control circuit stores it long enough to let a mercury lamp expose the paper. After the copy is made, a timing circuit reduces the reference voltage to revert the machine to idle speed until the next sheet of paper is inserted.

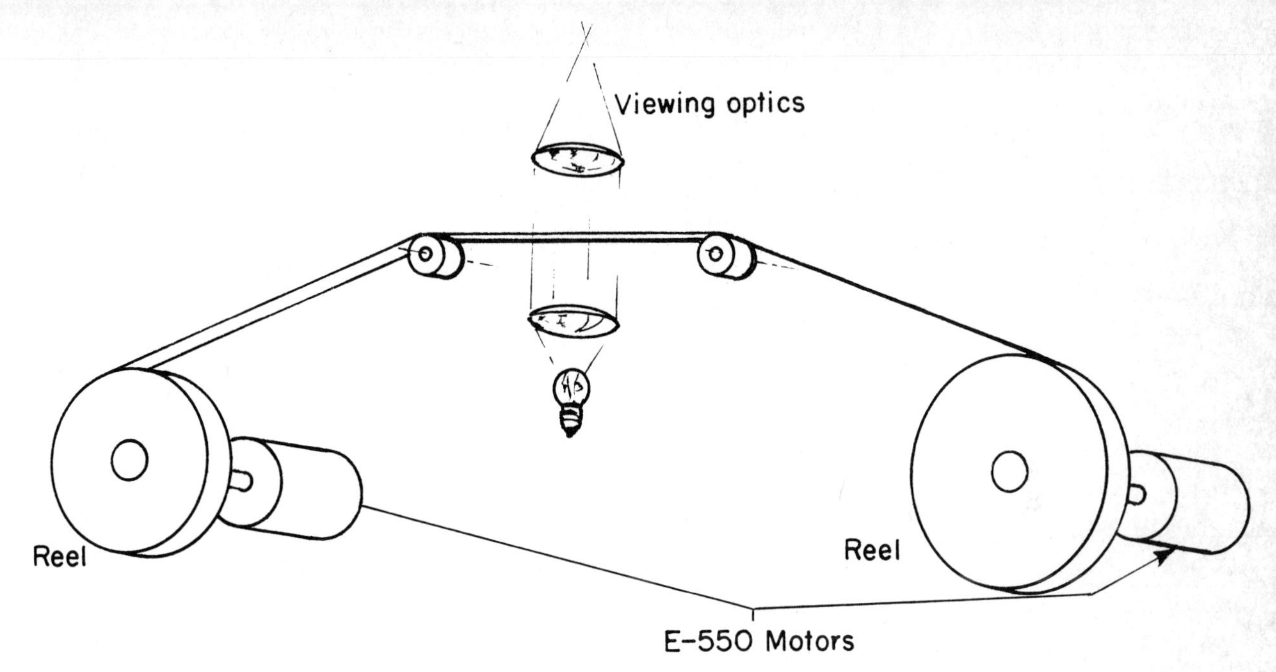

Fig. 5.5.34. Film viewer setup.

FILM SCANNING DEVICE

A customer wanted a wide range film scanning device, enabling the viewer to rapidly find the general area of interest, and then at a controlled low speed view the individual frames until the proper one is found and then stop. The film was stored on spools, and in the process of scanning and viewing, the film would be transferred from the supply spool to a take-up spool (Fig. 5.5.34).

A simple solution was achieved by using two E-550 motor-generators, driving each reel as shown in Fig. 5.5.34. However, since the reel sizes vary during a scan, the relationship between shaft speed and linear film speed varies in proportion to the reel size change. This would result in undesirable control characteristics. An unique solution to this problem is illustrated in Fig. 5.5.35, where the two tachometers are electrically

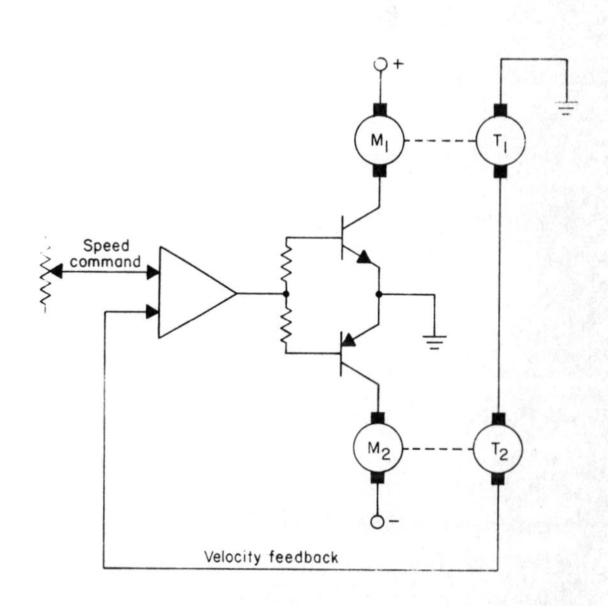

Fig. 5.5.35. Tandem reel drive system.

connected in tandem. Since the reel motors and tachometers are mechanically connected through the film, a pseudo-linear relationship between speed command and film speed is achieved, so that comfortable manual control of film motion results.

DRIVE FOR CHART RECORDER

The user wanted an infinitely variable speed paper drive for a strip-chart recorder. A speed range of 10 000:1 was required, and the acceleration time at the highest chart speed was to be less than 0.15 s.

Electro-Craft proposed an E-576 motor-generator, with a special "ultra-wide range" velocity amplifier with high peak power availability. The result was a chart drive with a capability of chart speed between 5 m/s to 0.5 mm/s, and an acceleration time of less than 0.15 s from 0 to 5 m/s chart speed.

PAPER CUTTING CONTROL

In this application paper was unwound from a feed roll, then cut to size by a mechanical knife. The system used a clutch-brake combination which provided intermittent motion to position the paper properly at the cutting station. The clutch-brake system needed frequent maintenance and adjustment to maintain the required accuracy.

By replacing the clutch and brake with a Motomatic E-550 direct-drive system, the entire cutting operation was accelerated, cutting accuracy was improved, and maintenance was minimized.

In the direct-drive system, shown in Fig. 5.5.36, the motor is commanded to follow

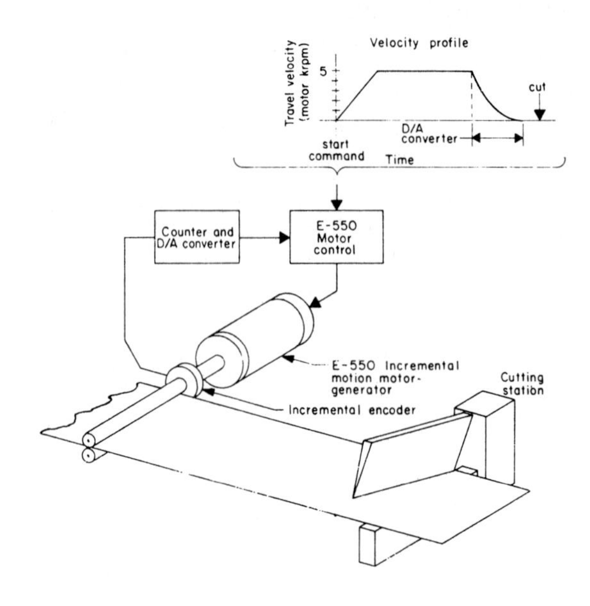

Fig. 5.5.36. Paper cutting control utilizing a Motomatic E-550 MG and control for incremental motion.

an input trapezoidal velocity command. The incremental encoder produces pulse counts proportional to distance of paper travel. The counter has a preset number stored, representing the proper paper size. When a predetermined number of binary counts remains in the memory, a digital-to-analog converter will be switched into the velocity feedback path. Thus, the final stop position will be approached in a controlled velocity manner, and when the last count has been received, the "stop" command can be followed instantly with no overshoot or error.

This hybrid analog-digital system has numerous applications in industrial and scientific applications.

BLOOD CELL SEPARATOR

The Celltrifuge™, a continuous flow experimental blood cell separator shown in Fig. 5.5.37, was developed in a program sponsored by the National Cancer Institute and IBM. The machine uses several Motomatic control systems powering eight pumps and a centrifuge to perform the task. The systems have been specially designed to provide ultrahigh reliability and very low ground leakage current for patient safety.

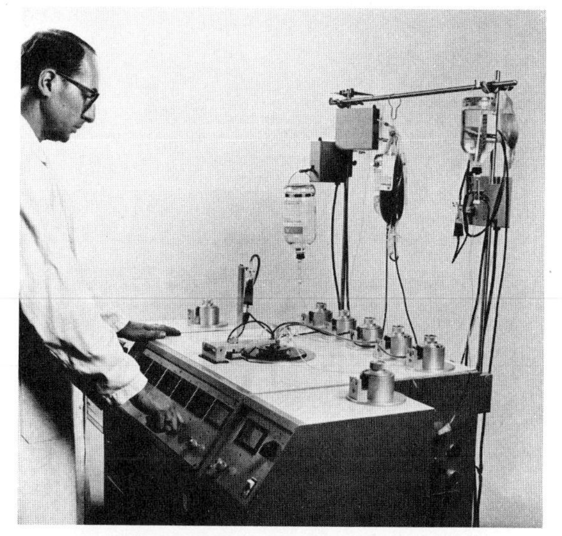

Fig. 5.5.37. Blood Cell Separator. (Photo courtesy of American Instrument Co., Silver Springs, Md.)

ADJUSTABLE LATHE FEED RATE CONTROL

Fig. 5.5.38 shows an E-650 motor-generator adapted to replace the quick-change gears in a lathe application. The variable speed, remote control feature was used to continuously alter the feed rate in a complex industrial turning process, keeping tool

Fig. 5.5.38. Adjustable speed lathe carriage feed.

pressure constant in spite of varying material thickness.

CONTOUR LATHE

The requirements for a contour lathe cutting tool position drive are high speed for slewing, low speed with high torque for the actual cutting operation, and a stable speed control over this range, with very good no-load to full-load speed regulation. A lathe manufacturer has found that an E-650 system met all these requirements and provided an economical and highly reliable solution, shown in Fig. 5.5.39.

THREE-AXIS MILLING MACHINE CONTROL

Fig. 5.5.40 shows an example of an industrial adaptation of Motomatic controls to a milling machine table drive.

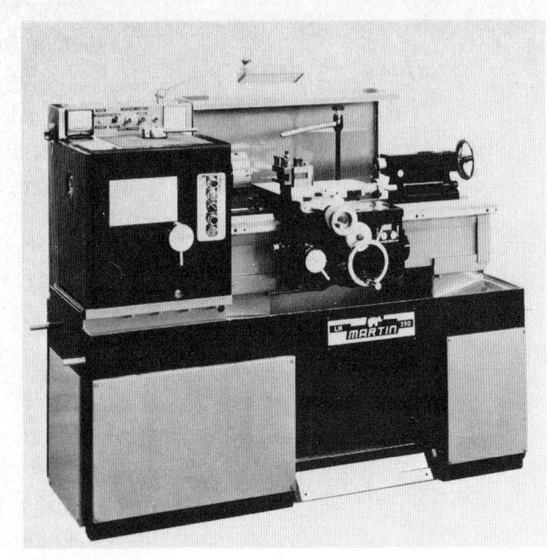

Fig. 5.5.39. A contour lathe, model LM 350. (Photo courtesy of LEFEBRE et MARTIN, France)

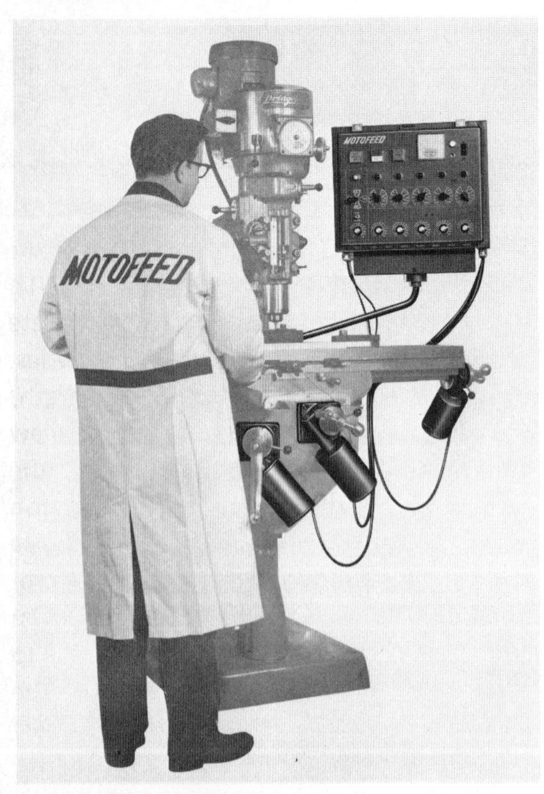

Fig. 5.5.40. Automatic three-axis milling machine control.

In this case the system was designed to provide a variable speed, three-axis position control for repetitious, automatic, routine milling applications. The control system is based on the wide range Motomatic control, with an individually adjustable speed for each program sequence. The position sensors work against adjustable micrometer stops, with a dual feature; as the control system approaches the final stop position, it goes into a "final approach" speed – a low speed mode – from which the control system can repeatedly stop within known tolerances. Thus, the MOTOFEED® system holds a repeatability of ±0.0005 in in all three axes, with infinitely variable speeds from 0.1 to 42 in/min. Using a specially designed gear reduction, the system can provide torque of 80 lb-in on the lead screw, even at the lowest speed, permitting heavy cuts at close tolerances. A uniquely designed clutch device can uncouple the drive system at any time, providing a manual operation feature without interference by motors or gears.

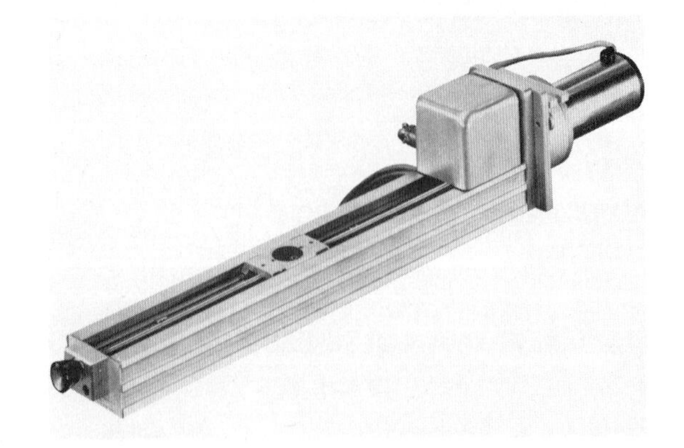

Fig. 5.5.41. A linear drive mechanism UNISLIDE, model B2515 CJ. (Photo courtesy of Velmex, Inc., Bloomfield, N.Y.)

LINEAR DRIVE SYSTEM

A manufacturer of linear motion mechanisms needed a drive system with infinitely variable speed to drive the lead screw for its line of modular length, linear motion devices.

An E-550 MGHP Master Control System was selected for this application because of its versatility, wide speed range and torque limiting features. Fig. 5.5.41 shows the finished linear drive mechanism, marketed under the trade name UNISLIDE (Velmex, Inc., Bloomfield, N.Y.).

PHOTOTYPESETTER APPLICATION

In a typical phototypesetter (see Fig. 5.5.42) the font, or character set, is in the form of negative silhouettes on a rotating disc. The particular character to be set is selected by flashing a strobe light behind the disc at the appropriate time. The character image is then projected via a moving mirror on to a photosensitive surface where it is recorded. A minicomputer in the machine calculates the exact letter spacing, word spacing and hyphenations to achieve the correct line end justification. The mirror is then moved to cause the characters to be positioned correctly on the surface. One manufacturer uses an Electro-Craft L5000 series servo control with the Digital Positioning Option and an M1438 moving coil motor to drive the lead screw on which the mirror is mounted. Use of this system enables direct interface with the minicomputer. The performance of the system is such that the motor and load can be moved and stopped through a typical step of 17 degrees in 5 ms and can make 90 such steps per second; positional accuracy is better than ±½ degree of rotation. This gives considerably faster operation than other systems.

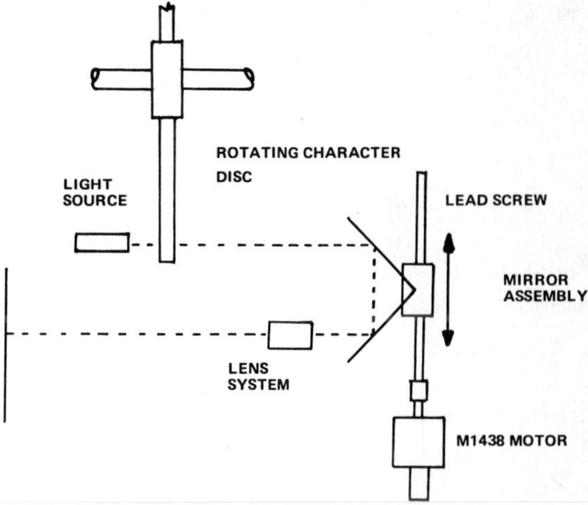

Fig. 5.5.42. Phototypesetter Application

DIAMOND SORTING SYSTEM

The system shown in Fig. 5.5.43 is designed to monitor the size of industrial diamonds and to automatically segregate them into ten (10) different categories. The diamonds (20 to 60 mils in length) are transported through the field of view of a Reticon line scan camera (first on the left), back illuminated using a Reticon light source (first on the right). The size is then displayed in mils on the microcomputer based controller (second from left) whose output is also transmitted to the motor controller (second from the right). An E576 servo motor (center) with ten bins on a rotary platform is then positioned to cause the various diamonds to be deposited into the appropriate bins.

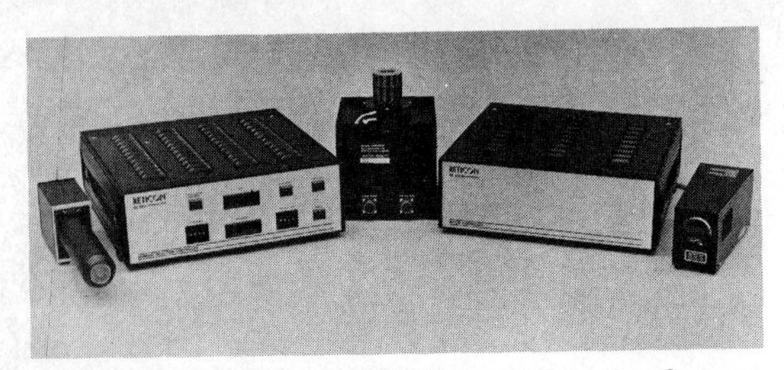

Fig. 5.5.43. Reticon RS825 Diamond Selection System

PRINTED CIRCUIT SOLDER FUSING MACHINE

The Research Inc. Solder Fusing System shown in Fig. 5.5.44 provides superior quality fusing on printed circuit board and multiple layer assemblies.

The solder fusing process converts the porous electroplated tin-lead on a circuit board into a homogeneous alloy with a strong intermetallic bond to the base copper.

The temperature achieved during the fusing process is critical to the formation of the bond. Since this temperature is to a large extent dependent on the length of time that the printed circuit is exposed to the heat source, the customer chose a Motomatic speed control to drive the conveyor which carries the circuit boards. This selection gave the advantage of smooth stable speed control over a wide range of speeds.

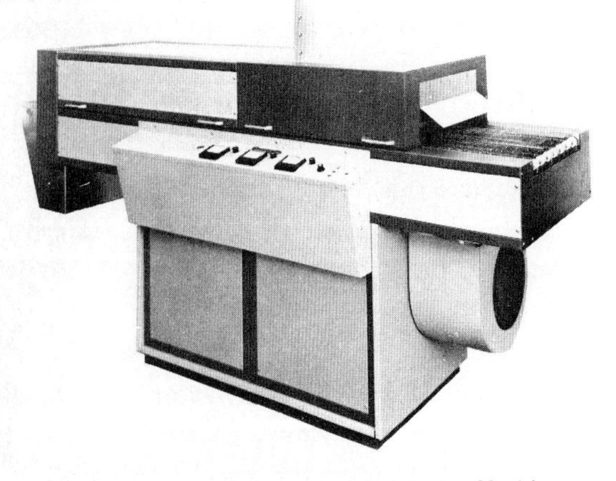

Fig. 5.5.44. Printed Circuit Solder Fusing Machine. (Courtesy of Research Incorporated)

COMPONENT MARKING SYSTEM

In this application a marking head had to be accurately synchronized with a conveyor chair carrying electronic components. The synchronization had to be such that the markings were accurately positioned on the component and that a minimum of components be missed during start up and shut down.

The solution is a P6200AP servo motor control, an E670/596 motor-gearhead and one of the PL100 series phase lock options. An analog tachometer generator mounted on the conveyor drive provides a reference signal which keeps the speed of the marking head approximately synchronized with that of the conveyor. The signals from the two pulse generators are fed into the phase lock system which adjusts the motor speed continuously to ensure that the speeds are matched identically and that the marker head is also correctly positioned so as to mark the component.

The resulting increased performance enables the marking of components to be increased in speed to about double that achieved by previous systems.

5.6. INCREMENTAL MOTION APPLICATIONS

5.6.1. INTRODUCTION

In this section we will discuss the principles of incremental motion operation, covering system dynamics and acceleration torques. We work through an example so that the reader will gain experience in calculating armature acceleration current, motor terminal voltage, motor power dissipation, maximum permissible repetition rate, cooling and slewing requirements, to aid in selecting a servomotor for his own application.

The application examples in this section illustrate the concepts brought forward in the preceding discussions of motor selection for incremental motion requirements. A comparison is made here between the rather tedious hand calculations with Electro-Craft computer solutions.

These application examples are drawn from Electro-Craft files, and serve to illustrate the motor and amplifier selection process. They should be of help to anyone who is contemplating a systems design. As we shall see, the computer design aids are of immense help in making a proper choice of system components. Electro-Craft Corporation Applications Engineering staff has available portable computer terminals, and can give you design help right in your office, since we can connect the terminal to our computer through the telephone lines. If you require assistance in your servomotor selection, our staff stands ready to be of help.

The *analysis forms* shown in next pages give the reader the basic information, what data, and in which form, are necessary to be supplied by the customer and Electro-Craft staff for exact and complex calculation of a given application.

The velocity and displacement data of an analysis form should be described in more detail by means of velocity profile and displacement profile graphs, as shown in Fig. 5.6.1.

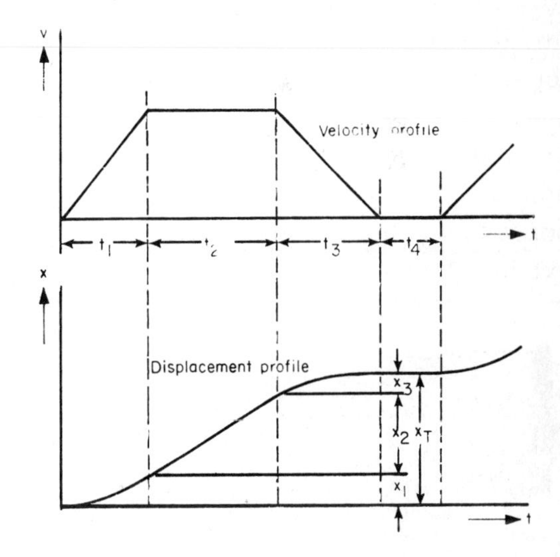

Fig. 5.6.1. Velocity and displacement profiles.

ELECTRO-CRAFT CORPORATION

MOTOMATIC System Application Data

Power Supply

AC Voltage______V,______Hz

Max. Current_______ A

Other (describe):______________________

DC Voltage_______V, rectified and filtered

_______% Ripple (p-p)

or

_______ V battery

Maximum DC voltage______ V and maximum DC current_____ A available

Velocity Profile

Acceleration Time__________ ms

Run Time __________ ms

Deceleration Time__________ ms

Dwell Time __________ ms

Reversing:

☐ No ☐ Yes, from max. speed____ rpm

Running Speed_______ rpm

Slewing Speed_______ rpm

Loading

Radial Load_____oz, applied_____ in from shaft end

Axial Load _____oz, applied ☐ away ☐ toward mounting face

Load Friction Torque _____ oz-in

Load Moment of Inertia_____oz-in-s^2 (if unknown, sketch the driven load, giving dimensions and material used)

Gear Reduction (motor-to-load)________________________

Tachometer	Regulation
Gradient (Voltage Constant)_____V/krpm	Long Term Regulation_____% of set speed
Ripple______ V p-p max. at_____ rpm	Short Term Accuracy _____% of set speed

Special Requirements

Ambient Temperature _____ °C max., _____ °C min.

Other Ambient Characteristics:______________________________

Shaft Requirements (sketch); Mounting Face; Base (sketch)

Shaft Position: ☐ horizontal ☐ shaft up ☐ shaft down

Space Available: length_____in, width_____ in, height_____ in

Forced Air Cooling Available: ☐ Yes ☐ No

Others (describe):______________________________

ELECTRO-CRAFT CORPORATION

LEAD SCREW APPLICATION

Customer ______________________ Date__________

Program Genrl*
Application ______________________ Contact Engineer______________

Line No.		Parameter	Units
50		Input Data Units =	/British/ or /Metric/
		Output Data Units =	/British/ or /Metric/ or /Neither/
		Load Type =	/Lead Screw/
		Motor Data Required =	/Full Data/ or /Temp. only/
51	*V1	Peak Velocity (Run)	/__/ cm/s /__/ in/s
	V2	Slew Velocity	/__/ cm/s /__/ in/s
	*S	Total Displacement	/__/ cm /__/ in

*Specify V1 or S, not both, equate other one to zero.

Line No.		Parameter	Units
52	T1	Acceleration Time	/__/ ms
	T2	Run Time	/__/ ms
	T3	Deceleration Time	/__/ ms
	T4	Dwell & Settle Time	/__/ ms
53	J1	Lead Screw Moment of Inertia	/__/ kg m^2 /__/ oz-in-s^2
	M	Carriage/Load Mass	/__/ kg /__/ oz
	P	Screw Pitch	/__/ turns/cm /__/ turns/in
	F1	Friction Torque	/__/ Nm /__/ oz-in
	G	Gear Ratio	/__/ : 1
	T7	Ambient Temperature (Max.)	/__/ °C
54 thru 99	M	Motor Model	/__/
	K1	Torque Constant	/__/ oz-in/A
	R2	Armature Resistance @25°C	/__/ Ω
	J2	Armature Moment of Inertia	/__/ oz-in-s^2
	K8	Damping Constant	/__/ oz-in/krpm
	F2	Static Friction Torque	/__/ oz-in
	R3	Thermal Resistance	/__/ °C/W

ELECTRO-CRAFT CORPORATION

CAPSTAN/DRIVE ROLLER APPLICATION

Program Genrl* Application   Customer ____________________ Date ____________

____________________ Contact Engineer ____________________

Line No.		Parameter	Units					
50		Input Data Units =	/British/	or	/Metric/			
		Output Data Units =	/British/	or	/Metric/	or	/Neither/	
		Load Type =	/Capstan/					
		Motor Data Required =	/Full Data/	or	/Temp. only/			
51	*V1	Peak Velocity (Run)	☐	cm/s	☐	in/s		
	V2	Slew Velocity	☐	cm/s	☐	in/s		
	*S	Total Displacement	☐	cm	☐	in		

*Specify V1 or S, not both, equate other one to zero

Line No.		Parameter	Units			
52	T1	Acceleration Time		☐ ms		
	T2	Run Time		☐ ms		
	T3	Deceleration Time		☐ ms		
	T4	Dwell & Settle Time		☐ ms		
53	J1	Capstan/Drive Roller Moment of Inertia	☐ $kg\ m^2$		☐ $oz\text{-}in\text{-}s^2$	
	R1	Capstan/Drive Roller Radius	☐ mm		☐ in	
	F1	Friction Torque	☐ Nm		☐ oz-in	
	G	Gear Ratio		☐ : 1		
	T7	Ambient Temperature (Max)		☐ ^{o}C		
54 thru 99	M	Motor Model		☐		
	K1	Torque Constant		☐ oz-in/A		
	R2	Armature Resistance @ $25^{o}C$		☐ Ω		
	J2	Armature Moment of Inertia		☐ $oz\text{-}in\text{-}s^2$		
	K8	Damping Constant		☐ oz-in/krpm		
	F2	Static Friction Torque		☐ oz-in		
	R3	Thermal Resistance		☐ $^{o}C/W$		

ELECTRO-CRAFT CORPORATION

REEL SERVO ANALYSIS

Program Name: REEL 5* Customer: ______

Application: ______ Engineer: ______

Date: ______

Input Data (supplied by customer) :

At line 10	(1)	V1	Operating tape speed (in/s)	
	(2)	V2	Rewind speed (in/s)	
At line 11	(1)	B	Tape width (in)	
	(2)	R4	Min. reel radius (no tape) (in)	
	(3)	R5	Max. reel radius (full tape) (in)	
	(4)	X	Max. +/- motion of sense point (in)	
At line 12	(1)	J3	Moment of inertia, reel & hub (oz-in-s^2)	
	(2)	G	Gear reduction (ratio) from motor shaft to reel	
	(3)	N	Number of tape loops	
	(4)	F	Friction torque of reel and tape (oz-in)	
	(5)	T7	Tape tension (oz)	

Motor Data (supplied by ECC)

At line 20 through 99	(1)	M	Motor model number (any number of motors can be input to program up to 80)			
	(2)	R2	Motor resistance @ 25°C (Ω)			
	(3)	K1	Motor torque constant (oz-in/A)			
	(4)	J2	Motor moment of inertia (oz-in-s^2)			
	(5)	K8	Motor viscous damping constant (oz-in/krpm)			
	(6)	W1	Motor max. permissible armature temperature (°C)			

Note: To assist ECC Applications Engineering in the selection of the best available motor, it is helpful if the customer can supply the following additional information:

1. Max. ambient operating temperature for the system (°F or °C)
2. Is forced-air cooling available?
3. Max. voltage and current available with proposed (or existing) power amplifier.

□

5.6.2. SELECTING THE PROPER SERVOMOTOR FOR AN INCREMENTAL MOTION SYSTEM

The most important class of servo systems in which Moving Coil Motors have been successfully applied is generally described as the *intermittent motion* or *incrementer systems.* In these applications, a low inertial load is required to be started and stopped rapidly and repeatedly, either in fixed increments, or in a random but programmable manner.

Typical examples are:

	Typical rate [steps/s]
a) Feed mechanisms	12-50
b) Line-by-line reading and punching	50-250
c) Incremental operation in digital magnetic tape transports	50-500
d) Punch card readers/ punchers	20-50
e) Magnetic head actuators	1-25

It may be seen intuitively in each example shown, that a low moment of inertia/high troque motor is required to successfully drive the load at the high repetition rates involved. Moreover, it is frequently necessary to run the motor at high speeds in order to *slew* through large shaft angular displacements. For example, document feed devices are required to step the paper for a predetermined number of lines, followed by a jump from page to page. Since all motion time is lost time, the slewing operation must be completed at the highest possible speed compatible with drive servo capability.

Principles of Operation

The incremental drive is basically a high-performance DC velocity servo system, as shown in block diagram form in Fig. 5.6.2. The motor is fitted with a standard DC tachometer, and is driven by a DC power amplifier.

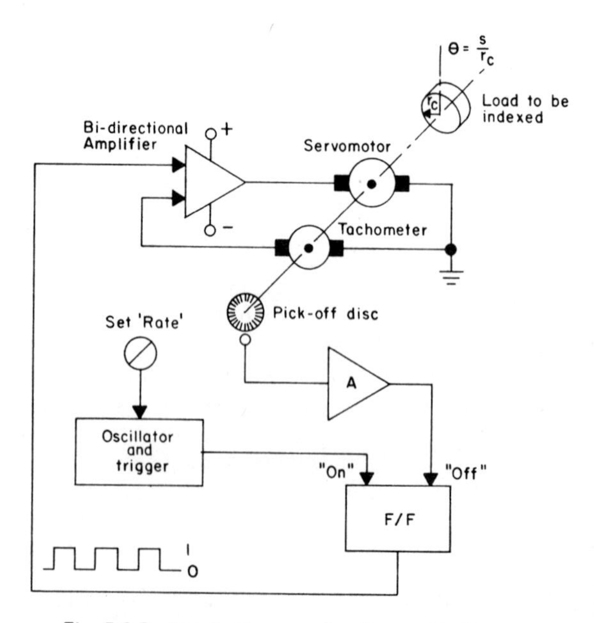

Fig. 5.6.2. Block diagram of an incremental servo system.

Incremental motion is obtained by operating the servo system in a non-linear mode, and is easily explained by reference to the waveforms in Fig. 5.6.3. Suppose that with the system at rest, we apply a step voltage to the input. Since the motor cannot attain any speed in *zero* time, there is no counter voltage from the tachometer.

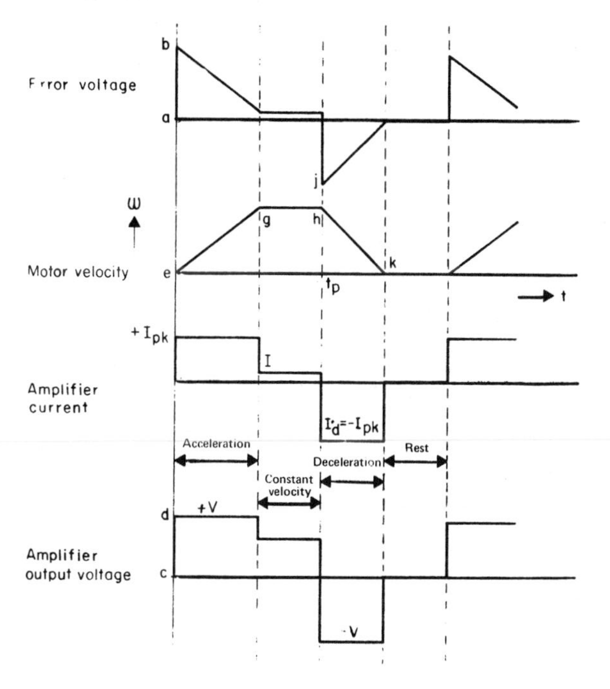

Fig. 5.6.3. Incremental motion waveforms.

Therefore, the effect of the step input signal is to produce a large error signal, **a–b**, at the amplifier input. The error signal exceeds the saturation level of the amplifier.

As a result, the full power supply voltage, **c–d**, is applied to the motor and acceleration begins following the line **e–g**. The rising tachometer voltage reduces the error signal during this time until the error signal falls within the acceptance band of the amplifier, and the system stabilizes in the constant velocity mode **g–h**. Only a nominal current, **I**, is required to supply the friction and load torque.. The operating speed will be such that tachometer voltage is very nearly equal to the input voltage.

If now the input voltage step is terminated at the time $\mathbf{t_p}$, the process reverses itself. The input change now produces a large error signal **i–j** in the opposite polarity causing a deceleration current $\mathbf{I_d}$. The re rulting torque, in turn, drives the velocity back to zero along the slope **h–k** which is nearly identical to the slope **e–g**, unless the friction torque is significant.

An analysis of the motion performance capabilities of an incrementer gets down to the equations of classical mechanics, with the acceleration values as the critical param eter.

In applications of this type, the incremental motion is almost wholly dissipative; i.e., little useful work is extracted from the motor since virtually the entire input energy is dissipated as heat. Such systems are therefore *power limited* in the sense that perfor mance is largely determined by the maximum armature dissipation capability. The calculations are, therefore, aimed at deter mining the power dissipation at "worst case" operating conditions.

System Dynamics

It should already be apparent that the incremental servo is a sophisticated piece of electromechanical engineering. In particular, the motor must operate satisfactorily in three quite distinct modes: a high speed start-stop mode, a good speed regulation mode for constant rates, and a high rewind speed mode.

Let us first look at the basic dynamic equation for motor output torque in an incremental motion application.

$$T_g = K_T I_a = J \frac{d\omega}{dt} + T_f + D\omega \quad (5.6.1)$$

From this we can see that first of all we must know what $\omega(t)$, the velocity profile, is. Let us assume that $\omega(t)$ is an incremental step motion which is repeated with a certain frequency — a profile many practical systems would be required to follow.

In the velocity profile of Fig. 5.6.4, ω_1 is the running velocity, and the angle of increment is $\theta = \int_0^{t_T} \omega(t)\, dt$. The acceleration time is t_g; t_r is the running time; t_s is the deceleration time; t_d is the dwell time; and the total time $t_T = t_g + t_r + t_s + t_d$.

EXAMPLE: A magnetic tape transport is to operate at linear speed $v = 75$ in/s; we assume a capstan radius of 1 in. Therefore, the motor run speed will be:

$$\omega_1 = \frac{75}{1} = 75 \text{ rad/s} \cong 716 \text{ rpm}$$

In typical digital tape transports, the capstan is required to perform very rapid start-stop (accelerate-decelerate) motions for repetitive block read-write operation. Fig. 5.6.4 shows the velocity profile for the short block, where t_r tends towards zero as the amount of data recorded per block becomes smaller. Thus, a worst case occurs when, for all practical purposes, the entire motion time t_T is made up of entirely of t_g and t_s.

Mathematically, as $t_r \to 0$ and $t_d \to 0$:

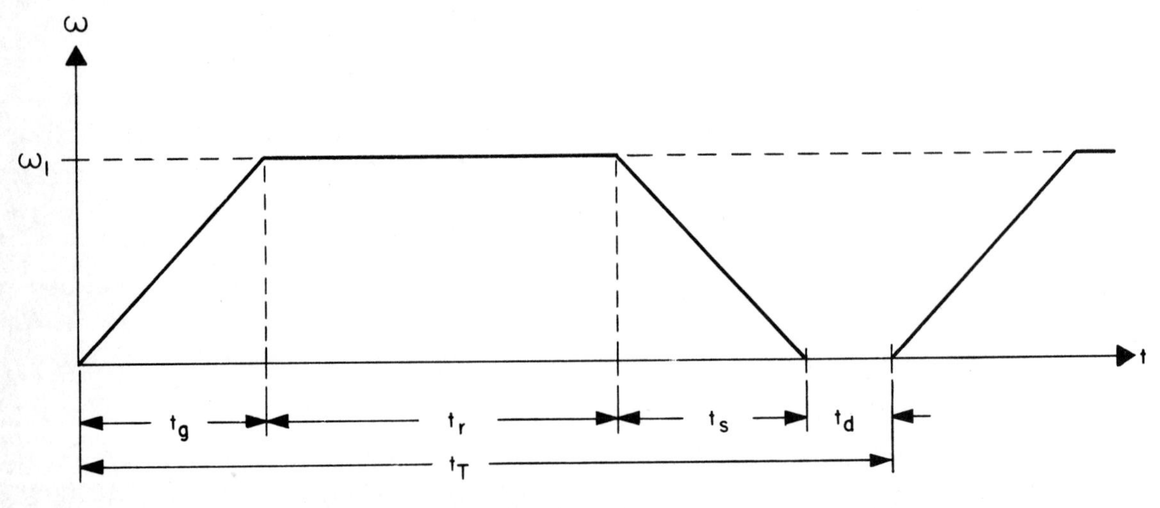

Fig. 5.6.4. Velocity profile of a typical incremental motion system.

$$t_T = t_g + t_s \qquad (5.6.2)$$

This is the condition which we now will investigate.

It is first necessary to establish the tape acceleration necessary to insure that the inter-record gap will not be exceeded in a single start-stop cycle. Let total displacement $d = d_g + d_s = 0.375$ in, allowing sufficient margin for overshoot, etc., then:

$$d_g = d_s = \frac{0.375}{2} = 0.1875 \text{ in}$$

Assuming the acceleration of the tape remains constant during the time t_g, we have:

$$a = \frac{v^2}{2d_g} = \frac{75^2}{0.375} = 15\,000 \text{ in/s}^2 \qquad (5.6.3)$$

The acceleration time t_g is given by:

$$t_g = \sqrt{\frac{2\,d_g}{a}} = \sqrt{\frac{0.375}{1.5 \times 10^4}} = 5 \text{ ms} \qquad (5.6.4)$$

The total motion time is therefore $t_T = 10$ ms when $t_g = t_s$.

Acceleration Torque

Since the capstan radius $r_c = 1$ in, then the capstan angular acceleration is

$$\alpha = \frac{a}{r_c} = \frac{15\,000 \text{ in/s}^2}{1 \text{ in}}$$

$$= 15\,000 \text{ rad/s}^2 \qquad (5.6.5)$$

Acceleration torque is given by the familiar expression:

$$T_a = J_T\,\alpha \qquad (5.6.6)$$

where

$$J_T = J_m + J_L$$

(J_T is the sum of the moments of inertia of the motor and the load, including tachometer)

Unfortunately, there is no direct analytic method whereby we can put in the numbers and come out with the motor model number for the solution. In fact, Electro-Craft's computerized analysis simply starts with the lowest cost servomotor available and works up the list until a motor is found which meets all requirements, since the equations apply to the entire performing system including the motor and load. Therefore, the analysis is an automated process of elimination based on acceptable performance and lowest possible cost.

After several attempts, we find that a suitable motor for operation in this type of capstan application is the Electro-Craft M-1400, which has the following parameters:

Moment of inertia $J_m = 0.002$ oz-in-s^2

Torque constant $K_T = 12$ oz-in/A

Armature Resistance $R = 1.6\ \Omega$

Total system moment of inertia is therefore:

Motor armature	**0.002 oz-in-s^2**
Capstan (typical)	**0.002**
Tachometer (typical)	**0.0005**
J_T =	**0.0045 oz-in-s^2**

The system acceleration torque becomes:

$$T_a = 4.5 \times 10^{-3} \times 1.5 \times 10^4$$

$$= 67.5 \text{ oz-in}$$

The *deceleration torque* is, for all practical purposes, *equal* and *opposite,* since friction is generally negligible in systems of this type.

Armature Acceleration Current

The armature peak current I_{pk} is simply obtained from the expression:

$$I_{pk} = \frac{T_a + T_f}{K_T} \qquad (5.6.7)$$

where T_a is the acceleration torque and T_f is the total system friction torque.

As we have noted, static friction in systems of this type is very small. Also, in a well-designed DC servomotor the damping loss can be neglected. Hence:

$$I_{pk} = \frac{67.5}{12} = 5.6 \text{ A}$$

Motor Terminal Voltage

The maximum *terminal voltage* **V** is required at the end of acceleration interval and may be calculated from the standard equation:

$$V = R\,I_a + K_E\,n \qquad (5.6.8)$$

In this case we must consider the "hot" resistance of the motor armature (20% increase) so that

$$R = 1.2 \times 1.6\ \Omega = 1.92\ \Omega$$

$$I_a = I_{pk} = 5.6 \text{ A}$$

$$K_E = \frac{K_T}{1.3524} \quad [\text{V/krpm; oz-in/A}]$$

[see (2.1.24)]

$$K_E = 8.873 \text{ V/krpm}$$

$$n = 716 \text{ rpm}$$

Substituting into (5.6.8) we get

$$V = 17.1 \text{ V}$$

Thus, a suitable pulse amplifier required to drive this capstan servomotor should be capable of delivering a peak current pulse of **6 A** at approximately **20 V** in the described repetition mode.

Calculation of Motor Power Dissipation

We will now proceed to calculate *motor power dissipation* for the example at hand.

The maximum repetition rate is chosen as the basis for "worst case" calculation. Although the power dissipation during the velocity mode portion of the velocity profile is very small compared to the acceleration/deceleration dissipation, it has been included in the calculation below. In some examples, it can be safely ignored.

The peak acceleration/deceleration current was calculated to be 5.6 A. The power dissipation — heat loss — in any device is given by $R\ I^2_{RMS}$. We can calculate the RMS value of the current as follows (refer to Fig. 5.6.5):

$$I_{RMS} = \sqrt{\frac{\Sigma I_i^2\ t_i}{\Sigma t_i}} \tag{5.6.9}$$

For the current profile of Fig. 5.6.5 we get

$$I_{RMS} = 4.02\ A$$

Armature heat loss is therefore

$$P_L = R\ I^2_{RMS} = 1.92 \times 4.02^2$$

$$= 31.03\ W$$

In order to see whether the motor can handle this power dissipation, we can check the temperature rise of the uncooled Electro-Craft M-1400 motor under the conditions calculated above.

The M-1400 has the thermal resistance

$$R_{th} = 2.4\ ^oC/W$$

The motor armature temperature rise will then be

$$\Theta_r = P_L\ R_{th} = 74.47\ ^oC \tag{5.6.10}$$

Assuming a maximum ambient temperature of 40 ^oC (and a maximum safe armature temperature of 155 ^oC), we will reach a final armature temperature of:

$$\Theta_a = \Theta_A + \Theta_r = 114.47\ ^oC \tag{5.6.11}$$

which is safely inside the maximum temperature limit of the armature.

If we were to use forced-air cooling, the thermal resistance would decrease drastically, and the armature temperature would be much lower than the foregoing example shows. Conversely, much higher dissipation rates could be accommodated.

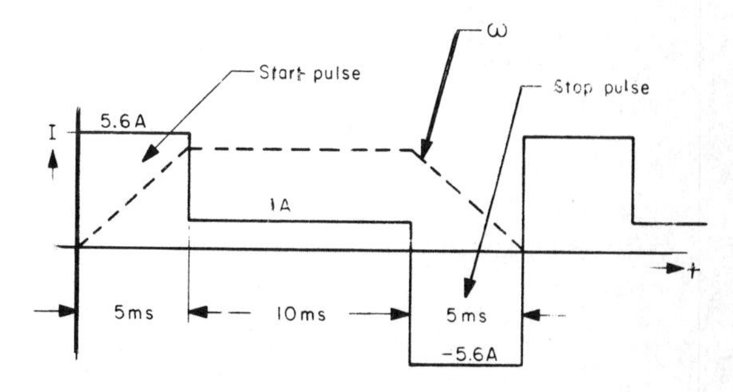

Fig. 5.6.5. Current profile for a typical incrementing application.

Slewing Requirement

One of the most interesting features of the incrementer servo — as distinct from the stepper motor — is its ability to perform large displacements at very high *slewing* rates without the need to worry about unstable velocity regions or programmed acceleration-deceleration. In the system under discussion, for example, line printers are required to skip large areas of printout paper as, for example, when printing out on standard size forms. Check writing is a typical example.

The problem reduces to one of allowing the servo drive motor to accelerate to much higher slewing velocities (typically a few thousand rpm) and controlling motion so that the system will be decelerated to a stop within permissible displacement tolerances.

A typical velocity versus time plot of a line printer application is shown in Fig. 5.6.6. We consider the system acceleration as being more or less linear; and, with a sufficiently high amplifier voltage and good power supply regulation, the mechanism and paper can be brought up to 100 in/s in less than 25 ms.

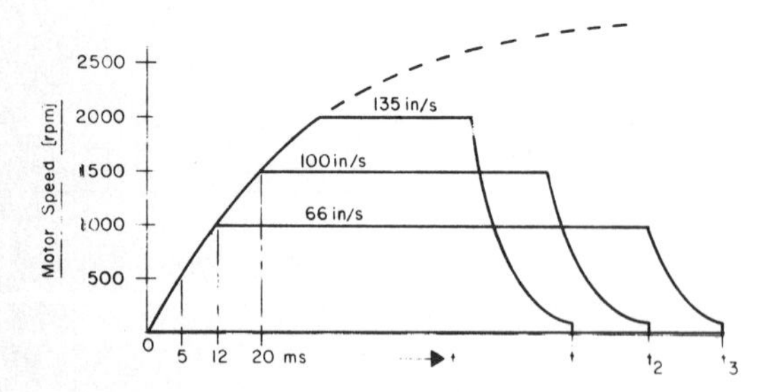

Fig. 5.6.6. Velocity vs. time plot for a line printer servo slewing through large displacements.

Conclusion

Electro-Craft Moving Coil Motors are ideally suited for use as incremental motion drive servomotors, because they have inherently good acceleration (high torque-to-moment of inertia ratios). Of equal significance in practical systems is the economy of using direct-drive DC servo systems with their low maintenance costs and high reliability.

Extensive life testing of the Electro-Craft Moving Coil Motors over a five-year period has indicated that there is no observable deterioration in the performance of the motors when they are operated within specification limits of pulse current and armature dissipation. *These factors should be given very serious consideration when specifying Moving Coil Motors for use in incremental motion and high acceleration drives.*

The Electro-Craft concept of the Moving Coil Motor undoubtedly has considerable development potential, especially in industrial equipment where the performance limits of mechanical indexing or clutch-brake devices are approached. It is well worth the design engineering effort required to investigate the potential performance improvement — sometimes an order of magnitude improvement in motion time — of which the Moving Coil Motor is now capable. Nevertheless, it is important to realize that the *total system's concept* must be considered for optimum performance. Electro-Craft is unique in being *the only manufacturer of low inertia motors with an equivalent capability in the electronic servo control field.* Many customers have benefited from our wide experience in *control systems design.* □

5.6.3. TAG PRINTER FEED DRIVE

System Requirement

The drive mechanism positions sales tags incrementally during printing and coding operations. The tags are driven by a 0.75 in diameter wheel which is driven directly by the motor. The tags are to be positioned 0.167 in in 21 ms and remain at rest for 29 ms. In the worst case, the cycle is repeated continuously. The power supply is limited to 10 V and 4 A. Roller moment of inertia is 0.001 oz-in-s^2, and load friction torque is 10 oz-in. The maximum ambient temperature is 65 ^oC.

Motion Profile

Since there are no restrictions on the motion profile expect for motion time, a trapezoidal velocity profile with equal acceleration, run and deceleration times will be selected to minimize power dissipation as illustrated in Fig. 5.6.7.

The basic data and variables of the system are:

roller radius r = 0.375 in

linear displacement per step x = 0.167 in

θ_1 **angular displacement during acceleration time t_1**

θ_2 **angular displacement during run time t_2**

θ_3 **angular displacement during deceleration time t_3**

$\theta_T = \theta_1 + \theta_2 + \theta_3$ **total step angular displacement**

ω_o = **run velocity**

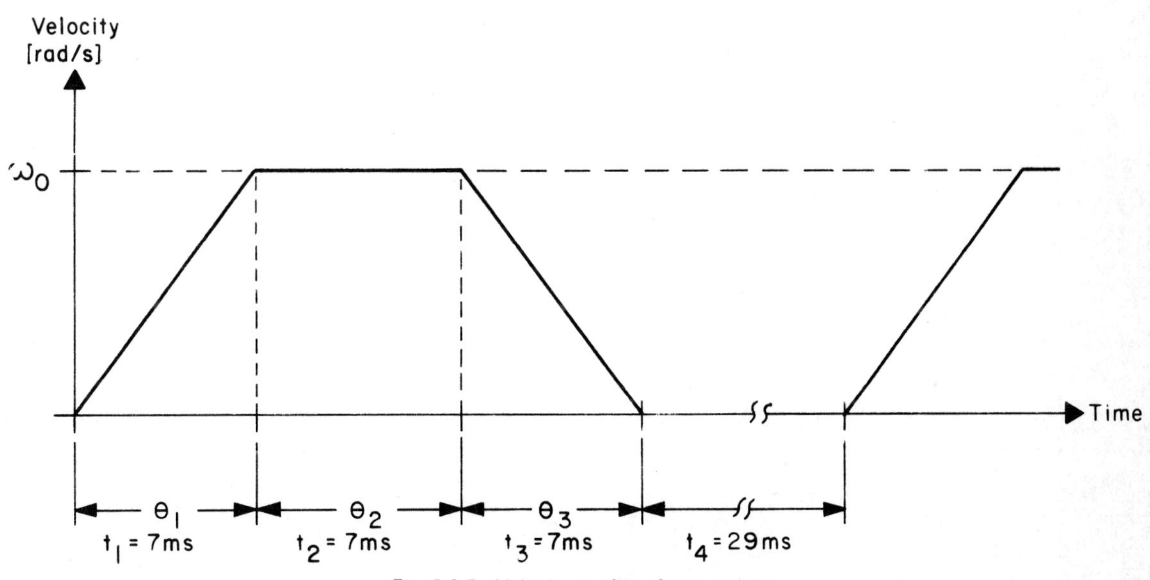

Fig. 5.6.7. Velocity profile of a tag printer.

For a trapezoidal velocity profile

$$t_1 = t_2 = t_3 = 0.007 \text{ s}$$

$$\theta_1 = \theta_3 = \frac{\omega_o t_1}{2}$$

$$\theta_2 = \omega_o t_1$$

$$\theta_T = 2\,\omega_o t_1 = \frac{x}{r}$$

$$\omega_o = \frac{x}{2rt_1} = \frac{0.167}{2 \times 0.375 \times 0.007}$$

$$= 31.81 \text{ rad/s}$$

Load Calculations

The next step in the system analysis is to compute the load at the motor shaft with the following inertia and friction values:

$$J_L = 0.001 \text{ oz-in-s}^2$$

$$T_L = 10 \text{ oz-in}$$

We can calculate the load torques during time intervals t_1, t_2 and t_3 and also the RMS load torque as follows:

a) load torque during time t_1

$$T_{L1} = \frac{J_L \omega_o}{t_1} + T_L = 14.54 \text{ oz-in}$$

b) load torque during time t_2

$$T_{L2} = T_L = 10 \text{ oz-in}$$

c) load torque during time t_3

$$T_{L3} = \frac{J_L \omega_o}{t_3} - T_L = -5.46 \text{ oz-in*}$$

d) RMS load torque (for t_1 thru t_4)

$$(T_L)_{RMS} = \sqrt{\frac{\Sigma T_{Li}^2 t_i}{\Sigma t_i}} \qquad (5.6.12)$$

Motor Selection

The initial selection of a first motor candidate is based on educated guesswork. When the results of the first selection are examined, the second or third choices (if needed) are usually close to the optimum. In this case, a high-performance Moving Coil Motor is not required because acceleration rates are relatively low. Therefore, we can accept a certain mismatch between the conventional motor armature moment of inertia and the relatively low load moment of inertia in order to choose the *most economical* motor for the application. Then we can also expect that the RMS generated torque required will be 1.5 to 2 times greater than the RMS load torque. We will try a Motomatic E-540 with the following parameters:

$$K_T = 10.02 \text{ oz-in/A}$$

$$R = 1.55\ \Omega \text{ @ } 25\ ^oC$$

*The negative sign here indicates that the friction torque is higher than the torque required to stop the load in time t_3. However, when the motor inertia is later added, we shall see that a positive torque is, indeed, needed to stop the motor and load in time.

J_m = 0.0038 oz-in-s^2

D = 0.1 oz-in/krpm

T_f = 3 oz-in (maximum)

R_{th} = 5.0 °C/W

We will first go through a hand calculation of the motor and amplifier requirements, and then follow with a computer analysis of the same problem.

The torque values T_{L1}, T_{L2}, and T_{L3} are functions of the load conditions only — no attention has yet been paid to the effect of a motor and its associated friction and inertia contributions. In the following section we will now include the constants of the selected motor to get values for the generated torque, which is the total torque required to run the motor and the load under specified conditions.

Generated torque required during time t_1

$$T_1 = \frac{(J_m + J_L)\,\omega_o}{t_1} + T_L + T_f$$

$$= 34.81 \text{ oz-in} \quad (5.6.13)$$

Damping torque is neglected here since motor speed is low and the damping constant is small.

Generated torque required during time t_2

$$T_2 = T_L + T_f = 13 \text{ oz-in} \quad (5.6.14)$$

Generated torque required during time t_3

$$T_3 = \frac{(J_m + J_L)\,\omega_o}{t_3} - T_L - T_f$$

$$= 8.81 \text{ oz-in} \quad (5.6.15)$$

Total RMS generated torque required (for t_1 thru t_4)

$$T_{RMS} = \sqrt{\frac{\Sigma T_i^2\, t_i}{\Sigma t_i}}$$

$$= 14.29 \text{ oz-in} \quad (5.6.16)$$

Armature Temperature Calculation

Temperature rise [see (5.3.7)]

$$\Theta_r = \frac{R\,R_{th}\,I_{RMS}^2}{1 - R\,R_{th}\,\psi\,I_{RMS}^2} \quad (5.6.17)$$

where

$$I_{RMS} = \frac{T_{RMS}}{K_T} = \frac{14.29}{10.02} \cong 1.43 \text{ A}$$

Thus,

$$\Theta_r = \frac{1.55 \times 5 \times 1.43^2}{1 - 1.55 \times 5 \times 0.00393 \times 1.43^2}$$

$$= 16.9\ ^\circ\text{C}$$

Remembering that Θ_A was established at 65 °C, we get

$$\Theta_a = \Theta_A + \Theta_r \cong 82\ ^\circ\text{C}$$

Thus, we notice that the biggest contributor to motor heat is the ambient temperature, but the motor temperature is well below the maximum limit of 155 ^{o}C.

Amplifier Requirement

During time t_1

$$I_1 = \frac{T_1}{K_T} = \frac{34.81}{10.02} = 3.474 \text{ A}$$

$$V_1 = R_h I_1 + K_E n_o \qquad (5.6.18)$$

where the "hot" resistance

$$R_h = R\,(1 + \psi_{cu}\,\Theta_r)$$

$$= 1.55\,(1 + 0.00393 \times 57)$$

$$= 1.9\ \Omega$$

Note that Θ_r must be the temperature rise above 25 ^{o}C (which is the temperature at which R is established).

$$K_E = \frac{K_T}{1.3524} = 7.409 \text{ V/krpm}$$

$$n_o = 9.5493\ \omega_o = 303.8 \text{ rpm}$$

Thus,

$$V_1 = 8.85 \text{ V}$$

During time t_2

$$I_2 = \frac{T_2}{K_T} = \frac{13}{10.02} \cong 1.3 \text{ A}$$

$$V_2 = R_h I_2 + K_E n_o = 4.73 \text{ V}$$

During time t_3

$$I_3 = \frac{T_3}{K_T} = \frac{8.81}{10.02} = 0.88 \text{ A}$$

$$V_3 = R_h I_3 + K_E n_o = 3.92 \text{ V}$$

Thus, we can stay within the power supply limitation of **10 V** and **4 A.**

Next, we will show the same analysis performed by a computer program developed by Electro-Craft Corporation. Fig. 5.6.8 shows the input data sheet, displaying not only input data but also the symbols accompanying it on the printout. These symbols may vary somewhat from the ones used throughout this book, due to the absence of suitable symbols on the printer.

The computer program (Fig. 5.6.9) simplifies the calculation chores and allows easy comparison of power dissipation and power supply demand by various motor models.

Worst Case Motor Analysis

In the following example we have evaluated the tag printer application power dissipation and amplifier requirement, using the worst case torque constant and armature resistance. Assume K_T to be 8% lower than nominal and **R** to be 8% higher than nominal for this evaluation. The result for this situation is given in Fig. 5.6.10. We can see that the armature temperature rose by 6 ^{o}C, and that the voltage and current requirements rose less than 10% from the values computed using nominal motor constants. The results shown here may not seem dramatic, but this is due to the relatively low motor power dissipation in this application. In cases where the motor is operating close to its dissipation limit, it is very important to perform the worst-case analysis to prevent unexpected dissipation problems from limiting the operational performance of a system.

INCREMENTAL SERVO; COMPUTER ANALYSIS

ELECTRO-CRAFT CORPORATION

Program Name INCR5* Customer: ______________ Date: ______________

Application: TAG PRINTER FEED DRIVE Contact (Engineer): ______________

Velocity Profile: Telephone: ______________

Vel. V_1 V_2

S_1 S_2 S_3

$S = S_1 + S_2 + S_3$

T_1 T_2 T_3 T_4

Problem Input Data (supplied by customer):

At line 10 (1) V1, Forward/Reverse velocity | 31.9
(2) V2, Rewind/Slew velocity |

At line 11 (1) T1, Acceleration time, mS | 7
(2) T2, Run time, mS | 7
(3) T3, Deceleration, mS | 7
(4) T4, Dwell time, mS | 29

(NOTE: Lines 10 and 11 define the velocity profile: it may be necessary to satisfy a given value for 's'; however, before entering the data into the computer, it is a preferred method to define the velocity profile in terms of velocity and time values.)

At line 12 (1) J1, Load inertia, oz.-in.-sec.^{2} | .001
(2) R1, Drive roll or capstan radius | .375
(3) F, Friction Torque at load shaft, oz.-in. | 10
(4) G, Reduction gear, motor to load | 1
(5) T_a, Ambient temperature °C. | 65

(NOTE: Line 12 defines the load conditions.)

Input Data (supplied by ECC):

At line 20 (1) M, Motor Model No. | 540
(2) K1, Motor torque constant (oz.-in./amp.) | 10.02
(3) R2, Motor armature resistance (at 25°C.) | 1.55
(4) J2, Motor inertia (oz.-in.-sec.^{2}) | .0038
(5) K8, Motor viscous damping constant (oz.-in./KRPM) | .1
(6) T_F, Friction Torque oz.-in. | 3
(7) R_{th}, Thermal Resistivity °C./watt | 5

(NOTE. To assist ECC Applications Engineering in the selection of the best available motor, it is helpful if customer can supply the following additional information:

1. Max. ambient operating temperature for the system (°F. or °C.)
2. Is air cooling available?
3. Max. voltage and current available with proposed (or existing) power amplifier.

Fig. 5.6.8. Incremental servo computer analysis for tag printer feed drive. (Not corrected for this third edition)

```
RUN VELØCITY (RAD/SEC) = 31.9

T1 = 7       S2 = Ø.Ø42     ACC, RAD/S↑2 = 4552
T2 = 7       S2 = Ø.Ø84
T3 = 7       S3 = Ø.Ø42     DEC, RAD/S↑2 = 4552
DWELL TIME = 29      TØTAL DISPLACEMENT = Ø.167
MØTØR RUN RPM = 3Ø4
MAX REP RATE (S-S CYCLES/SEC) = 2Ø
AMBIENT ØPERATING TEMP (DEGREES C) = 65
J1 = Ø.ØØ1    R1 = Ø.375   F1 = 1Ø    G = 1

JMØTØR MØDEL #54Ø        ARMATURE TEMP (DEGREES C) = 8.

K1 = 1Ø.??   R2=1.55   J2 =Ø.ØØ38   K8=Ø.1   R2 = 3   R3=5

ARMATURE PØWER DISSIPATIØN ( WATTS) = 3.9
RMS CURRENT (AMPS) = 1.4
TERMINAL RESISTANCE AT MAX TEMP (ØHMS) = 1.91
ACC TØRQUE (IN-ØZ) = 34.9

I-ACC = 3.5      V-ACC = 8.9
I-RUN = 1.3      V-RUN = 4.7
I-DEC = Ø.9      V-DEC = ?.9
```

Fig. 5.6.9. Computer printout of tag printer feed drive analysis. (Not corrected for this third edition)

```
T1 = 7      S1 = Ø.Ø42      ACC, RAD/S↑2 = 4552
T2 = 7      S2 = Ø.Ø84
T3 = 7      S3 = Ø.Ø42      DEC, RAD/S↑2 = 4552
DWELL TIME = 29      TØTAL DISPLACEMENT = Ø.167
MØTØR RUN RPM = 3Ø4
MAX REP RATE (S-S CYCLES/SEC) = 2Ø
AMBIENT ØPERATING TEMP (DEGREES C) = 65
J1 = Ø.ØØ1   R1 = Ø.375   F1 = 1Ø    G =1

MØTØR MØDEL #54Ø        ARMATURE TEMP (DEGREES C) = 9Ø

K1 = 9.2   R2=1.67   J2=Ø.ØØ38   K8=Ø.1   R2 = 3   R3 =5

ARMATURE PØWER DISSIPATIØN (WATTS) = 5.1
RMS CURRENT (AMPS) = 1.6
TERMINAL RESISTANCE AT MAX TEMP (ØHMS) = 2.Ø9
ACC TØRQUE (IN-ØZ) = 349

I-ACC = 3.8      V-ACC = 1Ø
I-RUN = 1.4      V-RUN = 5
I-DEC = 1         V-DEC=4.1
```

Fig. 5.6.10. Tag printer drive worst case analysis. □

5.6.4. CONVEYOR DRIVE SYSTEM

System Requirements

A motor must be selected to mate with existing drive components. The conveyor and load must be accelerated from rest to 200 ft/min in 0.5 s, driven at constant speed for 14 s and decelerated to rest in 0.5 s. For worst case analysis, the cycle is repeated continuously. The drive train description and motion profile are shown in Figs. 5.6.11 and 5.6.12, respectively.

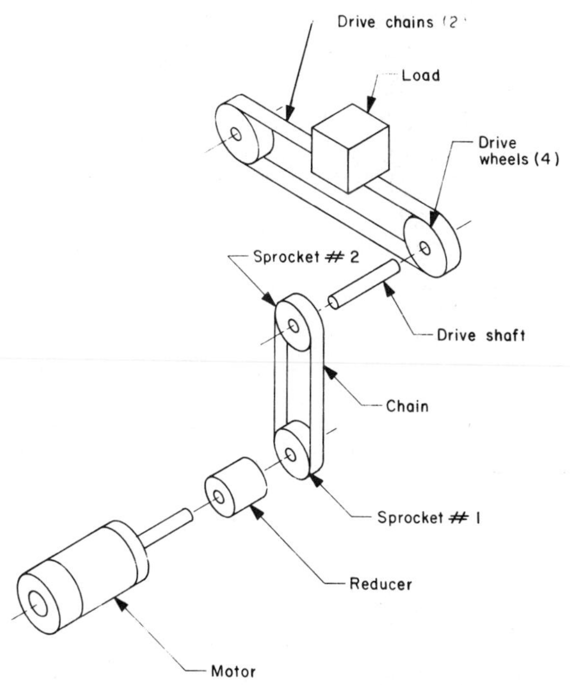

Fig. 5.6.11. Drive train mechanism for conveyor drive system.

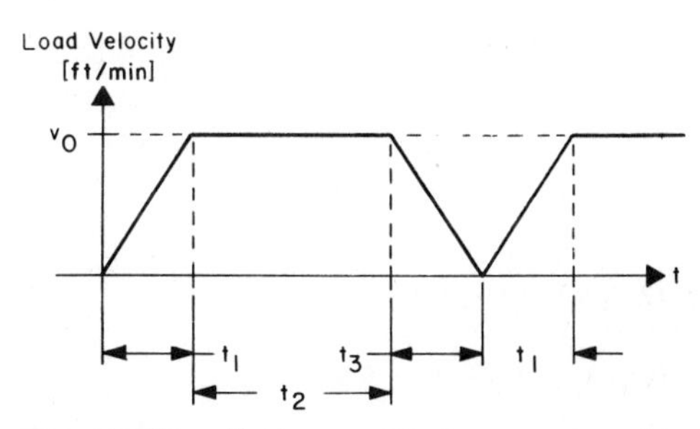

Fig. 5.6.12. Motion profile for conveyor drive system.

Load Calculations

The first step in the system analysis is to calculate the load at the motor shaft. In this case, all the inertial loads were not known by the customer, but mechanical dimensions and weights of each were furnished. From this data, the moments of inertia can be calculated. All moments of inertia are calculated here as equivalent moments of inertia at the motor shaft.

Note: In this example, the mass m of individual parts of the drive mechanism is calculated in units [oz-s^2/in], not in [slug] (see description of the British system of units in the Appendix A.2.). This is more convenient because the moments of inertia, J, are calculated in units [oz-in-s^2] and no conversion factors have to be used. The mass m is thus calculated from given weight w [oz] as follows:

$$m = \frac{w}{g} \qquad [oz\text{-}s^2/in;\ oz,\ in/s^2] \tag{5.6.19}$$

where

g = **386.0878 in/s^2** is the standard gravitational acceleration.

Load moments of inertia are as follows:

1. *Reducer*

$$J_1 = 0.04 \text{ oz-in-s}^2$$

(manufacturer's list value)

N = 10 – gear ratio

2. *Sprocket #1*

$r_2 = 2$ in — radius

$w_2 = 2$ lb $= 32$ oz

$$m_2 = \frac{w_2}{g} = 0.0829 \text{ oz-s}^2/\text{in}$$

$$J_2 = \frac{m_2 r_2^2}{2 N^2} = 0.00166 \text{ oz-in-s}^2$$

3. *Sprocket #2*

$r_3 = 6$ in

$w_3 = 20$ lb $= 320$ oz

$m_3 = 0.829$ oz-s^2/in

$$J_3 = \frac{m_3 r_3^2}{2\left(N \frac{r_3}{r_2}\right)^2} = 0.0166 \text{ oz-in-s}^2$$

4. *Chain*

$w_4 = 24$ lb $= 384$ oz

$m_4 = 0.995$ oz-s^2/in

$$J_4 = \frac{m_4 r_2^2}{N^2} = 0.0398 \text{ oz-in-s}^2$$

5. *Drive Shaft*

$r_5 = 1.5$ in

$w_5 = 80$ oz

$m_5 = 0.207$ oz-s^2/in

$$J_5 = \frac{m_5 r_5^2}{2\left(N \frac{r_3}{r_2}\right)^2} = 0.000259 \text{ oz-in-s}^2$$

6. *Drive Wheels (4 pcs)*

$r_6 = 8$ in

$w_6 = 20$ lb $= 320$ oz
$m_6 = 0.829$ oz-s^2/in
} one wheel

$$J_6 = 4 \frac{m_6 r_6^2}{2\left(N \frac{r_3}{r_2}\right)^2} = 0.118 \text{ oz-in-s}^2$$

7. *Load*

$w_7 = 100$ lb $= 1600$ oz

$m_7 = 4.144$ oz-s^2/in

$$J_7 = \frac{m_7 r_6^2}{\left(N \frac{r_3}{r_2}\right)^2} = 0.295 \text{ oz-in-s}^2$$

8. *Drive Chains (2 pcs)*

$w_8 = 150$ lb $= 2400$ oz
$m_8 = 6.216$ oz-s^2/in
} one chain

$$J_8 = 2 \frac{m_8 r_6^2}{\left(N \frac{r_3}{r_2}\right)^2} = 0.884 \text{ oz-in-s}^2$$

Total load moment of inertia

$$J_L = J_1 + \cdots + J_8 = 1.395 \text{ oz-in-s}^2$$

Load Friction Torque

Next, we calculated the friction torque at the motor shaft. T'_{Lf} = **40 lb-ft** is the load friction torque measured at the drive shaft. Therefore, load friction torque at the motor shaft is given by

$$T_{Lf} = \frac{T'_{Lf}}{N \frac{r_3}{r_2}} = 1.333 \text{ lb-ft} = 256 \text{ oz-in}$$

RMS Load Torque

We now have sufficient information to calculate the RMS load torque at the motor shaft. This can be computed from torque values during t_1, t_2 and t_3.

Load torque during time t_1

Load torque during t_1 is acceleration torque plus load friction torque:

$$T_1 = T_{a1} + T_{Lf}$$

where

$$T_{a1} = J_L \frac{\omega_o}{t_1}$$

and

$$\omega_o = \frac{v_o}{r_6} N \frac{r_3}{r_2}$$

$$v = 200 \text{ ft/min} = 40 \text{ in/s}$$

Then

$$\omega_o = 150 \text{ rad/s}$$

$$T_{a1} = 418.5 \text{ oz-in}$$

$$T_1 = 674.5 \text{ oz-in}$$

Load torque during time t_2

Load torque during time t_2 is simply the load friction torque

$$T_2 = T_{Lf} = 256 \text{ oz-in}$$

Load torque during time t_3

Load torque during time t_3 is the difference between acceleration and friction load torques (see the velocity profile in Fig. 5.6.12):

$$T_3 = T_{a1} - T_{Lf} = 162.5 \text{ oz-in}$$

RMS load torque is then given by

$$(T_L)_{RMS} = \sqrt{\frac{\Sigma T_i^2 t_i}{\Sigma t_i}} = 277.87 \cong 278 \text{ oz-in}$$

Motor Selection

In this system, acceleration torques contribute a relatively small part to the RMS load

INCREMENTAL SERVO; COMPUTER ANALYSIS

ELECTRO-CRAFT CORPORATION

Program Name INCR 5** Customer: ______________ Date: ______________

Application: CONVEYOR DRIVE SYSTEM Contact (Engineer): ______________

Velocity Profile: Telephone: ______________

Problem Input Data (supplied by customer):

At line 10	(1) V1, Forward/Reverse velocity, ips	150
	(2) V2, Rewind/Slew velocity, ips	
At line 11	(1) T1, Acceleration time, mS	500
	(2) T2, Run time, mS	14000
	(3) T3, Deceleration, mS	500
	(4) T4, Dwell time, mS	0

(NOTE: Lines 10 and 11 define the velocity profile: it may be necessary to satisfy a given value for 's'; however, before entering the data into the computer, it is a preferred method to define the velocity profile in terms of velocity and time values.)

At line 12	(1) Jl, Load inertia, oz.-in.-sec.^{2}	1.4
	(2) Rl, Drive roll or capstan radius	
	(3) F, Friction Torque at load shaft, oz.-in.	256
	(4) G, Reduction gear, motor to load	
	(5) T_a, Ambient temperature °C.	25

(NOTE: Line 12 defines the load conditions.)

Input Data (supplied by ECC):

At line 20	(1) M, Motor Model No.	702	
	(2) Kl, Motor torque constant (oz.-in./amp.)	26.9	
	(3) R2, Motor armature resistance (at 25°C.)	.43	
	(4) J2, Motor inertia (oz.-in.-sec.^{2})	.15	
	(5) K8, Motor viscous damping constant (oz.-in./KRPM)	7	
	(6) T_F, Friction Torque oz.-in.	15	
	(7) R_{th}, Thermal Resistivity °C./watt	1.6	

(NOTE: To assist ECC Applications Engineering in the selection of the best available motor, it is helpful if customer can supply the following additional information:

1. Max. ambient operating temperature for the system (°F. or °C.)
2. Is air cooling available?
3. Max. voltage and current available with proposed (or existing) power amplifier.

Fig. 5.6.13. Analysis form for conveyor drive system motor application. (Not corrected for this third edition)

torque. Therefore, it is estimated that a motor capable of providing the calculated acceleration torque and delivering in excess of 278 oz-in RMS torque will be adequate. A Motomatic E-702 seems to be a logical choice.

Motor parameters:

K_T = **26.9 oz-in/A**

R = **0.43 Ω @ 25 °C**

J_m = **0.15 oz-in-s²**

D = **7 oz-in/krpm**

T_f = **15 oz-in (maximum)**

R_{th} = **1.6 °C/W**

Motor Performance Calculation — a computer input data sheet is shown in Fig. 5.6.13, and the corresponding solution printout appears as Fig. 5.6.14. For examples of the calculations, hand computations will also be carried out below, permitting a comparison to be made.

```
RUN VELOCITY (RAD/SEC) = 150
T1 = 500        ACC, RAD/S↑2 = 300
T2 = 14000
T3 = 500        ACC, RAD/S↑2 = 300
DWELL TIME = 0
MAX REP RATE (S-S CYCLES/SEC) = .67
AMBIENT OPERATING TEMP (DEGREES C) =25
J1 =1.4   F1 = 256
MOTOR MODEL #702        ARMATURE (DEGREES C)=160
K1 = 26.9  R2 = .43  J2 = .15  K8 = 7  F2 = 15  R3 = 1.6

ARMATURE POWER DISSIPATION (WATTS) = 84.2
RMS CURRENT (AMPS) = 11.3
TERMINAL RESISTANCE AT MAX TEMP (OHMS) .65
ACC TORQUE (IN-OZ) = 741
I-ACC = 27.5        V-ACC = 46.6
I-RUN = 10.4        V-RUN = 35.3
I-DEC = 7           V-DEC = 33.2
```

Fig. 5.6.14. Conveyor drive system motor selection printout.

Manual Calculations

Generated RMS Torque

The required motor torques in time intervals t_1, t_2 and t_3 can be calculated from load torque data by adding the motor parameters J_m, D and T_f. The individual generated torques then become

T_1 = **741 oz-in**

T_2 = **281 oz-in**

T_3 = **189 oz-in**

and the required RMS generated torque is T_{RMS} = **305.3 oz-in.**

RMS value of motor current is then

$$I_{RMS} = \frac{T_{RMS}}{K_T} = \frac{305.3}{26.9} = 11.35 \text{ A}$$

Amplifier Requirements

Current and voltage requirements during times t_1, t_2 and t_3 must be calculated based on temperature stabilized armature resistance.

Armature temperature rise Θ_r is given by

$$\Theta_r = \frac{R\,R_{th}\,I^2_{RMS}}{1 - R\,R_{th}\,\psi\,I^2_{RMS}} \tag{5.6.17}$$

The E-702 motor has

R = **0.43 Ω**

R_{th} = **1.6 °C/W**

so that

$$\Theta_r = 136\ ^oC$$

The motor resistance at operating temperature is then

$$R_h = R\,(1 + \psi\Theta_r) = 0.66\ \Omega$$

a) During time t_1

$$I_1 = \frac{T_1}{K_T} = \frac{741}{26.9} = 27.55\ A$$

$$V_1 = R_h I_1 + K_E n_o \qquad (5.6.18)$$

where

$$K_E = \frac{K_T}{1.3524} = 19.9\ V/krpm$$

$$n_o = 9.5493 \times 10^{-3}\ \omega_o = 1.432\ krpm$$

so that

$$V_1 = 46.7\ V$$

b) During time t_2

$$I_2 = \frac{T_2}{K_T} = \frac{281}{26.9} = 10.45\ A$$

$$V_2 = R_h I_2 + K_E n_o = 35.4\ V$$

c) During time t_3

$$I_3 = \frac{T_3}{K_T} = \frac{189}{26.9} = 7.03\ A$$

$$V_3 = R_h I_3 + K_E n_o = 33.14\ V$$

Motor temperature

$$\Theta_a = \Theta_r + \Theta_A = 136 + 25 = 161\ ^oC$$

Notice that armature temperatures, 161 °C and 160 °C (calculated by computer — see Fig. 5.6.14) exceed the temperature rating of the motor. Let's evaluate the *next larger unit* available, a model E-703 with the following characterics:

$$K_T = 24.5\ oz\text{-}in/A$$

$$R = 0.26\ \Omega\ @\ 25\ ^oC$$

$$J_m = 0.2\ oz\text{-}in\text{-}s^2$$

$$D = 10\ oz\text{-}in/krpm$$

$$T_f = 15\ oz\text{-}in\ (maximum)$$

$$R_{th} = 1.2\ ^oC/W$$

The computer solution is given in Fig. 5.6.15.

```
RUN VELOCITY (RAD/SEC) = 150
T1 = 500          ACC RAD/S↑2 = 300
T2 = 1400
T3 = 500          ACC, RAD/S↑2 = 300
DWELL TIME = 0
MAX REP RATE (S-S CYCLES/SEC) = .67
AMBIENT OPERATING TEMP (DEGREES C) = 25
J1 = 1.4    F1 = 256
MOTOR MODEL #703          ARMATURE TEMP (DEGREES C)  = 87
K1 = 24.5  R2=0.26  J2 = 0.2  K8 =10  F2 =15  R3 =1.2

ARMATURE POWER DISSIPATION (WATTS) = 51.8
RMS CURRENT (AMPS) = 12.7
TERMINAL RESISTANCE AT MAX TEMP (OHMS) = 0.32
ACC TORQUE (IN-OZ) = 758.2

I-ACC = 30.9   V-ACC = 36
I-RUN = 11.6    V-RUN=29.7
I-DEC = 8.2    V-DEC - 28.7
```

Fig. 5.6.15. Re-run of the conveyor drive system problem, applying the E-703 motor.

Note that in the printout of Fig. 5.6.15 the *armature temperature* (87 °C) is well within the temperature rating of the motor. This is due to the added stack length of the E-703, which provides a comfortable margin for temperature consideration.

Therefore, the Motomatic E-703 can be used in this application, meeting all system requirements without overheating. □

5.6.5. COMPUTER TAPE TRANSPORT REEL MOTOR

The function of a reel motor in a computer tape system is to hold the amount of tape stored in the buffer bins constant (see Fig. 5.6.16). The reels are wound and unwound as required by the motion of the drive roll or capstan motor. Rotational speed and direction of the reel motors is controlled by the amount of tape hanging in the buffer bin.

The tape transport under consideration has 32 photocell sensors placed on the wall of the buffer bin. They produce a voltage which is proportional to the tape loop position in the buffer bin. As the tape moves up and down in the buffer bin, a greater or lesser number of photocells are shaded. If a sensor is shaded by tape, it turns a switching circuit off. Summing the voltages produced by the switching circuits yields a voltage signal proportional to tape position in the bin (Fig. 5.6.17). When the position signal is positive, the reel drive functions so as to provide the buffer bin with tape. When it is negative, the reel drive functions so as to remove tape from the bin.

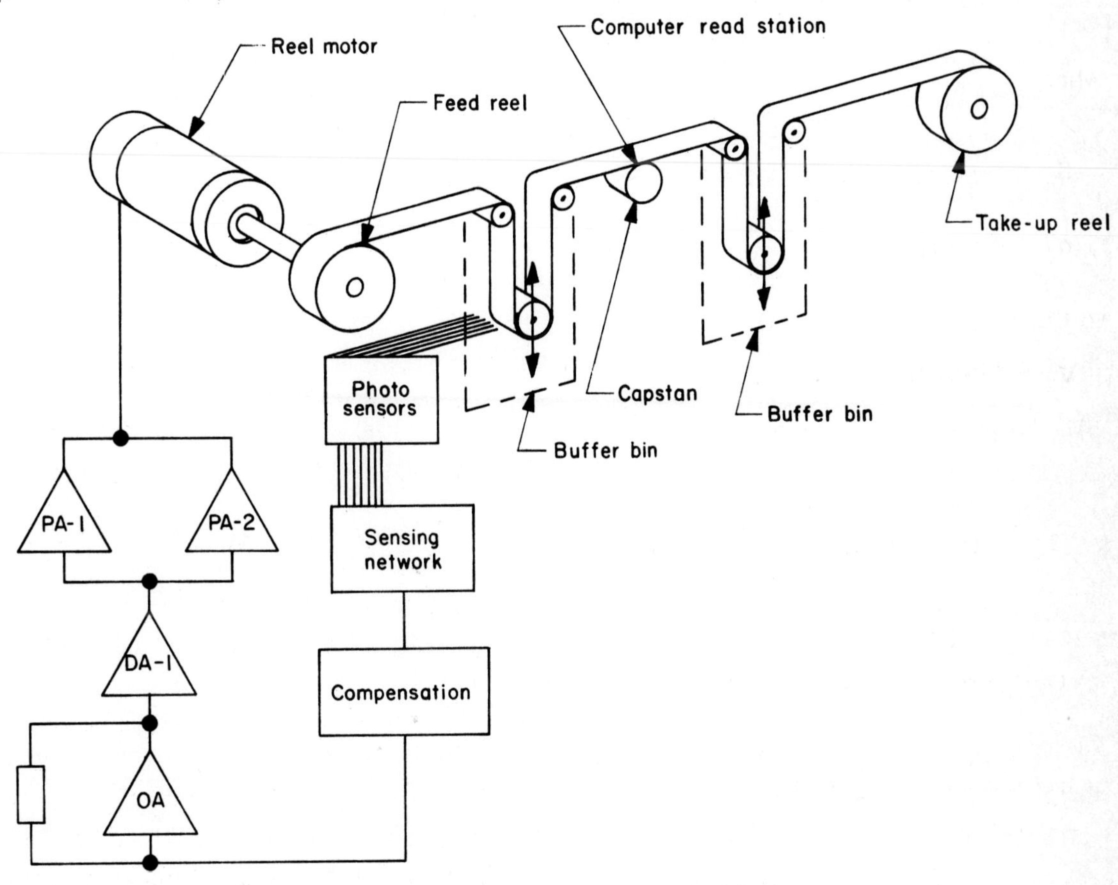

Fig. 5.6.16. Computer tape transport reel drive system.

The computer run for an application of this type is presented in Figs. 5.6.18 and 5.6.19. The results show that the motor selected, under specified conditions, will achieve a maximum armature temperature of 108.5 °C, with an armature dissipation of nearly 70 W.

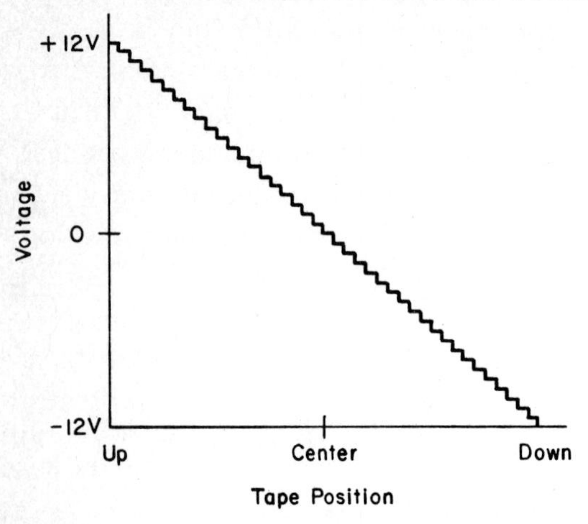

Fig. 5.6.17. Voltage vs. tape position in the bin of Fig. 5.6.16.

```
REEL SERVO ANALYSIS COMPLIMENTS OF ELECTRO CRAFT CORP

FORWARD OPERATING SPEED (IPS)= 125
REWIND SPEED (IPS)= 500

                          MOTOR MODEL NO.  E-703

                                  REEL EMPTY WO  REEL FULL WO   REWIND

MAX MOTOR SPEED (RPM)             459.334        229.667        1837.33
ARMATURE DISSIPATION (WATTS)      35.0157        69.3668
ARMATURE TEMP (DEG C)             67.0797        108.45
ACCEL TORQUE (OZ-IN RMS)          259.349        342.276
ACCEL CURRENT (AMPS RMS)          6.17497        8.14943


INPUT DATA CHECK
AMBIENT TEMP= 25
R4= 2.6 R5= 5.2 J1= 0.355 J2= 1.2
G1= 1 X1= 14 N1= 1 F1= 4 T7= 8

MOTOR DATA FOR MODEL NO.  E-703
R1= 42 R2= 0.79 J3= 0.2 K8= 10 F2= 15 K3= 1.2
```

Fig. 5.6.18. Computer printout of tape transport reel motor analysis.

ELECTRO-CRAFT CORPORATION

Reel Servo Analysis

Program Name: REEL 5* Customer: ________

Application: *REEL SERVO* Engineer: ________

Date: ________ Telephone: ________

Input Data (supplied by customer):

At line 10	(1)	V1= Operating tape speed, ips	125
	(2)	V2= Rewind speed, ips	500
At line 11	(1)	R4= Empty Reel Radius	2.6
	(2)	R5= Full Reel Radius	5.2
	(3)	X1= Max. overall motion of sense point (Max. +point to max. - point in inches)	14.0
	(4)	J1= Empty Reel Inertia	.355
	(5)	J2= Full Reel Inertia	1.2
At line 12	(1)	G1= Gear Reduction Ratio (from motor to reel)	1
	(2)	N1= No. of Tape Loops	1
	(3)	F1= Friction of Reel and Tape	4
	(4)	T7= Tape Tension	8
	(5)	ϕ1= Ambient Temperature (°C)	25

Motor Data (supplied by ECC):

At line 20 through 99	(1)	M$= Motor Model No. (Any number of motors can be input to program up to 80) (Enter Motor No. Using This Format: E-703)	E-703
	(2)	K1= Motor Torque Constant oz-in/amp	42
	(3)	R2= Motor Resistance at 25°C(ohms)	.79
	(4)	J3= Motor Inertia (oiss)	.2
	(5)	K8= Damping Constant (oz-in/KRPM)	10
	(6)	F2= Friction Torque (oz-in)	15
	(7)	R3= Thermal Resistance	1.2

NOTE: To assist ECC Applications Engineering in the selection of the best available motor, it is helpful if customer can supply the following additional information:

1. Max. ambient operating temperature for the system (°F. or °C.)
2. Is air cooling available?
3. Max. voltage and current available with proposed (or existing) power amplifier.

Fig. 5.6.19. Computer analysis input for the tape transport reel motor application. (Not corrected for this third edition)

5.7. APPLICATION OF P6000 SERIES SERVO MOTOR SYSTEMS

The P6000 systems are unique among the spectrum of servo motor controls currently on the market. Their uniqueness stems from the "total system" approach used during their conception and design. The design features incorporated in the series result in a wide range of highly flexible modules which can be interfaced with each other to enable the construction of custom systems tailored exactly to the needs of the user. Electro-Craft has had over 10 years of experience designing custom systems for specific applications; the know-how built up over those years has enabled the incorporation of many unique and previously unavailable features into the P6000 series.

5.7.1. GENERAL DESCRIPTION OF THE P6000 RANGE OF SERVO MOTOR CONTROLS

The heart of P6000 series is a fast response Pulse Width Modulated fully regenerative DC amplifier. The amplifier uses a fixed switching frequency of 20 kHz which eliminates the annoying audible noise generated by other types of PWM amplifiers. The series incorporates the following standard features:

Dual Stage Current Limit — Independently adjustable peak current and average current limits, together with full adjustment of the peak current limit duration. These adjustments can be made internally or from remotely mounted potentiometers.

Regenerative Energy Protection — The amplifiers are fully protected against the build-up of electrical energy in the power supply caused when high inertia loads are decelerated slowly. Units not having this kind of protection are susceptible to catastrophic failures under certain load conditions.

Crowbar Protected Power Supply — This feature makes the amplifier short circuit proof. If either output terminal on the amplifier is shorted to the other or to ground, or if the amplifier operation is disrupted by component failure or misuse, an SCR in the power supply is triggered blowing the power supply fuse and discharging the energy stored in the storage capacitors. This feature protects the output transistors and prevents catastrophic failure. Switching amplifiers not having this feature can fail in a literally "explosive" manner.

Amplifier Clamps — The amplifier has built-in clamping to prevent output surges during application or removal of AC power. Other circuits also provide independent clamps for clockwise and counterclockwise rotation for uses as end of travel limits. These clamps can be actuated by logic signals or contact closures.

Separate Power Supply — The power supply which enables operation from standard AC lines can be separated from the amplifier unit and mounted independently, or if desired, the amplifier can be purchased without the power supply and used in conjunction with existing unipolar DC supplies.

Flexibility – The amplifier has provision for up to three plug-in option p.c. cards. One of these is permanently used for the servo summing amplifier; the other two can be used to connect a wide range of standard or custom options. A complete list of options can be found in section 5.7.2.

Serviceability – The amplifiers have been designed to meet the maintainability requirements of both the "minimum downtime user" and the user who is working on a tight budget and cannot keep a stock of spare assemblies. For the "minimum downtime user", all major subassemblies and p.c. cards can be replaced by removing a maximum of four screws in each and unplugging the assembly. In fact, since all input/output connections are made through plugs and sockets, if preferred, the entire amplifier can be replaced in minutes.

For the user who prefers to troubleshoot and repair rather than replace, all test points are brought out to a 14-pin test point array for easy location; current loops are provided on the output stages to enable use of oscilloscope current probes; and all p.c. cards are hinged for access to components without removal.

Dual Axis Amplifiers – All models of the 6000 series are available in a dual axis configuration with two amplifiers sharing one power supply. This arrangement gives maximum efficiency and economy saving as much as 20% on the cost of a dual axis system.

5.7.2. STANDARD OPTIONS

This list represents all of the standard options for the P6000 series which were available at the publication date (August 1975). More standard options are planned, and in addition custom options to meet your exact needs are available.

R1000 Ramp Generator

The R1000 option permits the selection of one of two preset adjustable acceleration rates and one of two preset adjustable deceleration rates. The adjustments for ramp slopes are independent of each other and the different preset slopes can be selected by switch contacts or logic signals.

The R1000 also permits the selection of one of two preset adjustable motor speeds or a speed set by an external analog signal, thus the option can be used for digital selection of speed.

The ramp generator does not affect the closed-loop frequency response of the system to load disturbances.

The option is useful in any application where the acceleration and deceleration need to be controlled in a repeatable manner.

PL1000 Phase Lock System

The phase lock option enables speed control with zero long term speed drift and synchronization of a motor with any other moving part.

This option is particularly suited to applications where the speed of the motor forms a reference, such as chart drives or scanning drives. The system is also ideally suited to component marking applications where a motor driving a marking head has to be accurately synchronized with components on a moving conveyor.

DPS 6000 Digital Positioning System

The DPS 6000 is digital position system option which can be ordered with the standard P6000 Pulse Width Modulated servomotor controller series or the L5000 linear servo amplifiers. The unique design of the DPS 6000 and the high performance of the P6000 series controllers combine to provide a positioning system that can achieve higher accuracy and faster step times than stepper motors, clutch/brakes and other widely used systems.

A typical 200 step per revolution system will give single step times of 15 ms, including settling and slew speeds of 10,000 steps/second — even with load moments of inertia as high as 0.05 oz-in-s^2 — without loss of position.

Positional accuracy is typically in the order of 0.5°, even with high disturbing torques. Greater accuracy can be obtained by using systems having higher resolutions without deterioration of high speed slew and large step performance. Holding torques are on the order of 250 oz-in peak, 140 oz-in continuous.

The system accepts parallel binary or BCD step size inputs which enables direct control from thumb wheel switches and direct operation from minicomputers, eliminating the need for a translator/indexer or other interface logic.

An 8 bit speed input is provided so that the speed of the motor during stepping can be controlled. This feature, or the alternative analog speed input, enables the DPS 6000 to be used for contouring as well as point to point positioning applications.

For specifications see Chapter 7.

5.7.3. STANDARD DRIVE PACKAGES

The SD series of Standard Drive Packages consist of specially matched combinations of Electro-Craft servo motors and servo controllers. Both the motor and control are manufactured by Electro-Craft and are covered by a single warranty. Thus Electro-Craft can offer total system responsibility.

All of the SD series can be fitted with any of the standard or custom plug in options.

Full specifications of the SD series can be found in Chapter 7.

5.7.4. AMPLIFIER INSTALLATION

Portions of the amplifier and power supply produce heat which must be vented away from the components. Thus it is important to allow sufficient space around and above the amplifier for proper air circulation. Electro-Craft recommends at least 1 in clearance above and on both sides of the amplifier and power supply. If the amplifier is located in a completely enclosed cabinet, louvers or vents must be provided on each side near the fan. In addition to this, enough space should be allowed in front of the amplifier so that the logic control motherboard may be tilted down for servicing. For this to be done, 5.5 in of clearance should be provided in front of the amplifier.. Another 2 in of clearance should be left behind the power supply to allow replacement of fuses and disconnection of rear panel connectors.

Holes have been provided in the chassis rails to secure the amplifier and power supply to a sub-chassis or other surface. If the power supply is being mounted remotely and the rails are not being used, the holes in the chassis may be used. These components should be secured using #10-24 machine bolts of sufficient length.

5.7.5. CONTROL ADJUSTMENTS

Motherboard Controls

The controls that regulate the current limiting circuitry of the amplifier are located on the motherboard. Because the peak current adjustment will affect the average current adjustment, the peak current control should be set first. Unless a model P6042 Tektronix DC current probe is available, the peak current adjustment should be set by percent of rotation. The P6100 peak current range is between 3.5 and 15 A. Thus, when the control is fully clockwise, the peak current limit will be 15 A, while one-half rotation will allow only 9.25 A, or one-half the difference between the upper and lower limits. The P6200 amplifier should be adjusted using a range between 7 and 25 A, and then following the same procedure. The same method may be used when setting the peak current duration, or the time to which the current limiter will allow peak current. The range of adjustment is between 0.01 s and 0.1 s. With the control set at maximum clockwise rotation, the peak current duration will be 0.1 s. A central position will allow about one-half this time.

The next step is to set the average current limit. This setting will not effect the peak current limit. Set the control fully counterclockwise. With a DC ammeter in series with either motor lead, operate the motor under a stall condition. Slowly turn the average current limit control clockwise until a current reading is obtained which is considered the safe maximum continuous current of the motor to be used with the amplifier.

An alternate and more precise method of current adjustment would be to make use of the Test Point number 6 on the Motherboard. This point is the output of the current sensing amplifier and will give a voltage reading proportional to the amplifier current. An oscilloscope can be carefully connected to this point and the motor current set using the following proportion:

P6200 –	0.2 Volts = 1 Amp
P6100 –	0.4 Volts = 1 Amp

Summing Amplifier

The gain control on the summing amplifier (plug-in card) should be turned as far clockwise as is practical to insure stable operation. This will provide maximum gain. Set the input signal level control fully counterclockwise. After this is done, set the balance control so that there is no rotation of the motor. Then adjust the input level control for the proper speed range for the particular application in which the system is to be used. This must be done in conjunction with the external signal speed command. See section 5.7.7. for further details.

5.7.6. CUSTOMER CONNECTIONS

Motor Tachometer Hook-up

The 6-pin connector located at the rear panel of the power supply is used to interface the motor and tachometer with the amplifier. If the amplifier is purchased without a power supply the connector will be located on the rear of the amplifier chassis. To ensure that the motor is properly phased, the following procedure should be followed.

1. Connect a DC voltmeter to the motor leads. Set the meter on the 10 V range.
2. As viewed from the motor shaft end, rotate the motor shaft clockwise (CW) by hand.
3. Observe the voltmeter deflection and determine and mark the positive (+) lead of the motor.
4. Repeat steps 2 and 3 with the voltmeter connected to the tachometer leads.
5. Connect the motor and tachometer to the amplifier as indicated (Figure 5.7.1)

CW Rotation with Positive Command (into summing amplifier)

Motor positive (+)	pin 1
Motor negative (–)	pin 4
Tach positive (+)	pin 6
Tach negative (–)	pin 3

Counterclockwise (CCW) Rotation with Positive Command

Motor positive (+)	pin 4
Motor negative (–)	pin 1
Tach positive (+)	pin 3
Tach negative (–)	pin 6

If either the motor housing or a tach shield must be grounded, this can be done by

SINGLE AMPLIFIER

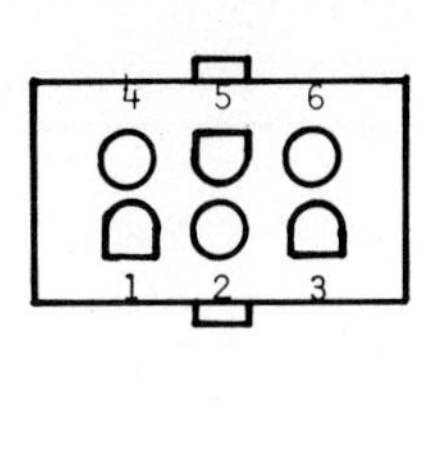

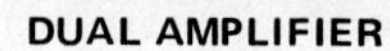

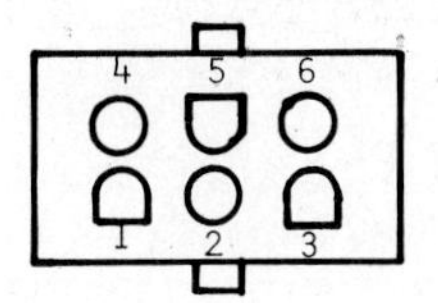

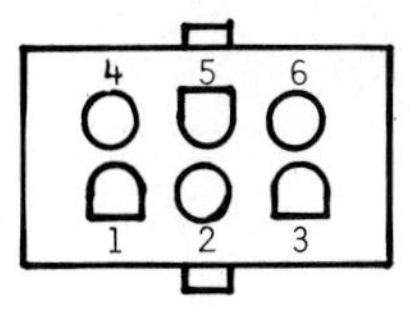

Fig. 5.7.1. Motor Connector as viewed from rear panel.

making the following connections to the six pin motor connector:

Motor Housing Ground	pin 5
Tachometer shield	pin 2

Control Signal Connections

The connections necessary to provide for signals to control the operation of the motor are made through the 36-pin connector located on the back panel of the power supply or amplifier chassis, if the power supply is not used or is used remotely. Some of the inputs are designated as auxiliary inputs. These auxiliary inputs are connected to one of the optional plug-in card connectors on the motherboard but they may be used to interface with other cards by the use of jumper wires on the motherboard. The functions of the inputs shown in Fig. 5.7.2 are as follows:

Pin 1: *External Command,* input to the summing amplifier to drive the motor. This input is not used with the ramp generator.

Pin 2: *External Command Common,* the common to be used in conjunction with the command signal. This point is common to the power supply and may be used as the common for any low level signal.

Pin 3: *Inhibit, A Drive* removes drive signal from one side of the output stage bridge. To use the inhibit, connect a normally closed switch between pin 3 and pin 5. When the switch opens, the motor will stop rotating in one direction (the motor does not brake to stop). If the inhibit is not used, pins 3 and 5 (A Drive) and pins 4 and 6 (B Drive) must be jumpered together.

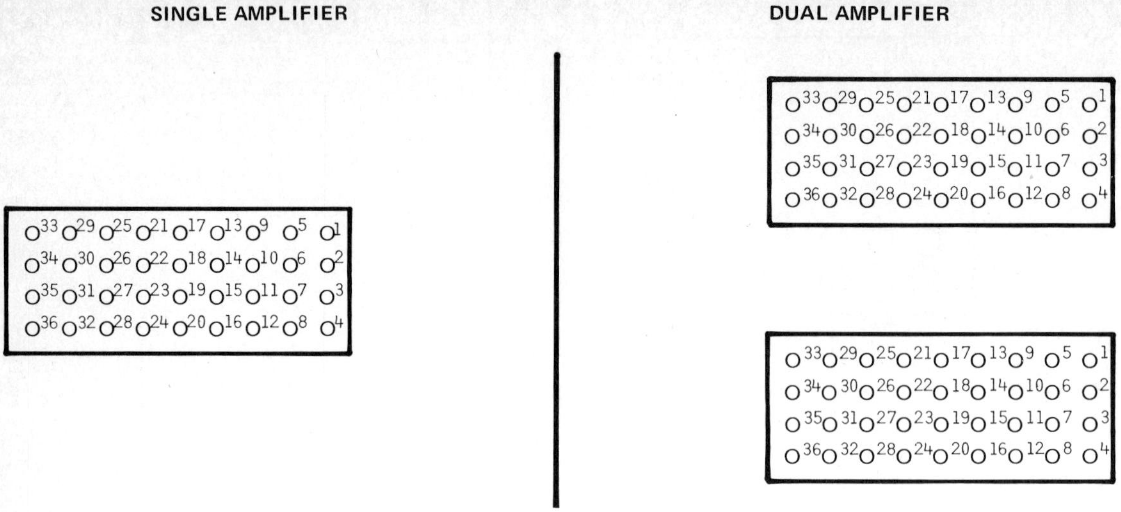

SINGLE AMPLIFIER

O33 O29 O25 O21 O17 O13 O9 O5 O1
O34 O30 O26 O22 O18 O14 O10 O6 O2
O35 O31 O27 O23 O19 O15 O11 O7 O3
O36 O32 O28 O24 O20 O16 O12 O8 O4

DUAL AMPLIFIER

O33 O29 O25 O21 O17 O13 O9 O5 O1
O34 O30 O26 O22 O18 O14 O10 O6 O2
O35 O31 O27 O23 O19 O15 O11 O7 O3
O36 O32 O28 O24 O20 O16 O12 O8 O4

O33 O29 O25 O21 O17 O13 O9 O5 O1
O34 O30 O26 O22 O18 O14 O10 O6 O2
O35 O31 O27 O23 O19 O15 O11 O7 O3
O36 O32 O28 O24 O20 O16 O12 O8 O4

Fig. 5.7.2. Control Signal Connector as viewed from rear panel.

Pin 4: *Inhibit, B Drive* removes drive signal from the other side of the output stage bridge. When pin 4 is open, the motor is prevented from turning in one direction.

Pin 5: *Inhibit Common* provides the signal to be used with the inhibit inputs. When the A and B Drive inhibits are connected to the inhibit commons, the output stage drive signals are turned off whenever the regulated power supply is below a predetermined point. This prevents damage to the output transistors during turn-on and turn-off of the amplifier.

Pin 6: *Inhibit Common* is the same point as Pin 5 and provides the signal for the B Drive Inhibit.

Pin 7: *Logic Clamp, A Drive*, if a TTL logic "I" is applied to this point the A Drive is clamped or inhibited.

Pin 8: *Logic Clamp, B Drive,* if a TTL logic "I" is applied to this point the Drive is clamped or inhibited.

Pin 9: *Reference Voltage, +15 Volts,* provides a regulated voltage through a 1,000 ohm, 1% resistor that may be used with a control potentiometer for a drive signal.

Pin 10: *Reference Voltage, −15 Volts,* provides the same feature as Pin 9 but with a negative voltage.

Pin 11: *Reference Voltage Common* is the common point to be used in conjuction with the reference voltage or any other low level signal.

Note: The following are used only with the Ramp Generator option.

Pin 12: *Ramp Generator External Input,* allows for control of the motor speed through the Ramp Generator with an external analog signal.

Pin 13: *Slope B Select* selects the B slope by application of a TTL logic "0" or a switch closure to ground.

Pin 14: *Slope A Select* provides the same functions as Pin 13 for the A Slope.

Pin 15: *Ramp Number 2 Initiate* selects one of two adjustable final motor speeds. This is controlled by applying a logic "0" or a switch closure to ground.

Pin 16: *Ramp Number 1 Initiate,* performs the same function as Pin 15 for Ramp Number 1.

Pins 17-36: are used for the DPS Option or any special inputs that may be required for specialized applications. If these inputs are not required they will not be wired into the harness.

NOTE: Pin numbers and pin configurations may vary depending on the particular options used. Verify the above connections by consulting the instruction manual.

5.7.7. SUMMING AMPLIFIER COMPENSATION

The components on the summing amplifier card must be selected for the type of motor and load conditions that are expected to be used with the amplifier. Assuming that this has already been done at the factory, the adjustments on the card will allow for proper compensation for an efficiently operating servo system. The first step in compensating the summing amplifier is to adjust the controls to a starting point that will ensure a stable, damped response. This can be done by adjusting the controls according to the following table. Refer to Fig. 5.7.3 for control location.

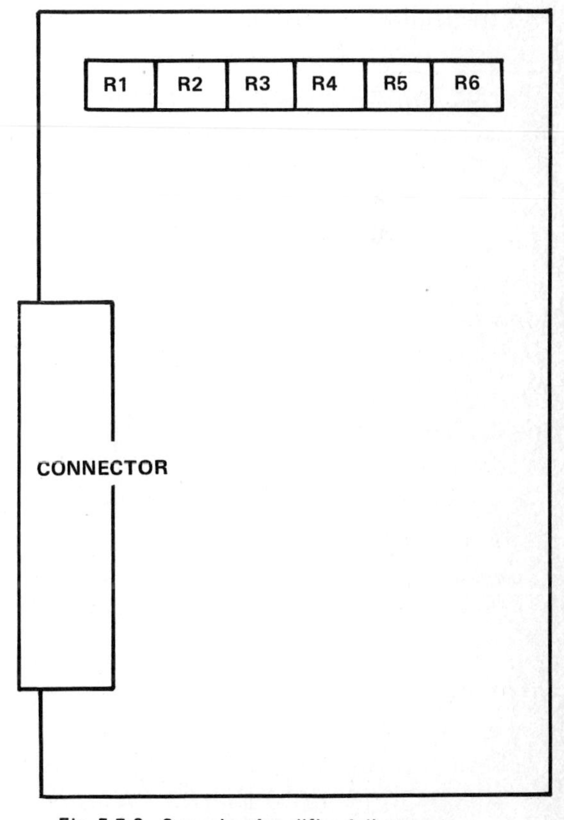

Fig. 5.7.3. Summing Amplifier Adjustments.

R1	Offset	Midrange
R2	Lag-Lead Compensation	Fully CCW
R3	DC Gain	Fully CCW
R4	Input Signal Magnitude	Fully CCW
R5	Lead-Lag Compensation	Fully CW
R6	Tach Feedback	Midrange

Offset Adjustment

With the amplifier turned on and the motor connected, adjust R1 so that the motor shaft does not turn. This will adjust for any amplifier offset. R4 has previously been set to minimum so that external input signals will have no effect.

Test Instrument Connection

Connect an oscilloscope to the output of the tachometer at pin 10 of the summing amplifier card. The ground of the scope should be connected to the point marked "GND" on the lead of the electrolytic capacitor C39 near the summing amplifier. Connect a DC voltage source of appropriate magnitude at pin 7 of the control signal connector. (The ramp generator, if one is being used, should not be plugged in at this time). The signal common of the voltage source should be connected to pin 8. This voltage source should be a step function (square wave generator), capable of being switched on and off.

Servo System Compensation

The objective of this compensation is to get a high DC gain (R6 and R3 CW as far as possible) consistent with stability, and a wide bandwidth (R2 CW as far as possible). The waveforms shown in Fig. 5.7.4 are typical of those obtained during the adjustment procedure.

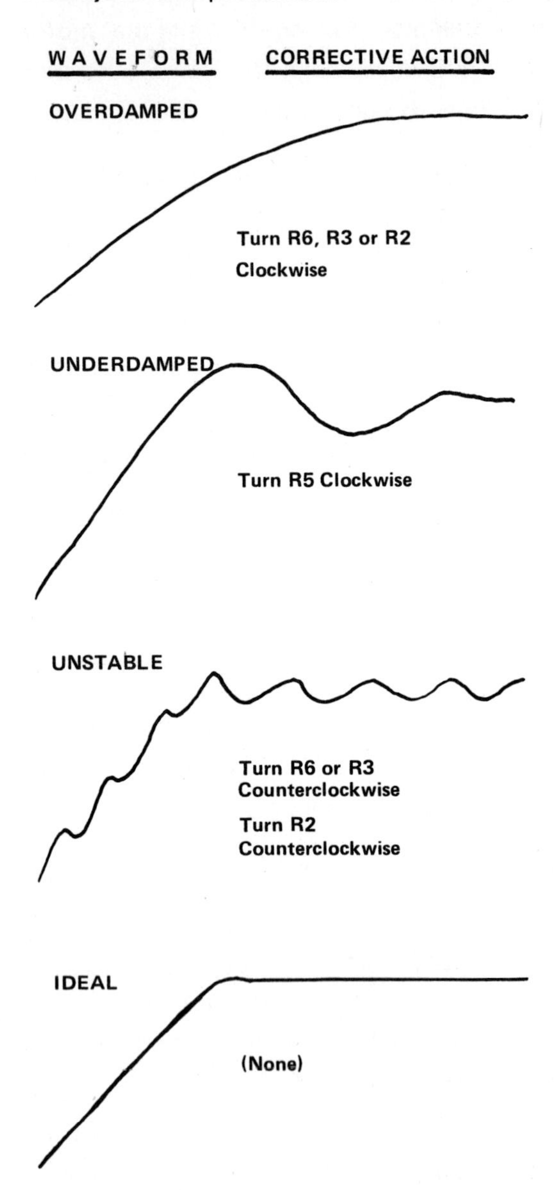

Fig. 5.7.4. Waveforms and Corresponding Corrective Action.

The "ideal" waveform should be the end result of the servo system compensation. This, of course, should be done with the proper load conditions that are to be used with the motor. The following procedure for adjustment is recommended for best results.

1. Turn R4 CW far enough to drive the motor and tachometer to achieve about a 10 V deflection on the oscilloscope (or any level consistent with the end application of the motor).

2. Turn R3 CW as far as possible without causing a sustained oscillation.

3. Turn R5 CCW until an underdamped waveform is obtained.

4. Correct the overshoot by turning R2 CW.

5. Repeat steps 3 and 4 until a waveform as close to ideal as possible is obtained.

6. Turn R6 CW in about 10° increments while adjusting R2 and R5 to compensate for any under or overshoot until the best response is achieved. (Note: As R6 is turned CW the motor speed will reduce. This may be corrected by turning R4 CW).

7. The input command can then be scaled for the correct output speed per volt by adjusting R4. If the potentiometer does not allow enough adjustment, a new range can be obtained by replacing the resistor R21. A larger resistance will require a higher input voltage to produce a given speed, and conversely a smaller resistance will require a lower input voltage.

This procedure should make it possible for individuals with little experience to get a working system. However, there are difficulties such as "gear backlash", or "dead band" which may be encountered, for which this procedure may not be adequate. In these cases it may be necessary to design a special compensation scheme to fit the application. When using a ramp generator, the basic summing amplifier should be adjusted for a slight under-damping or slight overshoot to assure sharp corners on the ramp. If the ramp generator is being used it should now be inserted into the correct option connector (with power to the amplifier off) and the set-up procedure for the ramp generator performed. Refer to RAMP GENERATOR ADJUSTMENT for the correct procedure.

5.7.8. RAMP GENERATOR ADJUSTMENT

Before the ramp generator can be adjusted, the summing amplifier must be properly compensated. The following procedure assumes that this has already been done. The input command signal to the control connector will have to be moved from the summing amplifier input to the proper input of the ramp generator unless the on-board ramp amplitude signal is used, in which case only a logic input signal is required. The ramp generator is designed to provide linear ramp inputs to the command input of the summing amplifier to provide the required servo system velocity profile. Both the ramp amplitudes and ramp slopes are independently adjustable. Two discrete (adjustable) ramp amplitudes plus zero (adjustable) are initiated by two TTL inputs to the board. Two other TTL inputs select one of two positive slopes or one of two negative slopes, each individually adjustable over a decade range. Ranges may be changed by changing the integrating capacitor to obtain ramp slopes ranging from 10V/ms to 0.667 V/s.

(Refer to Fig. 5.7.5.)

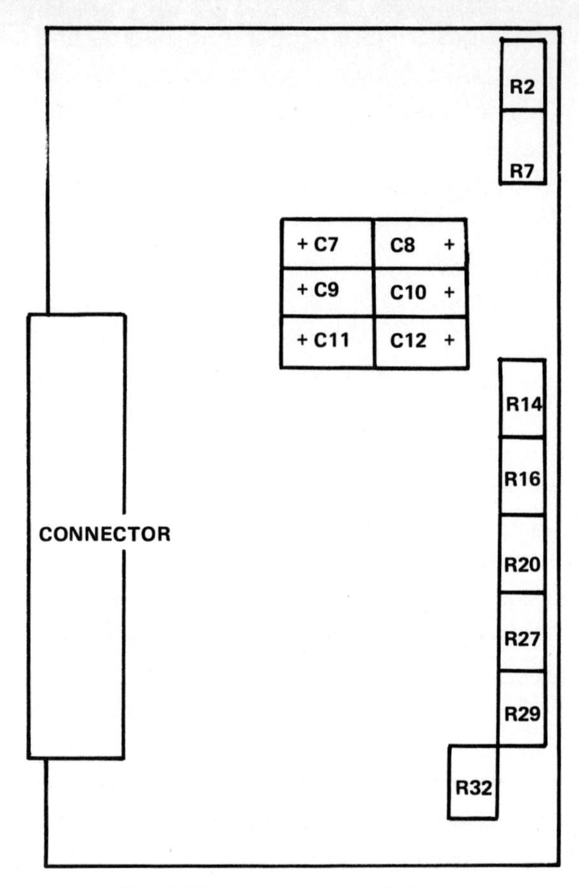

Fig. 5.7.5. Ramp Generator Adjustments.

1. Ramp Initiate

One of the two ramp amplitudes (final values) can be selected by a TTL low signal on *either* pin 11 *or* pin 12. The final ramp amplitudes are individually adjustable via R1 and R2 on the ramp generator board. R2 adjusts overall gain. Jumpers on the ramp generator board allow the selection of either a 0 to +10 V ramp or a 0 to −10 V ramp on either input. A typical ramp sequence is as follows. A ramp is initiated by a TTL low on *either* pin 11 *or* pin 12 (a low on both is an undefined state). The output will ramp at the selected slope to a final amplitude and polarity determined by pots R29 and R32 and their appropriate jumpers. A high signal on *both* inputs will cause the unit to ramp to zero. Zero is offsettable ± mV by R7.

2. Slope Select

A TTL low on pin 10 selects one set of ramp slopes, one positive slope adjusted by R16 and one negative slope adjusted by R20. A TTL low on pin 9 selects a second independent set of slopes; the positive slope adjusted by R14 and the negative By R27. A TTL high on both pin 10 and pin 9 is not normally used. It will cause the ramp integrator to hold the value it was at when both inputs went high until the integrator drifts due to input bias currents. A TTL low on both inputs is not defined.

3. Adjustment Procedure

With a low on pin 10 or pin 9 and both pin 11 and pin 12 high, adjust R7 for 0 VDC at pin 1 (ramp generator output). Apply a low to pin 12 and adjust R29 for the correct ramp amplitude. Release the low on pin 12 and apply a low to pin 11 and adjust R32 for the other ramp amplitude. Ramp polarities are selected by jumpers. To adjust ramp slopes apply a low to pin 10, apply a periodic low to pin 11 or pin 12 (whichever causes a negative ramp) and adjust R16 for the correct slope. With a periodic low on pin 11 or pin 12 (whichever causes a positive ramp) adjust R20 for the correct slope. Release the low on pin 10, apply a low to pin 9, and repeat the above adjusting R14 and R27 for the correct slope. Adjust overall gain pot R2 if necessary. Ranges may be changed by changing integrating capacitor(s). If polarized capacitors are used, they must be placed back to back i.e. (+) to (+) or (–) to (–). The currents adjusted by R16, R14, R20 and R27, vary between 10 and 100μA. The ramp slope is given by

$$\frac{dV}{dt} = \frac{I}{C} \quad [V, s; \mu A, \mu F]$$

An external voltage command, such as a square wave input, may be used with the ramp generator by feeding this signal into pin 7 of the ramp generator connector. This input is normally connected to pin 12 of the control connector.

5.7.9. IN CASE OF DIFFICULTY

The best procedure for troubleshooting is to isolate the source of the problem and then replace the section that is at fault. Try to determine what is the most likely thing to cause the difficulty. This troubleshooting guide consists of only the most common problems. It may be necessary to refer to two or more areas to correct a problem in some cases. Time spent analyzing a problem is time well spent. As you gain experience working with the amplifier and power supply, it will become very easy to isolate the source of any problem.

Never overlook the obvious, no matter how ridiculous it may seem. The following is a list of items that should be considered before looking to more difficult possibilities.

1. Line cord unplugged.
2. Circuit or connectors loose.
3. Fuses removed or blown out.
4. Shorted or open wires.
5. Defective motor or tachometer.
6. No power at AC outlet.

The following "Troubleshooting Guide" should help solve almost any difficulty that may arise in the operation of the amplifier.

5.7.10. TROUBLESHOOTING GUIDE

PROBLEM	POSSIBLE CAUSE
Amplifier and fan inoperative, no DC voltage present.	1. Line cord not plugged in. 2. No power at AC outlet. 3. Line fuse at back panel defective. NOTE: If the line fuse continues to blow out, one of the rectifiers on the rectifier bracket may be shorted. Disconnect transformer secondary and check rectifier for short using an ohmmeter.
Amplifier inoperative, fan works but no DC voltage at output stage or filter capacitor on power supply chassis.	1. Loose connection to transformer or rectifier bracket. 2. Defective rectifier on rectifier bracket.
Amplifier inoperative, fan works and DC voltage is present at the filter capacitor but not at the output stage.	1. Burned out or defective crowbar fuse. CAUTION: If the crowbar fuse continues to burn out, this may indicate a fault in the output stage. It should be determined if there is a defective component or shorted output before reapplying power to the amplifier. The problem may be isolated by disconnecting the power connector to the amplifier, replacing the fuse and reapplying power to the power supply. If the crowbar does not fire, this indicates the source of the problem is in the amplifier. 2. Loose or unplugged connector on output stage circuit board.
Amplifier inoperative, fan works and DC voltage present at both filter capacitor and the stage, motor shaft stiff in both directions of rotation.	1. Speed command signal missing.

PROBLEM	POSSIBLE CAUSE
	2. External command connector loose or disconnected. 3. External command level control potentiometer located on summing amplifier, set at minimum.
Amplifier inoperative, fan works and DC voltage present at both the filter capacitor and the output stage, motor shaft not stiff in one or both directions.	1. Clamp circuit activated due to loss of plus or minus 15 V supply voltages. 2. Open connection between inhibit output and either or both of the inhibit inputs. NOTE: At this point it would be beneficial to isolate the problem to either the output or the motherboard. To do this, it is recommended that an oscilloscope be used to check the test points on the motherboard. Start at the drive to the output stage. If these signals match those in Table 5.7.1., the problem is in the output stage. Continue comparing signals and note all differences. Then consult the Electro-Craft service representative. When spare parts are available, the isolated problem component should be replaced. □

Table 5.7.1. Test Point Waveforms

TEST POINTS		
NO.	DESCRIPTION	WAVEFORM
1	Current Limit Amp	Proportional to I
2	B Trip Point	–0.5V
3	IC Power Supply	–15V
4	Power Supply Comm.	0V
5	IC Power Supply	+15V
6	Current Diff. Amp	Proportional to I
7	Peak Current Mono.	←PK I→ 30V P-P
8	Tri. Gen. Drive	12V P-P 20 kHz
9	Tach Input	
10	A Trip Point	+0.5V
11	Triangle Generator	7V P-P 20 kHz
12	No Connection	
13	Output Drive A	25V P-P 20 kHz
14	Output Drive B	25V P-P 20 kHz
N.B.	I = Amplifier Output Current	

Chapter 6

Brushless DC Motors

6.1. INTRODUCTION

Since the performance of brushless DC motors is intimately tied to the commutation of current in motor windings, it may be appropriate to briefly review the commutation of conventional DC permanent magnet motors. The main reason for dividing a conventional DC motor winding system into segments is to minimize the effect of winding inductance as it relates to the turn-on and turn-off behavior of current in a segment. A secondary reason for the choice of the appropriate number of winding segments is to control the torque ripple. Thus we find that a fractional horsepower DC motor may have anywhere from 7 to 32 commutator bars per armature, and an integral horsepower motor below 5 to 10 hp may have less than 100 bars, and larger versions may have more than 100 bars per armature. The transient event of a coil commutation is generally well hidden inside the rotating armature of a conventional DC motor. The commutation event is not usually viewed on an oscilloscope screen, except as seen from outside the motor brush connections – the reason being that it is not easy to reach proper internal test points in a rotating structure. Much of the design effort in such motors, then, has gone into proper brush and commutator choice, and the results have been viewed from the observer's point by recording emitted radiation, brush and commutator heating, and erosion of commutation surface or brush surface. A great variety of brush compositions have enabled designers to cope with the widely varying operational and environmental conditions that exist in today's technology.

As we contemplate the desirability of using semiconductor devices for commutation of a DC permanent magnet motor we are faced with a set of new problems, and a simple translation of the brush-type motor designed to a brushless type is not practical. For example, if we take a 16 bar, two-pole DC motor and were to substitute semiconductor devices for the brush and commutator assembly to achieve the same function, we would end up with 32 power transistors, two of which would be conducting at any one given time. It would then result in a very inefficient use of semiconductors – a utilization factor of about 6%. Obviously the problem has to be solved in a new way to achieve cost effective performance.

Before we discuss brushless motor principles, let us remember the varying design constraints which conventional DC motors have to face and the varying applications which exist today. Examples of the variety of performance requirements are:

- open loop, on-off applications
- unidirectional speed-control motor systems

- bidirectional speed-control motor systems
- unidirectional speed-control motor systems able to handle over-running loads
- torque motors able to operate at stalled conditions
- low inertia servo motors able to handle acceleration torque demands one order of magnitude larger than the continuous torque
- servo motors with low torque-ripple specifications, in either low inertia or high inertia versions

The listing could go further, but these examples illustrate the unique design demands which exist for brush type DC motors.

The same conditions hold for brushless DC motor systems, and therefore the following discussion of brushless DC motors will treat some of the more common configurations which serve their intended purposes in a cost-optimal way.

Although the following describes brushless DC motors controlled by transistor systems, the reader should realize that thyristors (SCR's) can as well be used for the same purpose, although the turn-off of the latter may be complicated due to the unique characteristics of the thyristor. □

6.2. DEFINITION OF A BRUSHLESS DC MOTOR SYSTEM

Since the control circuits for a step motor or frequency controlled AC motor at first glance may appear to be similar to some brushless DC motor controllers, it is appropriate at the outset to clarify the distinctions between them. If we first define what a brushless DC motor system characteristic should be, we can then realize the differences between that type of system and the others.

A brushless DC motor system should have the torque-speed characteristics of the conventional DC permanent magnet motor.

Fig. 6.2.1 illustrates the torque-speed characteristics of a conventional DC motor. By our definition above, the brushless DC motor should have the same basic characteristics. From these basic relationships we can derive mathematical expressions which enable us to apply simple mathematical analysis and achieve solutions to servo system problems. These conditions are discussed in chapter 2, and apply generally to brushless DC motor systems.

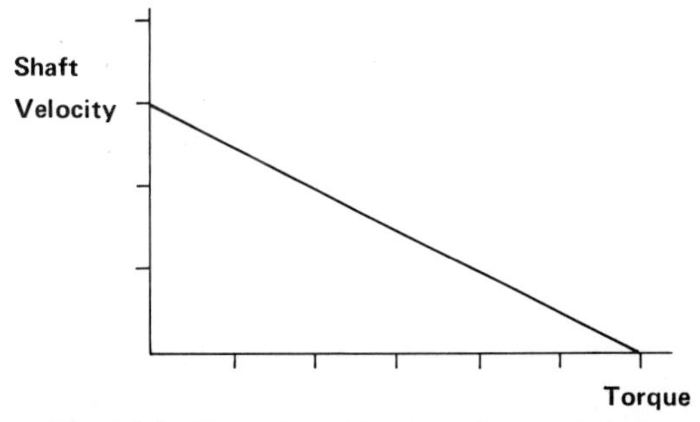

Fig. 6.2.1. Typical speed-torque characteristic for a conventional permanent magnet DC motor. Armature voltage is held constant.

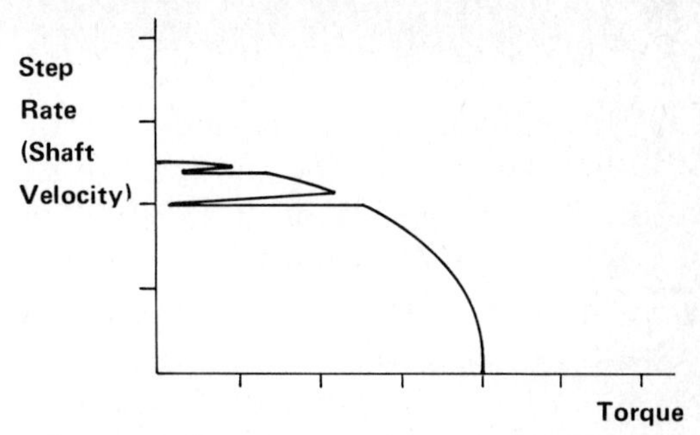

Fig. 6.2.2. Typical speed-torque characteristic for a step motor system. Excitation voltage is held constant, step frequency varied.

If we examine a step motor system speed-torque characteristic (Fig. 6.2.2) we observe non-linear relationships between the two variables. The discontinuous part of the curve is due to a resonant oscillatory mode inherent in the motor design. The discontinuous nature of the step motor low-speed performance limits its usefulness in velocity control applications. Thus the step motor is a unique device, which differs from a brushless DC motor in fundamental performance aspects.

The AC motor characteristic in Fig. 6.2.3 shows an entirely different relationship between speed and torque. The useable

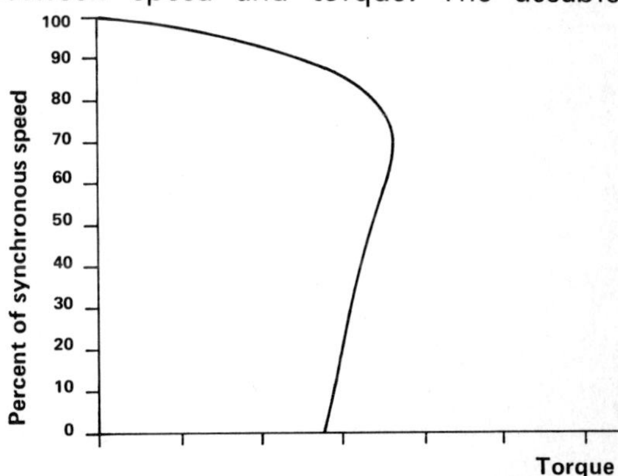

Fig. 6.2.3. Typical speed-torque curves of a fractional-horsepower polyphase induction motor. Excitation voltage and frequency is held constant.

part of the curve lies in the region between 90 and 100% of synchronous speed – and the synchronous speed depends on the excitation frequency. The controllable AC motor system depends on a variable frequency, variable AC voltage, which has to be coordinated with the shaft velocity to produce a controlled "slip frequency" current in the rotor windings. Because the rotor-stator structure can be considered a transformer, it does not work well at low frequencies (which corresponds to low shaft speeds). This is a fundamental difference between the AC motor and the brushless DC motor, since in the latter torque is produced by the interaction of a magnetic field produced by a permanent magnet rotor, and magnetic field due to a DC current in a stator structure. □

6.3 PRACTICAL SOLUTIONS TO BRUSHLESS COMMUTATION

The first general comment about brushless motors and controllers is that the design philosophy should point to a motor design which would require the least number of power semiconductor switches to adequately meet the performance requirements. The second assumption is that wherever possible one should use permanent magnet materials in order to eliminate the need for slip rings in the rotor assembly. In a brushless DC motor it is usually most practical to provide a stator structure as shown in Fig. 6.3.1 where the windings are placed in an external, slotted stator. The rotor consists of the shaft and a hub assembly with a magnet structure. The picture shows a two-pole magnet. For contrast Fig. 6.3.2 shows the equivalent cross-sectional view of a conventional DC motor where the permanent magnets are situated in the stator structure and the rotor carries the various winding coils. We can see that there are significant differences in winding and magnet locations. The conventional DC motor has the active conductors in the slots of the rotor structure, and in contrast, the brushless DC motor has the active conductors in slots in the outside stator. The removal of heat produced in the active windings is easier in the brushless DC motor, since the thermal path to the environment is shorter. Since the permanent magnet rotor does not contribute any heating, the result is that the brushless DC motor is a more stable mechanical device from a thermal point of view.

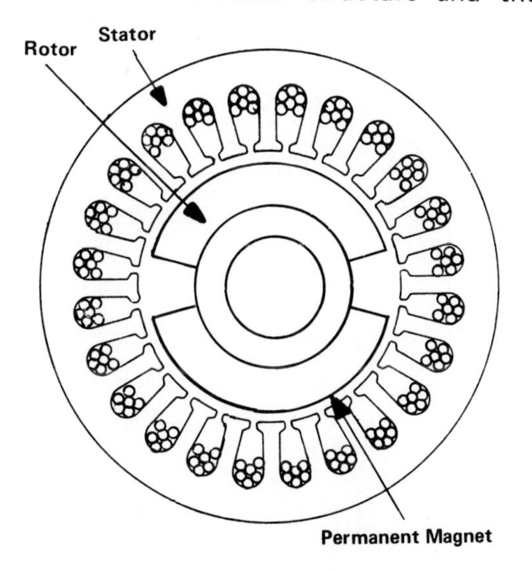

Fig. 6.3.1. Cut-away view of brushless DC motor assembly.

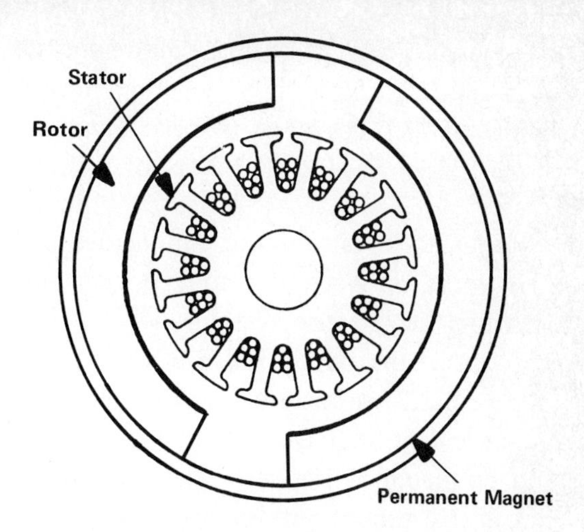

Fig. 6.3.2. Cut-away view of a conventional permanent-magnet DC motor assembly.

In spite of the advantages discussed above, there are cases where a brushless DC motor uses the configuration shown in Fig. 6.3.1, but then the roles of the two parts are reversed so that the permanent magnet outside structure rotates and the wound lamination part is the stator. The result of such a design is a high rotor moment of inertia, which sometimes is of interest to the user who needs a high mechanical time constant in order to filter any existing torque ripple. In such cases the thermal advantage of the design in Fig. 6.3.2 is not utilized.

In order to illustrate the similarities and differences between the conventional and brushless DC motor systems, we can see sketches of the two in Fig. 6.3.3 and 6.3.4. In Fig. 6.3.3 we have the system elements of a conventional DC motor and control. The connections between the rotor windings and the commutator are shown in an oblique sketch for the sake of clarity. A bidirectional controller and driver stage is shown together with a power supply and servo compensation and interlock logic.

The equivalent brushless DC motor system is shown in Fig. 6.3.4, where the main differences are seen to be static windings, permanent magnet rotor, four transistors instead of the two found in the conventional motor, and a shaft position encoder. The latter generates logic signals which control the commutation of the windings. As we shall see, many different commutation configurations exist, and next we will examine the most commonly encountered models.

One of the simplest practical brushless DC motor circuits is shown in Fig. 6.3.5 (SEE NOTE IN FIGURE) This is a "half-wave" control circuit with a conduction angle of

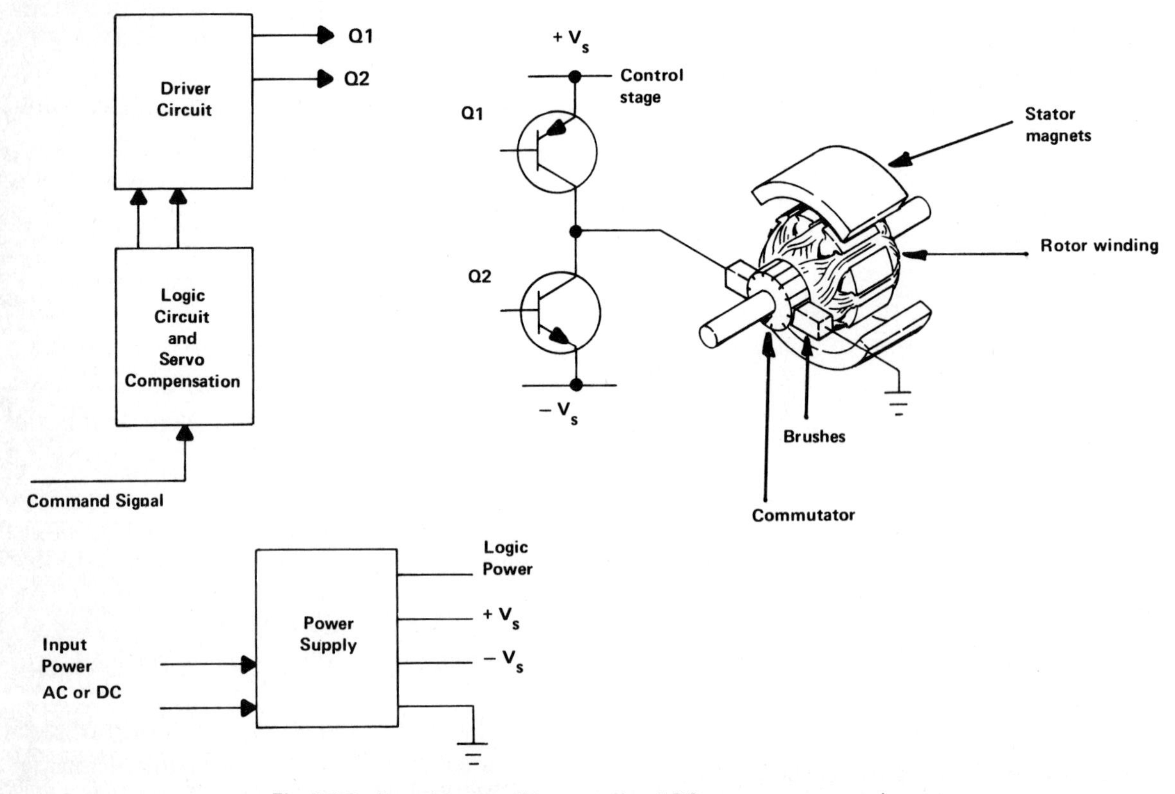

Fig. 6.3.3. Essential parts of a conventional DC motor servo control.

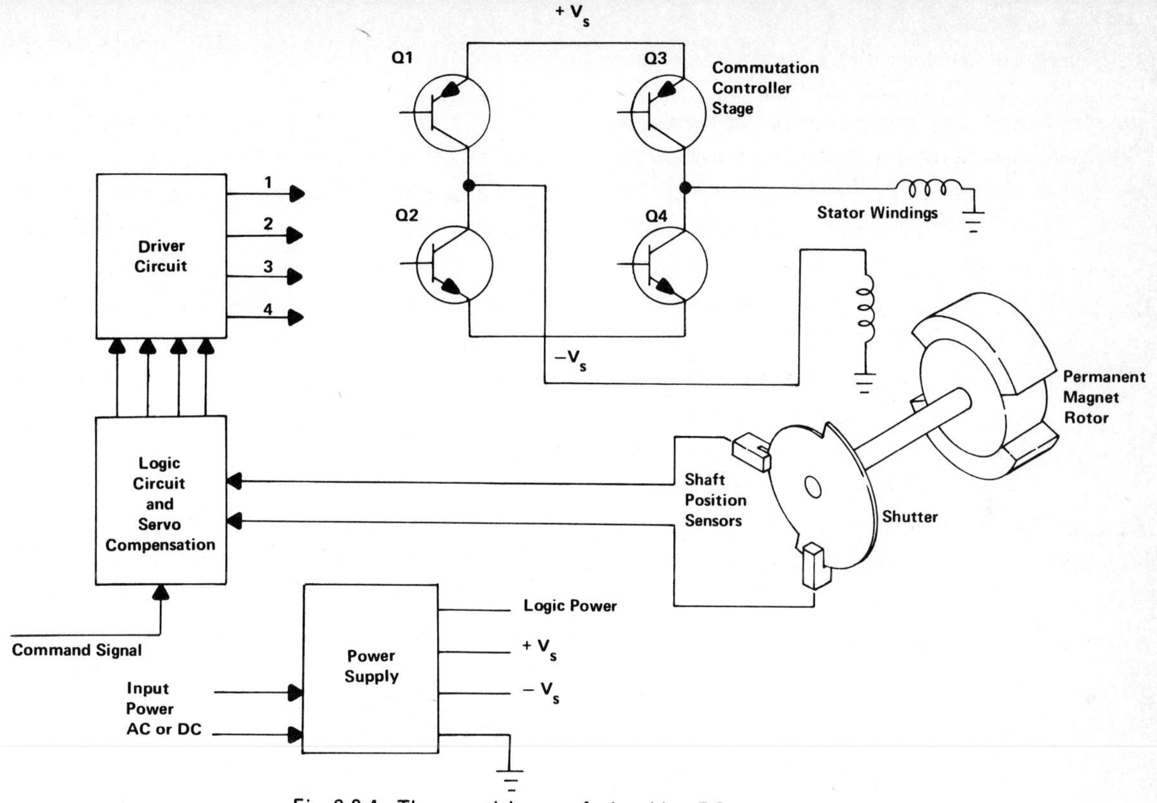

Fig. 6.3.4. The essential parts of a brushless DC motor control.

120° el. As we see in the accompanying diagram each winding is used one third of the time and the logic control of the system is rather simple. The speed and torque output of the motor can be controlled by varying the power supply voltage V_s. In the lower part of the diagram we see the same system with a reversed torque. The torque reversal is achieved not by reversing the power supply voltage as in a conventional DC motor, but instead by shifting all logic functions 180° el. This example illustrates one of the basic differences between brush type and brushless type DC motors.

The example just given ignores one very important point in brushless motor commutation performance, namely the handling of the inductive transient current in each winding as it is commutated. The circuit shown in Fig. 6.3.5 would experience a forward voltage breakdown of each transistor as its conduction would be discontinued. This is due to the voltage produced by the stored energy in each winding. Such breakdown conditions can be tolerated in low-power systems, where the stored energy is low. However, if any significant amounts of current and voltage are handled in such a system, a point is reached where breakdown conditions would cause damage to the semiconductor junctions. Therefore other methods are employed to assure proper commutation of the inductive energy in each winding.

NOTE: The diagram of Fig. 6.3.5 shows the "torque function" [T(I)] of each winding. This function shows that winding's contribution of torque at various shaft angles when a constant current I is flowing in the winding. The unit value of torque over the conduction angle then becomes the torque constant K_T of the winding.

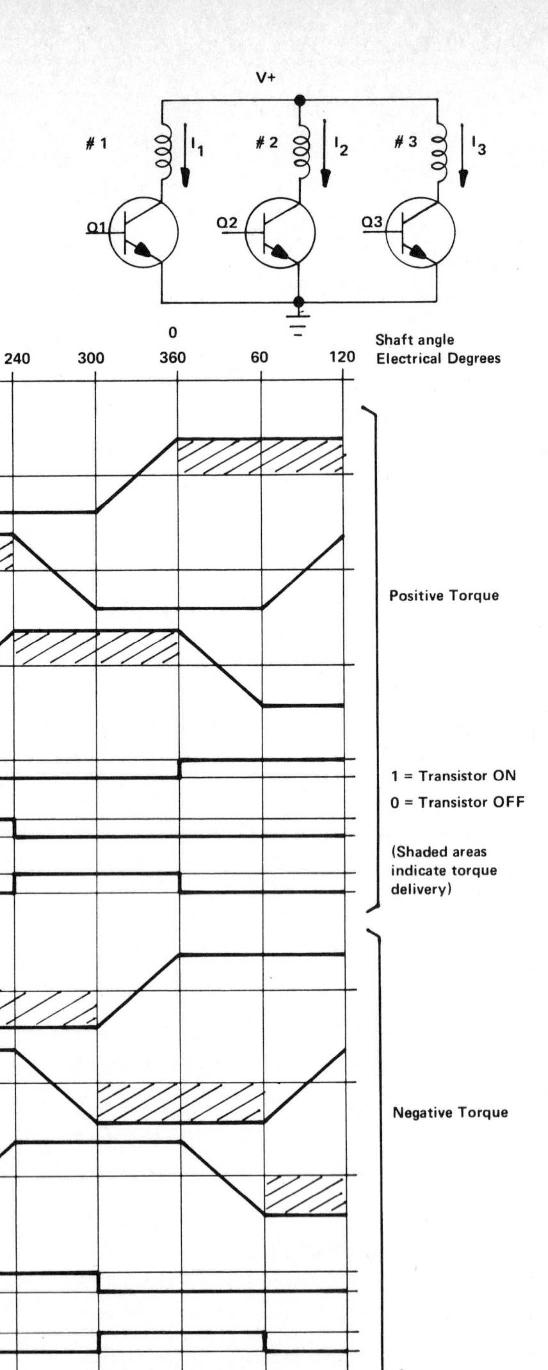

Fig. 6.3.5. A three-phase, half-wave brushless motor controller.

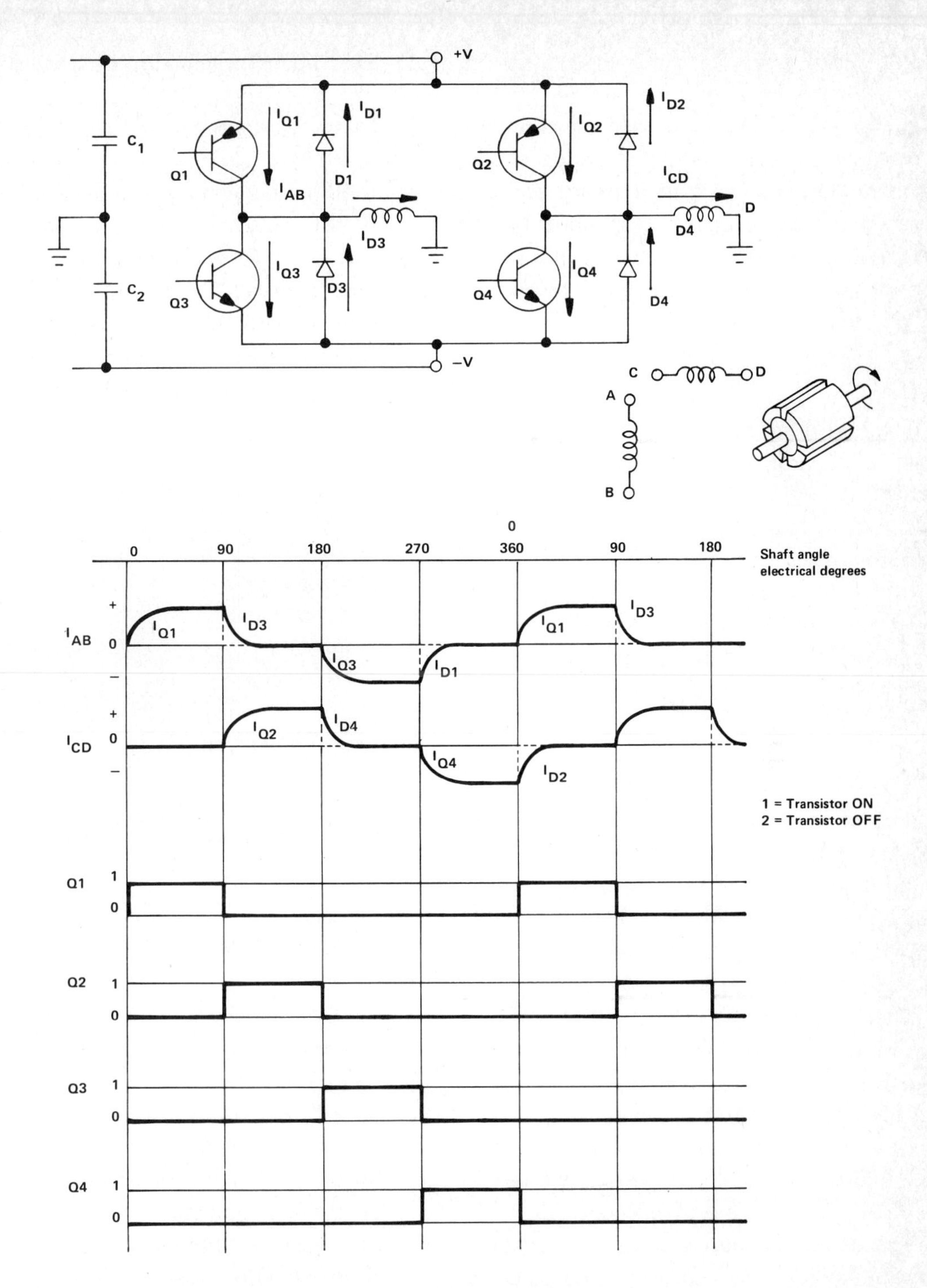

Fig. 6.3.6. Controller configuration, current wave-form and logic sequence for a two-phase 4-pole brushless DC motor.

Instead of applying the remedy for the inductive energy problem to the circuit we have just reviewed, let us instead for the moment look at another type of circuit where the commutation transient concept can be easily visualized without having to worry about other problems which exist in the circuit of Fig. 6.3.5 (which we will discuss later).

(See Fig. 6.3.6 on preceding page 6-9)

If we turn to Fig. 6.3.6, which shows a two-phase brushless motor using two power supplies, $+V_s$ and $-V_s$, we note that we now have four power transistors and four diodes. Each half of the circuit controls its own winding, and the two are essentially independent from each other. The accompanying diagram shows phase currents and logic signals for one direction of operation. The current waveforms in the diagram are intended to by typical of real-time response of current at a given shaft velocity. The diagram, therefore, shows not only current response with respect to rotor position, but also current versus time at a given shaft velocity. If we, for instance, look at current I_{Q1} we see that it has an exponential initial increase to a steady state value which is maintained until the 90° position has been reached. Then **Q1** is switch to the OFF condition. The stored energy is now dissipated through the power supply and returned through diode **D2**, and the exponential decline is shown in I_{D2}, while the current rise now is progressing in **Q3.** Thus there is a continuous torque production maintained in the motor as one stage is turned off and the next is turned on.

The diodes provide an ideal transient path for the stored conducted energy. The energy is either stored in one of the filter capacitors or transferred to the other winding, depending on the logic sequence. If fast recovery diodes are used, the commutation event is carried out without any significant RFI emission. An oscilloscope picture of the current waveforms in a two-pole full-wave brushless motor is shown in Fig. 6.3.7. The pictures were taken during continuous motor operation at a torque level of 80 oz-in and 3600 rpm.

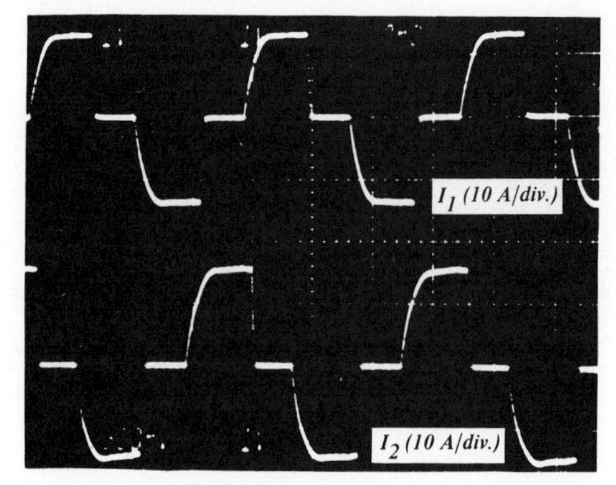

Fig. 6.3.7. Current waveforms in the phase windings of a four-pole, two-phase, full-wave brushless motor. Motor running at 3600 rpm, delivering 80 oz-in. Power supply voltage ±30V.

Fig 6.3.8 shows an extended version of the two-phase full-wave brushless motor system. The winding configuration is based on a "star" connection stator arrangement where each winding is oriented 120° from the other. The six transistors are connected to the end points of each stator leg, and thus form a three-phase full-wave motor control. This system differs from the previous one in that conduction is always

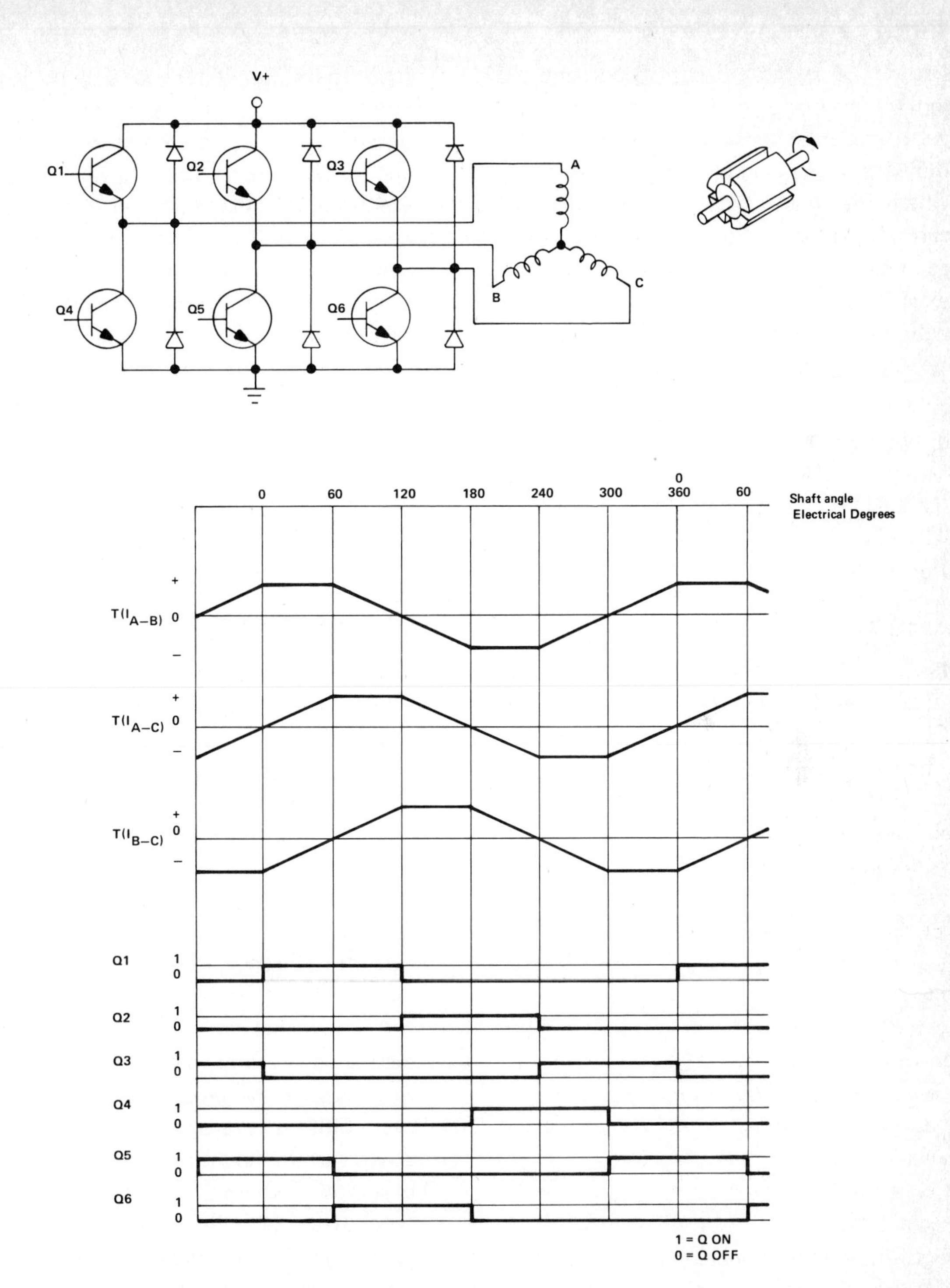

Fig. 6.3.8. Controller configuration, torque function, and logic sequence for a three-phase, full-wave brushless DC motor.

continuous in one leg when the other is being commutated. We can see that when **Q1** is energized between 0 and 60°, **Q5** is also conducting, and current is thus flowing from point A to point B. In the next sequence (60-120°) **Q6** is energized and the current will then flow from point A to point C. In the meantime the current through leg B declines to zero by conduction through D2. The conduction angle per phase is 120°, as compared to 90° in the two-phase circuit. Since this circuit requires only a single power supply it tends to be more compact than the two-phase circuit; and further, since 67% of the available windings are used at any one time compared to 50% in the former circuit, the three-phase circuit is more efficient than the two-phase circuit of Fig. 6.3.6.

We can now return to the half-wave three-phase circuit in Fig. 6.3.5, which was shown without any means for handling the stored inductive energy. If we were simply to provide a diode path for each winding, such as shown in Fig. 6.3.9a — typical of relay coil circuits — we would

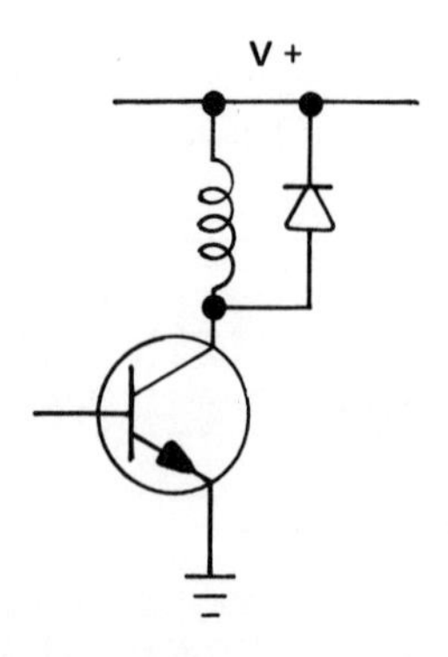

Fig. 6.3.9a. A conventional "fly-back" diode arrangement.

find that the inductive energy is well handled by the diode arrangement. But because the diode is located across the coil and not across the transistors as in Figures 6.3.6 and 6.3.8, we find that the back EMF will also flow in the diode circuit when the polarity of the back EMF is reversed (180-360° in coil 1). In order to prevent current flow due to the back EMF during this "off" period of each cycle, one has to let diode conduction occur to a voltage source which has a voltage at least two times the power supply voltage. One arrangement accomplishing this is shown in Fig. 6.3.9b. As the conduc-

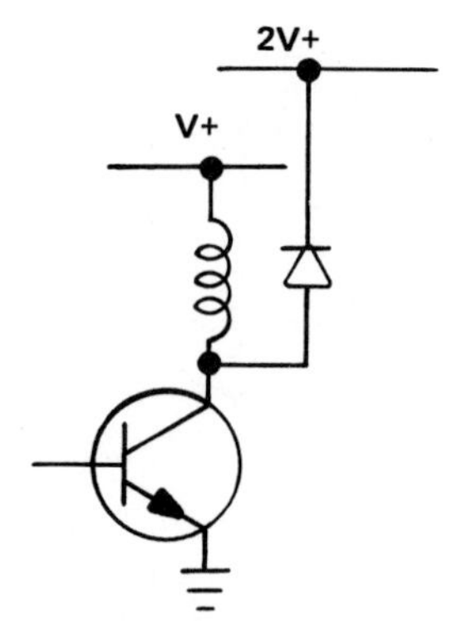

Fig. 6.3.9b. A biased "fly-back" diode arragement.

tion of **Q1** ceases, the winding inductance will generate a voltage equal to $2V_s$, which will allow the inductive transient to be dissipated in the power supply circuit. But because the diode is biased, when the back EMF is reversed and could normally cause current flow, it cannot overcome the bias, and therefore no current flows. This type of circuit will thus allow for proper handling of the stored inductive energy, but at the expense of extra components and power dissipation in the bias supply.

The controller circuits we have seen so far represent the basic forms which may be encountered. Often due to special requirements, other variants of these circuits will be used. For instance in Fig. 6.3.10 we have a four-phase, half-wave controller. This circuit features a 90° conduction angle and is at times used where low torque ripple is important. While it would seem that the two-phase full-wave circuit in Fig. 6.3.6 basically fulfills the same purpose, the reason for using this circuit may

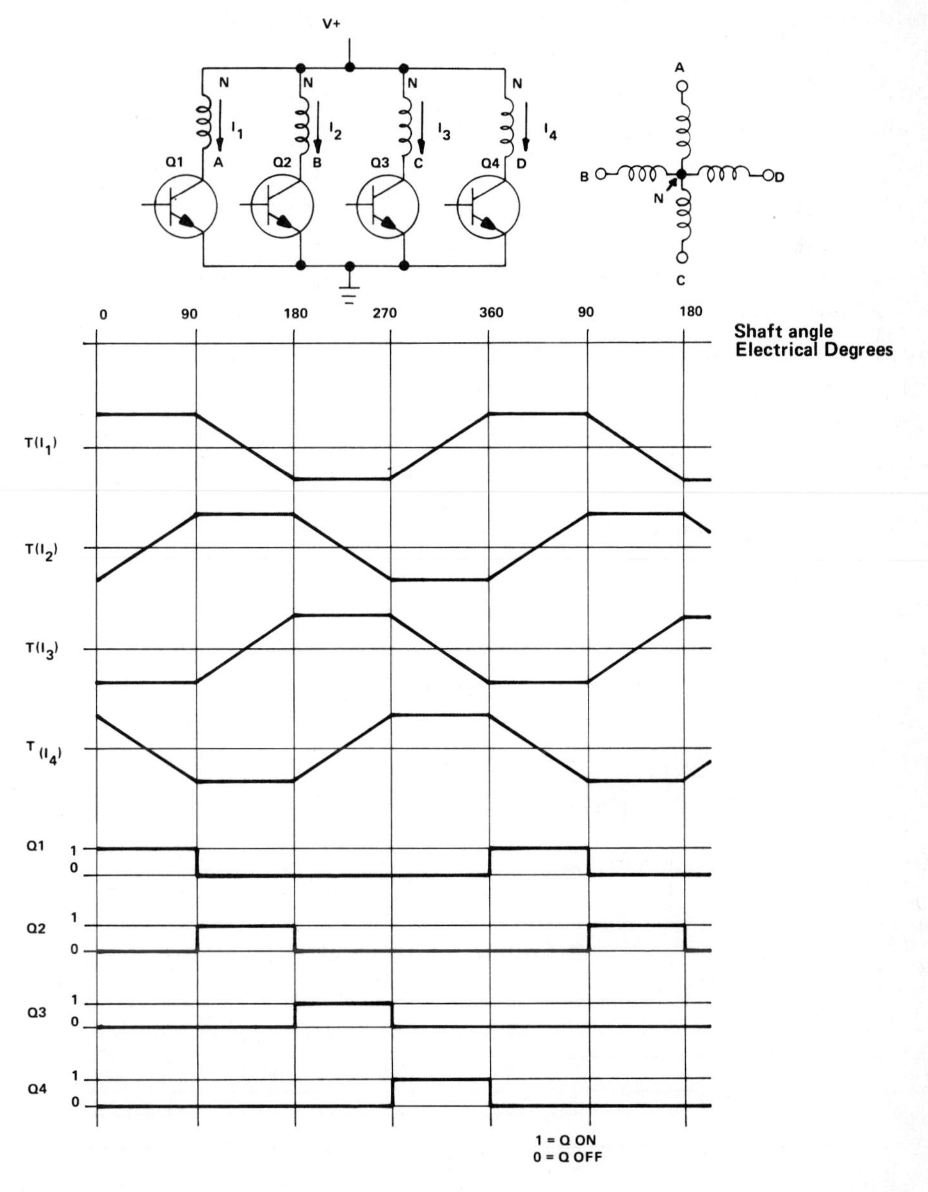

Fig. 6.3.10. Four-phase, half-wave circuit.

be that it gives similar performance with a single power supply and does not need cross-fire prevention interlocks. It can therefore be used in relatively low power circuits where cost is important and where the fly-back current problem can be handled in a simple way.

Fig. 6.3.11 shows a three-phase full-wave circuit with a grounded neutral. This circuit

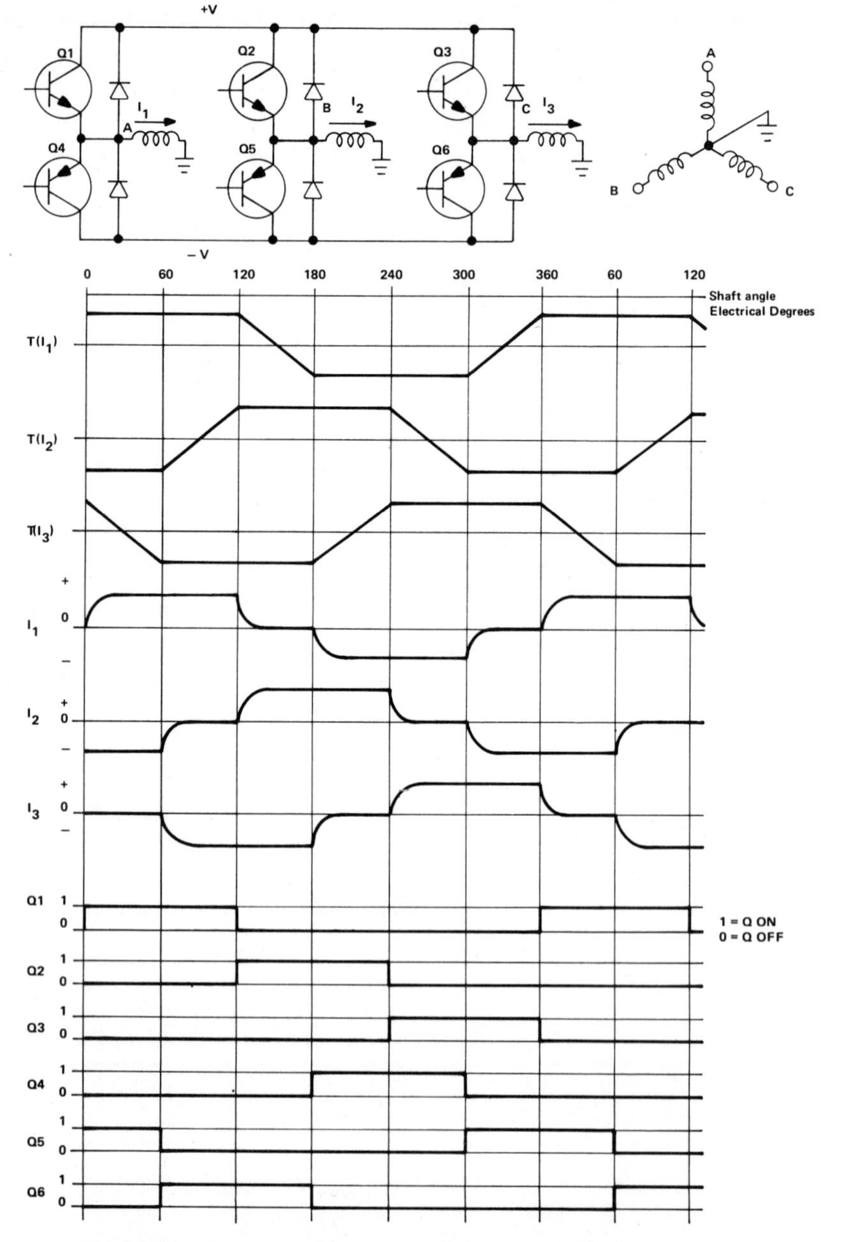

Fig. 6.3.11. Three-phase, full-wave circuit with grounded neutral.

offers some advantages when a constant current drive method is necessary. The symmetrical location of transistors with respect to ground offers advantages in the linear amplifier driving each transistor pair. We notice that at any given time two stages are conducting simultaneously although commutation occurs at alternate angular intervals. The significance of this is that the load current is shared between two transistors, which may affect the choice of their characteristics. On the other hand, due to the dual power supply, each transistor has to withstand the sum of the two supply voltages. This is in contrast with the three-phase circuit shown in Fig. 6.3.8, where two transistors conduct at any given time, but they are always in series. Therefore no load sharing occurs – but the transistors see only one supply voltage instead. Returning to Fig. 6.3.11, we should be aware of another potential problem, namely that the dual supply type circuits (this includes the circuit in Fig. 6.3.6) are susceptible to power supply "pump-up" during certain operational conditions such as start and stop conditions driving high inertial loads, especially if pulse-width or pulse-frequency modulation is utilized. It is also true that such pump-up situations may occur in the single supply type circuits, but they are somewhat easier to control in this respect.

If we now turn our attention to the output stage in Fig. 6.3.12, we see a six-phase full-wave circuit with grounded neutral. The timing diagram for this circuit is shown in Fig. 6.3.13, and we notice a 120° conduction angle and that four transistors are conducting at any one time. We thus have excellent load sharing, and since commutation occurs every 30°, we have very low torque ripple conditions.

In addition, since the windings are divided into six independent coils, the electrical time constant is lower than in the equivalent three-phase circuit. Therefore, this type of motor performs excellently in ultra high performance requirements, rivaling that of low inertia moving coil motors. The price for this, of course, is that 12 power transistors are used in the output stage circuit. On the other hand, since there is load sharing between four transistors at any one time, the ratings for each transistor can

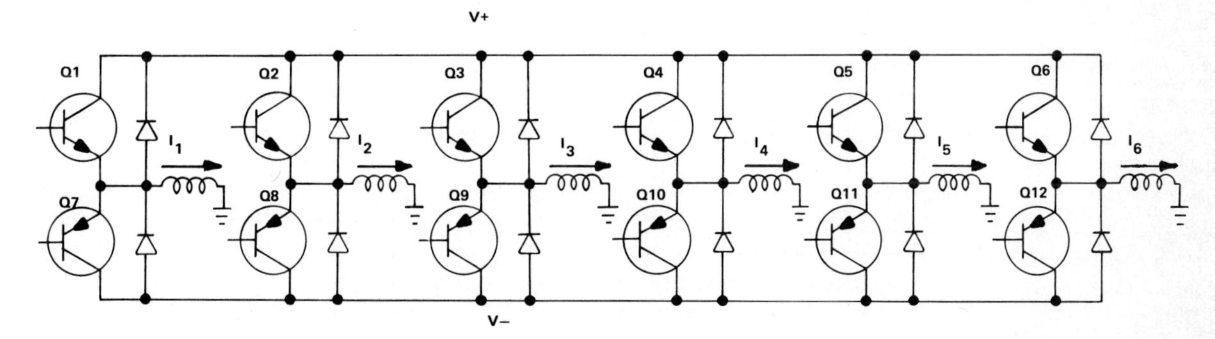

Fig. 6.3.12. Six-phase, full-wave commutation circuit with grounded neutral, and two power supplies.

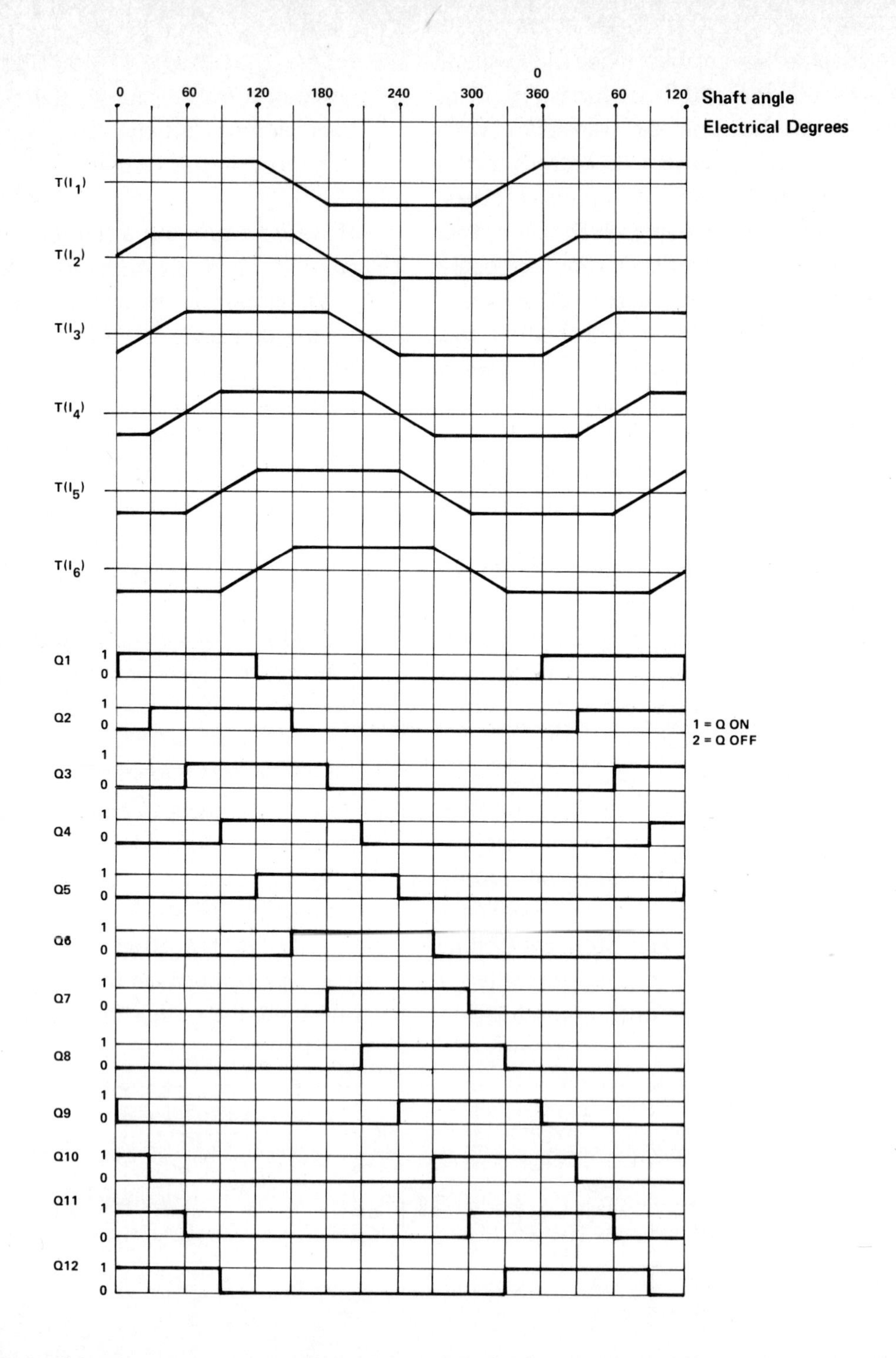

Fig. 6.3.13. Torque function and logic sequence for unidirectional operation of six-phase, full-wave circuit of Fig. 6.3.12.

be held at a lower level than for the equivalent three-phase circuit.

Other variants of these circuits can be used, such as a pair of three-phase full-wave star or delta connected three-phase bridges, each displaced 30° el from the other. The resultant system would then use a single power supply, but two independent commutation circuits.

The foregoing examples illustrate the large variety of circuit possibilities for commutation control of brushless DC motors. If we recall the statements at the beginning of this discussion citing the various application requirement which DC motors – and therefore brushless DC motors – encounter, we can realize that there is no single "good" circuit which can economically serve all purposes. Since the functional aspects of a brushless DC motor system are also intimately tied into the system cost, we have to keep in mind that an independent choice of the best motor and control has to take many factors into account. □

6.4. TORQUE GENERATION BY VARIOUS CONTROLLER CONFIGURATIONS

We have now discussed some general concepts in the area of brushless motor commutation without specific regard for the control of torque and proper utilization of available options in the magnetic circuit. We can distinguish two principal control configurations which are used in today's brushless motors. The first method is based on the use of a *sinusoidal torque* generation scheme, and the second utilizes a *trapezoidal torque* generation principle.

6.4.1. THE SINUSOIDAL CONTROL SCHEME

This system uses a magnet and field coil arrangement which produces a torque function due to current in each of two coils which has a sine-cosine relationship as shown in Fig. 6.4.1. The torque contribution by each coil is:

Coil 1 $$T_1 = I_{m1} K_T \sin\theta \qquad (6.4.1)$$

Coil 2 $$T_2 = I_{m2} K_T \cos\theta \qquad (6.4.2)$$

where

T_1 and T_2 are instantaneous torque values

K_T is the torque constant of each winding

The current in each coil is controlled as a function of shaft angle as follows:

$$I_{m1} = I \sin\theta \qquad (6.4.3)$$

$$I_{m2} = I \cos\theta \qquad (6.4.4)$$

with I being the amplitude of desired motor current. The total output torque is

$$T_o = T_1 + T_2 \qquad (6.4.5)$$

and combining (6.4.1) thru (6.4.4) we get

$$T_o = K_T I (\sin^2\theta + \cos^2\theta) \qquad (6.4.6)$$

and since

$$\sin^2\theta + \cos^2\theta = 1$$

we finally arrive at

$$T_o = K_T I \qquad (6.4.7)$$

and we have a motor which has essentially no torque ripple and has linear characteristics similar to a conventional DC motor. However, the generation of sinusoidal, amplitude controlled currents requires a complex sine-cosine generator (resolver)

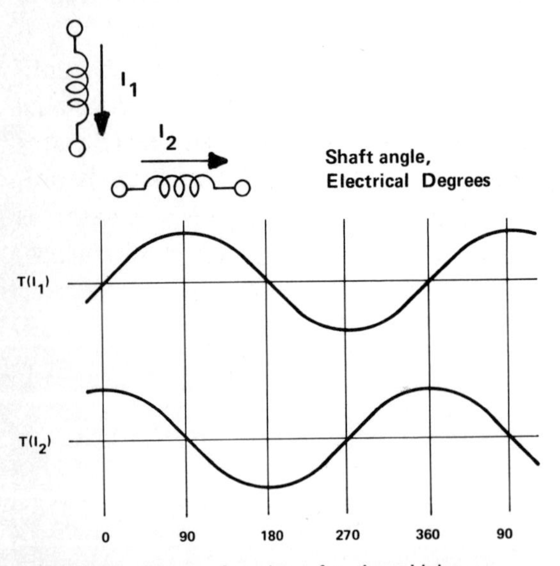

Fig. 6.4.1. Torque function of a sinusoidal type two-phase brushless DC motor.

either by analog or digital design, and a linear amplifier. In addition, certain compromises must be made in the motor design to achieve the sinusoidal torque relationship, which does not give optimum motor efficiency. Thus this method has limited use because of technical and economic reasons.

6.4.2. TRAPEZOIDAL TORQUE FUNCTION

The other control method is based on a *trapezoidal torque function* such as shown in Fig. 6.3.8 (see this illustration in the previous discussion). The principle in this case is to design the motor to provide a trapezoidal torque function and to utilize the unvarying portion of the torque function to produce useful torque. One has to choose a conduction angle and a number of coils such that the resultant torque is independent of shaft position and the motor output torque is then directly proportional to the input current. The torque function in Fig. 6.3.8 has a trapezoidal shape. Proper use of modern high coercive force magnets makes such torque functions possible. While the torque in some motors does not quite conform to this simple form, the relationship is sufficiently close to be useful for purposes of analysis. Thus, for example, if a controlled current I is allowed to flow from A to B for 0 - 60° and from B to A for 180-240°, and if other coils will similarly contribute torque at the intermediate angular intervals, we then conclude that the resultant torque is

$$T_o = K_T I \qquad (6.4.8)$$

where K_T is the value of torque per unit current at the flat portion of the trapezoid. Consideration has to be given to the rise and fall times of the current in the various coils, so that T_o is essentially constant.

The two basic control principles discussed above are most commonly used in brushless motor systems today. The sinusoidal control scheme requires that both the magnetic circuit design and the control current will have sinusoidal shapes to fulfill the motor function. On the other hand, the trapezoidal torque function scheme requires that the magnetic flux is distributed such that the torque function is near constant over the desired conduction angle. Since the controller can commutate the motor windings with square wave control inputs, the control circuit can be of digital nature. This fact makes the control logic for the trapezoidal system rather simple. In addition, this type of circuit utilizes current *only* when it can be optimally effective. Therefore the trapezoidal control scheme tends to yield lower motor losses than the sinusoidal scheme. Hence, the trapezoidal commutation system is generally preferred for high performance, high power control systems. □

6.5. COMMUTATION SENSOR SYSTEMS

Several methods are available today for the angular position sensing system. The most commonly used methods are *Hall effect sensors, electro-optical sensors,* and *radio frequency (RF) sensors.*

The *Hall effect sensing system* utilizes a sensor which detects the magnitude and polarity of a magnetic field. The signals are amplified and processed to form logic compatible signal levels. The sensors are usually mounted in the stator structure, where they sense the polarity and magnitude of the permanent magnet field in the air gap. The outputs of these sensors control the logic functions of the controller configuration to provide current to the proper coil in the stator, and the system can provide some compensation for armature reaction effects which are prominent in some motor designs. One drawback with such a location of the angular position sensor is that it is subject to stator temperature conditions, which may at times be rather severe in high performance applications. It is not unusual, for instance, to allow a winding temperature to reach 160-180 ^{o}C for peak load conditions. Such a temperature may affect the Hall effect switching performance, and can therefore be a system performance limitation.

The Hall effect device can, of course, be located away from the immediate stator structure and may use a separate magnet for angular sensing. In such a case the sensor is not necessarily subject to the severe operating conditions mentioned above; but on the other hand, neither does it compensate for armature reaction problems.

The second alternative for angular sensing is the *electro-optical switch*, most commonly a combination of a light emitting diode (LED) and a phototransistor. A shutter mechanism controls light transmission between the transmitter and the sensor. The sensor voltages can be processed to supply logic signals to the controller. The electro-optical system lends itself well to generation of precise angular encoding signals.

The third angle sensing method *(radio frequency sensing)* is based on inductive coupling between RF coils. Several varieties of such devices have been used with varying degrees of circuit complexity. The angular sensing accuracy of such devices depends on several design factors, which may limit their use in high performance systems. In addition, their switching time (which depends on oscillator frequency and sensor circuit damping) may cause switching time delays which can be undesirable at higher shaft speeds.

The three systems discussed have their own advantages and drawbacks, and the requirements of each application usually dictate which is best suited. □

6.6. POWER CONTROL METHODS

So far we have discussed commutation control without reference to *power control* of the brushless DC motor. One method of power control is to vary the supply voltage to the commutation system. An example of this method is shown in Fig. 6.6.1. The six switching transistors will control commutation at the proper angular intervals, and the series-connected power transistor will handle velocity and current control of the brushless motor. This can be accomplished either by linear (Class A) control or by pulse-width or pulse-frequency modulation. If directional control is needed, the commutation sequence must be adjustable 0 or 180° el. In effect, then, we have a series regulator controlling the power supply voltage for a switching stage commutation controller.

Another way of controlling voltage and current in the brushless motor is to let the commutation transistors control the motor current by either linear control means or by pulse-width or pulse-frequency modulation. Such a control method results in better utilization of available semiconductor devices, but proper attention must be paid to power dissipation in the controller stage. In the case of *linear transistor control* the control stage must be operated in a constant current mode rather than constant voltage mode, since otherwise the transistor stage held at a zero output voltage would tend to conduct back EMF induced currents during the inactive part of the cycle, causing a viscous damping effect which is detrimental to motor operation.

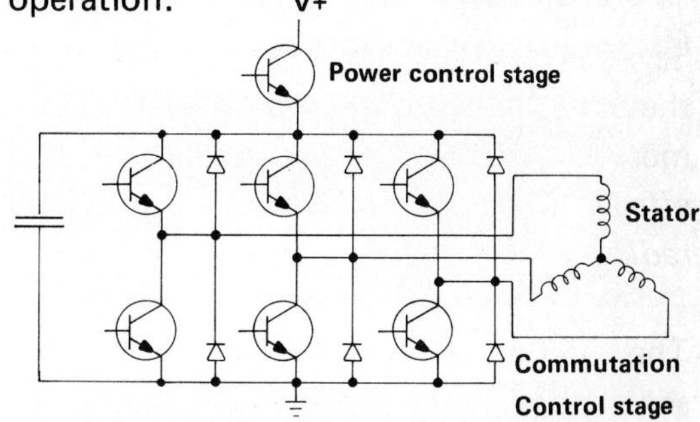

Fig. 6.6.1. Series regulator control of power for three-phase brushless DC motor.

The *pulse-width* or *pulse-frequency control scheme* is well suited for control of voltage and current to a brushless motor. Since logic circuitry is already in place capable of switching the appropriate transistors on and off, the implementation of such control is possible. The switching rate has, of course, to be compatible with a proper current form factor (see the discussion of pulse-width modulations and form factor in section 3.3), and also has to be within the switching capability of the power transistors so as not to cause undue dissipation losses during the transistor turn-off and turn-on times. An example of the variation in current form factor at two different pulse-width frequencies is shown in Fig. 6.6.2. Although the two oscilloscope pictures are taken at two different motor speeds, the improvement in current form factor is readily apparent. In addition, proper attention has to be focused on safety interlocks to prevent "cross-firing" between adjacent transistor stages. Such switching schemes are based on the principles discussed in Chapter 4, and many of the advantages (and problems) discussed in this chapter apply equally well to the switching type commutation controller for brushless DC motors.

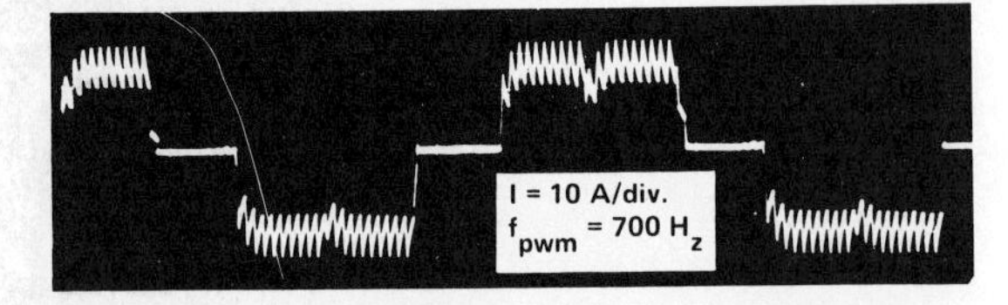

Fig. 6.6.2. Examples of poor current form factor (upper) and improved current form factor (lower) in a three-phase, full-wave brushless DC motor winding. Output torque = 225 oz-in.

Any high performance brushless DC motor will require some form of current limit control, either to protect the controller stage or to protect the magnetic circuit. With either of the controller schemes discussed above it is easy to apply such current limit control. However, the switching type controller can sustain current limit conditions without significant circuit power dissipation — as compared to a linear type of control system, where the excess power is dissipated in the power transistors. The reasons given above point clearly toward pulse-width or pulse-frequency modulation as a superior control scheme for brushless DC motors where any significant amount of DC power is being controlled. Since the intention of this chapter on brushless motors is to familiarize the reader with the various parts of the brushless motor system, and not to be a design manual for brushless motors and controls, it would be beyond the scope of this chapter to go into detail of the variety of logic schemes, power control circuits, and current limit options which are available today.

Power control systems used in Electro-Craft brushless DC motors utilize pulse-width modulation controls for all size motors except ultra-high performance, low inertia motors for incremental motion, where constant-current techniques are used. All systems are provided with current limit circuits, some of which work on a multi-level, time based current-limit principle. Thus the motor control systems can provide extremely high acceleration rates and yet have adequate safety provisions for moderate or long term overload. □

6.7. MOTOR CONSTANTS

Since the same basic principles are used in both the conventional permanent magnet DC motor and the brushless DC motor, it would seem that the basic motor constants also would be the same. This is true if we use some caution in applying the constants.

A simplified electrical equation of a brushless DC motor is:

$$V = IR + L\frac{dI}{dt} + K_E\omega \qquad (6.7.1)$$

where

- I is the sum of the phase currents.
- R is the resistance of a phase winding
- L is the inductance of a phase winding
- K_E is the voltage constant of a phase winding over the conduction angle
- ω is the angular velocity of the motor shaft

These relationships hold well for the common brushless motor structures, although there are lower order effects due to mutual inductance between windings, overlapping conduction angles and unequal rise and fall times of current (due to differing charge and discharge paths). For practical applications eq. (6.7.1) is adequate, however.

The dynamic equation for a motor coupled to a load is:

$$K_T I = (J_m + J_L)\frac{d\omega}{dt} + D + T_f + T_L \qquad (6.7.2)$$

where

- K_T = torque constant of motor winding
- J_m = motor moment of inertia
- J_L = load moment of inertia
- D = viscous damping coefficient
- T_f = motor friction torque
- T_L = load friction torque

In a brushless DC motor T_f is small, usually only due to bearing drag, the viscous damping coefficient is also very small, and both items can usually be ignored in dynamic performance calculations. (For derivation of the motor transfer function see section 2.3.3.) □

6.8. BRUSHLESS DC TACHOMETERS

DC analog output tachometers are often necessary in high performance servo applications, where they provide velocity feedback for speed control purposes or servo system stability. The requirements of a brushless DC tachometer are the same as for a brush-type DC tachometer: an output voltage proportional to shaft velocity and the polarity corresponding to the direction of the shaft rotation. The quality of the tachometer depends on the stability of the voltage constant and the magnitude and frequency of voltage ripple.

Digital-type tachometers can be used successfully where the controlled motor speed range is in a region where the pulse rate is high enough to provide a sufficient servo bandwidth after the required filtering following D/A conversion. However, when servo control is required over a wide range of speeds, down to a stop position, then rate information is required over the entire range; therefore the digital tachometer is often not sufficient for servo system operation.

The Electro-Craft brushless DC tachometer is based on a permanent magnet rotor and a multi-coil stator structure, which is commutated by a MSI (medium scale integration) circuit. The output voltage of the system behaves exactly as the brush-type DC tachometer. The ripple voltage depends on tachometer design, but can be controlled to a level better than 1%. Depending on the rotor moment of inertia and the temperature stability required, ceramic, alnico or rare-earth materials are used to provide the flux for the tachometer. The block diagram of the system is shown in Fig. 6.8.1. When the brushless DC tachometer is mounted directly on the motor shaft, the motor shaft encoder output can also be used for tachometer commutation. Since the motor and tachometer rotors are both located on the same shaft, a very stiff mechanical structure is realized. Thus the unavoidable torsional resonances can be kept at very high frequency levels, where they do not influence servo system behavior. □

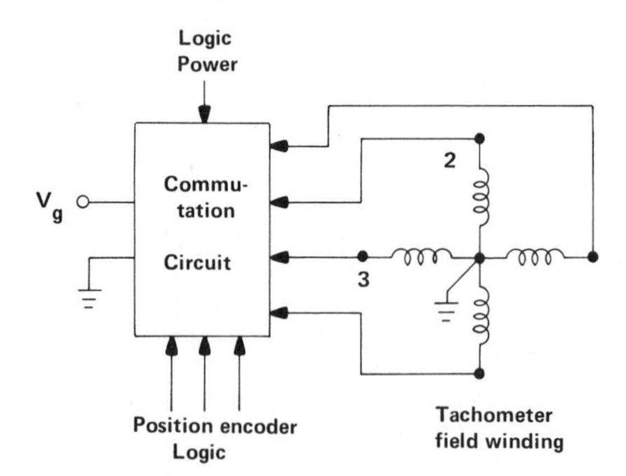

Fig. 6.8.1. Brushless DC tachometer system.

6.9. EXAMPLES OF BRUSHLESS MOTORS

In Fig. 6.9.1 we see a picture of a high performance incremental motion brushless DC motor, Electro-Craft Corporation Model 17. This motor was designed to be light in weight in addition to having low armature moment of inertia. The rotor configuration includes a four-pole magnet structure using rare-earth magnet materials. The performance parameters are shown in Table 6.9.1. Shown are two alternative connections, 3-phase and 6-phase full-wave systems. Slightly better characteristics are achieved by the 6-phase connection, because of better winding utilization.

The system performance is shown in Fig. 6.9.2, where we see an example of the output current of one "constant current" stage during operation in a velocity mode with 75 oz-in torque at 500 rpm.

We see one complete commutation cycle of one of the six stages. The current is kept constant during the positive and negative commutation cycles, each 120°. In this system four phases are always conducting. Since each phase contributes one-fourth of the torque, and the current per phase is 3.6 A, we can calculate the torque constant:

$$K_T = \frac{75}{4\text{x}3.6} = 5.2 \text{ oz-in/A}$$

In order to demonstrate incremental motion performance, the servo motor was connected to a load with $J_L = 0.5\text{x}10^{-3}$ oz-in-s^2, and the system was compensated for the added inertia. The amplifier was adjusted for a peak current limit of 3.8 A per phase (this is the continuous current rating of the motor, with 5 cfm of air cooling). The measured system servo bandwidth with the load was 270 Hz.

Fig. 6.9.1. Electro-Craft Corporation Brushless DC Motor Model 17.

Fig. 6.9.3 shows a torque-speed characteristic of the motor-tachometer in a closed-loop velocity mode. The current limit causes the fall-off of speed as the torque requirements exceed 70 oz-in, and at

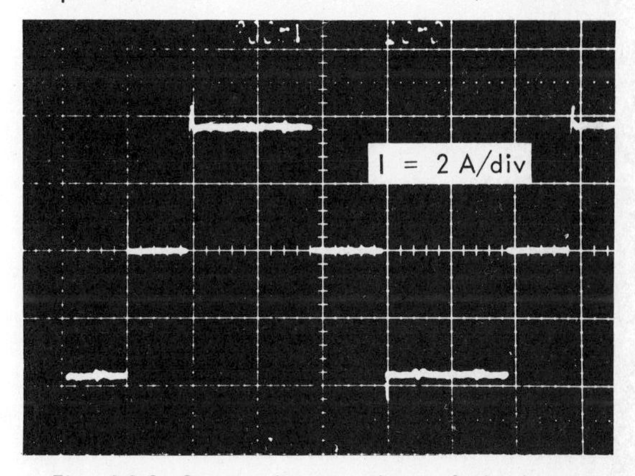

Fig. 6.9.2. Current in one phase of six-phase low inertia servo motor in a "constant current" drive mode. Constant velocity of 500 rpm at T = 75 oz-in.

TABLE 6.9.1. Preliminary Motor Characteristics of Model 17 Brushless DC Motor

Parameter	Unit	Connection		
		3 phase	6 phase	
			−1	−2
K_T (per phase)	oz-in/A	9.6	5.3	10.6
K_E (per phase)	V/krpm	7.5	4.1	8.2
R (per phase)	Ω	5.4	2.7	10.8
τ_e (per phase)	μs	700	350	350
J_a	oz-in-s^2	$.65\times10^{-3}$	$.65\times10^{-3}$	$.65\times10^{-3}$
T cont (free air)	oz-in	30	30	35
T cont (12″ x 12″ heat sink)	oz-in	45	45	50
T cont (5 cfm @ 0.2 in H_2O)	oz-in	60	60	70
T max	oz-in	120	120	140
T max/J_a	rad/s^2	185,000	210,000	210,000
Max. winding temp. = 160°C				
Motor dimensions: Diameter 2.75 in, Length 3.0 in				
Weight 16 oz				

speeds above 3000 rpm the effects of power supply voltage limits are evident.

In order to assess the worst-case output torque ripple characteristics, a measurement of torque output of a stall motor was made under various constant current conditions. The torque was measured by a strain gage torque arm, as the motor housing was slowly rotated in a mechanical "dividing head". An X-Y recorder plotted the varying torque as one complete electrical revolution was made (Fig. 6.9.4). The torque ripple under these conditions was

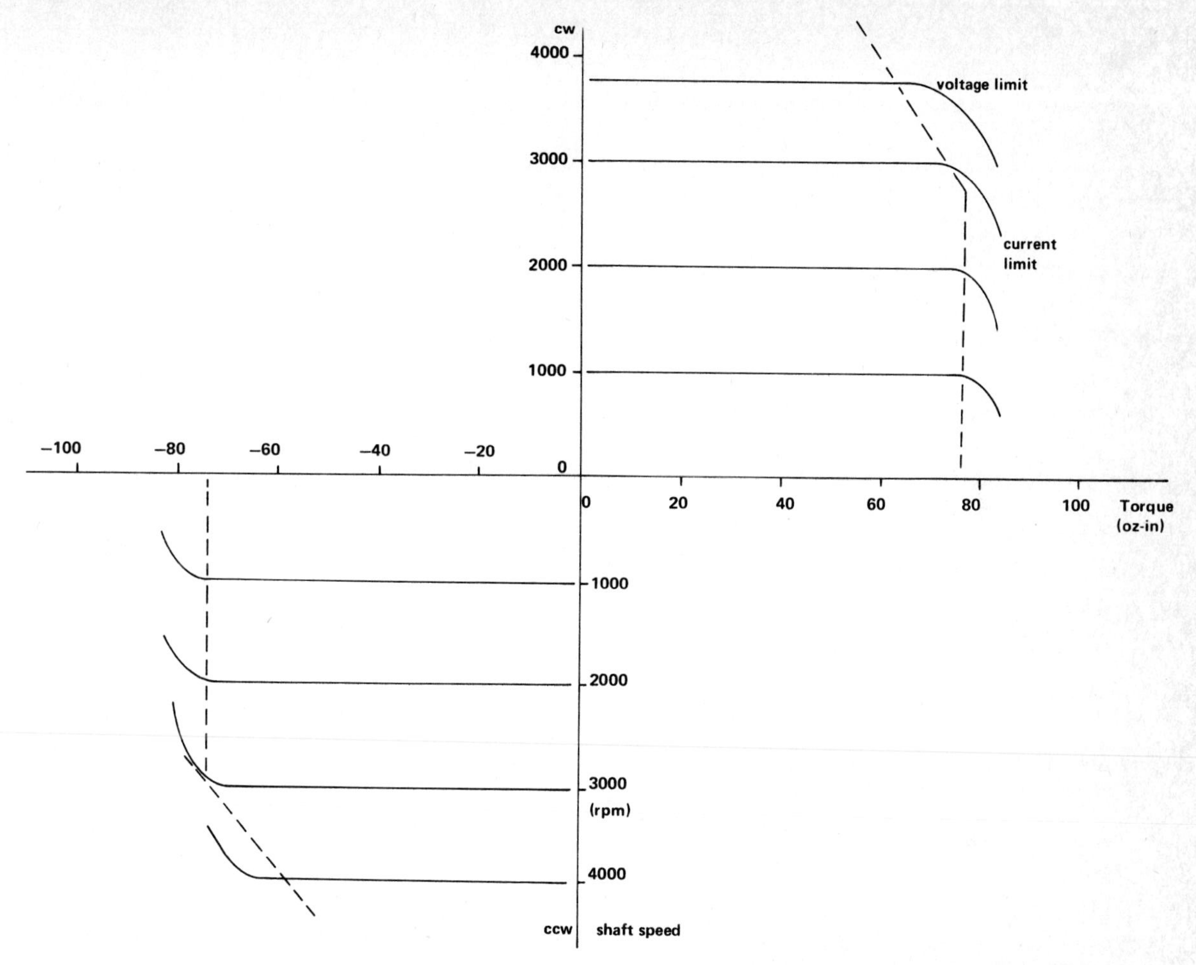

Fig. 6.9.3. Torque-speed characteristics of a brushless motor-tachometer in a closed-loop velocity mode.

measured to be less than ±5% of average output torque. Since this motor was designed as a low inertia motor, it does not necessarily represent the best attainable torque ripple performance in brushless DC motors. It may reflect the trade-off between low rotor moment of inertia and torque ripple.

A demonstration of incremental motion performance characteristics of Model 17 is displayed in Fig. 6.9.5, where the system is excited with a larger command signal. Thus we see the motor velocity switched from +5850 rpm to −5850 rpm. The linear portion of the velocity trace shows the acceleration, in this case limited by the current limit of 3.8 A per phase. If we measure the slope, we obtain

$$\frac{d\omega}{dt} = \frac{1220 \text{ rad/s}}{19\text{x}10^{-3}\text{s}} = 64 \text{ krad/s}^2$$

which is the acceleration rate at the set current limit of 3.8 A and a load moment of inertia of

$$J_L = 0.5 \times 10^{-3} \text{ oz-in-s}^2$$

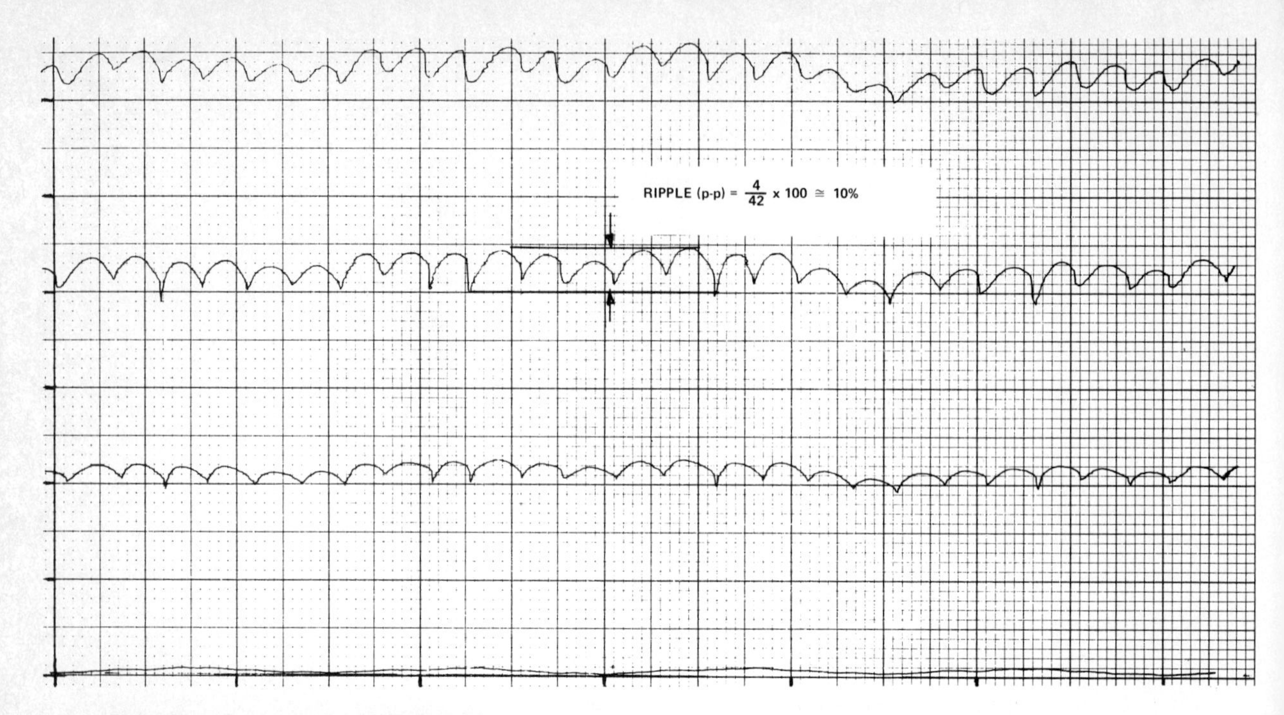

Fig. 6.9.4. Static torque ripple test of Mod. 17, six-phase brushless DC motor and constant-current type controller.

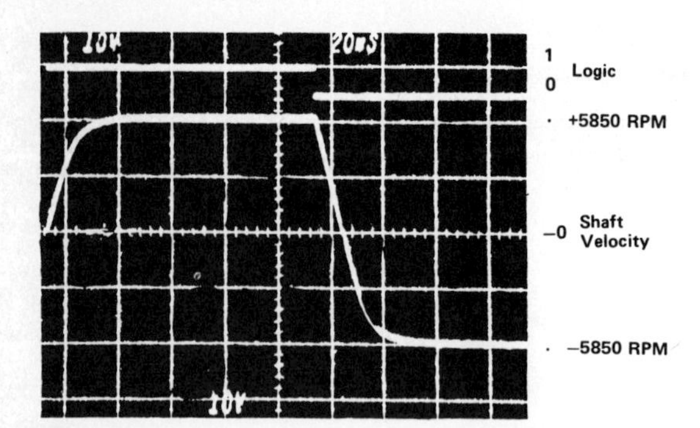

Fig. 6.9.5. Constant acceleration at a current of 3.8A, and a load moment of inertia of $J_L = .5 \times 10^{-3}$ oz-in-s^2.

A cross-sectional view of a brushless DC motor-tachometer is shown in Fig. 6.9.6. The motor rotor, tachometer rotor and the shaft encoder shutter are all mounted on the same shaft, which contributes to a high torsional stiffness between all elements. A special brushless DC motor designed for constant speed applications is shown in Fig. 6.9.7. The motor is intended for use in disc memory systems, where it drives the discs at a constant speed of 3600 rpm and a torque of 50 oz-in. The run-out of the shaft end is held to a tolerance of less than 50 millionths of an inch.

An example of a two-phase, full-wave controller for a 0.25 hp brushless DC motor is shown in Fig. 6.9.8. The circuit board contains analog velocity loop summing amplifier, pulse-width modulation circuit, commutation logic, driver circuits and power output stage. DC power supply is not included.

Fig.6.9.9 shows an example of a brushless DC motor, displayed with two optional stator assemblies of different stack lengths and a typical ceramic magnet rotor assembly. The commutation encoder is located in the rear portion of the housing assembly. □

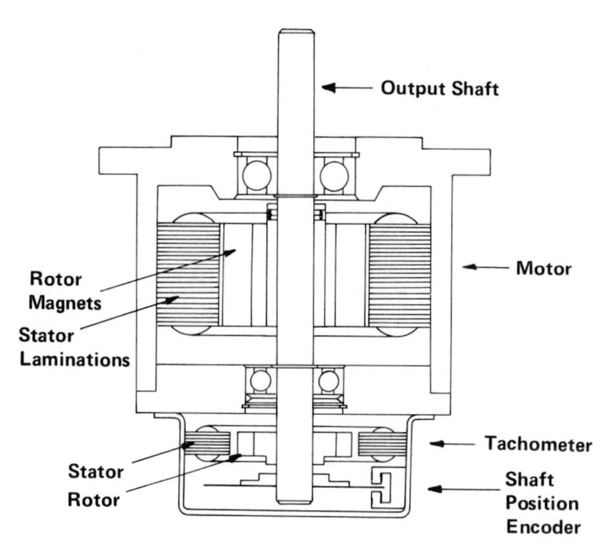

Fig. 6.9.6. Brushless DC motor-tachometer.

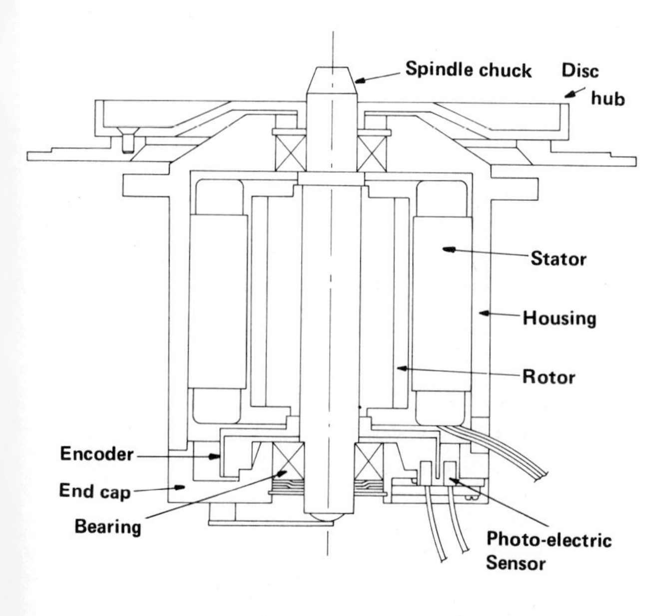

Fig. 6.9.7. Brushless DC motor disc memory spindle drive package.

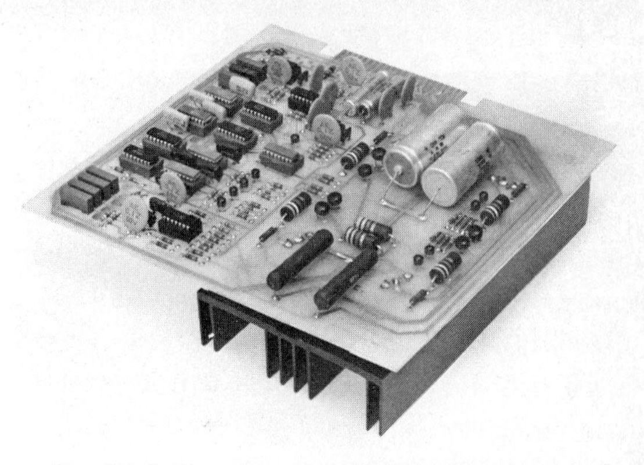

Fig. 6.9.8. Complete two-phase, full-wave amplifier-controller for a ¼ HP brushless DC servo motor.

Fig. 6.9.9. Brushless DC motor housing together with two different stator assemblies and a ceramic magnet rotor.

6.10. SUMMARY

Brushless DC motors can be used for a variety of applications. The advantages of state-of-the-art magnet materials have made possible brushless DC motor designs which have very high torque-to-inertia ratios. Since commutation is performed in circuit elements external to the rotating parts, there are no items which would suffer from mechanical wear, except the motor bearings. The brushless DC motor will, therefore, have a life expectancy limited only by mechanical bearing wear, and the reliability of the electronic controller.

The brushless DC motor can be controlled with very efficient amplifier configurations. In cases of severe environmental conditions the controller can be located remotely from the motor. The control system can easily interface with digital and analog inputs, and is therefore well suited for incremental motion (see the following discussion 6.11) and for phase-locked speed control systems. The motor and control have a lower level of radio frequency emission than the conventional DC motors and controls.

The brushless DC motor controller has, in general, a more complex configuration than the controller for an equivalent conventional DC motor, but may be similar in size and complexity to a closed-loop step motor controller. The commutation sensor system can in some cases provide some incremental information of shaft angle for incremental motion applications, but usually an encoder system is added to the commutation system to suit the application.

Table 6.10.1 shows a brief comparison of the conventional DC motor and the brushless DC motor. Due to the many variants available of the two kinds, the comparison has to be very general. However, it may give more insight into the basic characteristics of the two.

TABLE 6.10.1.

A brief comparison between conventional DC motors and brushless DC motors

CONVENTIONAL DC MOTORS

GOOD POINTS:

1. Controllability over a wide range of speeds.
2. Capable of rapid acceleration and deceleration.
3. Convenient control of shaft speed and position by servo amplifiers.

PROBLEM AREA:

1. Commutation (brushes) causes wear and electrical noise.
   This problem can be kept under control by selection of best materials for each application.

BRUSHLESS DC MOTORS

GOOD POINTS:

1. Controllability over a wide range of speeds.
2. Capable of rapid acceleration and deceleration.
3. Convenient control of shaft speed and position.
4. No mechanical wear problem due to commutation.
5. Better heat dissipation arrangement.

PROBLEM AREA:

1. Requires more semiconductor devices than the brush type DC motor for equal power rating and control range. □

6.11. STEP MOTORS VERSUS BRUSHLESS DC MOTORS IN DIGITAL POSITION SYSTEMS

6.11.1. INTRODUCTION

Step motors are, in their proper market, very useful and economical. The designer of a system only has to send a given number of pulses to a controller and the motor will move an appropriate number of steps and then stop. This works well as long as the load moment of inertia and friction levels are within design limits, and the step rate is within the capability of the motor and controller. The brushless DC motor has, by our earlier definition, no inherent capability of moving in discrete steps. However, with the addition of very few parts the brushless DC motor can perform better than step motors in incremental motion systems.

6.11.2. STEP MOTOR PERFORMANCE LIMITATIONS

With the ever-increasing technological demands on incremental motion devices, step motor systems are continually called upon to provide higher step rates and slew speeds, and to handle larger inertial loads. While the step motor industry is continually improving their products to handle increased demands, the problems inherent in the motor design principles come to the foreground as step rate demands increase. Following is a list of the major problems with currently available step motor systems:

1) *Settling time.* The open-loop step motor controller cannot cope well with the damped oscillatory settling behavior of the step motor shaft into a final position.
2) *Lack of availability of a variety of step angles.* Step motors have fixed step angles, which may not always fit given applications. A change in step angle involves re-tooling the rotor and stator laminations, a very expensive procedure.
3) *Slew speed problems.* Step motors are limited in high speed slew rates and have "forbidden zones" of operation, where the system can lose synchronism.
4) *"Jerky" speed.* In some applications the step motor's inherent reluctance cogging causes speed modulations which are inconsistent with acceptable performance.
5) *Inability to handle large inertial loads.* Because of the pulse excitation control, large inertial loads present difficulties for step motors.

The most common method of solving some of these problems in *high performance systems* is to provide an incremental shaft angle encoder to aid in the step switching, and the result is improved step rate and settling time. The user then has the following equipment: a power supply, a digital logic controller, a multi-phase switching controller, an ecoder and a motor. The parts are shown in block diagram form in Fig. 6.11.1.

6.11.3. BRUSHLESS DC MOTOR INCREMENTAL MOTION CONTROL

In order to get high performance "step motor" action with controllable damping

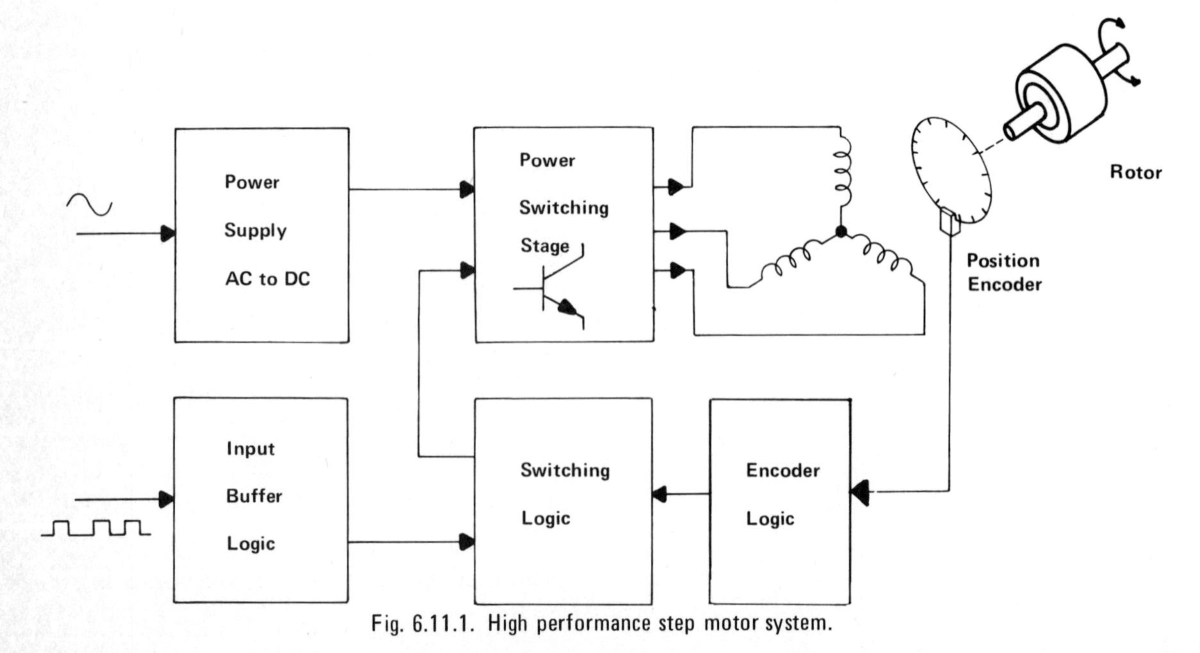

Fig. 6.11.1. High performance step motor system.

and with slew rates limited only by mechanical strength of the motor and counting rate limitations of the logic circuit, we can demonstrate the system as shown in Fig. 6.11.2. Included in the components, in addition to the parts shown in Fig. 6.11.1 (the closed-loop step motor system) are the commutation encoder, the DC brushless tachometer and the DC servo compensation network. The latter part is adjustable to accommodate varying load moments of inertia and other factors affecting system stability. There are, of course, other differences in signal processing and logic, but the purpose of this discussion is to demonstrate, in a general way, how this is accomplished. This system has performance parameters exceeding the closed-loop step motor system in Fig. 6.11.1. It is more versatile in that by choice of proper position encoder one can have any desired number of steps per revolution. In addition, the system will have the high torque-to-inertia ratio of a DC motor, will behave as a DC motor at high speeds and therefore will have higher operational range than the equivalent step motor.

In Table 6.11.1 we see a brief summary of the main features of each system. A photograph of a system demonstrator is shown in Fig. 6.11.3.

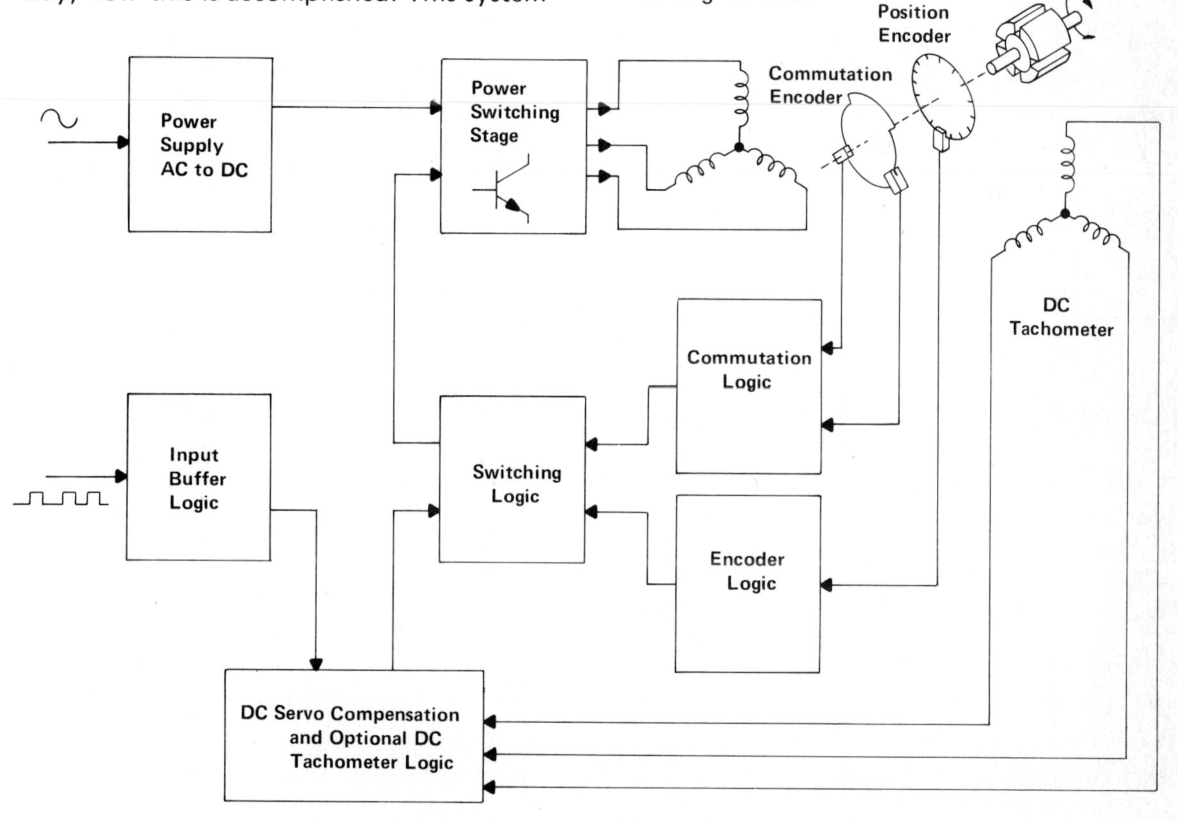

Fig. 6.11.2. High performance BLM "electronic" step motor.

6.11.4. CONCLUSION

Step motors used in their "natural" operational range are well suited for step applications. Whenever high performance step requirements are encountered it is worthwhile to investigate the advantages of brushless "electronic step" DC motor systems. □

TABLE 6.11.1

BRIEF COMPARISON OF CHARACTERISTICS OF CLOSED-LOOP STEP MOTORS AND BRUSHLESS MOTORS FOR INCREMENTAL MOTION

STEP MOTOR	BRUSHLESS DC MOTOR
Has an inherent number of discrete shaft stop positions.	Objective is to provide shaft rotation without any preferred positions due to motor construction.
Requires a high number of winding switching sequences.	Requires a low number of winding switching sequences.
Winding inductance is a dominant factor which may be limiting high speed performance.	Due to a lower number of switching sequences the winding inductance is a lesser problem at high shaft speeds.
Has non-linear torque-speed characteristics.	Has linear torque-speed characteristics.
Requires incremental shaft angle encoder.	Requires incremental shaft angle encoder.
The number of encoder lines is dependent on motor step angle.	The number of encoder lines is independent of motor parameters.
The incremental encoder and circuitry provide improved position damping over the open-loop motor system.	Has optimum position damping response if analog tachometer is used.
Position stiffness dependent on magnetic detent force and step angle.	Position stiffness dependent on encoder line-to-line angle, motor torque.

Fig. 6.11.3. Brushless DC motor and controller for demonstration of "electronic step motor" performance.

Chapter 7

Electro-Craft Corporation Products

This chapter groups generic types of Electro-Craft products as follows:

The products presented are only a small sampling of the complete Electro-Craft product line. For further details, contact:

Electro-Craft Corporation
1600 Second Street South
Hopkins, Minnesota 55343
(612) 935-8226
Telex: 29-0677

NOTE: Most product data in this chapter were retained intact from the first edition of this handbook. Some symbols of units, etc., differ, therefore, from standard form used elsewhere in the book.

7.1. DC SERVOMOTORS

E-250

The Model E-250 is a high performance DC permanent magnet servomotor. It is especially suited for applications in cassette tape drives, printer ribbon drives and other general purpose servo applications.

SPECIFICATIONS

MOTOR RATINGS

Maximum Safe Speed (no load)	10 krpm
Rated Torque (continuous at stall)	4.5 oz-in
Maximum Rated Armature Power Dissipation	10 W
Maximum Rated Armature Temperature	155 ^{o}C
Thermal Resistance (Armature to Ambient)	13.0 ^{o}C/W
Armature Moment of Inertia	0.00055 oz-in-sec^2
Electrical Time Constant	.85 ms
Mechanical Time Constant	21 ms
Damping Factor	0.01 oz-in/krpm
Static Friction Torque	0.5 oz-in

WINDING DATA

Torque Constant K_T (oz-in/A) $\pm$ 10%	4.45
Voltage Constant K_E (V/krpm) $\pm$ 10%	3.3
Armature Winding* (Ω) $\pm$ 15%	5.4
Armature Inductance (mH)	4.6
Max. Pulse Current (A)	8

*Add 0.5 ohms for terminal resistance. Contact factory for alternate windings.

SPEED-TORQUE CURVE

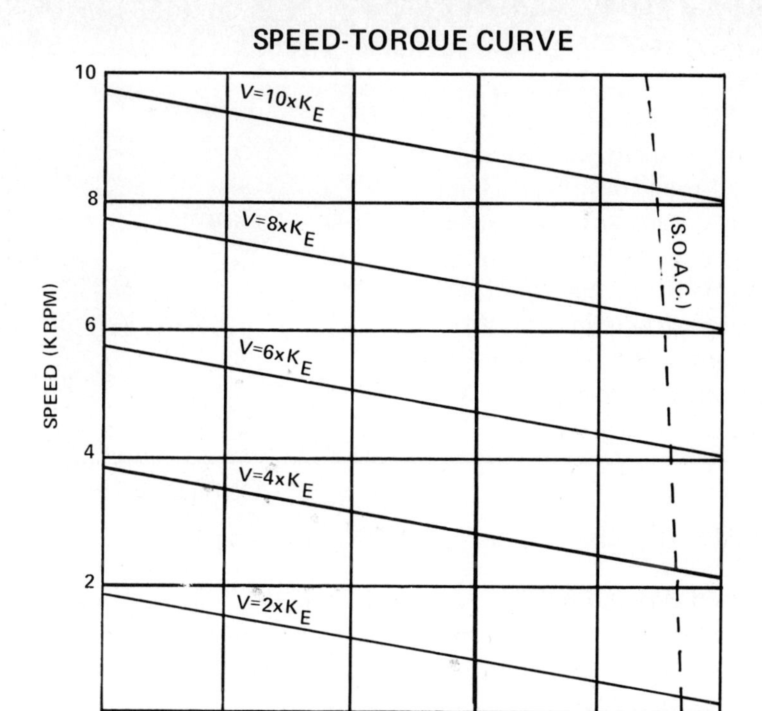

OUTLINE DRAWING

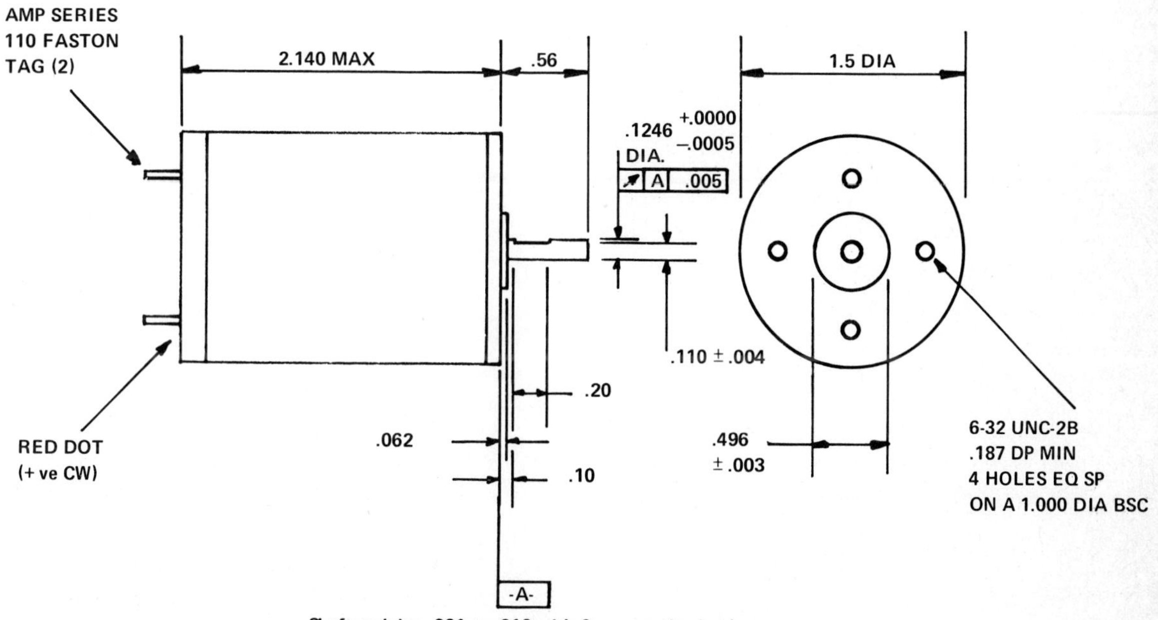

Shaft endplay .001 to .010 with 8 oz reversing load.

E-508/512

The E-508 - E-512 series is a moderate cost line of FHP D.C. permanent magnet motors designed for a broad range of industrial applications, such as tape transport reel drives or position control systems. Reliability and long life have been designed into these motors by the use of corrosion-resistant shaft and housing materials and permanently lubricated ball bearings. Pictured at the right is the E-508 and the longer E-512.

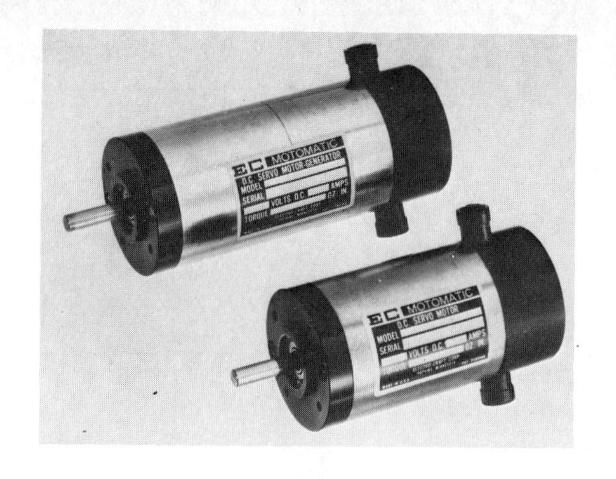

SERIES MOTOR RATINGS

Maximum Rated Armature Temp. 155°C.
Maximum No Load Speed 6 KRPM.
Rated Brush Life (Minimum) 5,000 HRS @ 1 KRPM
Maximum Friction Torque 3 oz. in.

Maximum Shaft Radial Load 10 lb., .5 in. from front bearing continuous for a min. 1000 Hours

CONSTRUCTION FEATURES

Winding Insulation Class F.
ABEC Class 1 Ball Bearings
PVC Lead Insulation
Totally Enclosed Housing
Available with Integrally Mounted Tachometers, ECC E-595 or M-110
Consult Electro-Craft Factory for Special Shaft and Mounting Configurations

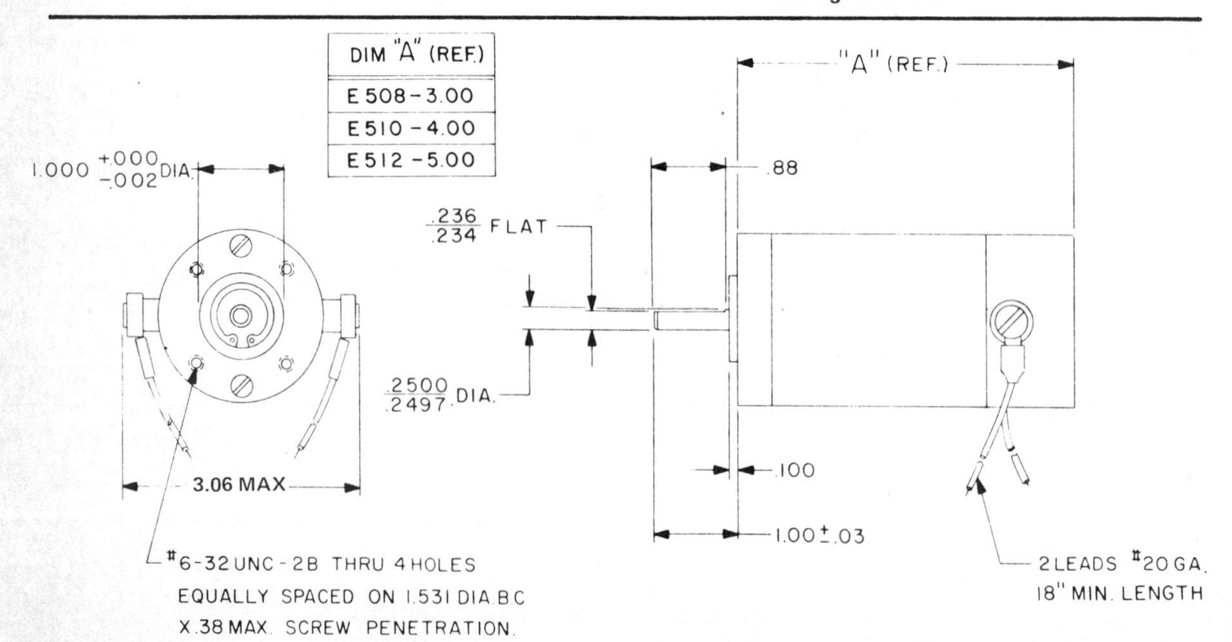

DIM "A" (REF.)
E 508 - 3.00
E 510 - 4.00
E 512 - 5.00

NOMINAL SPECIFICATIONS AND RATINGS FOR TYPICAL E.C.C. SIZE 20 MOTORS

Model 508

Max. Rated Continuous Torque (oz. in.) 10
Armature Inertia (oz. in. sec.^{2}) .0015
Damping Factor (oz. in./KRPM) .1
Thermal Resistance (^{o}C./Watt) 6.84
(Armature-to-ambient)
Elect. Time Constant (msec) 1.65
Mech. Time Constant (msec) 24.7

Winding Number	Torque Constant (K_T) (oz. in./amp) ± 10%	Voltage Constant (K_E) (Volts/KRPM) ± 10%	Winding Res. (Ohms)* at 25°C. ± 15%	Armature Inductance (millihenries)	Max. Pulse Current (Amps) [To avoid demagnetization]
01	2.79	2.07	.51	.84	39
02	3.53	2.61	1.05	1.73	31
03	4.47	3.31	1.94	3.20	25
04	5.58	4.14	3.25	5.36	20

MODEL 510

Max. Rated Continuous Torque (oz. in.) 19
Armature Inertia (oz. in. sec.^{2}) .0038
Damping Factor (oz. in./KRPM) .1
Thermal Resistance (^{o}C./Watt) 5.0
(Armature-to-ambient)
Elect. Time Constant (msec) 2.06
Mech. Time Constant (msec) 24.7

Winding Number	Torque Constant (K_T) (oz. in./amp) ± 10%	Voltage Constant (K_E) (Volts/KRPM) ± 10%	Winding Res. (Ohms)* at 25°C. ± 15%	Armature Inductance (millihenries)	Max. Pulse Current (Amps) [To avoid demagnetization]
01	5.80	4.29	1.15	2.37	24
02	7.34	5.43	2.09	4.31	19
03	9.26	6.85	3.56	7.33	15
04	11.60	8.58	5.78	11.91	12

Model 512

Max. Rated Continuous Torque (oz. in.) 25
Armature Inertia (oz. in. sec.^{2}) .0062
Damping Factor (oz. in./KRPM) .2
Thermal Resistance (^{o}C./Watt) 4.18
(Armature-to-ambient)
Elect. Time Constant (msec) 2.15
Mech. Time Constant (msec) 25.7

Winding Number	Torque Constant (K_T) (oz. in./amp) ± 10%	Voltage Constant (K_E) (Volts/KRPM) ± 10%	Winding Res. (Ohms)* at 25°C. ± 15%	Armature Inductance (millihenries)	Max. Pulse Current (Amps) [To avoid demagnetization]
01	8.78	6.50	1.86	3.99	20
02	11.15	8.25	3.24	6.96	16
03	14.06	10.40	5.42	11.65	12
04	17.56	13.00	8.64	18.58	10

*Add 0.40 Ohms for Terminal Resistance

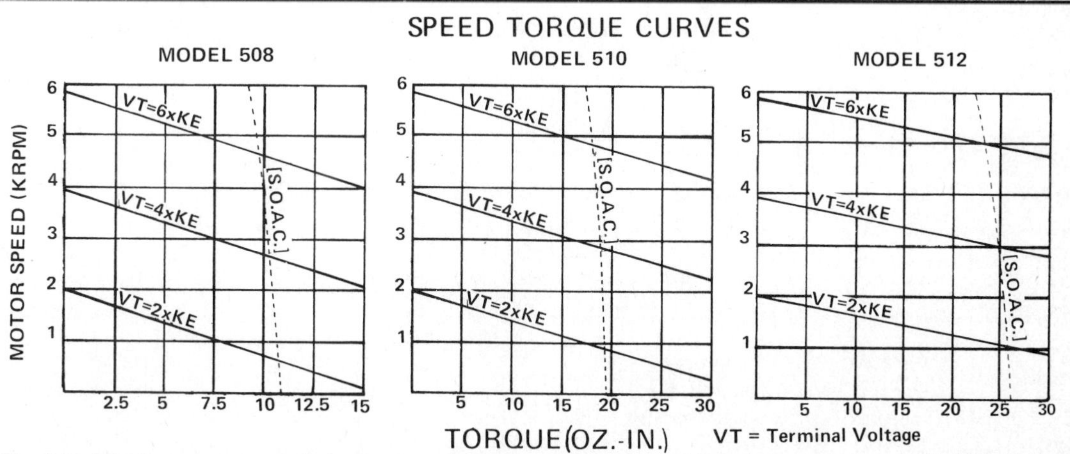

The S.O.A.C. line represents the safe operating area, continuous with unit mounted on 10"x10"x¼" heat sink.

E-540/E-542

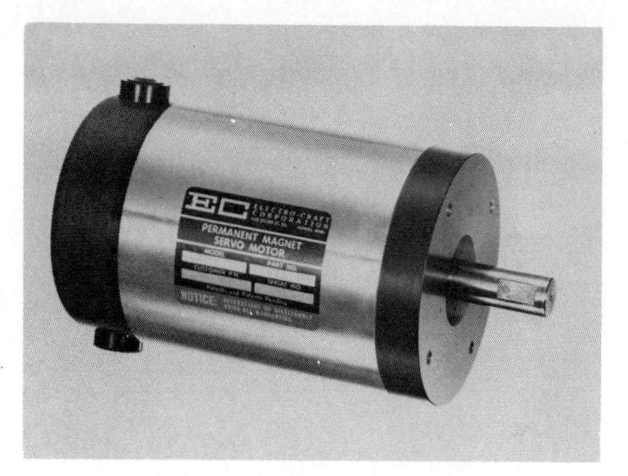

The E-540-E-542 series of FHP D.C. permanent magnet motors is derived from the E-508-E-512 series with increased output performance resulting from an enhanced magnetic circuit. This increased performance has substantially added to the cost effectiveness and versatility of these motors used in a variety of applications in professional and industrial products.

SERIES MOTOR RATINGS

Maximum Rated Armature Temp. 155°C

Maximum No Load Speed 6 KRPM

Rated Brush Life (Minimum) 5,000 HRS @ 1 KRPM

Maximum Friction Torque 3 oz.-in.

Maximum Shaft Radial Load 10 lb., 5 in. from front bearing continuous.

CONSTRUCTION FEATURES

Winding Insulation Class F.
ABEC Class 1 Ball Bearings
PVC Lead Insulation
Totally Enclosed Housing
Available with Integrally Mounted Tachometers, ECC E-596 or M-110
Consult Electro-Craft Factory for Special Shaft and Mounting Configurations.

OUTLINE DRAWING

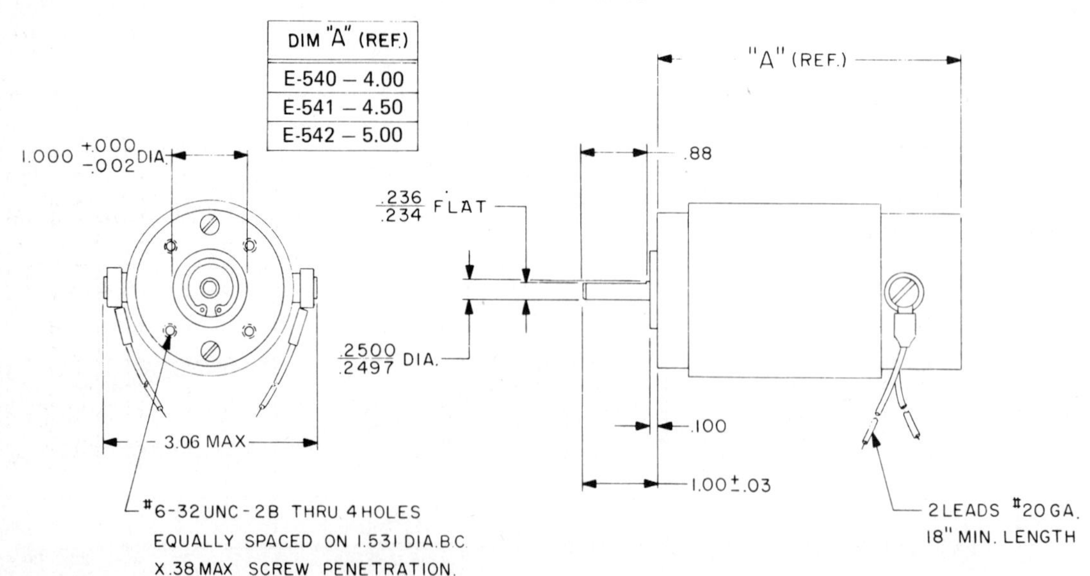

NOMINAL SPECIFICATIONS AND RATINGS

Model E-540
Max. Rated Continuous Torque (oz.in.) 29
Armature Inertia (oz.in.sec.^{2}) .0038
Damping Factor (oz.in./KRPM) .1
Thermal Resistance (^{o}C./Watt) 5.0
(Armature-to-ambient)
Elect. Time Constant (msec) 2.06
Mech. Time Constant (msec) 11

Winding Number	Torque Constant (K_T) (oz.in./amp) ±10%	Voltage Constant (K_E) (Volts/KRPM) ±10%	Winding Res. (Ohms)* at 25°C. ± 15%	Armature Inductance (millihenries)	Max. Pulse Current (Amps) (To avoid demagnetization)
01	9.20	6.81	1.36	3.63	26
02	11.65	8.63	2.27	6.43	21
03	14.72	10.90	4.33	9.74	16
04	18.40	13.63	6.81	14.85	13

Model E-541
Max. Rated Continuous Torque (oz.in.) 38
Armature Inertia (oz.in.sec.^{2}) .005
Damping Factor (oz.in./KRPM) .1
Thermal Resistance (^{o}C./Watt) 4.59
(Armature-to-ambient)
Elect. Time Constant (msec) 2.1
Mech. Time Constant (msec) 9.7

Winding Number	Torque Constant (K_T) (oz.in./amp) ±10%	Voltage Constant (K_E) (Volts/KRPM) ±10%	Winding Res. (Ohms)* at 25°C. ± 15%	Armature Inductance (millihenries)	Max. Pulse Current (Amps) (To avoid demagnetization)
01	11.79	8.73	1.51	4.01	23
02	14.94	11.06	2.48	6.05	18
03	18.87	13.97	4.0	8.4	14
04	23.87	17.68	6.35	13.34	11

Model E-542
Max. Rated Continuous Torque (oz.in.) 50
Armature Inertia (oz.in.sec.^{2}) .0062
Damping Factor (oz.in./KRPM) .2
Thermal Resistance (^{o}C./Watt) 4.18
(Armature-to-ambient)
Elect. Time Constant (msec) 2.15
Mech. Time Constant (msec) 7.6

Winding Number	Torque Constant (K_T) (oz.in./amp) ±10%	Voltage Constant (K_E) (Volts/KRPM) ±10%	Winding Res. (Ohms)* at 25°C. ± 15%	Armature Inductance (millihenries)	Max. Pulse Current (Amps) (To avoid demagnetization)
01	14.86	11.00	1.53	4.15	21
02	18.82	13.94	3.07	7.46	16
03	23.78	17.61	4.87	11.33	13
04	29.72	22.00	7.67	17.35	10

* Add 0.40 ohms for terminal resistance

SPEED TORQUE CURVES

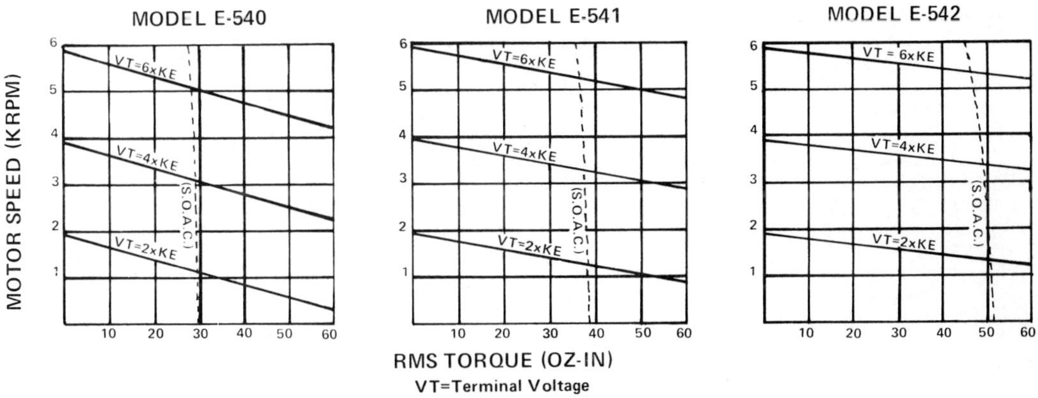

MODEL E-540 | MODEL E-541 | MODEL E-542

VT=6xKE
VT=4xKE
VT=2xKE
(S.O.A.C.)
MOTOR SPEED (KRPM)
RMS TORQUE (OZ-IN)

VT=Terminal Voltage

The S.O.A.C. line represents the safe operating area, continuous with unit mounted on 10"x10"x¼" heat sink.

E-660/E-670

The Model E-660-E-670 series permanent magnet D.C. motors are designed to provide high cost effectiveness in a variety of professional and industrial equipment. Applications include direct drive for reel motors in tape transports and web handling systems and material handling systems.

SERIES MOTOR RATINGS

Maximum Rated Armature Temp. 155°C.

Rated Brush Life (Minimum) 5,000 HRS @ 1 KRPM.

Maximum Friction Torque 7 oz.-in.

Maximum Shaft Radial Load 30 lbs., 1 in. from front bearing, continuous.

CONSTRUCTION FEATURES

Winding Insulation Class F.
ABEC Class 1 Ball Bearings
PVC Lead Insulation
Totally Enclosed Housing
Available with Integrally Mounted Tachometers, ECC E-596 or M-110
Consult Electro-Craft Factory for Special Shaft and Mounting Configurations

OUTLINE DRAWING

DIM "A" (REF.)
E-660 – 5.5
E-670 – 6.7

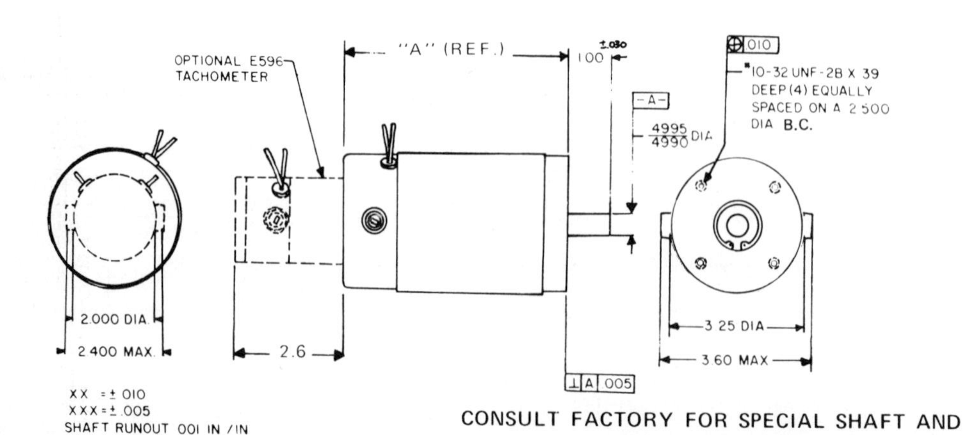

CONSULT FACTORY FOR SPECIAL SHAFT AND MOUNTING CONFIGURATIONS

NOMINAL SPECIFICATIONS AND RATINGS

Model E-660
Max. Continuous Stall Torque (oz-in) 150
Armature Inertia (oz.in.sec ^{2}) .03
Rotational Losses (oz.in./KRPM) .68
Thermal Resistance (^{o}C./Watt) 2.8
(Armature-to-ambient)
Elect. Time Constant (msec) 3.5
Mech. Time Constant (msec) 10

Winding Number	Torque Constant (K_T) (oz.in./amp) ±10%	Voltage Constant (K_E) (Volts/KRPM) ±10%	Winding Res. (Ohms)* at 25°C. ± 15%	Armature Inductance (millihenries)	Max. Pulse Current (Amps) (To avoid demagnetization)
01	12.9	9.6	.35	1.5	54
02	27.6	20.4	1.56	4.4	26
03	43.2	32.0	2.94	10.5	15
04	63.4	47.0	7.00	24.8	10
05	79.8	59.2	11.12	39.9	8

Model E-670
Max. Continuous Stall Torque (oz-in) 225
Armature Inertia (oz.in.sec.^{2}) .05
Rotational Losses (oz.in./KRPM) 1.0
Thermal Resistance (^{o}C./Watt) 2.1
(Armature-to-ambient)
Elect. Time Constant (msec) 3.5
Mech. Time Constant (msec) 11.0

Winding Number	Torque Constant (K_T) (oz.in./amp) ±10%	Voltage Constant (K_E) (Volts/KRPM) ±10%	Winding Res. (Ohms)* at 25°C. ± 15%	Armature Inductance (millihenries)	Max. Pulse Current (Amps) (To avoid demagnetization)
01	18.9	14.0	.42	1.75	60
02	36.8	27.3	1.40	5.95	28
03	60.7	45.0	3.70	14.00	18
04	82.0	60.6	7.70	22.40	13
05	102.5	75.9	9.20	35.00	11

*Add 0.30 Ohms for Terminal Resistance

SPEED TORQUE CURVES

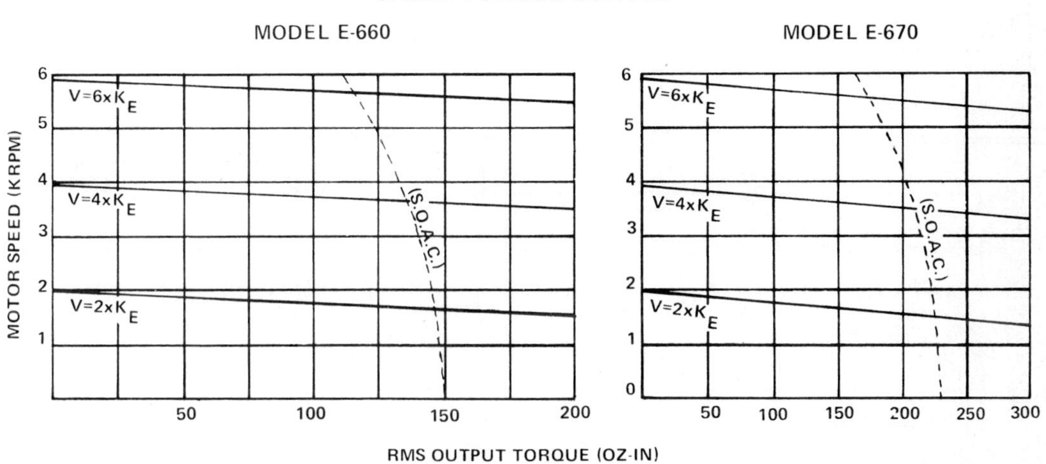

V = Terminal Voltage

The S.O.A.C. line represents the safe operating area, continuous with unit mounted on 12"x12"x½ Heat Sink.

E-701/E-703

The E-701-E-703 series permanent magnet D.C. motors are designed to be used in a broad range of industrial servo applications where high torque output at low speeds, or high pulse torque output is particularly useful. Applications include direct drive for reel motors in tape transports and web handling systems, material handling systems and machine tool drives. Shown in the photos are the standard square flange (above) end cap and the optional "in line" round end cap.

SERIES MOTOR RATINGS

Maximum Rated Armature Temp. 155°C.

Rated Brush Life (Minimum) 5,000 HRS @ 1 KRPM

Maximum Friction Torque 15 oz.-in.

Maximum Shaft Radial Load 30 lbs., 1 in. from front bearing, continuous.

CONSTRUCTION FEATURES

Winding Insulation Class F.
ABEC Class 1 Ball Bearings
PVC Lead Insulation
Totally Enclosed Housing
Available with Integrally Mounted Tachometers, ECC E-596 or M-110
Consult Electro-Craft Factory for Special Shaft and Mounting Configurations

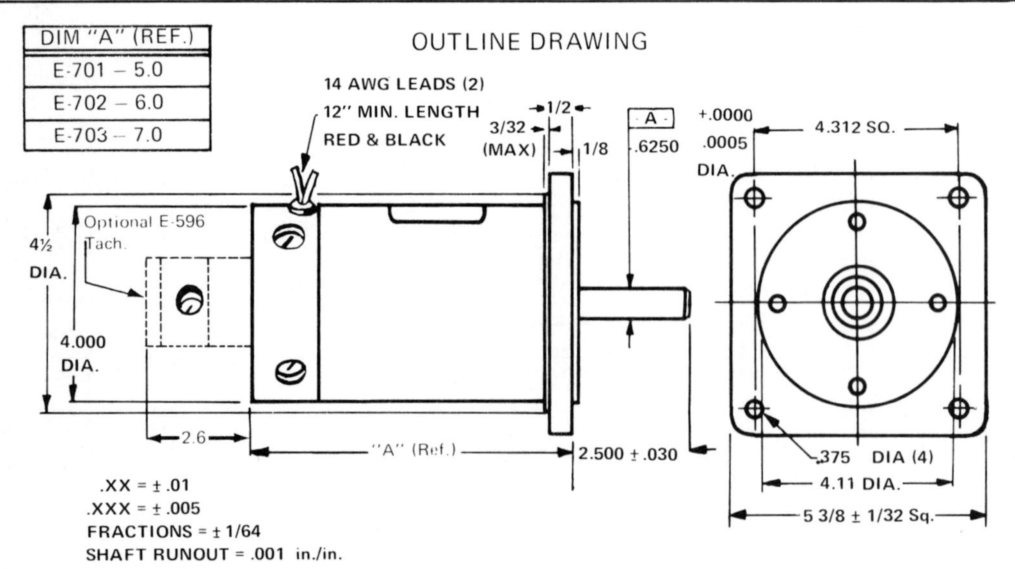

DIM "A" (REF.)
E-701 – 5.0
E-702 – 6.0
E-703 – 7.0

CONSULT FACTORY FOR SPECIAL SHAFT AND MOUNTING CONFIGURATIONS

NOMINAL SPECIFICATIONS AND RATINGS

Model E-701
Max. Continuous Stall Torque (oz-in) 150
Armature Inertia (oz.in.sec.^{2}) 0.1
Rotational Losses (oz.in./KRPM) 5
Thermal Resistance (^{o}C./Watt) 2.2
(Armature-to-ambient)
Elect. Time Constant (msec) 2.2
Mech. Time Constant (msec) 19.5

Winding Number	Torque Constant (K_T) (oz.in./amp) ± 10%	Voltage Constant (K_E) (Volts/KRPM) ± 10%	Winding Res. (Ohms)* at 25°C. ± 15%	Armature Inductance (millihenries)	Max. Pulse Current (Amps) (To avoid demagnetization)
01	10.7	8	0.16	0.35	100
02	21.6	16	0.64	1.92	50
03	32.4	24	1.52	3.34	33
04	43.2	32	2.60	5.72	25

Model E-702
Max. Continuous Stall Torque (oz-in) 300
Armature Inertia (oz. in.sec.^{2}) 0.15
Rotational Losses (oz.in./KRPM) 7
Thermal Resistance (^{o}C./Watt) 1.6
(Armature-to-ambient)
Elect. Time Constant (msec) 2.7
Mech. Time Constant (msec) 11

Winding Number	Torque Constant (K_T) (oz.in./amp) ± 10%	Voltage Constant (K_E) (Volts/KRPM) ± 10%	Winding Res. (Ohms)* at 25°C. ± 15%	Armature Inductance (millihenries)	Max. Pulse Current (Amps) (To avoid demagnetization)
01	25.9	19.2	0.37	1.0	71
02	53.6	39.7	1.51	4.1	34
03	79.6	59.0	3.60	9.7	23
04	107.0	79.3	5.95	16.1	17

Model E-703
Max. Continuous Stall Torque (oz-in) 410
Armature Inertia (oz.in.sec.^{2}) 0.20
Rotational Losses (oz.in/KRPM) 10
Thermal Resistance (^{o}C./Watt) 1.5
(Armature-to-ambient)
Elect. Time Constant (msec) 3
Mech. Time Constant (msec) 9

Winding Number	Torque Constant (K_T) (oz.in./amp) ± 10%	Voltage Constant (K_E) (Volts/KRPM) ± 10%	Winding Res. (Ohms)* at 25°C. ± 15%	Armature Inductance (millihenries)	Max. Pulse Current (Amps) (To avoid demagnetization)
01	24.2	17.9	0.18	0.54	110
02	48.6	36.0	0.72	2.16	55
03	77.2	57.2	1.84	5.52	35
04	100.0	74.1	3.00	9.03	26

*Add 0.15 ohms for 4 brush motor; 0.25 ohms for 2 brush motor.

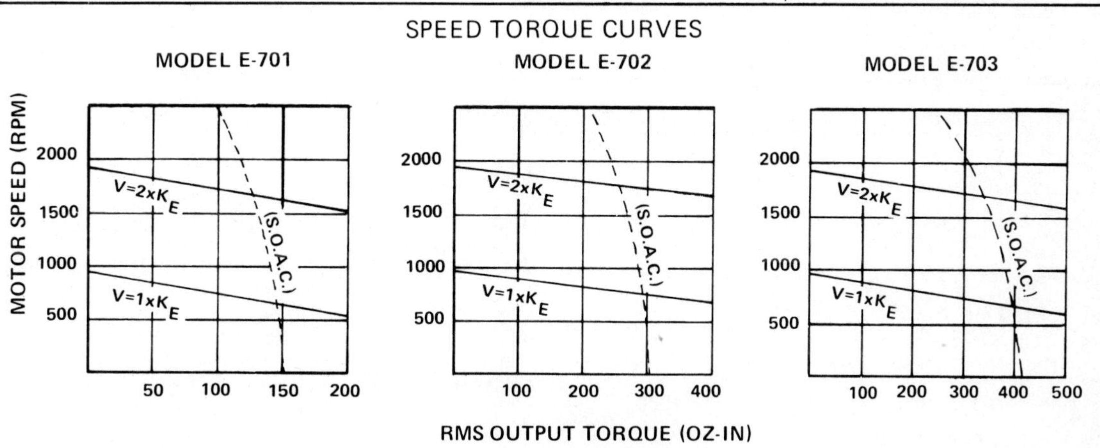

V=Terminal Voltage

The S.O.A.C. line represents the safe operating area, continuous with unit mounted on 12"x12"x½ Heat Sink.

7.2. SERVOMOTOR GENERATORS

E-576

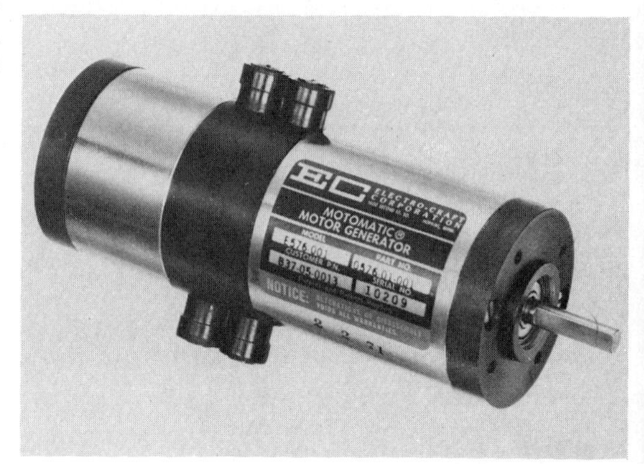

The E-576 Motor-tachometer represents a significant development in high performance dc instrument servos, designed for use in a wide range of precision velocity and positioning control systems. Inherent torque ripple or slot lock in the motor is minimized by careful design, and when used in a medium to high gain regulator type servo, velocity modulation is reduced to a level entirely determined by residual tachometer-generator ripple. The E-576 servomotor-tachometer is designed with both motor and generator armatures assembled onto the same shaft, but electrically and magnetically isolated. This motor/tach can be used as the drive elements in high response position control systems with bandwidths up to 200 Hz. Applications are found in low speed digital tape capstan servos and positioning drives in microfilm data retrieval systems.

OUTLINE DRAWING

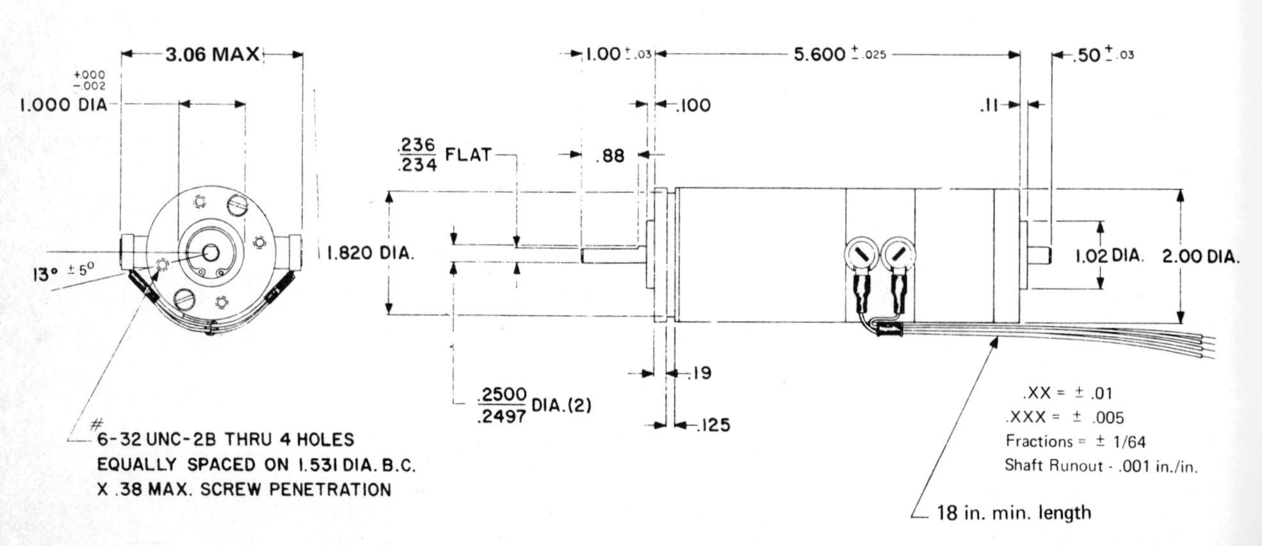

NOMINAL MOTOR SPECIFICATIONS

Max. Continuous Stall Torque (oz-in) 20
Armature Inertia (oz.-in-sec^2) .0055
Damping Factor (oz-in/KRPM) .4
Thermal Resistance (°C/Watt) 4.6
(Armature-to-ambient, with motor mounted on 10" x 10" x ¼" heat sink)

Electrical Time Constant (msec) 2.0
Mechanical Time Constant (msec) 35

Winding Number	Torque Constant (K_T) (oz. in./amp) ±10%	Voltage Constant (K_E) (Volts/KRPM) ±10%	Winding Res. (Ohms)* at 25° ± 15%	Armature Inductance (Millihenries)	Max. Pulse Current (Amps) [to avoid demagnetization]
01	5.80	4.29	1.15	2.3	24
02	7.34	5.43	2.10	4.2	19
03	9.26	6.85	3.56	7.1	15
04	11.60	8.58	5.78	11.6	12

* For Terminal Resistance add 0.40 Ohms.

NOMINAL TACHOMETER SPECIFICATIONS

Linearity .2% Maximum deviation either direction of rotation between 0 and 2000 RPM
Ripple at 1,000 RPM
5% peak to peak with 200 Hz filter
Dominant Ripple frequency 11 cycles per revolution
Temperature Stability .05% per °C maximum over temperature range of 0°C to 50°C

Winding Number	Output Voltage Gradient (V/KRPM) ±10%	Armature Resistance (ohms) (including temp. comp.) ±10%	Armature Inductance (mhy)	Optimum Load Impedance (ohms)
-01	3.0	570	6.2	5000
-02	7.0	570	33	5000
-03	14.2	720	138	5000
-04	21.0	950	255	10,000

MOTOR-TACH OPERATING LIMITS

Maximum Rated Armature Temp. 155°C
Maximum Rated Armature Power Dissipation 28 Watts
Rated Brush Life (minimum) 5,000 hrs. at 1 KRPM
Maximum No Load Speed 6 KRPM
Maximum Shaft Radial Load 10 lbs. at .5 inches from front bearing continuous for a min. 1000 hours
Maximum Friction Torque 3 oz-in.

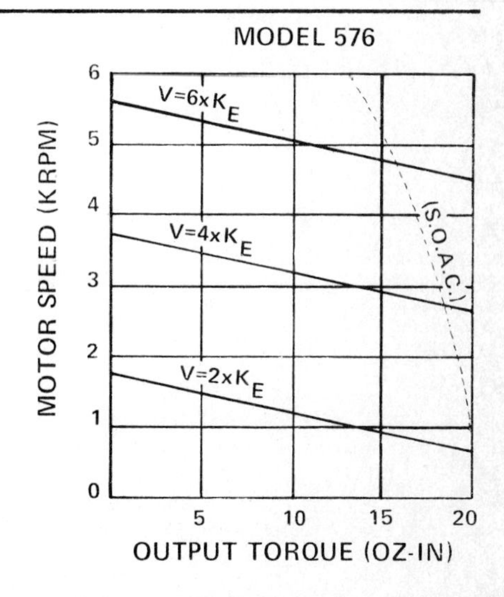

V=Terminal Voltage

The S.O.A.C. line represents the safe operating area, continuous with unit mounted on 10"x10"x1/4" Heat Sink

E-586/E-587

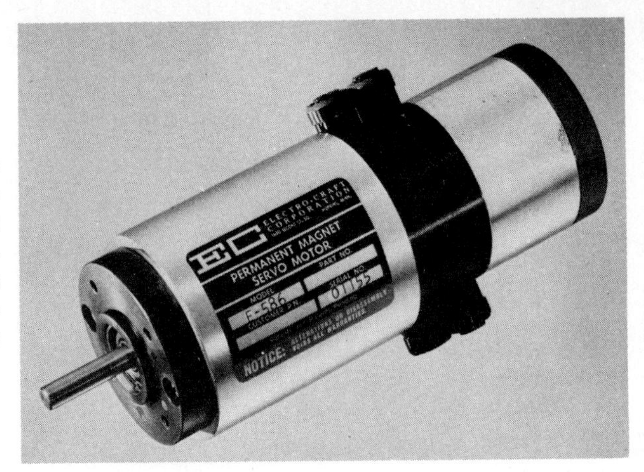

The Model E-586-E-587 motor-tachometer series is derived from the Model E-576 and is designed to add higher torque capability without increased inertia. With the increased performance and cost effectiveness, the E-586-E-587 has found applications in low speed digital tape transports, office equipment and a wide variety of velocity and position servo systems.

SERIES MOTOR RATINGS

Maximum Rated Armature Temp. 155°C.
Maximum Rated Armature Power Dissipation 28-31 Watts
Rated Brush Life (minimum) 5,000 HRS. at 1 KRPM
Maximum No Load Speed 6 KRPM
Maximum Shaft Radial Load 10 lbs at .5 inches from front bearings continuous for a min. 1000 hours
Maximum Friction Torque 3 oz.-in.

CONSTRUCTION FEATURES

Winding Insulation Class F.
ABEC Class 1 Ball Bearings
PVC Lead Insulation
Totally Enclosed Housing

Consult Electro-Craft Factory for Special Shaft and Mounting Configurations.

OUTLINE DRAWING

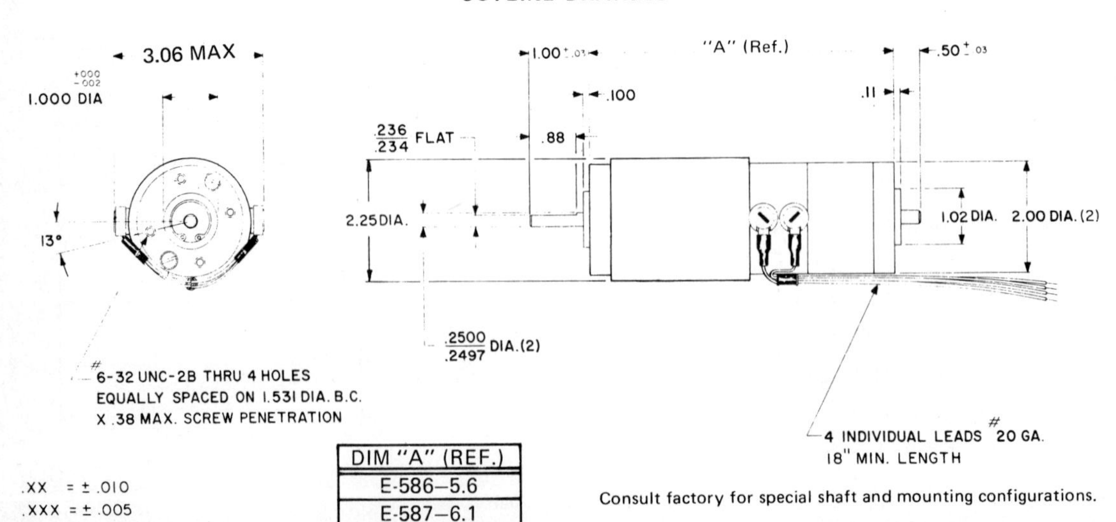

DIM "A" (REF.)
E-586–5.6
E-587–6.1

.XX = ± .010
.XXX = ± .005
SHAFT RUNOUT .001 IN./IN.

Consult factory for special shaft and mounting configurations.

NOMINAL SPECIFICATIONS AND RATINGS

	Winding Number	Torque Constant (K_T) (oz.in./amp) ± 10%	Voltage Constant (K_E) (Volts/KRPM) ± 10%	Winding Res. (Ohms)* at 25°C. ± 15%	Armature Inductance (millihenries)	Max. Pulse Current (Amps) (To avoid demagnetization)
Model E-586 Max. Continuous Stall Torque (oz-in) 30 Armature Inertia (oz.in.sec.^{2}) .0055 Damping Factor (oz.in./KRPM) 0.4						
Thermal Resistance (°C./Watt) 4.6	01	9.20	6.81	1.36	3.63	26
(Armature-to-ambient)	02	11.65	8.63	2.27	6.43	21
Elect. Time Constant (msec) 2.06	03	14.72	10.90	4.33	9.74	16
Mech. Time Constant (msec) 16	04	18.40	13.63	6.81	14.85	13
Model E-587 Max. Continuous Stall Torque (oz-in) 38 Armature Inertia (oz.in.sec.^{2}) .0067 Damping Factor (oz.in./KRPM) 0.4						
Thermal Resistance (°C./Watt) 4.1	01	11.79	8.73	1.51	4.01	23
(Armature-to-ambient)	02	14.94	11.06	2.48	6.05	18
Elect. Time Constant (msec) 2.1	03	18.87	13.97	4.00	8.40	14
Mech. Time Constant (msec) 13	04	23.87	17.68	6.35	13.34	11

*Add 0.04 ohms for terminal resistance.

NOMINAL TACHOMETER SPECIFICATIONS

	Winding Number	Output Voltage Gradient (V/KRPM) ± 10%	Armature Resistance (ohms) (including temp. comp.) ± 10%	Armature Inductance (mhy)	Optimum Load Impedance (ohms)
Linearity .2% Maximum deviation either direction of rotation between 0 and 2000 RPM Ripple at 1,000 RPM 5% peak to peak with 200 Hz filter					
Dominant Ripple frequency 11 cycles per revolution	–01	3.0	570	6	5000
Termperature Stability .05% per °C	–02	7.0	570	33	5000
maximum over temperature range	–03	14.2	720	138	5000
of 0°C to 50°C	–04	21.0	950	255	10,000

SPEED TORQUE CURVES

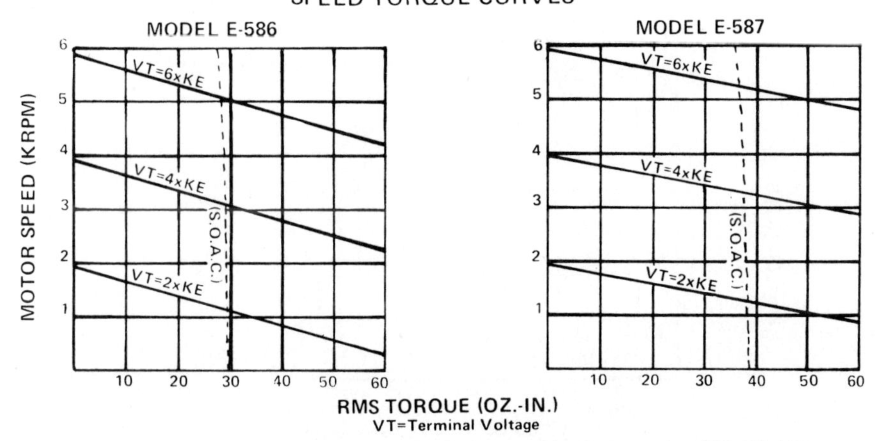

The S.O.A.C. line represents the safe operating area, continuous with unit mounted on 10"x10"x¼" heat sink.

S-586 Metric Motor-Tachometer

The model S-586 motor-tachometer is substantially the same as the E-586 motor-tachometer except that the shaft and certain other physical dimensions are metric standards. These motors will be available in small quantities from stocks located in the United States and Europe.

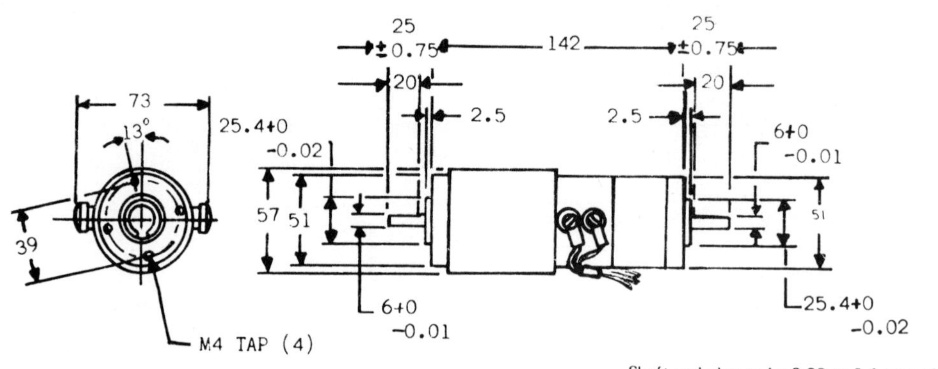

ALL DIMENSIONS IN MILLIMETRES

Shaft runout 0.008 mm/mm.
Shaft ⊥ to mounting surface within 0.127 TIR.

Shaft end play to be 0.06 to 0.4 mm with 5 kp reversing load.
Shaft radial play not to exceed 0.04 mm when measured 20 mm from front of pilot with 2.5 kp reversing load applied 6 mm from pilot.

S-586

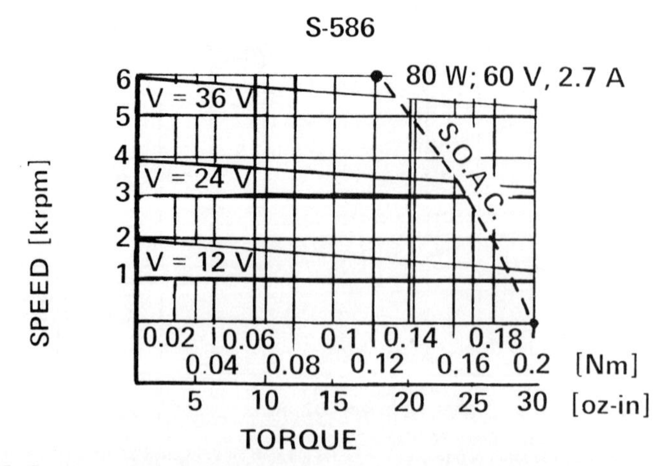

The S.O.A.C. line represents the safe operating area, continuous operation.

NOMINAL MOTOR-TACH RATINGS	UNITS BRITISH	SI (METRIC)
Output Torque	32 oz-in	0.225 Nm
Max. Speed	5000 rpm	523 rad/s
Armature Moment of Inertia	0.0055 oz-in-s^2	3.88 x 10^{-5} kg m^2
Max. Armature Temperature	155 °C	155 °C
Thermal Resistance — Armature to Ambient (Motor Heatsink Mounted)	4.6 °C/W	4.6 °C/W
Mechanical Time Constant	12 ms	12 ms
Static Friction Torque	3 oz-in max.	0.021 Nm max.
MOTOR SPECIFICATIONS		
Torque Constant K_T	7.9 ± 10% oz-in/A	0.056 ± 10% Nm/A
Voltage Constant K_E	5.85 ± 10%V/krpm	0.056 ± 10% V/rad·s^{-1}
Terminal Resistance R	1.25 ± 15% Ω	1.25 ± 15% Ω
Armature Inductance L_a	2.6 mH	2.6 mH
Electrical Time Constant τ_e	2.06 ms	2.06 ms
Max. Pulse Current	24 A	24 A
Damping Factor K_d	0.1 oz-in/krpm	0.0067 Nm/rad·s^{-1}
Maximum Armature Temperature	155 °C	155 °C
TACHOMETER SPECIFICATIONS		
Output Voltage Gradient	14.2 ± 10% V/krpm	0.136 ± 10% V/rad·s^{-1}
Armature Reisstance	750 ± 10% Ω	750 ± 10% Ω
Armature Inductance	138 mH	138 mH
Ripple Frequency	11 cycles/rev.	11 cycles/rev.
Temperature Coefficient	–0.05%/°C maximum	–0.05%/°C maximum

Ripple: 5% maximum peak to peak AC component in DC output voltage at 500 rpm.

Linearity: 0.2% maximum deviation from perfect linearity.

SUGGESTED TACHOMETER CIRCUIT

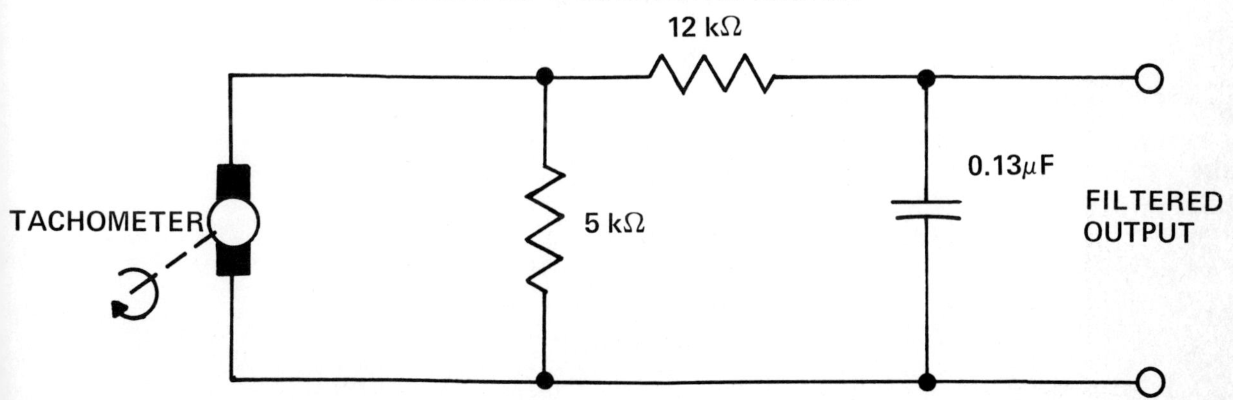

S-670/596 Metric Motor-Tachometer

The model S-670/596 motor-tachometer is substantially the same as the E-670/596 motor-tachometer except that the shaft and certain other physical dimensions are metric stardards. These motors will be available in small quantities from stocks located in the United States and Europe.

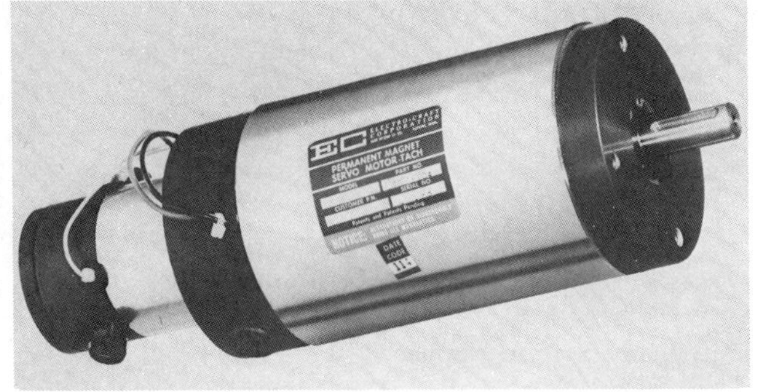

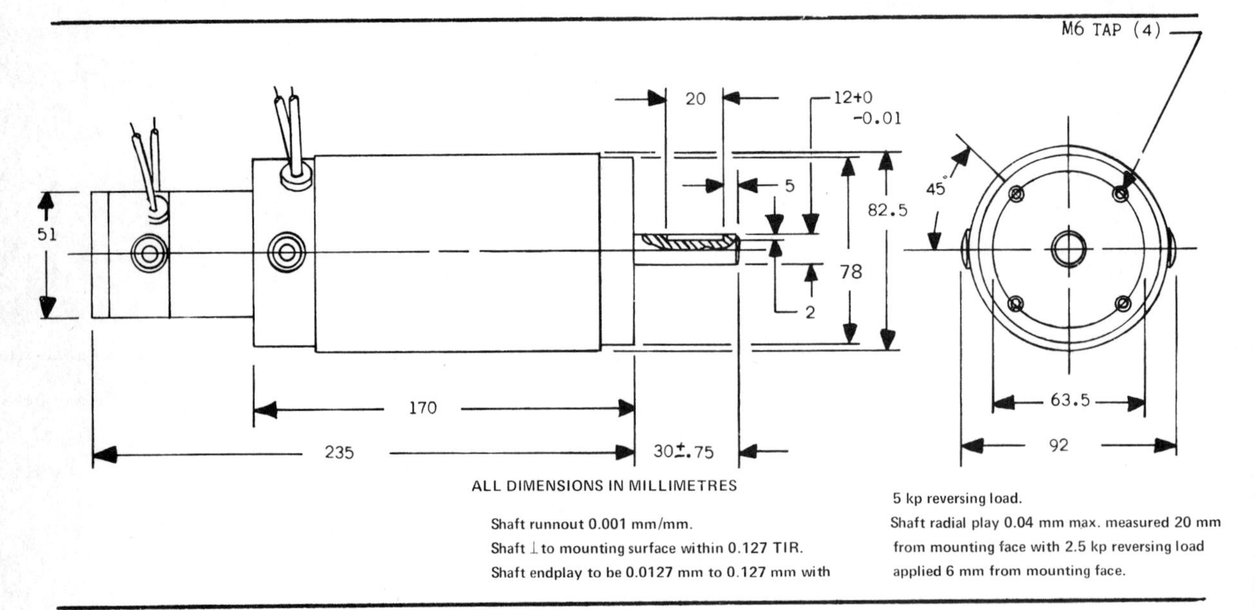

ALL DIMENSIONS IN MILLIMETRES

Shaft runnout 0.001 mm/mm.
Shaft ⊥ to mounting surface within 0.127 TIR.
Shaft endplay to be 0.0127 mm to 0.127 mm with 5 kp reversing load.
Shaft radial play 0.04 mm max. measured 20 mm from mounting face with 2.5 kp reversing load applied 6 mm from mounting face.

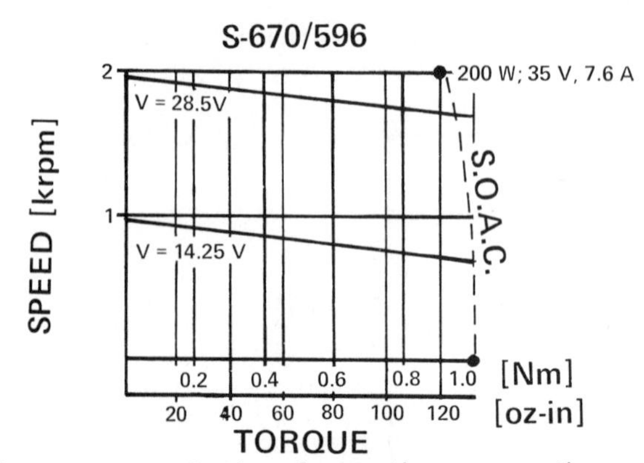

The S.O.A.C. curve represents the safe operating area, continuous operation.

NOMINAL MOTOR-TACH RATINGS	UNITS BRITISH	SI (METRIC)
Output Torque	150 oz-in	1.05 Nm
Maximum Speed	3000 rpm	314 rad/s
Armature Moment of Inertia	0.05 oz-in-s^2	3.53×10^{-4} kg m^2
Maximum Armature Temperature	155 °C	155 °C
Thermal Resistance-Armature to Ambient (Motor Heatsink Mounted)	2.1 °C/W	2.1 °C/W
Mechanical Time Constant	7.5 ms	7.5 ms
Static Friction Torque	8 oz-in	0.056 Nm

MOTOR SPECIFICATIONS		
Torque Constant K_T	18.9 ± 10% oz-in/A	0.133 Nm/A
Back emf Constant K_E	14 ± 10% V/krpm	0.133 ± 10% V/rad·s^{-1}
Terminal Resistance R	0.65 ± 15% Ω	0.65 ± 15% Ω
Armature Inductance L_a	1.91 mH	1.91 mH
Electrical Time Constant τ_e	2.2 ms	2.2 ms
Max. Pulse Current	54 A	54 A
Damping Factor K_d	0.68 oz-in/krpm	0.045 Nm/rad·s^{-1}
Maximum Armature Temperature	155 °C	155 °C

TACHOMETER SPECIFICATIONS		
Output Voltage Gradient	14.2 ± 10% V/krpm	0.136 ± 10% V/rad·s^{-1}
Armature Resistance	750 ± 10% Ω	750 ± 10% Ω
Armature Inductance	138 mH	138 mH
Ripple Frequency	11 cycles/rev.	11 cycles/rev.
Temperature Coefficient	–0.05%/°C maximum	–0.05%/°C maximum

Ripple: 5% maximum peak to peak AC component in DC output voltage at 500 rpm.

Linearity: 0.2% maximum deviation from perfect linearity.

SUGGESTED TACHOMETER CIRCUIT

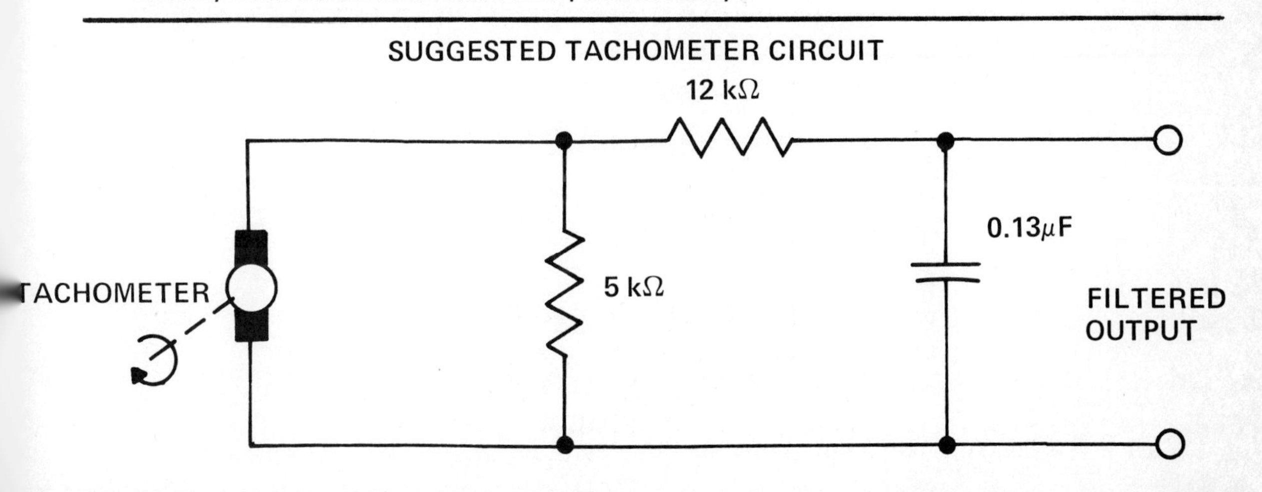

S-703/596 Metric Motor-Tachometer

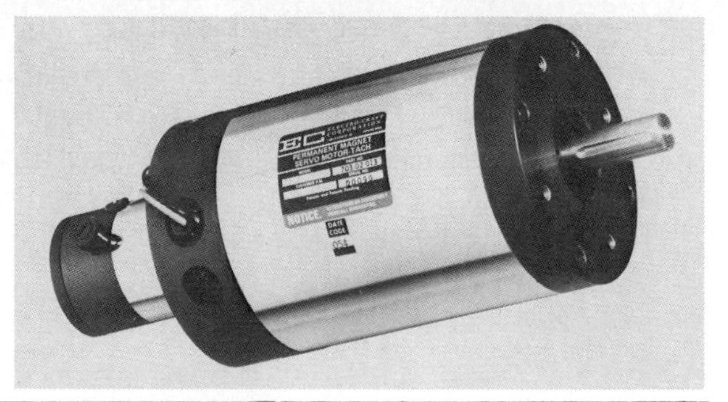

The model S-703/596 motor-tachometer is substantially the same as the E-703/596 motor-tachometer except that the shaft and certain other physical dimensions are metric standards. These motors will be available in small quantities from stocks located in the United States and Europe.

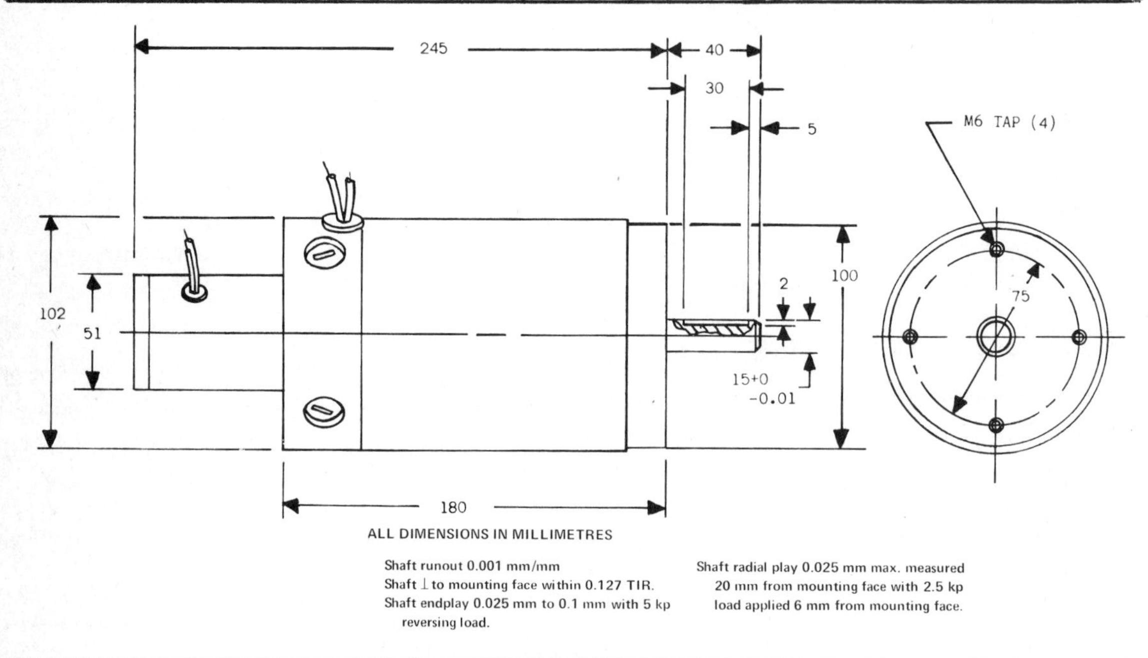

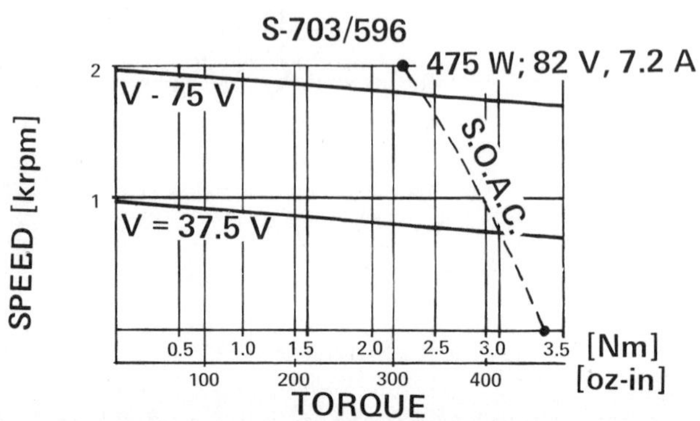

The S.O.A.C. curve represents the safe operating area, continuous operation.

S-703/596

NOMINAL MOTOR-TACH RATINGS	UNITS BRITISH	SI (METRIC)
Output Torque	400 oz-in	2.8 Nm
Max. Speed	2000 rpm	209 rad/s
Armature Moment of Inertia	0.2 oz-in-s^2	1.41×10^{-3} kg m^2
Max. Armature Temperature	155 ^oC	155 ^oC
Thermal Resistance—Armature to Ambient (Motor Heatsink Mounted)	1.2 $^oC/W$	1.2 $^oC/W$
Mechanical Time Constant	9.4 ms	9.4 ms
Static Friction Torque	14 oz-in	0.1 Nm

MOTOR SPECIFICATIONS		
Torque Constant K_T	49 ± 10% oz-in/A	0.345 Nm/A
Back emf Constant K_E	36.3 ± 10% V/krpm	0.345 V/rad·s^{-1}
Terminal Resistance R	0.85 ± 15% Ω	0.85 ± 15% Ω
Armature Inductance L_a	4.4 mH	4.4. mH
Electrical Time Constant τ_e	4.5 ms	4.5 ms
Max. Pulse Current	42 A	42 A
Damping Factor K_d	10 oz-in/krpm	0.67 Nm/rad·s^{-1}
Maximum Armature Temperature	155 ^oC	155 ^oC

TACHOMETER SPECIFICATIONS		
Output Voltage Gradient	14.2 ± 10% V/krpm	0.136 ± 10%V/rad·s^{-1}
Armature Resistance	750 ± 10% Ω	750 ± 10% Ω
Armature Inductance	138 mH	138 mH
Ripple Frequency	11 cycles/rev.	11 cycles/rev.
Temperature Coefficient	–0.05%/^oC maximum	–0.05%/^oC maximum

Ripple: 5% maximum peak-to-peak AC component in DC output voltage at 500 rpm.

Linearity: 0.2% maximum deviation from perfect linearity.

SUGGESTED TACHOMETER CIRCUIT

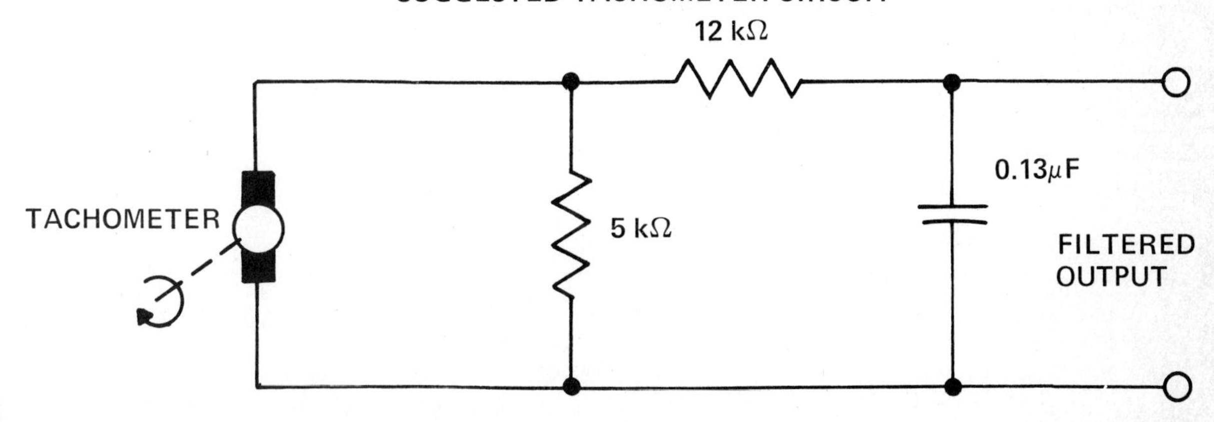

E-530/E-532

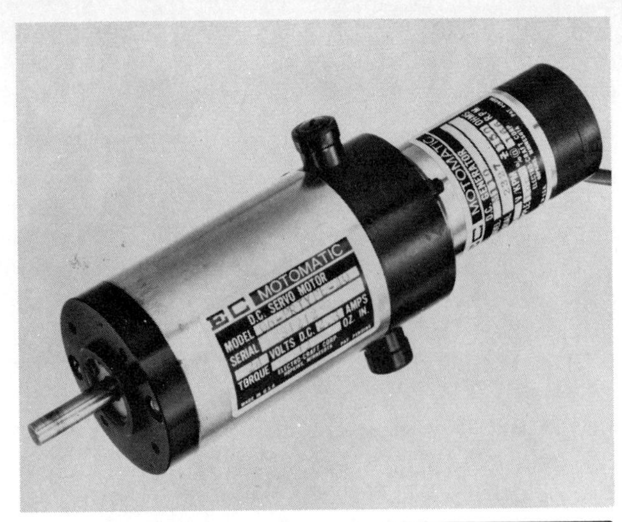

The E-530-E-532 motor-tachometer series is designed to offer high performance and low cost through the use of a moving coil tachometer for low system inertia and high quality low ripple tachometer output. The added responsiveness makes the E-530 series an excellent performer in applications such as digital tape transport capstans up to 45 ips and a variety of position servo systems.

SERIES MOTOR RATINGS

Maximum Rated Armature Temp. 155°C.

Maximum No Load Speed 6 KRPM

Rated Brush Life (Minimum) 5,000 HRS @ 1 KRPM.

Maximum Friction Torque 3 oz.-in.

Maximum Shaft Radial Load 10 lbs., .5 in. from front bearing continuous.

CONSTRUCTION FEATURES

Winding Insulation Class F.
ABEC Class 1 Ball Bearings
PVC Lead Insulation
Totally Enclosed Housing
Integrally Coupled Tachometer
NOTE: Refer to Electro-Craft publication PDS-804C for M-110 Tachometer Specifications.

OUTLINE DRAWING

DIM "A" (REF.)
E-530–4.0
E-531–4.5
E-532–5.0

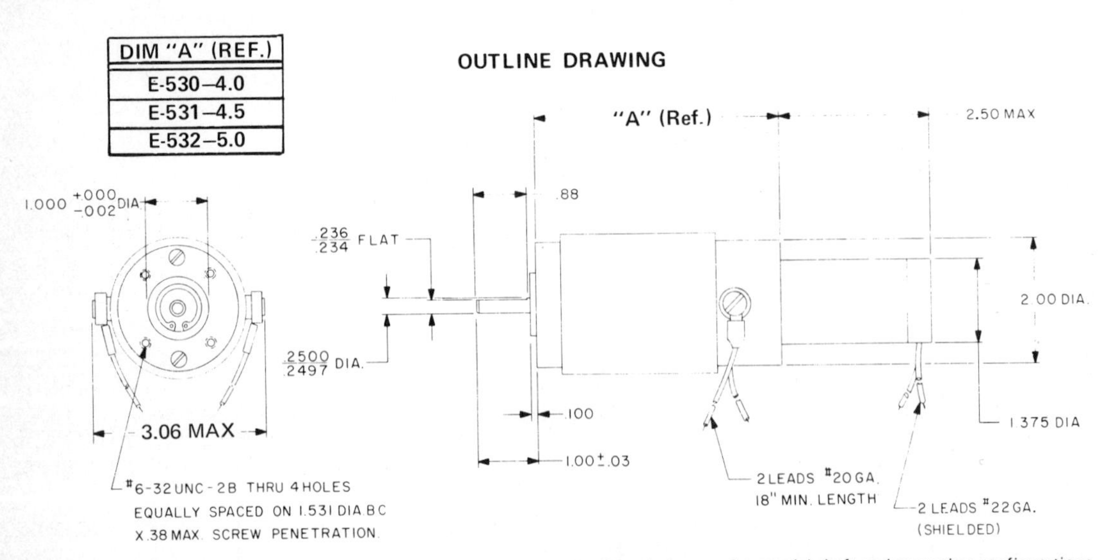

Consult factory for special shaft and mounting configurations.

NOMINAL SPECIFICATIONS AND RATINGS

Model E-530

Max. Continuous Stall Torque (oz-in) 29
Armature Inertia (oz.in.sec.^{2}) .0039
Damping Factor (oz.in./KRPM) .1
Thermal Resistance (^{o}C./Watt) 5.0
(Armature-to-ambient)
Elect. Time Contstant (msec) 2.06
Mech. Time Constant (msec) 8.3

Winding Number	Torque Constant (K_T) (oz.in./amp) ± 10%	Voltage Constant (K_E) (Volts/KRPM) ±10%	Winding Res. (Ohms)* at 25°C. ± 15%	Armature Inductance (millihenries)	Max. Pulse Current (Amps) (To avoid demagnetization)
01	9.20	6.81	1.36	3.63	26
02	11.65	8.63	2.27	6.43	21
03	14.72	10.90	4.33	9.74	16
04	18.40	13.63	6.81	14.85	13

Model E-531

Max. Continuous Stall Torque (oz-in) 38
Armature Inertia (oz.in.sec.^{2}) .0051
Damping Factor (oz.in./KRPM) .1
Thermal Resistance (^{o}C./Watt) 4.59
(Armature-to-ambient)
Elect. Time Constant (msec) 2.1
Mech. Time Constant (msec) 9.7

Winding Number	Torque Constant (K_T) (oz.in./amp) ± 10%	Voltage Constant (K_E) (Volts/KRPM) ±10%	Winding Res. (Ohms)* at 25°C. ± 15%	Armature Inductance (millihenries)	Max. Pulse Current (Amps) (To avoid demagnetization)
01	11.79	8.73	1.51	4.01	23
02	14.94	11.06	2.48	6.05	18
03	18.87	13.97	4.0	8.40	14
04	23.87	17.68	6.35	13.34	11

Model E-532

Max. Continuous Stall Torque (oz-in) 50
Armature Inertia (oz.in.sec.^{2}) .0063
Damping Factor (oz.in./KRPM) .2
Thermal Resistance (^{o}C./Watt) 4.18
(Armature-to-ambient)
Elect. Time Constant (msec) 2.15
Mech. Time Constant (msec) 8.5

Winding Number	Torque Constant (K_T) (oz.in./amp) ± 10%	Voltage Constant (K_E) (Volts/KRPM) ±10%	Winding Res. (Ohms)* at 25°C. ± 15%	Armature Inductance (millihenries)	Max. Pulse Current (Amps) (To avoid demagnetization)
01	14.86	11.00	1.53	4.15	21
02	18.82	13.94	3.07	7.46	16
03	23.78	17.61	4.87	11.33	13
04	29.72	22.00	7.67	17.35	10

*Add 0.40 ohms for terminal resistance.

SPEED TORQUE CURVES

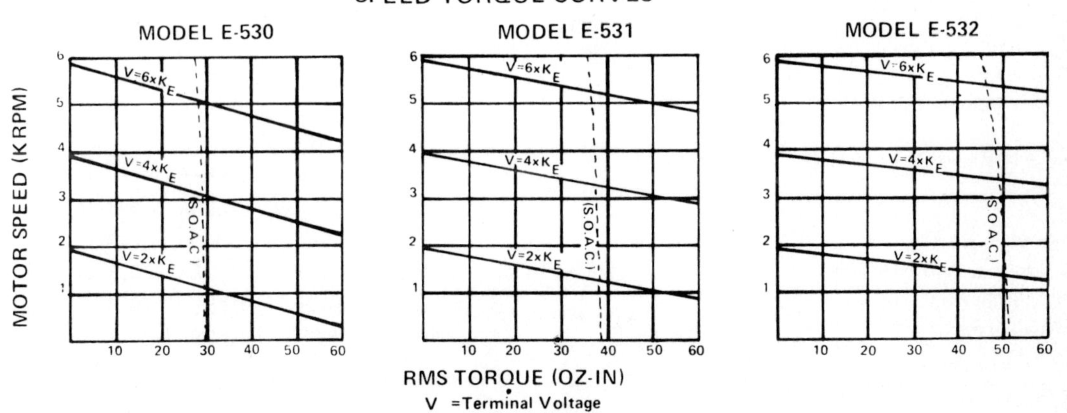

The S.O.A.C. line represents the safe operating area, continuous with unit mounted on 10"x10"x¼" heat sink.

ET-4000

The ET-4000 is a high performance permanent magnet servomotor which combines many of the characteristics of a "torque" motor with those of a low-inertia moving coil type. The motor as excellent regulation and low inherent ripple torque. It is therefore ideally suited for use in applications such as low-speed digital and incremental tape capstan and reel servo drives, film handling systems, and precision positioning servomechanisms.

SPECIFICATIONS

MOTOR RATINGS

Maximum Safe Speed (No Load)	6 krpm
Rated Torque (continuous at Stall)	25 oz.-in.
Maximum Rated Armature Power Dissipation	32 W
Maximum Rated Armature Temperature	155°C
Thermal Resistance (Armature-to-ambient)	**4.0 °C/W**
Armature Moment of Inertia	.0033 oz-in-sec^2
Electrical Time Constant	1.8 ms
Mechanical Time Constant	15 ms
Damping Factor	1.0 oz-in/krpm
Static Friction Torque	3.0 oz-in.

WINDING VARIATIONS

WINDING NUMBER	Torque Constant K_T (oz-in/A) ± 10%	Voltage Constant K_E (V/krpm) ± 10%	Armature Winding* (Ω) ± 15%	Armature Inductance (mH)	Max. Pulse Torque (oz-in) (1 sec)
00	5.6	4.1	0.85	1.8	75
01	8.0	5.9	1.45	2.9	75
02	10.0	7.4	2.25	4.3	75
03	20.0	14.8	10.85	19.8	75
04	40.0	29.6	43.85	79.2	75

*Add 0.4 ohms for terminal resistance.

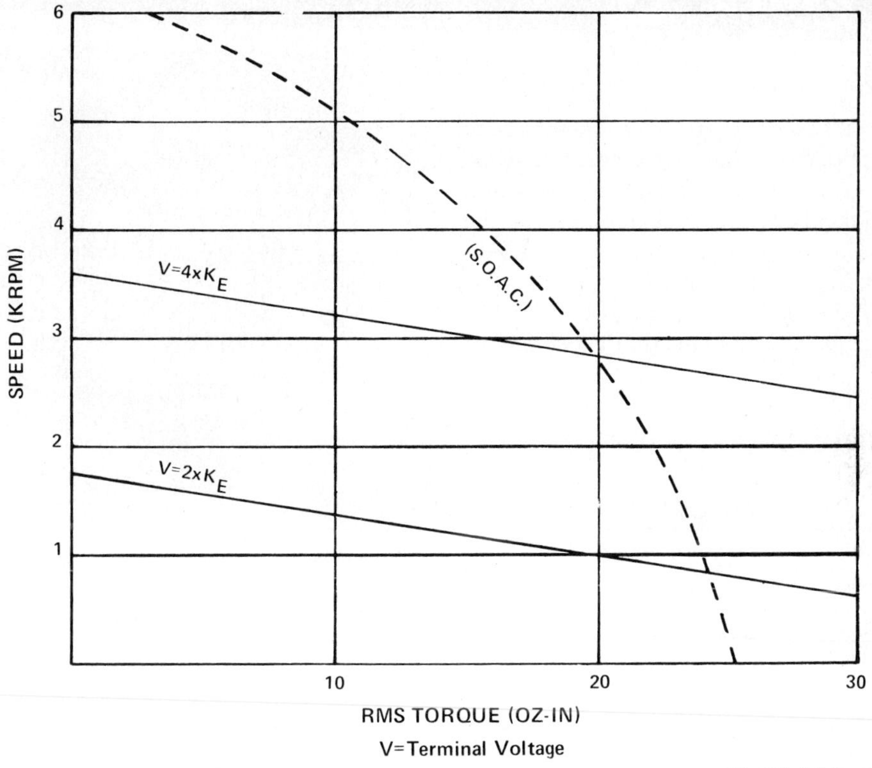

RMS TORQUE (OZ-IN)

V=Terminal Voltage

The S.O.A.C. line represents the safe operating area, continuous with unit mounted on 10"x10"x¼" Heat Sink.

OUTLINE DRAWING

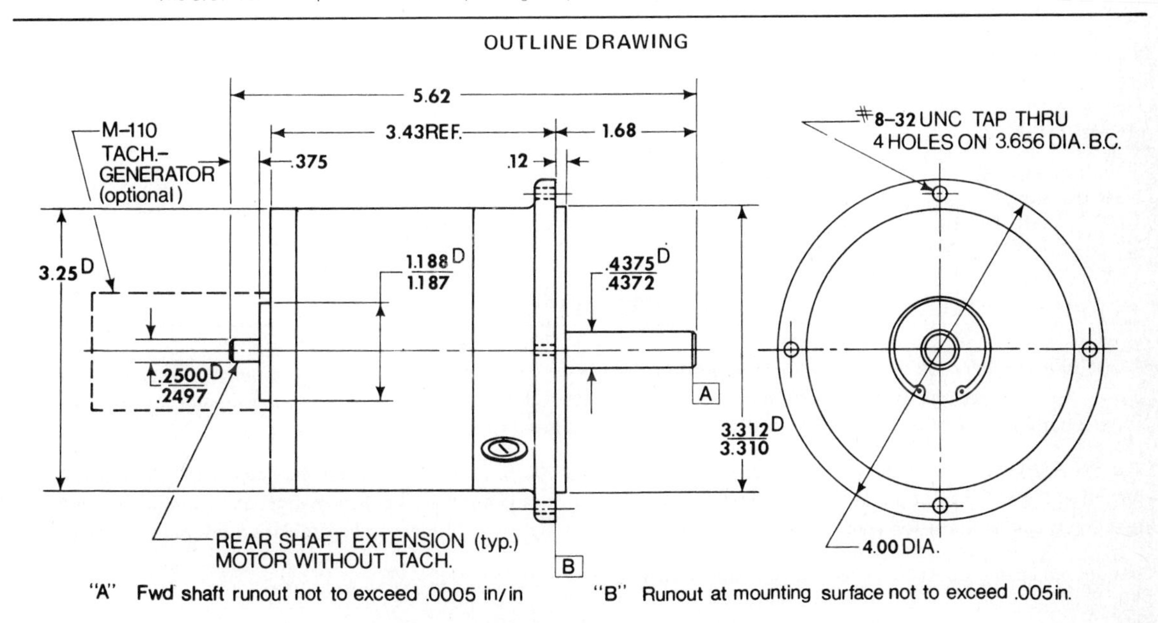

"A" Fwd shaft runout not to exceed .0005 in/in "B" Runout at mounting surface not to exceed .005in.

Note: Nominal ".500" diam. fwd. shaft available.

7.3. MOVING COIL MOTORS

M-1030

The Electro-Craft Model M-1030 is an intermediate performance D.C. moving coil servomotor designed for maximum economy over a broad range of applications. A significant range extension can be achieved by using forced air cooling due to the very efficient air baffle arrangement on this motor. The unit shown on the right has an Electro-Craft M-110 moving coil tachometer mounted on the rear end of the shaft.

SPECIFICATIONS

MOTOR RATINGS		Uncooled	W/Air Cooling (20 cfm)
Maximum Safe Speed (no load)		6 krpm	12 krpm
Maximum Speed at Rated Load		4 krpm	4 krpm
Rated Torque (continuous at stall)		32 oz-in	65 oz-in
Rated Power Output		70 W	170 W
Power Rate (continuous)		7.3 KW/sec	42 KW/sec
Power Rate (peak)		1570 KW/sec	1570 KW/sec
No Load Acceleration at Maximum Continuous Torque		43700 rad/sec	105000 rad/sec
No Load Acceleration at Maximum Pulse Torque		640000 rad/sec	640000 rad/sec
Maximum Rated Armature Temp.		155 °C	155 °C
Maximum Static Friction Torque		2.5 oz-in	2.5 oz-in
Electrical Time Constant		0.1 ms	0.1 ms
Mechanical Time Constant		2.3 ms	2.3 ms
Damping Factor		1 oz-in/krpm	1 oz-in/krpm
Armature Moment of Inertia (including bearings)			
1/2 in Dia. Shaft, Mtr. Only	0.00068	oz-in-s^2	
1/2 in Dia. Shaft, Mtr.-Tach	0.00075	oz-in-s^2	
3/8 in Dia. Shaft, Mtr.-Tach	0.00054	oz-in-s^2	
3/8 in Dia. Shaft, Mtr.-Tac	0.00061	oz-in-s^2	

WINDING VARIATIONS

WINDING NUMBER	Torque Constant K_T (oz-in/A) ± 10%	Voltage Constant K_E (V/krpm) ± 10%	Armature Winding* (Ω) ± 15%	Armature Inductance (mH)	Max. Pulse Current (A)
–00	5.8	4.3	0.7	0.09	60
–01	8.3	6.1	1.95	0.21	42
–02	11.6	8.6	3.35	0.35	30

*Add .15 ohms for terminal resistance.

MOTOR LIFE

Many users naturally ask what operational and brush life can be expected from a moving coil motor. There is no pat answer to this question as numerous application factors have a direct impact on motor life. Among these are peak current during acceleration, duty cycle, direction reversals, load inertia and method of heat removal.

For the M-1030, operated under normal conditions, a life expectancy of 1.5×10^9 stop-start cycles at 100 cycles per second or 2×10^8 revolutions at 1500 RPM with 15 oz-in. output can be achieved before maintenance is required. A life prediction can be made for your application after all pertinent factors are known and a simulation test has been made.

Let our application specialists examine your requirements.

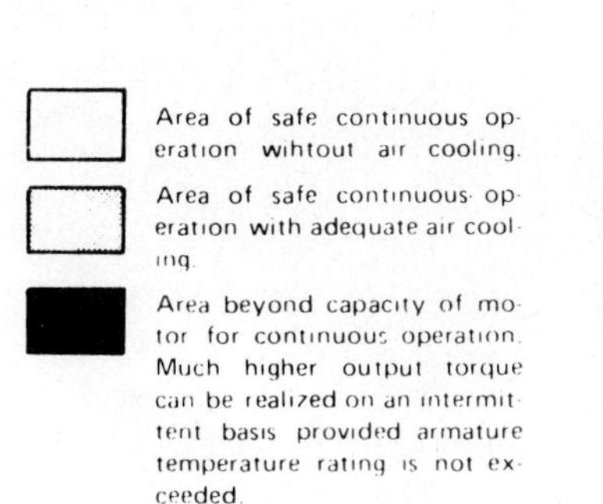
Area of safe continuous operation wihtout air cooling.
Area of safe continuous operation with adequate air cooling.
Area beyond capacity of motor for continuous operation. Much higher output torque can be realized on an intermittent basis provided armature temperature rating is not exceeded.

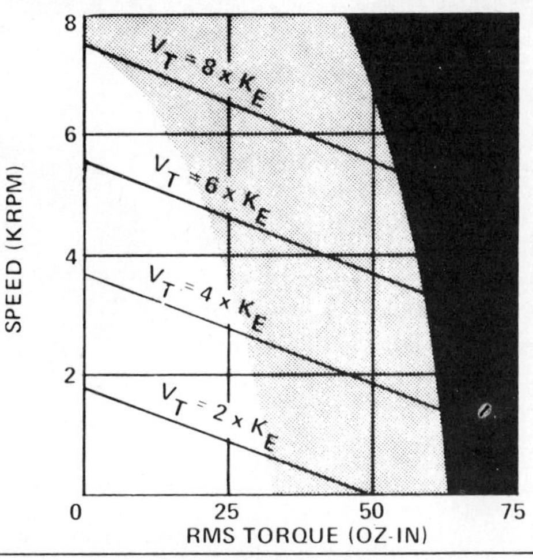
SPEED (KRPM)
RMS TORQUE (OZ-IN)
$V_T = 8 \times K_E$
$V_T = 6 \times K_E$
$V_T = 4 \times K_E$
$V_T = 2 \times K_E$
0
25
50
75
2
4
6
8

MODEL M-1030 THERMAL DATA

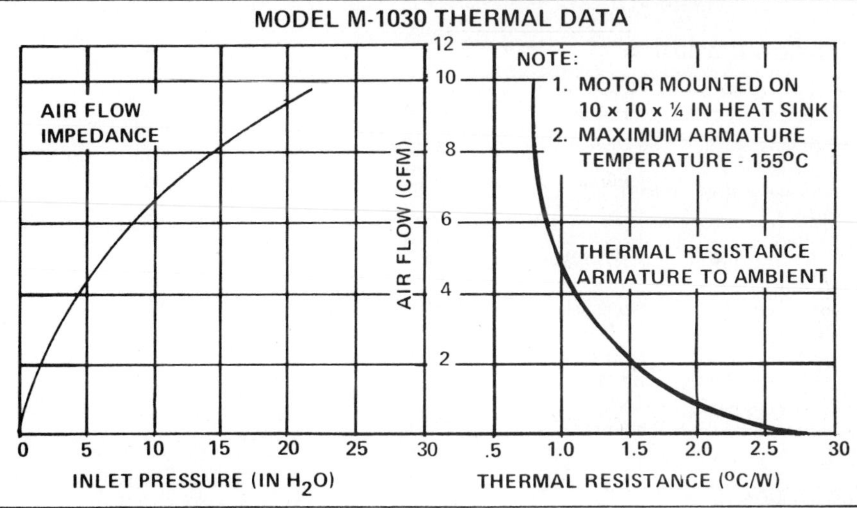
AIR FLOW IMPEDANCE
AIR FLOW (CFM)
NOTE:
1. MOTOR MOUNTED ON 10 x 10 x ¼ IN HEAT SINK
2. MAXIMUM ARMATURE TEMPERATURE - 155°C
THERMAL RESISTANCE ARMATURE TO AMBIENT
INLET PRESSURE (IN H_2O)
THERMAL RESISTANCE (°C/W)
0
5
10
15
20
25
30
.5
1.0
1.5
2.0
2.5
30
2
4
6
8
10
12

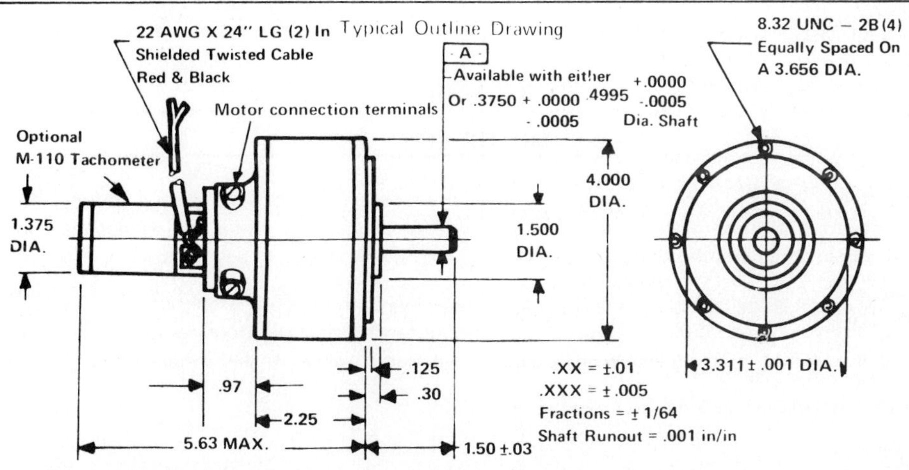
Typical Outline Drawing
22 AWG X 24" LG (2) In Shielded Twisted Cable Red & Black
Motor connection terminals
Optional M-110 Tachometer
A
Available with either .4995 +.0000 -.0005 Dia. Shaft
Or .3750 + .0000 - .0005
8.32 UNC – 2B (4) Equally Spaced On A 3.656 DIA.
4.000 DIA.
1.500 DIA.
1.375 DIA.
.97
2.25
5.63 MAX.
.125
.30
1.50 ±.03
3.311 ± .001 DIA.
.XX = ±.01
.XXX = ± .005
Fractions = ± 1/64
Shaft Runout = .001 in/in

M-1040

The Electro-Craft Model M-1040 is a medium performance DC moving coil servomotor designed to fill the gap between low cost iron core motors and the higher cost moving coil motors. A significant range extension can be achieved by using forced air cooling due to the very efficient air baffle arrangement on this motor.

SPECIFICATIONS

MOTOR RATINGS		Uncooled	W/Air Cooling (24 cfm)
Maximum Safe Speed (no load)		6 krpm	12 krpm
Maximum Speed at Rated Load		4 krpm	4 krpm
Rated Torque (continuous at stall)		35 oz-in	75 oz-in
Rated Power Output		80 W	200 W
Power Rate (continuous)		9.0 KW/sec	58 KW/sec
Power Rate (peak)		1570 KW/sec	1570 KW/sec
No Load Acceleration at Maxium Continuous Torque		49000 rad/sec^2	124000 rad/sec
No Load Acceleration at Maximum Pulse Torque		640000 rad/sec^2	640000 rad/sec^2
Maximum Rated Armature Temp.		155 °C	155 °C
Maximum Static Friction Torque		2.5 oz-in	2.5 oz-in
Electrical Time Constant		0.1 ms	0.1 ms
Mechanical Time Constant		2.3 ms	2.3 ms
Damping Factor		1.0 oz-in/krpm	1.0 oz-in/krpm
Armature Moment of Inertia (including bearings)			
1/2 in Dia. Shaft, Mtr. Only	0.00058	oz-in-s^2	
1/2 in Dia. Shaft. Mtr.-Tach	0.00072	oz-in-s^2	
3/8 in Dia. Shaft. Mtr. Only	0.00048	oz-in-s^2	
3/8 in Dia. Shaft, Mtr.-Tach	0.00055	oz-in-s^2	

WINDING VARIATIONS

WINDING NUMBER	Torque Constant K_T (oz-in/A) ± 10%	Voltage Constant K_E (V/krpm) ± 10%	Armature Winding* (Ω) ± 15%	Armature Inductance (mH)	Max. Pulse Current (A)
-00	5.8	4.3	0.70	0.09	60
-01	8.3	6.1	1.95	0.21	42
-02	11.6	8.6	3.35	0.35	30

*Add .15 ohms for terminal resistance

MOTOR LIFE

Many users naturally ask what operational and brush life can be expected from a moving coil motor. There is no pat answer to this question as numerous application factors have a direct impact on motor life. Among these are peak current during acceleration, duty cycle, direction reversals, load inertia and method of heat removal.

For the M-1040 operated under normal conditions a life expectancy of 1.5 x 10^9 stop-start cycles at 100 cycles per second or 1 x 10^8 revolutions at 1500 RPM with 15 oz-in output can be achieved before maintenance is required. A life prediction can be made for your application after all pertinent factors are known and a simulation test has been made.

Let our application specialists examine your requirements.

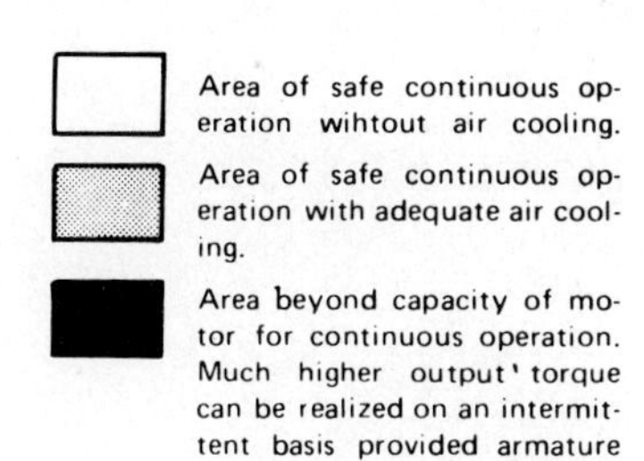

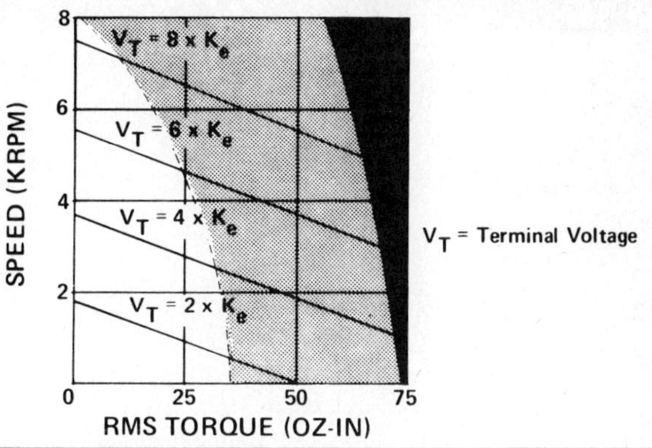

MODEL M-1040 THERMAL DATA

AIR FLOW (c.f.m.)

12
10
8
6
4
2

0 6 12 18 24 30

PRESSURE (in. H_2O)

.50 1.0 1.5 2.0 2.5 3.0

THERMAL RESISTANCE (°C/W)

OUTLINE DRAWING

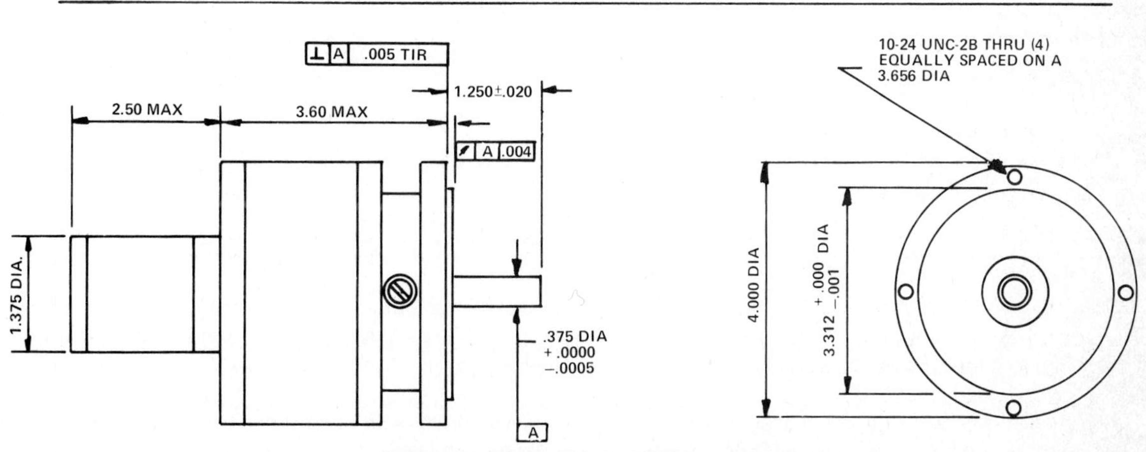

Shaft endplay .0005 min .004 max with 10 lb reversing load.
Shaft radial play .0015 max when measured with 5 lb reversing load.

M-1438

The Electro-Craft Model M-1438 is a high-performance DC moving coil servomotor designed to meet the requirements of a 200 IPS capstan drive. This motor has a theoretical acceleration from stall exceeding 10^6 rad/s^2 The unit shown on the right has a two-track bidirectional optical tachometer and a low inertia capstan mounted on the shaft.

SPECIFICATIONS

MOTOR RATING	Uncooled	W/Air Cooling (20 cfm)
Maximum Safe Speed (no load)	4.5 krpm	10 krpm
Maximum Speed at Rated Load	2.5 krpm	2.5 krpm
Rated Torque (continuous at stall)	95 oz-in	170 oz-in
Rated Power Output	130 W	280 W
Power Rate (continuous)	50 KW/sec	230 KW/sec
Power Rate (peak)	4400 KW/sec	4400 KW/sec
No Load Acceleration at Maximum Continuous Torque	99600 rad/sec^2	215000 rad/sec^2
No Load Acceleration at Maximum Pulse Torque	950000 rad/sec^2	950000 rad/sec^2
Maximum Rated Armature Temp	155°C	155°C
Maximum Static Friction Torque	5 oz-in	5 oz-in
Electrical Time Constant	.16 ms	.16 ms
Mechanical Time Constant	.69 ms	.69 ms
Damping Factor	4 oz-in/krpm	4 oz-in/krpm

Armature Moment of Inertia
(ref. only) (including bearings)

Motor only	.00070 oz-in-sec^2
Motor M 110 Tach	.00077 oz-in-sec^2
Motor/Optical Tach.	.00078 oz-in-sec^2

WINDING VARIATIONS

WINDING NUMBER	Torque Constant K_T (oz-in/A) ± 10%	Voltage Constant K_E (V/krpm) ± 10%	Armature Winding* (Ω) ± 15%	Armature Inductance (mH)	Max. Pulse Current (A)
00	9.5	7	.45	< .1	70
01	11.8	8.7	.80	< .15	57

*Add .1 ohms for terminal resistance.

MOTOR LIFE

Many users naturally ask what operational and brush life can be expected from a moving coil motor. There is no pat answer to this question as numerous application factors have a direct impact on motor life. Among these are peak current during acceleration, duty cycle, direction reversals, load inertia and method of heat removal.

For the M-1438 operated under normal conditions a life expectancy of 1.5×10^9 stop-start cycles at 150 cycles per second or 3×10^8 revolutions at 1750 RPM with 40 oz.-in. output can be achieved before maintenance is required. A life predition can be made for your application after all pertinent factors are known and a simulation test has been made.

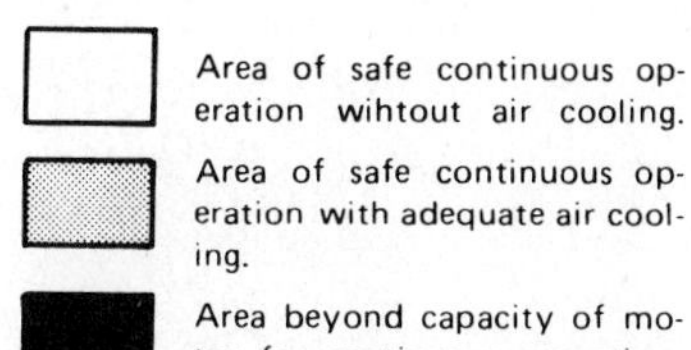

Area of safe continuous operation wihtout air cooling.

Area of safe continuous operation with adequate air cooling.

Area beyond capacity of motor for continuous operation. Much higher output torque can be realized on an intermittent basis provided armature temperature rating is not exceeded.

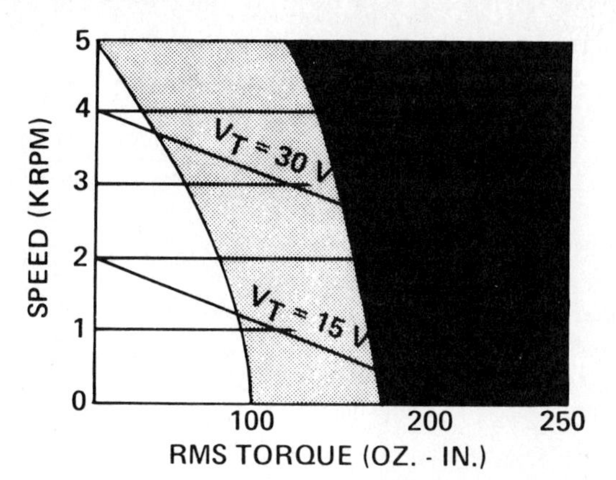

MODEL M-1438 THERMAL DATA

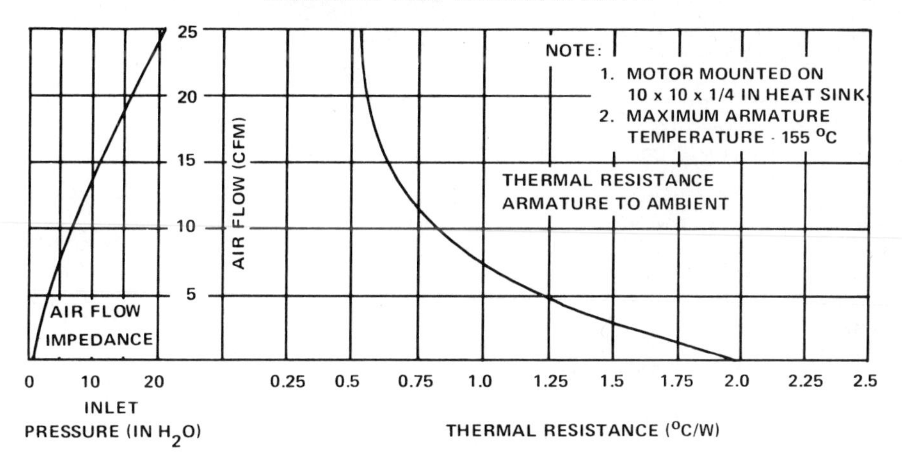

OUTLINE DRAWING

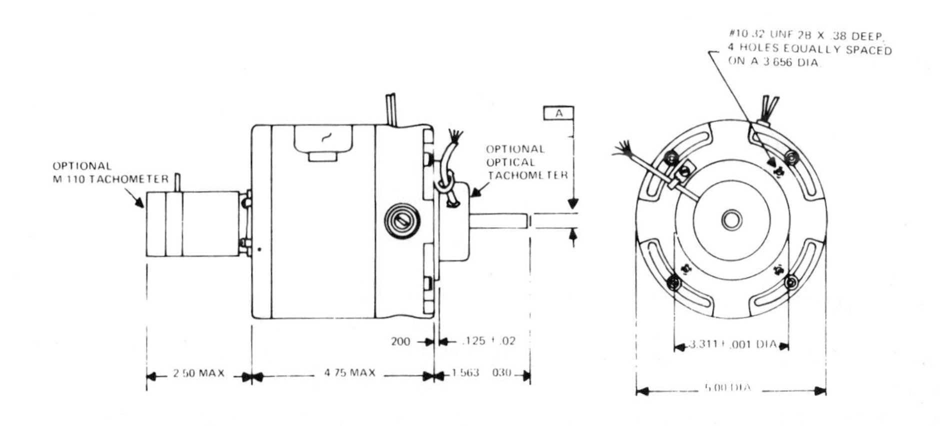

M-1450

The Electro-Craft Model M-1450 is a particularly powerful D.C. high-performance servo-motor designed to drive moderately high inertial loads with the fast response characteristic of moving coil motors. Typcial applications include high speed line printer drives, paper feed drives in high speed Optical Character recognition equipment and high speed material handling systems.

SPECIFICATIONS

MOTOR RATINGS	Uncooled	W/Air Cooling (25 cfm)
Maximum Safe Speed (no load)	4.5 krpm	7.5 krpm
Maximum Speed at Rated Load	2 krpm	2 krpm
Rated Torque (continuous at stall)	215 oz.-in.	355 oz.-in.
Rated Power Output	267 W	490 W
Power Rate (continuous)	22 KW/sec	76 KW/sec
Power Rate (peak)	1300 KW/sec	1300 KW/sec
No Load Acceleration at Maximum Continuous Torque	17900 Rad/sec^2	32900 Rad/sec^2
No Load Acceleration at Maximum Pulse Torque	139000 Rad/sec^2	139000 Rad/sec^2
Maximum Rated Armature Temp.	155° C	155° C
Maximum Static Friction Torque	9 oz.-in.	9 oz.-in.
Electrical Time Constant	.18 ms	.18 ms
Mechanical Time Constant	3 ms	3 ms
Damping Factor	7.6 oz.-in./krpm	7.6 oz.-in./krpm
Armature Moment of Inertia (including bearings) .01 oz-in-sec^2		

WINDING VARIATIONS

WINDING NUMBER	Torque Constant K_T (oz-in/A) ± 10%	Voltage Constant K_E (V/krpm) ± 10%	Armature Winding* (Ω) ± 15%	Armature Inductance (mH)	Max. Pulse Current (A)
00	24.9	18.4	1.15	<.3	56
02	18.8	13.9	.55	<.3	77

*Add .10 ohms for terminal resistance.

MOTOR LIFE

Many users naturally ask what operational and brush life can be expected from a moving coil motor. There is no pat answer to this question as numerous application factors have a direct impact on motor life. Among these are peak current during acceleration, duty cycle, direction reversals, load inertia and method of heat removal.

For the M-1450 operated under normal conditions a life expectancy of 10^9 stop-start cycles at 40 cycles per second or 2×10^8 revolutions at 1000 RPM with 100 oz.-in. output can be achieved before maintenance is required. A life prediction can be made for your application after all pertinent factors are known and a simulation test has been made.

Let our application specialists examine your requirement.

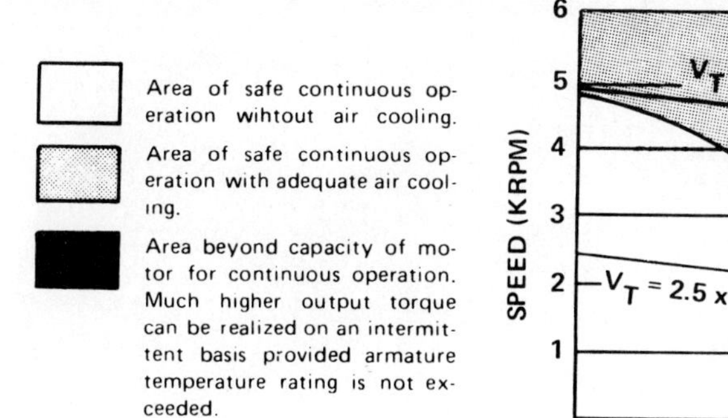

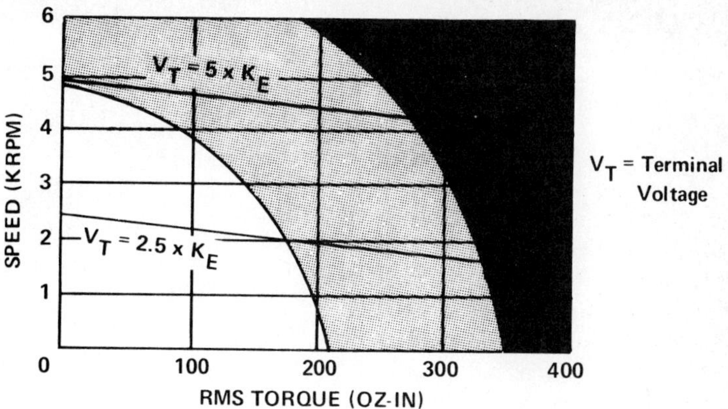

MODEL M-1450 THERMAL DATA

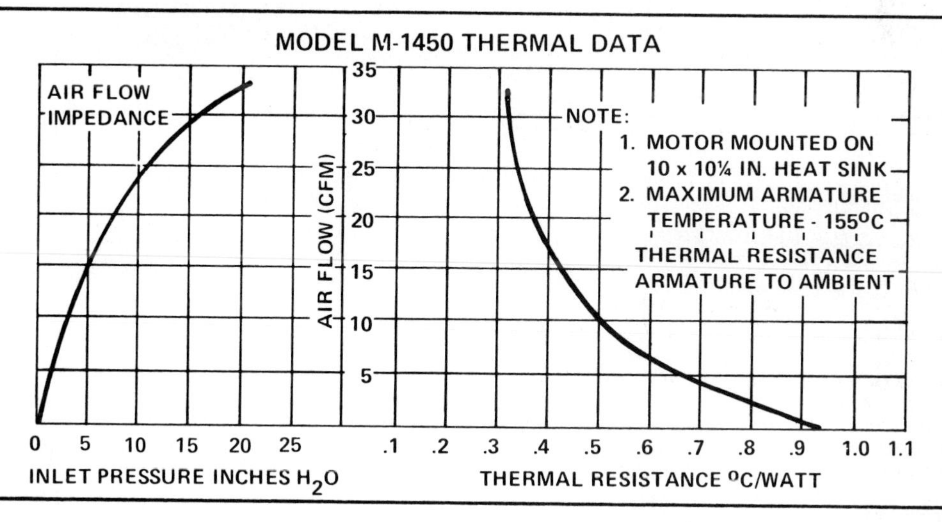

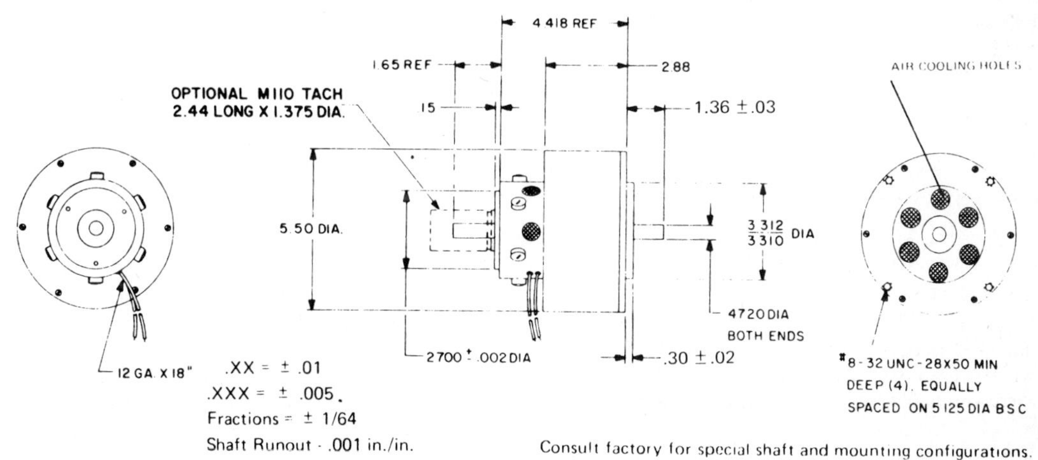

.XX = ± .01
.XXX = ± .005
Fractions = ± 1/64
Shaft Runout - .001 in./in.

Consult factory for special shaft and mounting configurations.

M-1500

The Electro-Craft Model M-1500 is a high-performance DC moving coil servomotor designed to meet the requirements of 250-300 IPS capstan drives. This motor has a theoretical acceleration from stall exceeding 10^6 rad/s^2.

SPECIFICATIONS

MOTOR RATINGS	Uncooled	W/Air Cooling
Maximum Safe Speed (no load)	5 krpm	9 krpm
Maximum Speed at Rated Load	2.5 krpm	2.5 krpm
Rated Torque (coninuous at stall)	85 oz.-in.	155 oz.-in.
Rated Power Output at Rated Speed	123 W	268 W
Power Rate (continuous)	50 KW/sec.	240 KW/sec
Power Rate (peak)	3400 KW/sec.	3400 KW/sec
No Load Acceleration at Maximum Continuous Torque	109000 rad/sec.^{2}	239000 rad/sec^2
No Load Acceleration at Maximum Pulse Torque	900000 rad/sec.^{2}	900000 rad/sec^2
Maximum Rated Armature Temp.	155°C	155°C
Maximum Static Friction Torque	5 oz.-in.	5 oz.-in.
Electrical Time Constant	0.12 ms	0.12 ms
Mechanical Time Constant	0.63 ms	0.63 ms
Damping Factor	4 oz-in/krpm	4 oz-in/krpm
Thermal Resistance	2.0 °C/W	0.6 °C/W
Armature Moment of Inertia (including bearings)		
Motor only .0006 oz-in-sec^2		

WINDING VARIATIONS

WINDING NUMBER	Torque Constant K_T (oz-in/A) ± 10%	Voltage Constant K_E (V/krpm) ± 10%	Armature Winding* (Ω) ± 15%	Armature Inductance (mH)	Max Pulse Current (A)
00	8.6	6.4	.35	<0.1	63
01	11.5	8.5	.45	<0.1	58

*Add .10 ohms for terminal resistance.

MOTOR LIFE

Many users naturally ask what operational and brush life can be expected from a moving coil motor. There is no pat answer to this question as numerous application factors have a direct impact on motor life. Among these are peak current during acceleration, duty cycle, direction reversals, load inertia and method of heat removal.

For the M-1500 operated under normal conditions a life expectancy of 1.5×10^9 stop-start cycles at 180 cycles per second or 3×10^8 revolutions at 1750 RPM with 50 oz.-in. output can be achieved before maintenance is required. A life predition can be made for your application after all pertinent factors are known and a simulation test has been made.

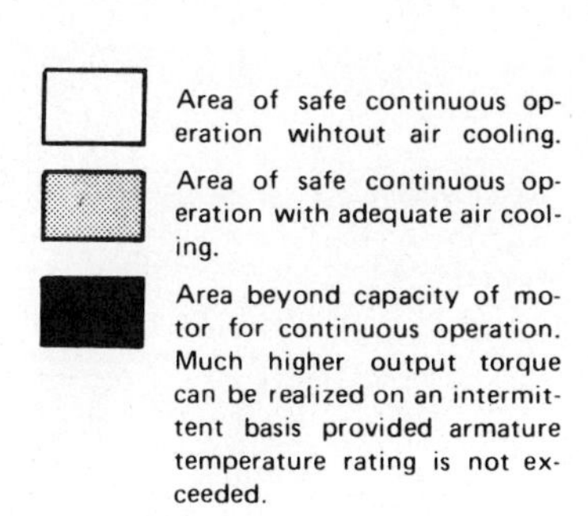

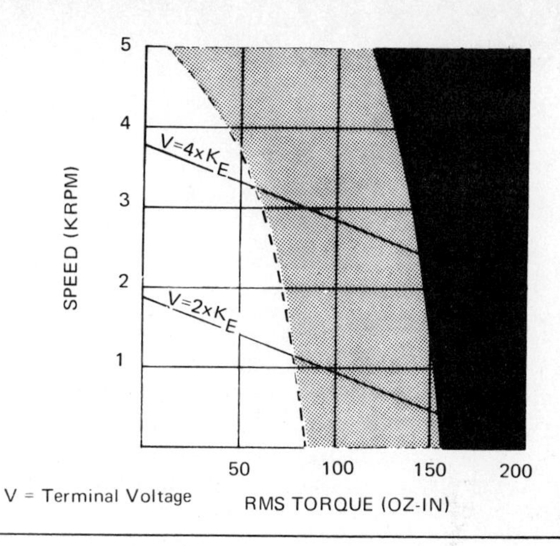

MODEL M-1500 THERMAL DATA

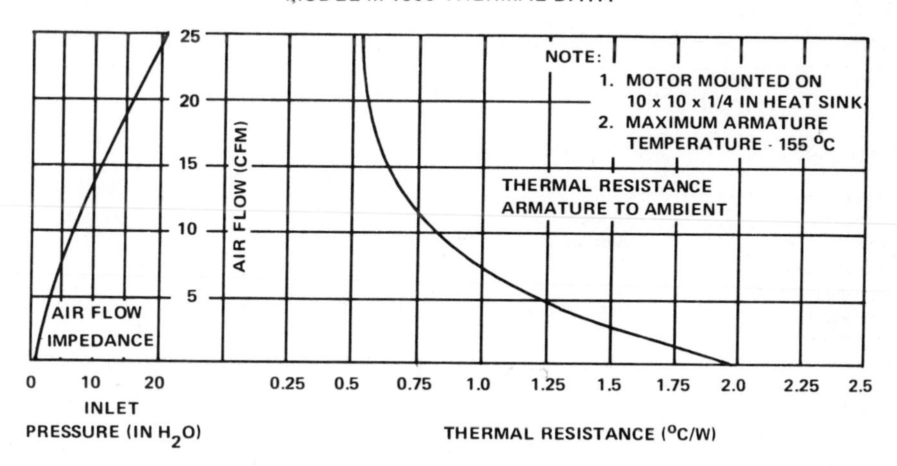

OUTLINE DRAWING

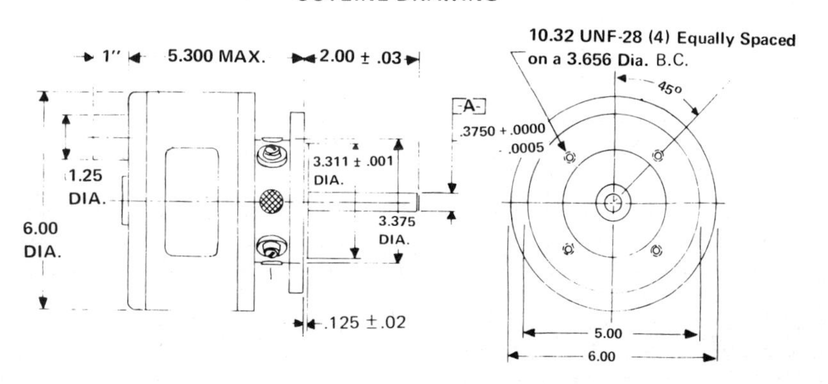

M-1600

The Electro-Craft Model M-1600 is a high-performance DC moving coil servomotor designed to meet the requirements of 250 IPS capstan drives. This motor has a theoretical acceleration from stall exceeding 10^6 rad/s^2.

SPECIFICATIONS

MOTOR RATINGS	Uncooled	W/Air Cooling
Maximum Safe Speed (no load)	7 krpm	7 krpm
Maximum Speed at Rated Load	4.5 krpm	4.5 krpm
Max. Continuous Stall Torque	60 oz-in	120 oz-in
Rated Power Output	150 Watts	375 Watts
Power Rate (continuous)	45 KW/sec	280 KW/sec
Power Rate (peak)	6630 KW/sec	6630 KW/sec
No Load Acceleration at Maximum Continuous Torque	144,000 rad/sec^2	360,000 rad/sec^2
No Load Acceleration at Maximum Pulse Torque	1,700,000 rad/sec^2	1,700,000 rad/sec^2
Maximum Rated Armature Temp.	155°C	155°C
Maximum Static Friction Torque	5 oz.-in.	5 oz.-in.
Electrical Time Constant	.1 ms	.1 ms
Mechanical Time Constant	.5 ms	.5 ms
Damping Factor	1 oz-in/krpm	1 oz-in/krpm
Armature Moment of Inertia	.00031 oz-in-sec^2	
Thermal Resistance	2.5 ^{o}C/Watt	.65 ^{o}C/Watt

WINDING DATA

Torque Constant K_T (oz-in/A) ± 10%	Voltage Constant K_E (V/krpm) ± 10%	Armature Winding* (Ω) ± 15%	Armature Inductance (mH)	Max. Pulse Current (A)
9.5	7	.8	.1	57

*Add .10 ohms for terminal resistance.

MOTOR LIFE

Many users naturally ask what operational and brush life can be expected from a moving coil motor. There is no pat answer to this question as numerous application factors have a direct impact on motor life. Among these are peak current during acceleration, duty cycle, direction reversals, load moment of inertia and method of heat removal.

For the M-1600 operated under normal conditions a life expectance of 1.5 X 10^9 stop-start cycles at 180 cycles per second or 3 X 10^8 revolutions at 1750 rpm with 30 oz-in output torque can be achieved before maintenance is required. A life prediction can be made for your application after all pertinent factors are known and a simulation test has been made.

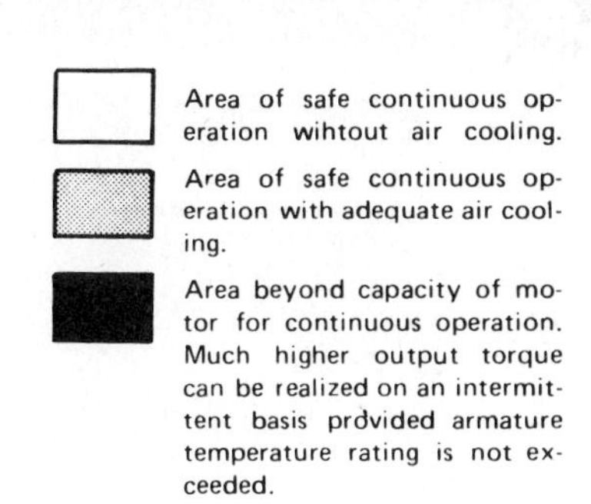

Area of safe continuous operation wihtout air cooling.
Area of safe continuous operation with adequate air cooling.
Area beyond capacity of motor for continuous operation. Much higher output torque can be realized on an intermittent basis prdvided armature temperature rating is not exceeded.
V = Terminal Voltage

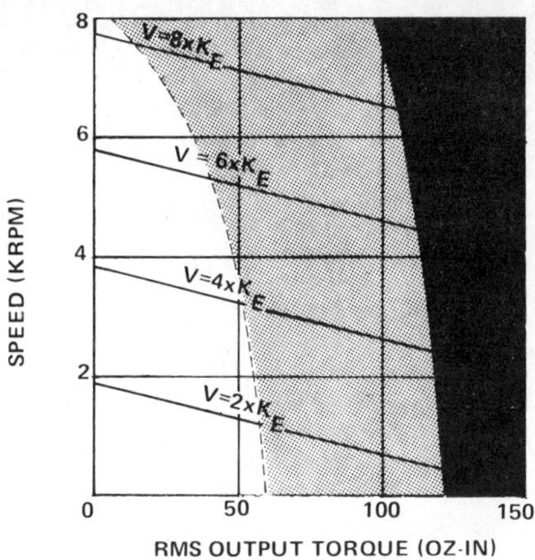

V=8xK_E
V = 6xK_E
V=4xK_E
V=2xK_E
SPEED (KRPM)
0
2
4
6
8
50
100
150
RMS OUTPUT TORQUE (OZ-IN)

MODEL 1600 THERMAL DATA

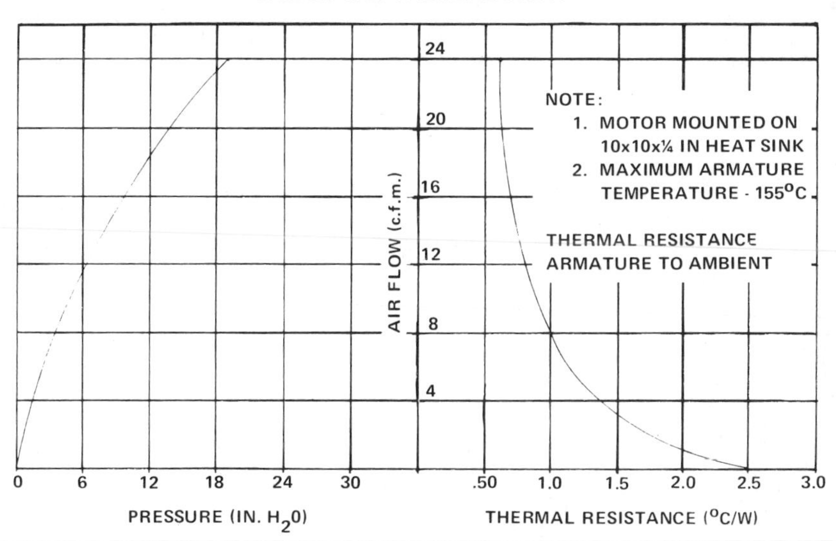

NOTE:
1. MOTOR MOUNTED ON 10x10x¼ IN HEAT SINK
2. MAXIMUM ARMATURE TEMPERATURE - 155°C
THERMAL RESISTANCE ARMATURE TO AMBIENT
AIR FLOW (c.f.m.)
4
8
12
16
20
24
0
6
12
18
24
30
.50
1.0
1.5
2.0
2.5
3.0
PRESSURE (IN. H2O)
THERMAL RESISTANCE (°C/W)

OUTLINE DRAWING

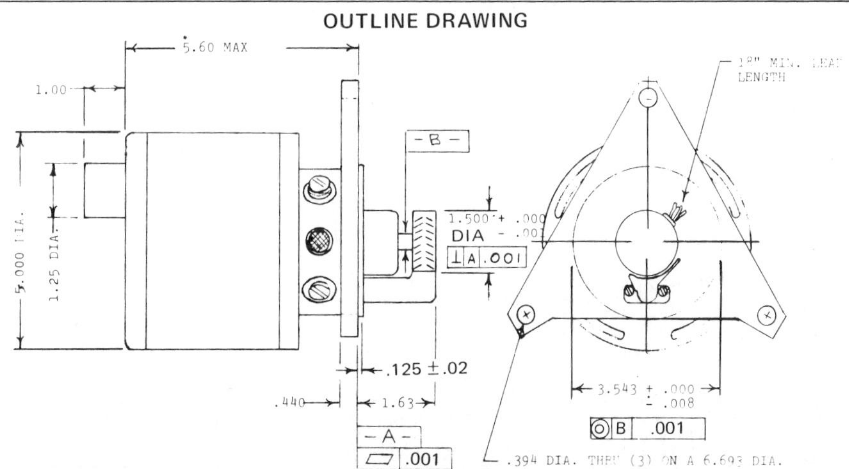

5.60 MAX
1.00
-B-
5.000 DIA.
1.25 DIA.
1.500 + .000 - .001 DIA
⊥ A .001
.125 ± .02
.440
1.63
-A-
▱ .001
3.543 + .000 - .008
◎ B .001
.394 DIA. THRU (3) ON A 6.693 DIA.

M-1700

The Electro-Craft model M-1700 is a particularly powerful high performance D.C. moving coil motor designed to drive high inertial loads with the fast response characteristics of D.C. servo-motors. The high energy permanent magnet field incorporates special windings which prevent demagnetization by pulse currents up to 800 amps. The model M-1700 is suitable for machine tool and other applications which require fast response, high accuracy and high torque.

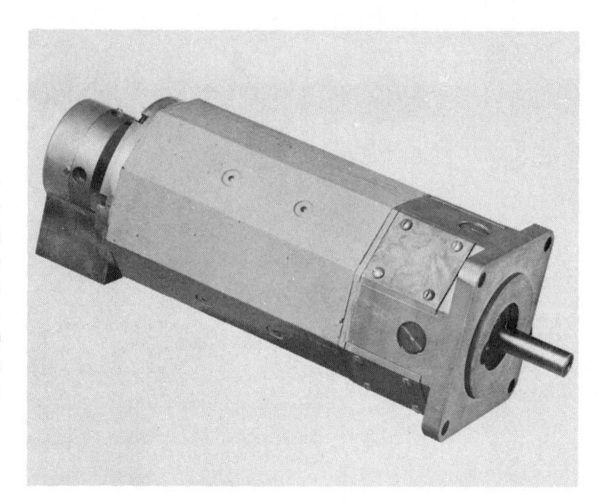

SPECIFICATIONS

MOTOR RATINGS	Uncooled	W/Air cooling
Maximum Safe Speed (no load)	4.5 krpm	6.0 krpm
Rated Torque (continuous at stall)	75 lb-in	150 lb-in
Maximum Rated Armature Power Dissipation	280 W	1060 W
Maximum Rated Armature Temperature	195°C	195°C
Thermal Resistance (Armature-to-ambient)	0.6°C/W	0.16°C
Aramture Moment of Inertia	0.172 lb-in-sec^2	
Electrical Time Constant	0.75 ms	
Mechanical Time Constant	7.75 ms	
Damping Factor	1.0 lb-in/krpm	
Static Friction Torque	2.2 lb-in.	

TACHOMETER CONSTANTS		
Output Voltage	14.5 V/krpm	±10%
Armature Resistance	15.0 ohms	±15%
Armature Inductance	12.0 mhy	±10%
Ripple (peak to average) at 200 rpm with a 1500 ohm load		

MOTOR WINDING CONSTANTS	
Torque Constant K_T (lb-in/A) ± 10%	2.97
Voltage Constant K_E (V/krpm) ± 10%	35.2
Armature Winding (Ω) ±15%	0.24
Armature Inductance (mH) ± 10%	0.34
Magnetizing Winding Resistance (Ω) ± 15%	0.097
Magnetizing Winding Inductance (mH) ± 10%	0.115

*Add 0.21 ohms for terminal resistance.

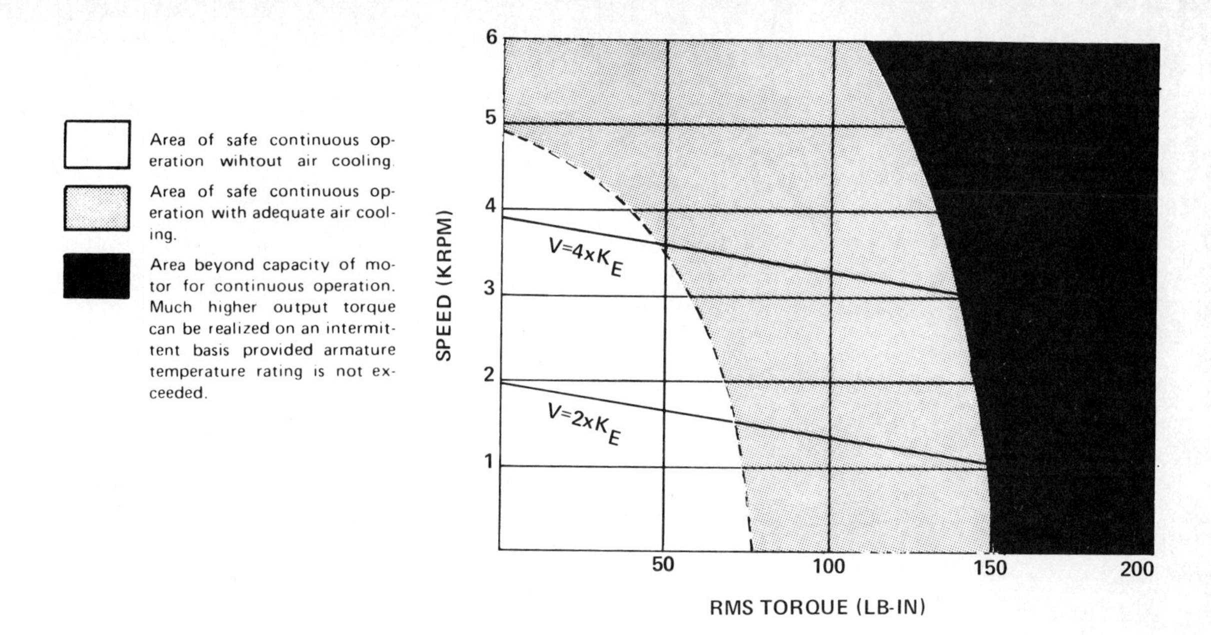

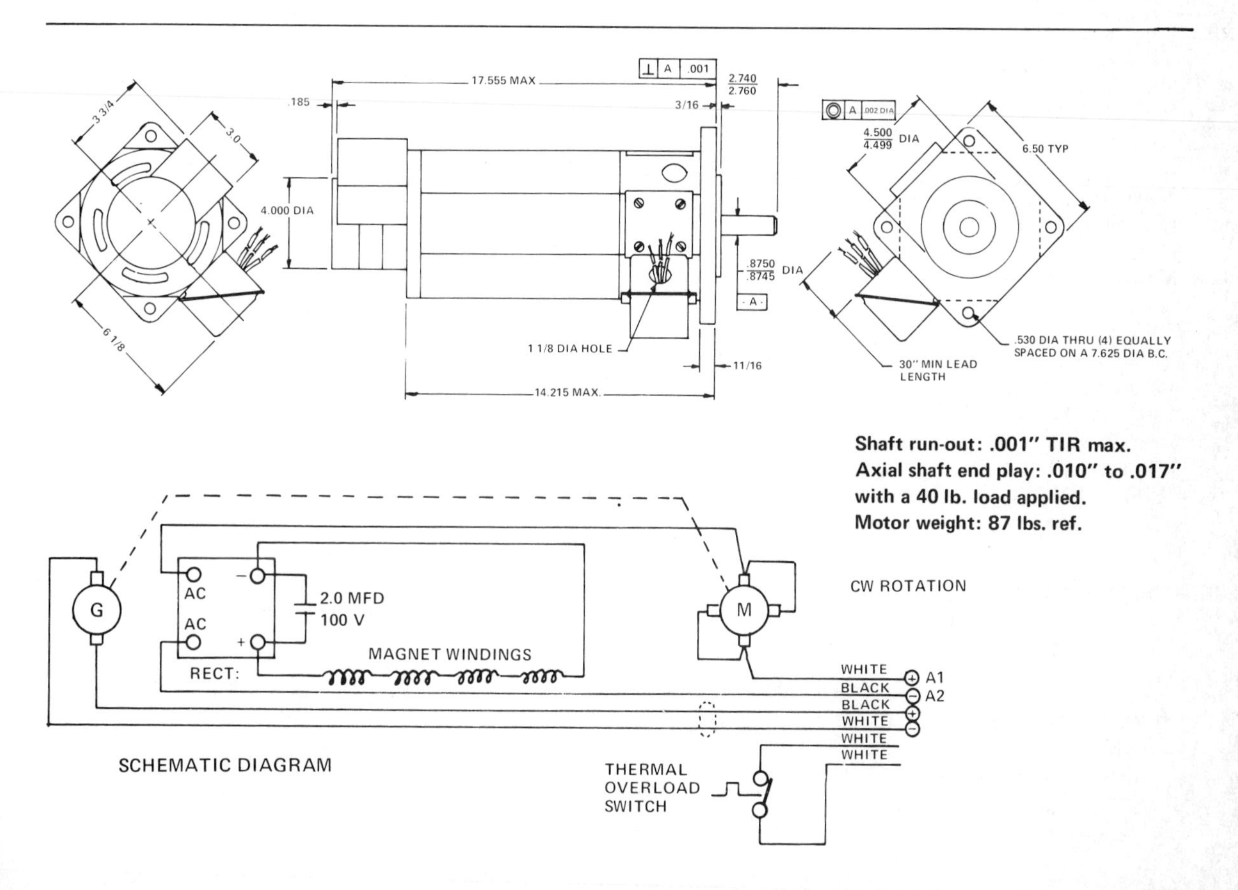

Shaft run-out: .001" TIR max.
Axial shaft end play: .010" to .017" with a 40 lb. load applied.
Motor weight: 87 lbs. ref.

7.4. TACHOMETERS-GENERATORS

E-591/596

The Model E-591 separate permanent-magnet tachometer is a precision D.C. rate generator suitable for use in a wide range of industrial velocity and position control systems where extremely low inertia or ripple is not required. The E-591 is supplied with standard base and face mounting.

The Model E-596 is the same tachometer integrally coupled to any Electro-Craft servomotor.

OUTLINE DRAWING

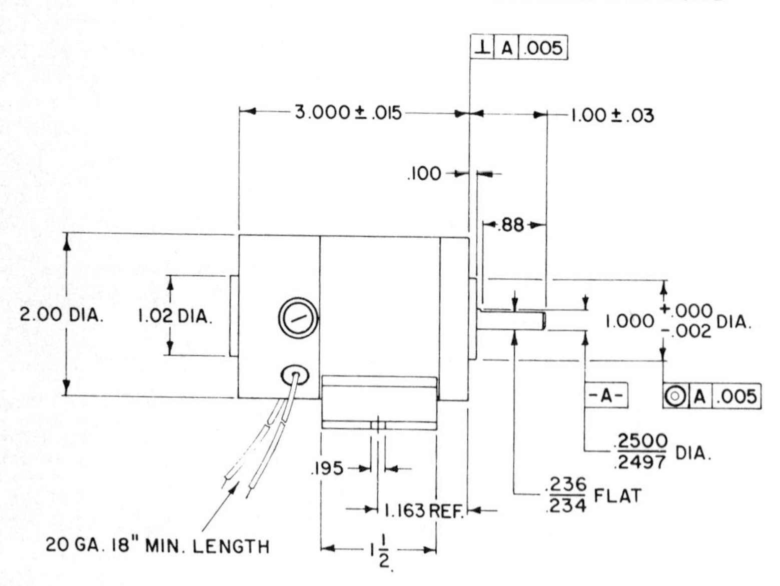

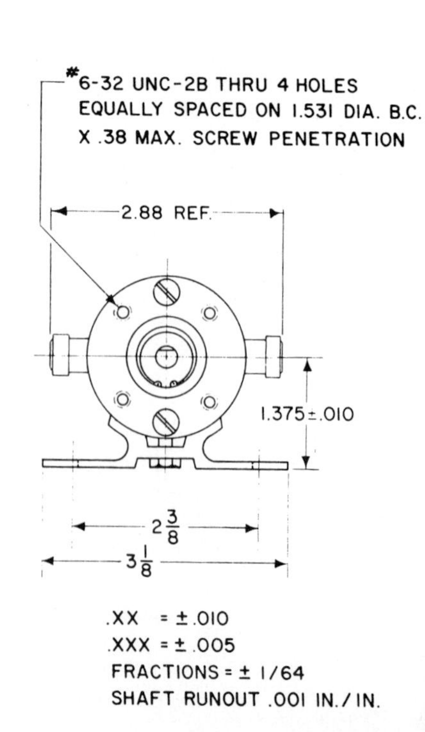

.XX = ± .010
.XXX = ± .005
FRACTIONS = ± 1/64
SHAFT RUNOUT .001 IN./IN.

PERFORMANCE SPECIFICATIONS

Parameter	Value	Units
Linearity	.2	% Max. Deviation from Perfect Linearity
Ripple	5	Max. Percentage Peak to Peak A.C. Component in D.C. output at 500 RPM.
Ripple Frequency	11	Cycles/Revolution
Speed Range	1-8,000	RPM
Temperature Coefficient	–.05	%/°C.
Armature Inertia	1.4×10^{-3}	oz.-in.-sec.^{2}
Insulation Resistance	10	Megohm
Friction Torque	3	oz.-in., max.
Rated Life	5,000	Hours at 3 KRPM
Operating Temperature Range	0-75	°C.

WINDING VARIATIONS

Winding Constants	Winding Number			
	-01	-02	-03	-04
Output Voltage Gradient (Volts/KRPM)	3.0	7.0	14.2	21
Armature Resistance (ohms) (includes temperature compensation network)	570	570	720	950
Armature Inductance (mhy)	6.2	33	138	255
Optimum Load Impedance (ohms)	5,000	5,000	5,000	10,000

STANDARD RIPPLE FILTER SCHEMATIC

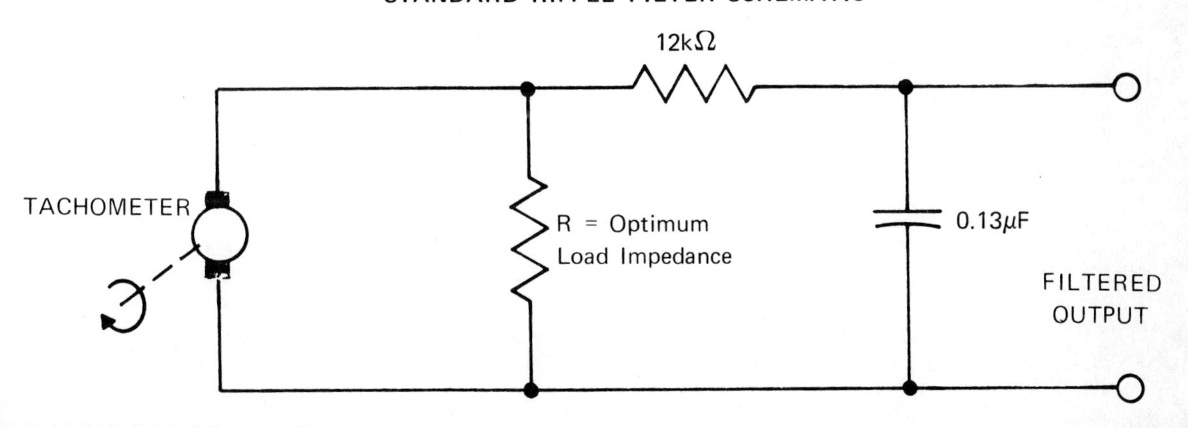

M-100/110

The model M-100 is a separate moving coil tachometer designed for use in applications requiring high quality velocity feedback is with minimum system inertia load. Commutators are manufactured from coil silver and are diamond cut after assembly to a surface finish better than .000020". Very long life – exceeding 10,000 hours at 3,000 RPM – may be obtained even in systems where high frequency shaft reversal and start-stop operation is a normal operating mode.

The model M-110 is the same generator integrally coupled to any Electro-Craft servomotor. Shaft coupling problems are totally eliminated and due to the very low inertia of the tachometer armature, the lowest shaft torsional resonant frequency exceeds 3.5 KHz.

OUTLINE DRAWING

1.37 REF.
3.0 REF.
.625
2.37 REF.
.062
.093
1.281 DIA.
.02 x 45°
$\frac{.5000}{.4997}$ DIA.
$\frac{.2500}{.2498}$ DIA. (STD.)
.093
WIRE EXIT – THIS SIDE ONLY
#2-56 x 1/8 DP. 3 HOLES ON .875 DIA. B.C.

Cable: 2 cond. shld., blk. and red, A.W.G. 22, 24 ± 1" lg., .21 O.D.
Polarity: W/C.W. Rotation viewed from shaft - Red = Pos.

PERFORMANCE SPECIFICATIONS

Parameter	Value	Units
Linearity	.2	% max. deviation from perfect linearity.
Ripple		
M-100	1.5	max. percentage peak to peak A.C.
M-110	2	Component in D.C. output at 167 to 6,000 RPM.
Ripple Frequency	19	Cycles/Revolution
Speed Range	1-6,000	RPM
Temperature Coefficient	−.01	%/°C.
Armature Inertia		
M-100	$9x10^{-5}$	$oz.-in.-sec.^2$
M-110	$7x10^{-5}$	$oz.-in.-sec.^2$
Insulation Resistance	10	Megohm
Friction Torque	.25	oz.-in., max.
Torsional Resonant Frequency When Coupled to an Electro-Craft MCM Servomotor	3.5	KHz, min.
Rated Operating Life	10,000*	Hours at 3,000 RPM
Operating Temperature Range	0-155	°C.

WINDING VARIATIONS

Winding Constants	Winding Number -00	-01	-02
Output Voltage Gradient (Volts/KRPM)	3.0	2.5	1
Armature Resistance (ohms)	150	150	150
Armature Inductance (mhy)	4	4	4
Recommended Minimum Load Impedance (ohms)	10,000	10,000	10,000

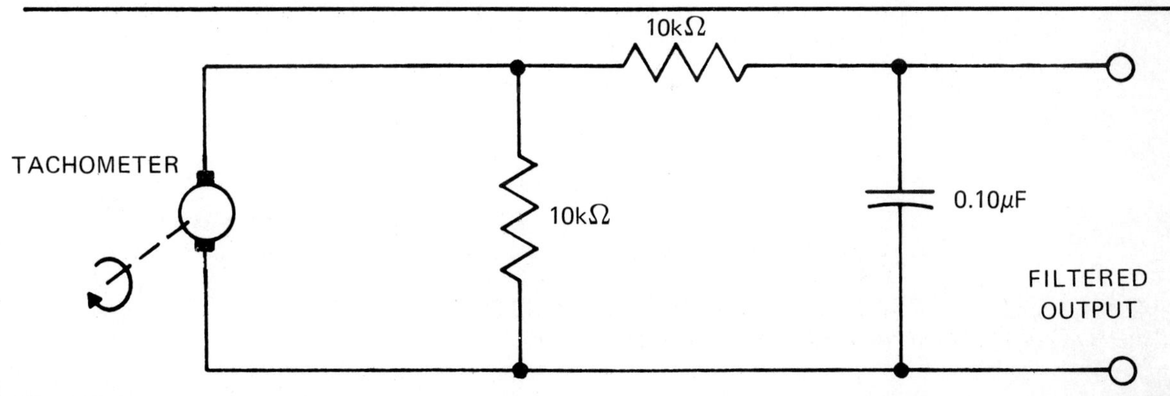

7.5. SPECIAL HYBRID MOTORS
H-5200 SERIES HYBRID MOTORS

The Electro-Craft H-5200 series are the first truly new development in digital tape transport reel motors. The series was designed with the objective of reducing total system costs by incorporating an innovative design development resulting in a "Hybrid" motor with the best qualities of both wound field and permanent magnet motors.

The digital tape transport designer of the part was forced to choose between a permanent magnet motor and a wound field motor. With the permanent magnet motor, the driving servo amplifier was required to supply high current for the normal incrementing mode and to be capable of high voltage to overcome the back e.m.f. during rewind; resulting in a very expensive amplifier. With the wound field motor, the torque constant (K_T) could be changed by controlling the field making rewind easier, but then an expensive field control system was required, often with the less than ideal results.

These motors bridge the gap by providing a 4:1 change in K_T while keeping the field control system simple and inexpensive. Test results to date have proven the H-5200 series to be a highly cost effective choice for the high speed digital tape transport designer.

RELATIONSHIP BETWEEN FLUX BIAS CURRENT AND TORQUE CONSTANT

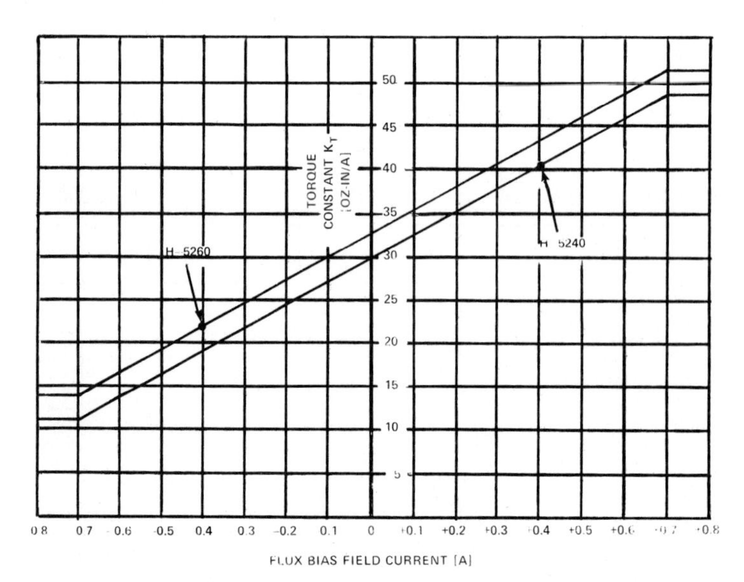

MODEL H-5240 THERMAL DATA

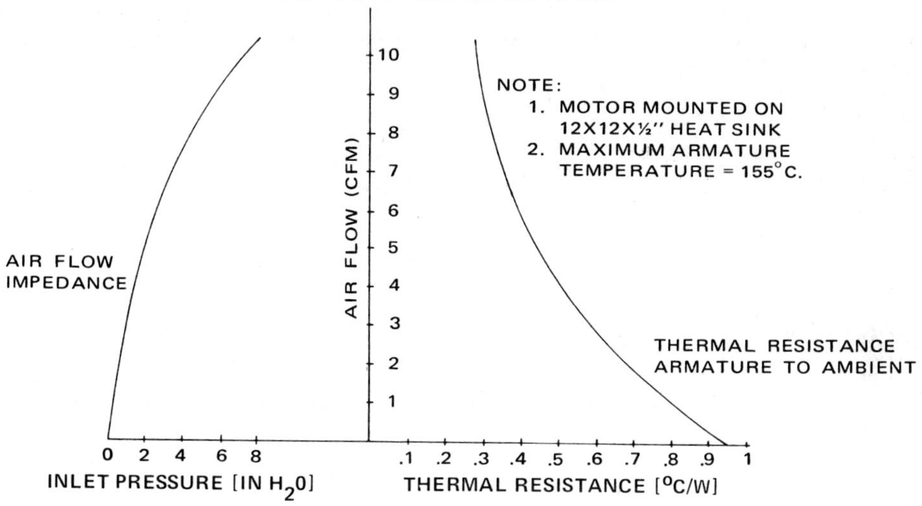

H-5240

SPEED TORQUE CURVES FOR $V_T = 40V$

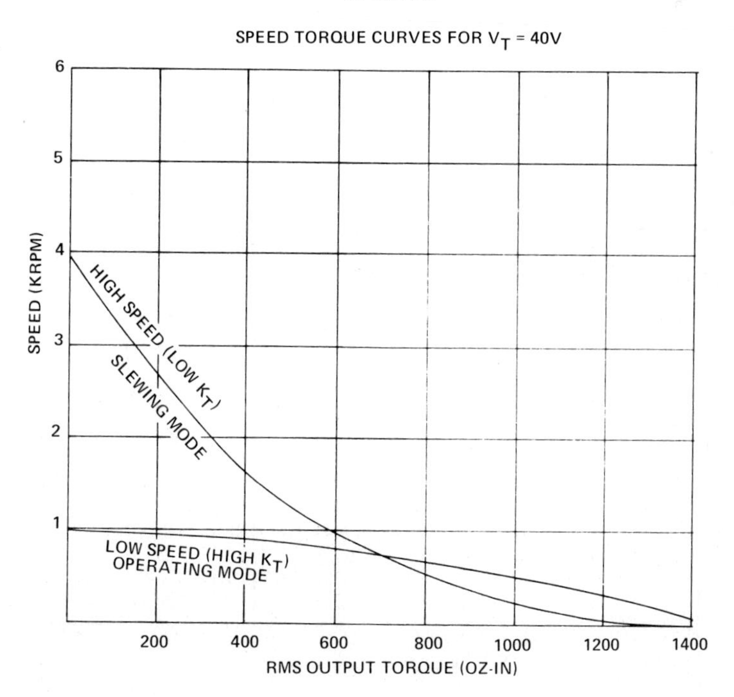

MODEL H-5260 THERMAL DATA

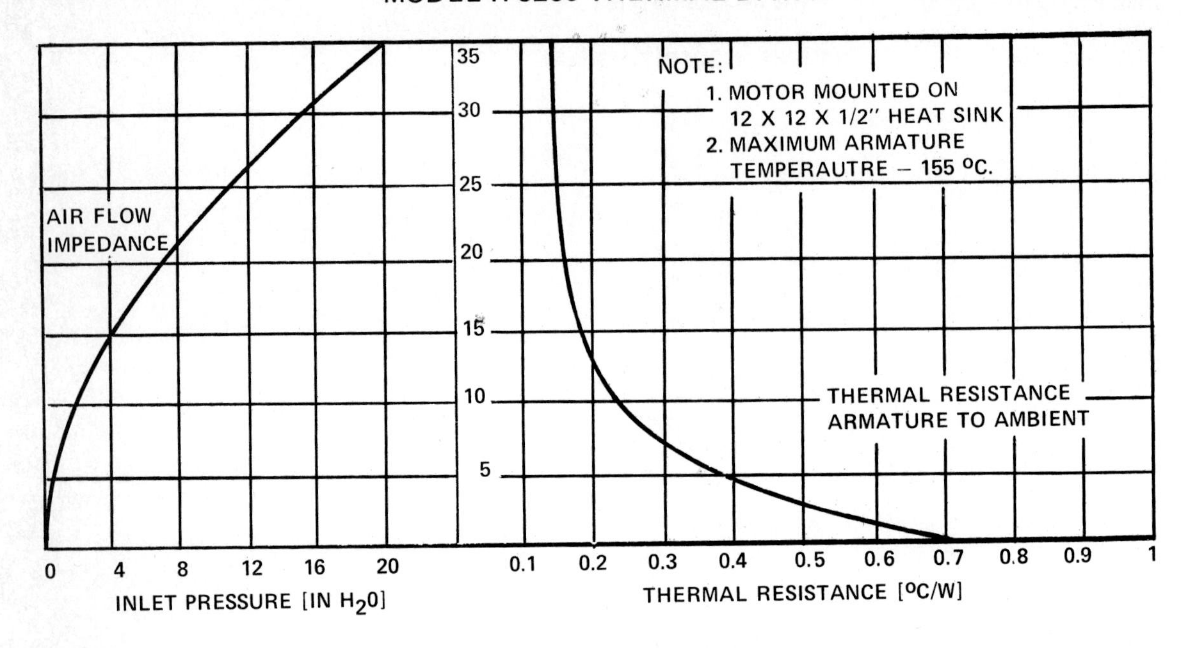

NOTE:
1. MOTOR MOUNTED ON 12 X 12 X 1/2" HEAT SINK
2. MAXIMUM ARMATURE TEMPERAUTRE – 155 °C.
AIR FLOW IMPEDANCE
THERMAL RESISTANCE ARMATURE TO AMBIENT
35
30
25
20
15
10
5
0
4
8
12
16
20
INLET PRESSURE [IN H20]
0.1
0.2
0.3
0.4
0.5
0.6
0.7
0.8
0.9
1
THERMAL RESISTANCE [°C/W]

H-5260

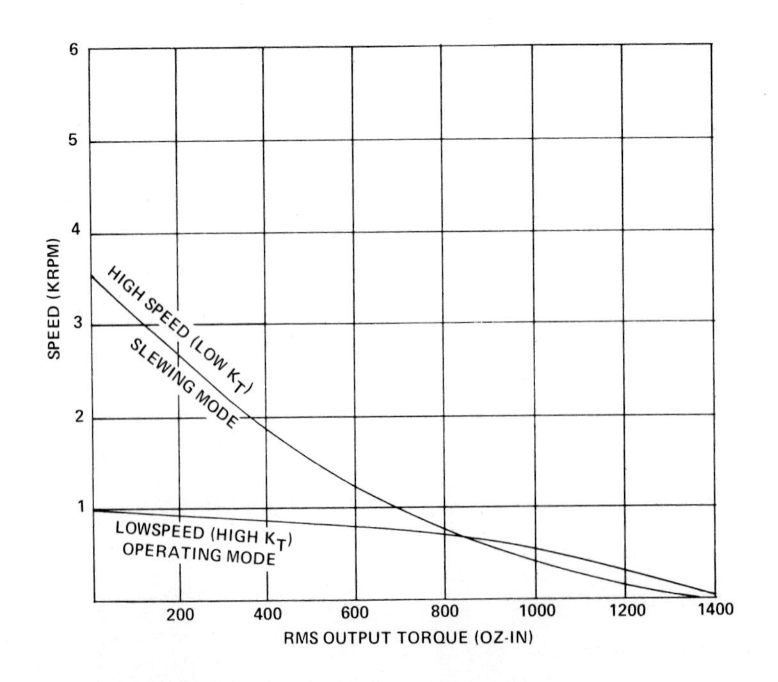

HIGH SPEED (LOW KT) SLEWING MODE
LOWSPEED (HIGH KT) OPERATING MODE
SPEED (KRPM)
1
2
3
4
5
6
200
400
600
800
1000
1200
1400
RMS OUTPUT TORQUE (OZ-IN)

MODEL	A	B
5240	8.042	3.590
5260	10.416	4.475

OUTLINE DRAWING

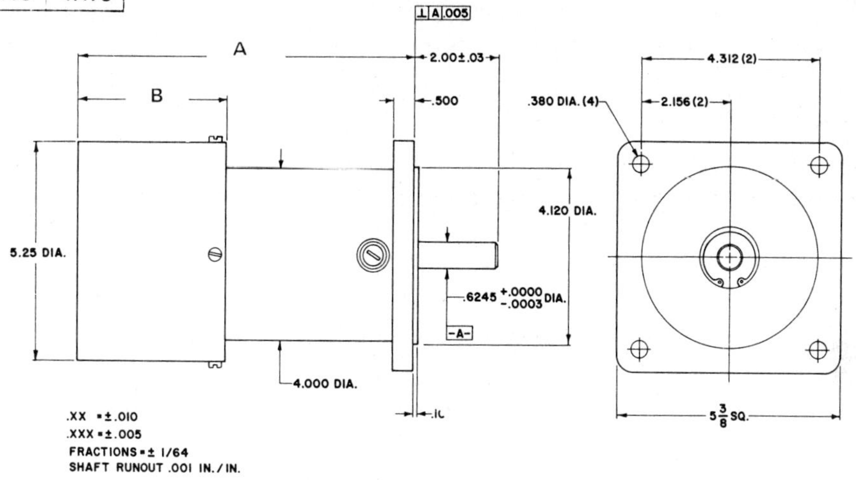

.XX = ±.010
.XXX = ±.005
FRACTIONS = ± 1/64
SHAFT RUNOUT .001 IN./IN.

NOMINAL SPECIFICATIONS

PARAMETER	5240	5260	UNITS
Torque Constant K_T			
Low Speed Drive	48	50	oz-in/A
High Speed Slew	11	14	oz-in/A
Voltage Constant K_E			
Low Speed Drive	36	37	V/krpm
High Speed Slew	8.2	12.6	V/krpm
Armature Resistance	0.6	0.5	Ω
Armature Inductance	4.2	4	mH
Armature Moment of Inertia	0.26	0.38	oz-in-s^2
Flux Bias Field Voltage	± 32	± 32	V
Flux Bias Field Current	± 0.7	± 0.8	A
Full Load Torque (Uncooled)	550	700	oz-in
Maximum Armature Temperature	155	155	^{0}C

THIS PAGE INTENTIONALLY LEFT BLANK

7.6.1. MOTOR SPEED CONTROLS

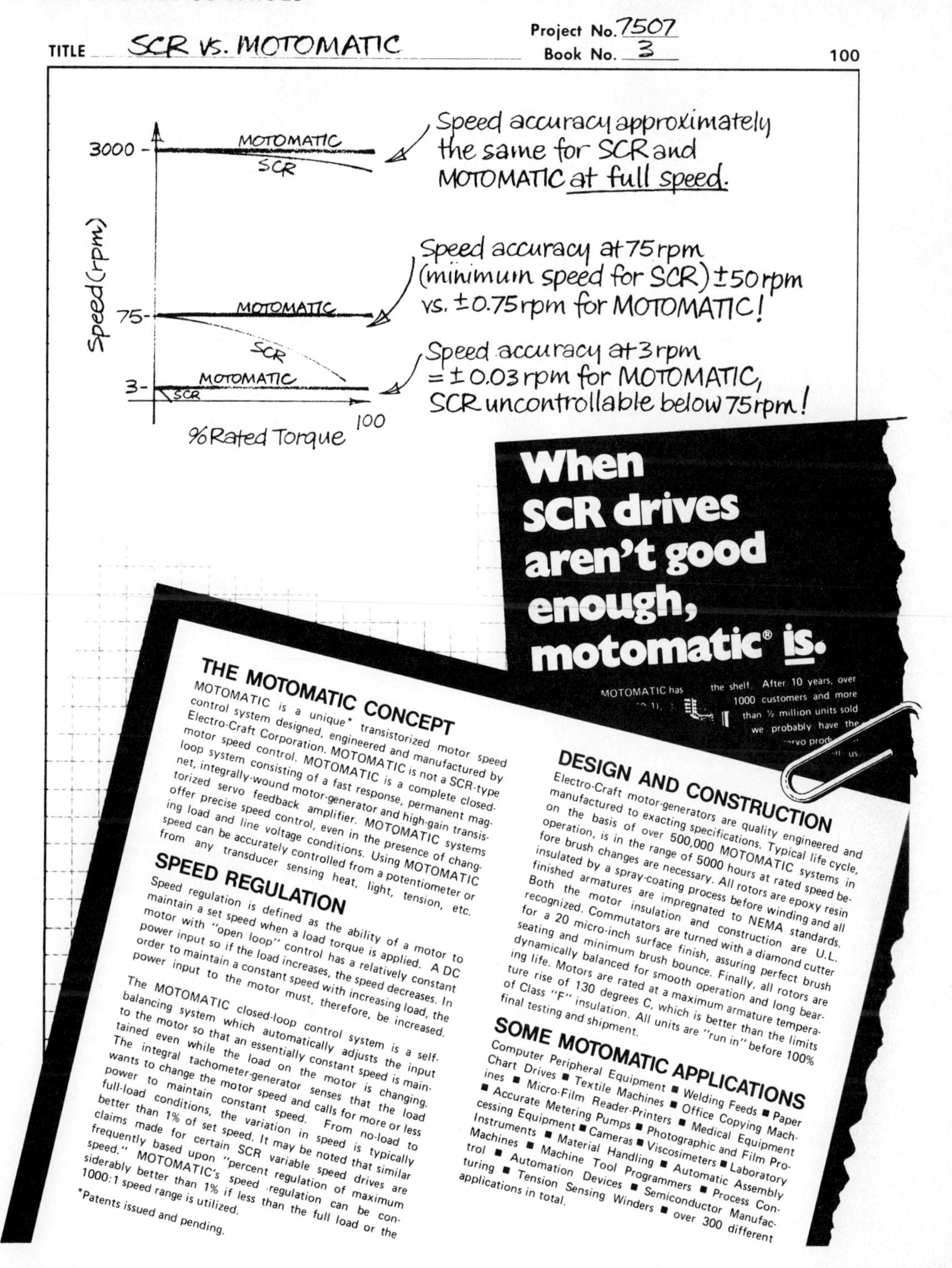

TITLE SCR vs. MOTOMATIC
Project No. 7507
Book No. 3
100
3000
75
3
MOTOMATIC
SCR
Speed (rpm)
% Rated Torque
100
Speed accuracy approximately the same for SCR and MOTOMATIC at full speed.
Speed accuracy at 75 rpm (minimum speed for SCR) ±50 rpm vs. ±0.75 rpm for MOTOMATIC!
Speed accuracy at 3 rpm = ±0.03 rpm for MOTOMATIC, SCR uncontrollable below 75 rpm!
When SCR drives aren't good enough, motomatic® is.
MOTOMATIC has
the shelf. After 10 years, over 1000 customers and more than ½ million units sold we probably have the
THE MOTOMATIC CONCEPT
MOTOMATIC is a unique* transistorized motor speed control system designed, engineered and manufactured by Electro-Craft Corporation. MOTOMATIC is not a SCR-type motor speed control. MOTOMATIC is a complete closed-loop system consisting of a fast response, permanent magnet, integrally-wound motor-generator and high-gain transistorized servo feedback amplifier. MOTOMATIC systems offer precise speed control, even in the presence of changing load and line voltage conditions. Using MOTOMATIC speed can be accurately controlled from a potentiometer or from any transducer sensing heat, light, tension, etc.
SPEED REGULATION
Speed regulation is defined as the ability of a motor to maintain a set speed when a load torque is applied. A DC motor with "open loop" control has a relatively constant power input so if the load increases, the speed decreases. In order to maintain a constant speed with increasing load, the power input to the motor must, therefore, be increased.
The MOTOMATIC closed-loop control system is a self-balancing system which automatically adjusts the input to the motor so that an essentially constant speed is maintained even while the load on the motor is changing. The integral tachometer-generator senses that the load wants to change the motor speed and calls for more or less power to maintain constant speed. From no-load to full-load conditions, the variation in speed is typically better than 1% of set speed. It may be noted that similar claims made for certain SCR variable speed drives are frequently based upon "percent regulation of maximum speed." MOTOMATIC's speed regulation can be considerably better than 1% if less than the full load or the 1000:1 speed range is utilized.
*Patents issued and pending.
DESIGN AND CONSTRUCTION
Electro-Craft motor-generators are quality engineered and manufactured to exacting specifications. Typical life cycle, on the basis of over 500,000 MOTOMATIC systems in operation, is in the range of 5000 hours at rated speed before brush changes are necessary. All rotors are epoxy resin insulated by a spray-coating process before winding and all finished armatures are impregnated to NEMA standards. Both the motor insulation and construction are U.L. recognized. Commutators are turned with a diamond cutter for a 20 micro-inch surface finish, assuring perfect brush seating and minimum brush bounce. Finally, all rotors are dynamically balanced for smooth operation and long bearing life. Motors are rated at a maximum armature temperature rise of 130 degrees C, which is better than the limits of Class "F" insulation. All units are "run in" before 100% final testing and shipment.
SOME MOTOMATIC APPLICATIONS
Computer Peripheral Equipment ■ Welding Feeds ■ Paper Chart Drives ■ Textile Machines ■ Office Copying Machines ■ Micro-Film Reader-Printers ■ Medical Equipment ■ Accurate Metering Pumps ■ Photographic and Film Processing Equipment ■ Cameras ■ Viscosimeters ■ Laboratory Instruments ■ Material Handling ■ Automatic Assembly Machines ■ Machine Tool Programmers ■ Process Control ■ Automation Devices ■ Semiconductor Manufacturing ■ Tension Sensing Winders ■ over 300 different applications in total.

E-350 Control
☐ Better than 1000:1 speed range
☐ Single polarity amplifier is easily adapted to DC power sources
☐ Complete closed loop transistorized velocity feedback amplifier
☐ Current limited
☐ IC regulated reference for stable operation
☐ Fuse protection
☐ 115/230 VAC - 50/60 Hz
☐ *
RPM
7500
6500
5500
4500
3500
2500
1500
500
100
50
10
5
1.25
1.50
1.75
2
TORQUE OZ-INCHES
SPEED REGULATION ZONES FOR E-350 CONTROL
E-350-MG
☐ 1/70 hp
☐ Centerless ground, nickel-chrome plated motor housing
☐ Shielded ball bearings
☐ Stainless steel shaft
☐ High-energy oriented ALNICO magnets
☐ Silver brushes for stable operation
☐ Unique negator spring design applies constant brush pressure
☐ *
E-350-MGH
☐ Spur gearhead
☐ Heavy duty construction
☐ Sintered bronze bearings
☐ *
RPM
5000
4000
3000
2500
2000
1500
1000
500
100
50
20
5
0
2
4
6
8
10
TORQUE OZ-INCHES
SPEED REGULATION ZONES FOR E-550 CONTROL
E-550-M/550-O Controls
☐ 1000:1 speed range (5-5000 rpm)
☐ Low drift differential input stage
☐ Complete closed loop transistorized velocity feedback amplifier
☐ Speed,torque readout†
☐ Reversing brake-switch†
☐ Fuse and thermal overload protection
☐ 115/230 VAC - 50/60 Hz
☐ *
HOW TO READ THE SPEED REGULATION ZONE GRAPHS:
Select the torque and rpm range; the data (% of regulation) is stated as a percentage of the set speed.
99% of set speed
98% of set speed
97% of set speed
<96% of set speed
Intermittent duty only—consult ECC Engineering Application Dept. for additional data.
E-550-MG
☐ 1/20 hp
☐ Fully-enclosed all metal housing
☐ Temperature compensated generator
☐ Shielded ball bearings
☐ Unique negator brush design applies constant brush pressure
☐ Stainless steel shaft
☐ Can be face or base mounted
☐ *
E-550-MGHP
☐ Precision low backlash (typical 1°) gearhead
☐ Base or flange mounting
☐ Output torque 3 to 50 inch-lbs (see buying guide)
☐ Bronze sleeve bearings
☐ Permanent lubrication - silicone grease
☐ *
E-550-MGHD
☐ Spur gearhead felt-wick lubrication
☐ Precision die cast gear housing
☐ Base or flange mounting
☐ Bronze sleeve bearings are oil impregnated and sintered
☐ Output torque 3 to 80 inch-lbs (see buying guide)
☐ *
* See buying guide for dimensions and additional information. † optional on E550-0.

E-586-O Control

- ☐ Dual tracking reference regulator insures better than .5% speed stability with line variations (based on nominal line of 115/230 VAC - 50/60 Hz ± 10%
- ☐ 1000:1 speed range (5-5000 rpm)
- ☐ Speed, torque readout (optional)
- ☐ Fuse and thermal overload protection
- ☐ Adjustable torque limit
- ☐ *

RPM 5000 4000 3000 2500 2000 1500 1000 500 100 50 20 5

0 5 10 15 20
TORQUE OZ-INCHES

SPEED REGULATION ZONES
FOR E-586 CONTROL*

E-586-MG

- ☐ 1/10 hp
- ☐ Die cast end caps
- ☐ Nickel-chrome plated motor and tach housings
- ☐ Stainless steel shaft
- ☐ Face or servo clamp mounting
- ☐ High energy oriented Ferrite magnets
- ☐ Shielded ball bearings
- ☐ Separate motor and tach-generator windings
- ☐ Temperature compensated generator
- ☐ *

E-586 BVC Control

- ☐ Velocity or position amplifier
- ☐ ± 15V @ 20 mA available for external supply
- ☐ Convection cooling
- ☐ Adjustable torque limit
- ☐ Fuse and circuit breaker protection
- ☐ Completely bi-directional
- ☐ *

RPM 5000 4000 3000 2500 2000 1500 1000 500 100 50 20 10

0 10 20 0 20 40
TORQUE OZ-INCHES
E-586/MG E-660/596

SPEED REGULATION ZONES
E-586 BVC CONTROL

E-660/596

- ☐ Rated 1/10 hp w/586 bi-directional control
- ☐ Die cast end caps
- ☐ Centerless ground nickel-chrome plated motor housing
- ☐ Bi-directional
- ☐ Adaptable to NEMA 42cz frame
- ☐ Shielded ball bearings
- ☐ Stainless steel shaft
- ☐ Separate motor and tach-generator windings.
- ☐ High energy oriented Ferrite magnets
- ☐ *

RPM 3000 2500 2000 1500 1000 500 100 50 20 10 5

5 10 20 30 40 50 60 70
TORQUE OZ-INCHES

SPEED REGULATION ZONES
E-650-0/S CONTROL

E-650-O/S Control

- ☐ 1000:1 speed range (3-3000 rpm)
- ☐ Low drift differential input stage
- ☐ Complete closed loop transistorized velocity feedback amplifier
- ☐ Fuse and circuit breaker protection
- ☐ 115 VAC - 50/60 Hz
- ☐ *

GENERAL INFORMATION: All items listed in this catalog are regularly stocked. For list prices and additional stock items see the buying guide. Duty Cycle: All items are designed for continuous duty in a 25°C ambient. The E650-MG may require a heat sink when operating continuously at full torque at 3000 rpm. Contact the Applications Engineering Department for additional information and data on operating in higher ambient temperatures.

E-650-M

E-650-O/M

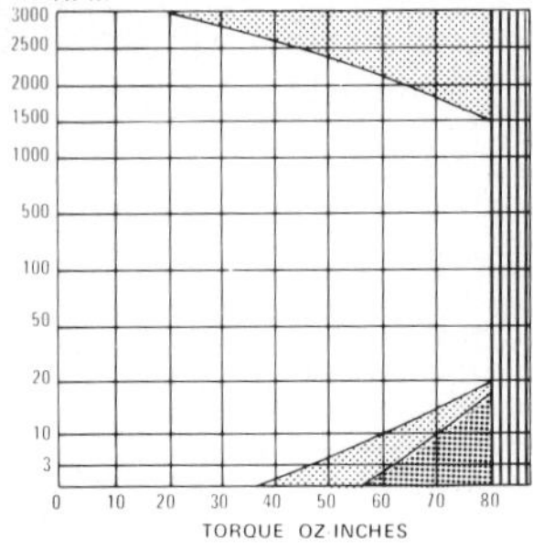

SPEED REGULATION ZONES
E-650-M/650-0/M CONTROLS

E-650-M•650-O/M Controls

- ☐ 1000:1 speed range (3-3000 rpm)
- ☐ Low drift differential input stage
- ☐ Complete closed loop transistorized velocity feedback amplifier
- ☐ Adjustable torque limit
- ☐ Speed torque readout†
- ☐ Reversing brake switch†
- ☐ Fuse and circuit breaker protection
- ☐ 115VAC - 50/60 Hz
- ☐ *

E-650-MG

- ☐ 1/5 hp
- ☐ Die cast end caps
- ☐ Centerless ground, nickel-chrome plate motor housing
- ☐ Temperature compensated generator
- ☐ Stainless steel shaft
- ☐ Bi-directional
- ☐ High energy oriented Ferrite magnets
- ☐ *

E-650-MGHR

- ☐ Output torque 12 to 115 inch-lbs. (see buying guide)
- ☐ Boston 310 series single reduction worm gear
- ☐ Precision machined cast iron housing
- ☐ Bi-directional
- ☐ Shipped oil filled
- ☐ *

E-650-MGHS

- ☐ Spur gearhead felt-wick lubricated
- ☐ Precision die cast gear housing
- ☐ Bi-directional
- ☐ Bronze sleeve bearings are oil impregnated and sintered
- ☐ Output torque 80 inch-lbs (see buying guide)
- ☐ Base or flange mounting
- ☐ *

HOW TO READ THE SPEED REGULATION ZONE GRAPHS:

Select the torque and rpm range; the data (% of regulation) is stated as a percentage of the set speed.

- **99% of set speed**
- **98% of set speed**
- **97% of set speed**
- **<96% of set speed**
- **Intermittent duty only—consult ECC Engineering Application Dept. for additional data.**

E-680-MG

- ☐ 680MG 1/4 hp
- ☐ Bi-directional
- ☐ Temperature compensated generator
- ☐ High energy oriented Ferrite magnets
- ☐ Die cast end caps
- ☐ Centerless ground, nickel-chrome plated motor housing
- ☐ Stainless steel shaft
- ☐ Shielded ball bearings
- ☐ *

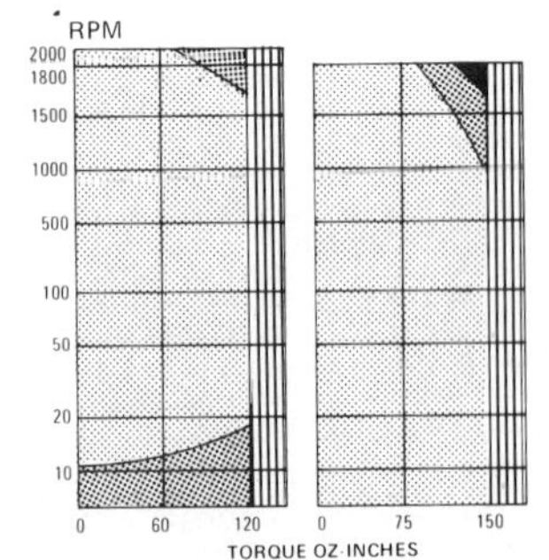

E-680-MG E-760-MG
SPEED REGULATION ZONES
E-650-M•650-0/M CONTROLS

E-760-MG

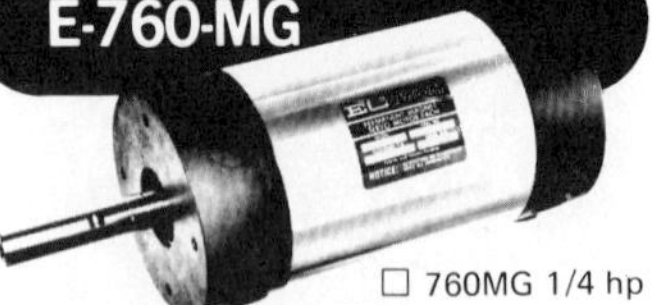

- ☐ 760MG 1/4 hp
- ☐ Bi-directional
- ☐ Temperature compensated generator
- ☐ High energy oriented Ferrite magnets
- ☐ Die cast end caps
- ☐ Centerless ground, nickel-chrome plated motor housing
- ☐ Stainless steel shaft
- ☐ Shielded ball bearings
- ☐ *

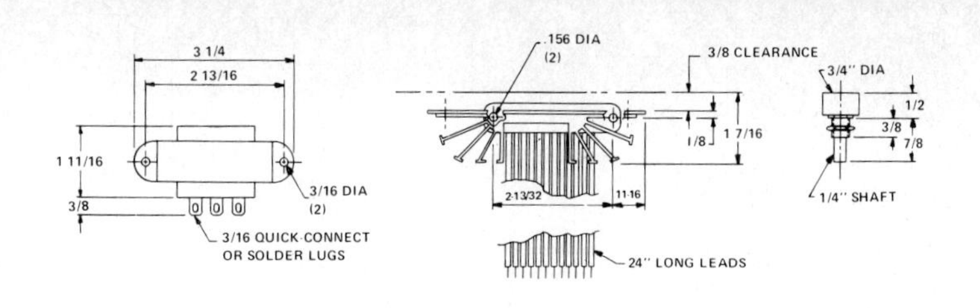

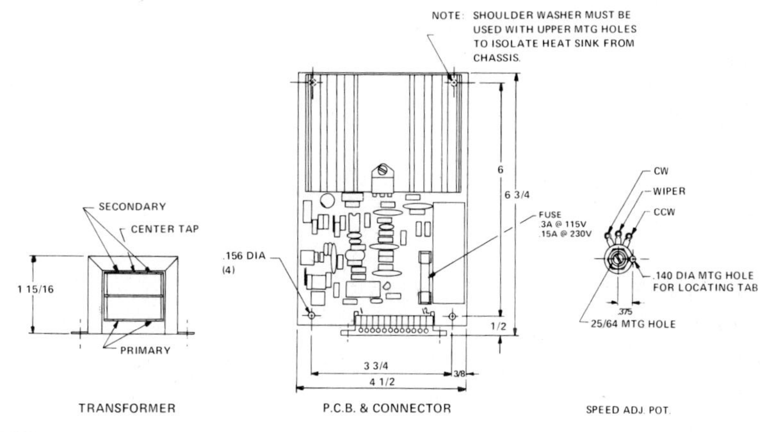

NOTES:
1. SEE SCHEMATIC FOR ELECTRICAL CONNECTORS.

CONTROLLER SPECIFICATIONS E350-0

Input Power

105-125 VAC 50/60 Hz
210-230 VAC

Line Regulation

1%

Maximum Output Voltage No Load

32 VDC

Output Current

.75 A

Part Number	Model Number	Approx. Shpg. Wt. lbs.	oz.
9086-0007 (115 VAC system)	350-0-115V	2	
9086-0008	350-0-220V	2	

Options

Speed & Torque Readout
0041-5011

Ten-Turn Potentiometer
Use Spectrol Model 534-10K potentiometer and Model 11 dial or Bourns Model 3507S-10K potentiometer and H492-3 dial.

Reversing Switch
Order schematic 9086-0009 for installation diagram and recommended hardware.

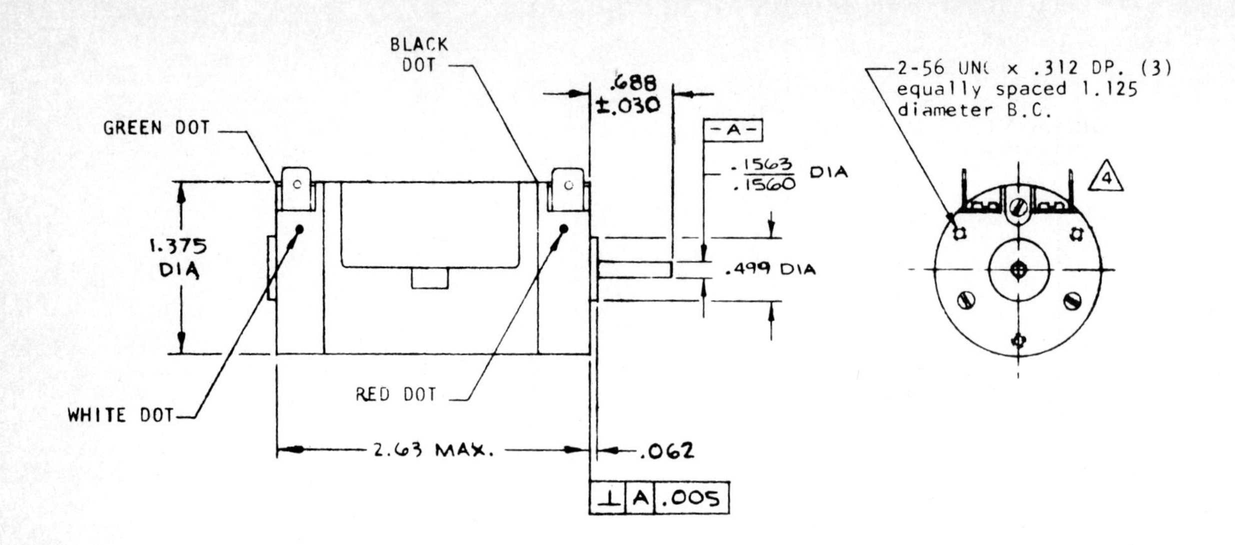

Part Number	Model Number	Approx. Shpg. Wt. lbs.	oz.
0350-00-000	E350-MG		6

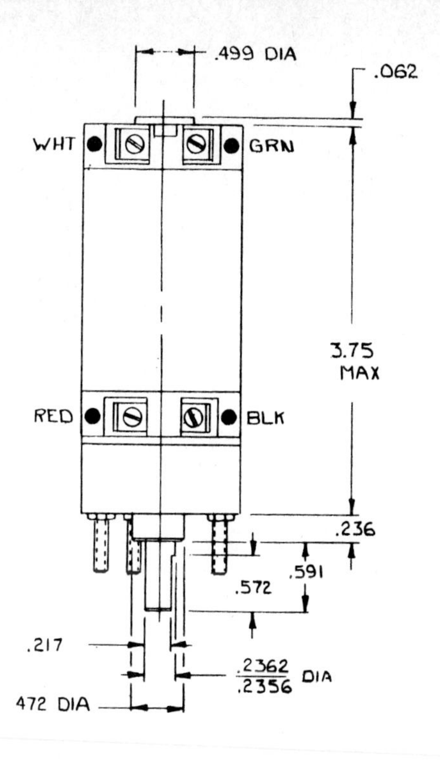

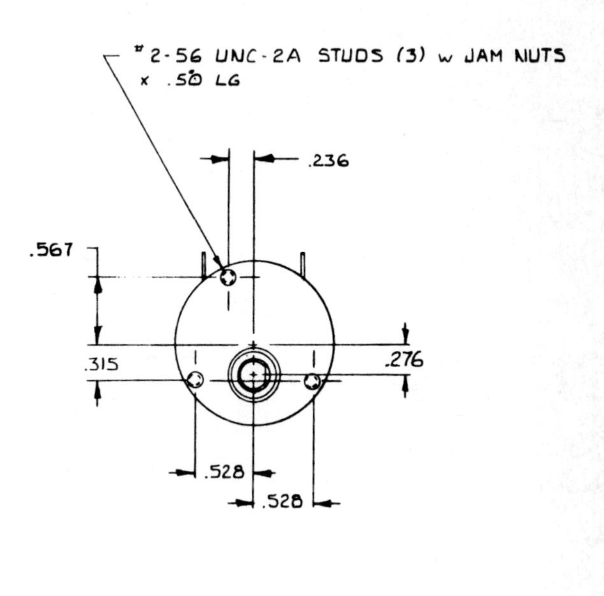

Gearhead Ratio	120:1
Maximum Output Torque inch-lbs.	2.5

Part Number	Ratio	Model Number	Approx. Shpg. Wt. lbs.	oz.
350-003-692	120:1	E350-MGH		9
0350-00-003		E350-MG w/o gearhead		6

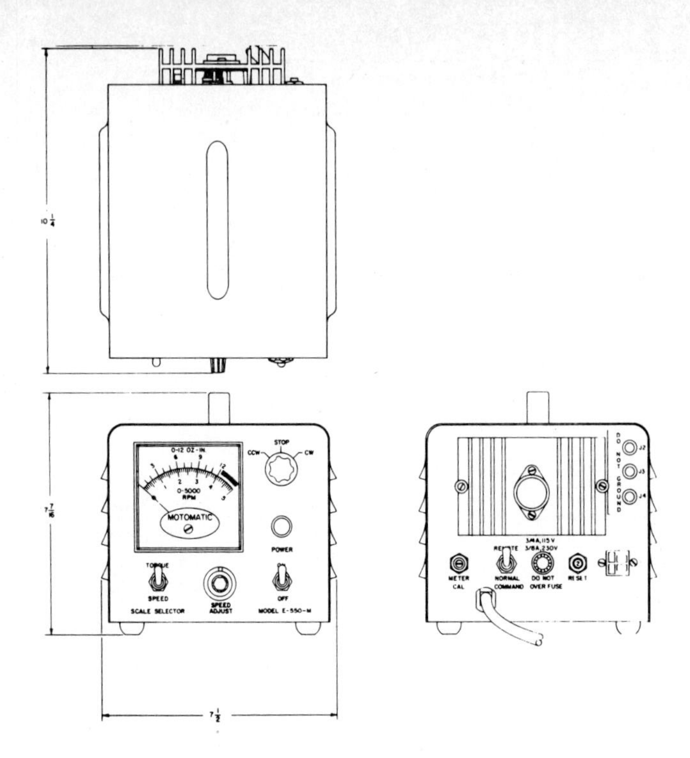

CONTROLLER SPECIFICATIONS E550-M, E550-0

Input Power

105-125 VAC 50/60 Hz
210-230 VAC

Line Regulation

1%

Maximum Output Voltage No Load

36 VDC

Output Current

5 A

Part Number	Model Number	Approx. Shpg. Wt. lbs.	Approx. Shpg. Wt. oz.
9022-0025	E550-M	7	14
9014-0025	E550-0	4	8

(See reverse side for outline drawing.)

Options

Option	lbs.	oz.
Speed & Torque Readout 0041-5006-1		9
Reversing Switch 0041-5005-1		3

Ten-Turn Potentiometer
Use Spectrol Model 534-5K potentiometer and Model 11 dial or Bourns 350S-5K potentiometer and H492-3 dial.

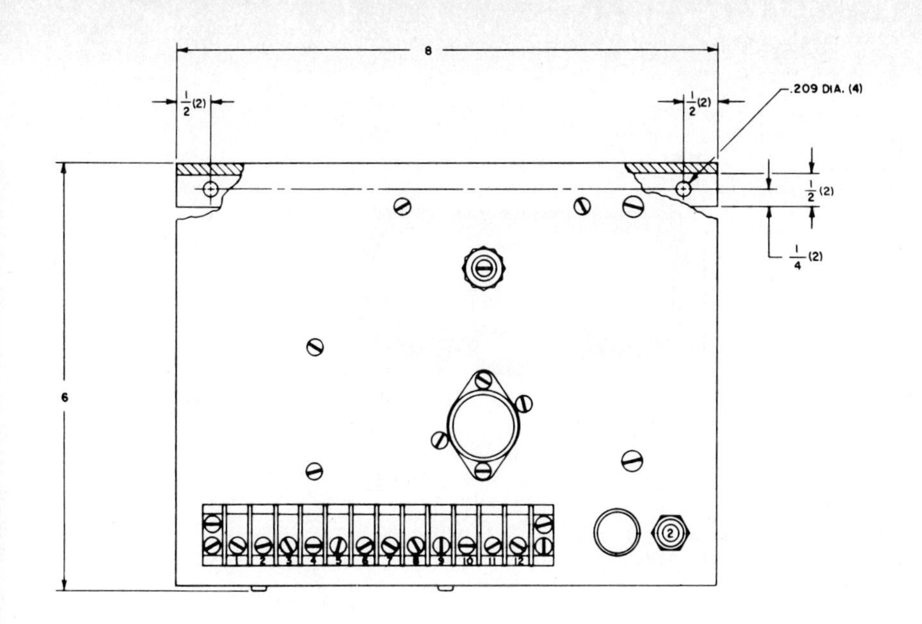

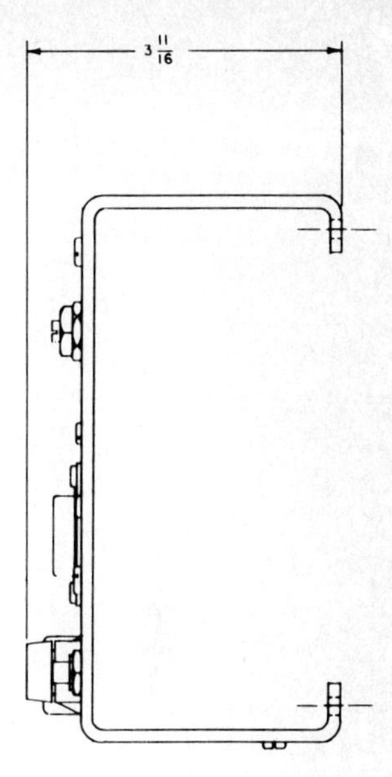

9014-0025

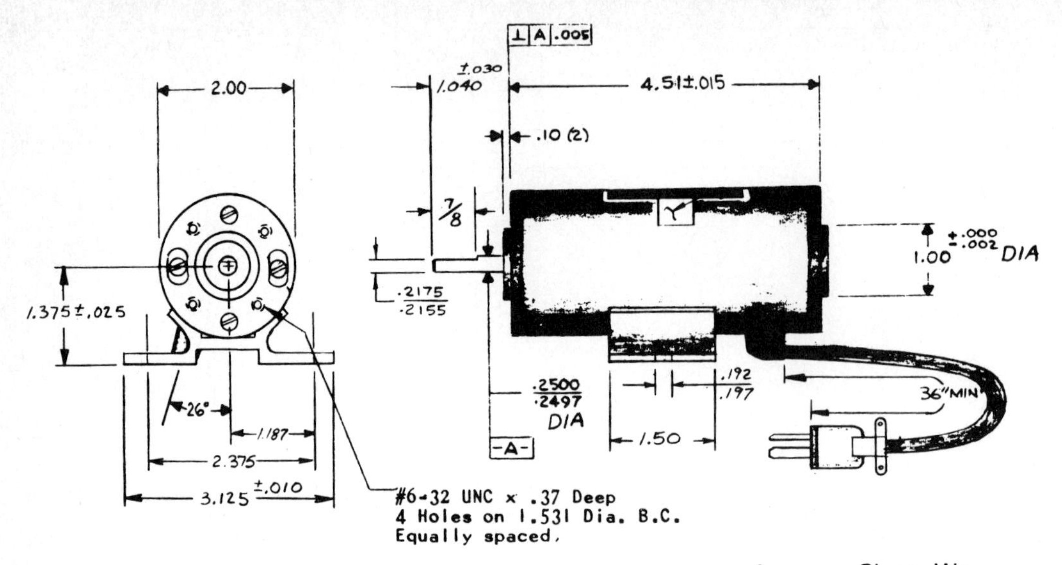

Part Number	Model Number	Approx. Shpg. Wt. lbs.	oz.
0550-00-000	E550-MG	1	12
0550-00-008 (double-ended shaft)	E550-MG	1	12

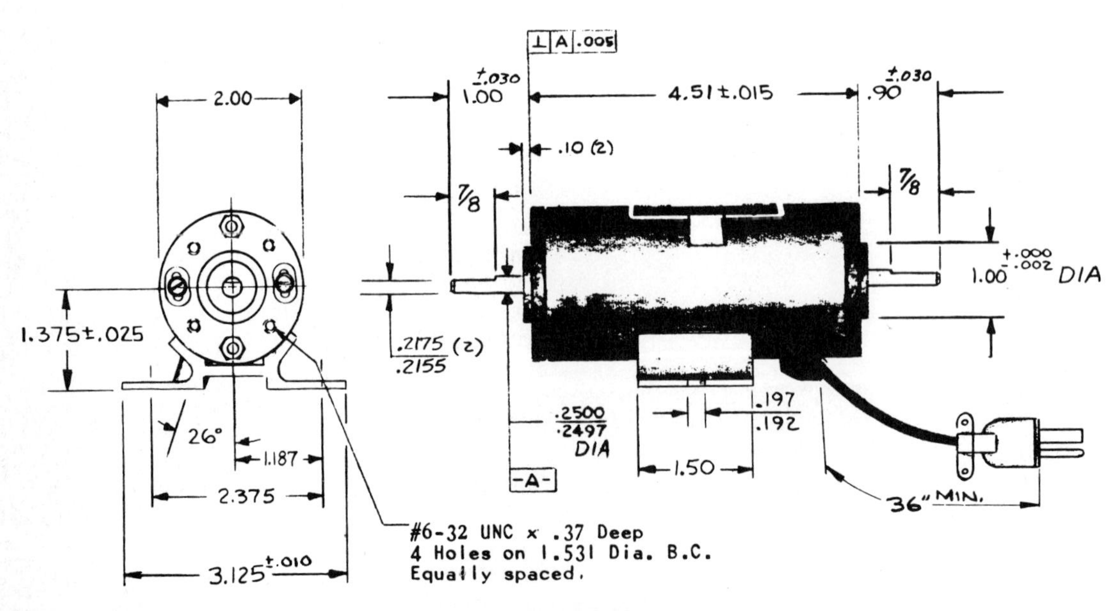

0550-00-008

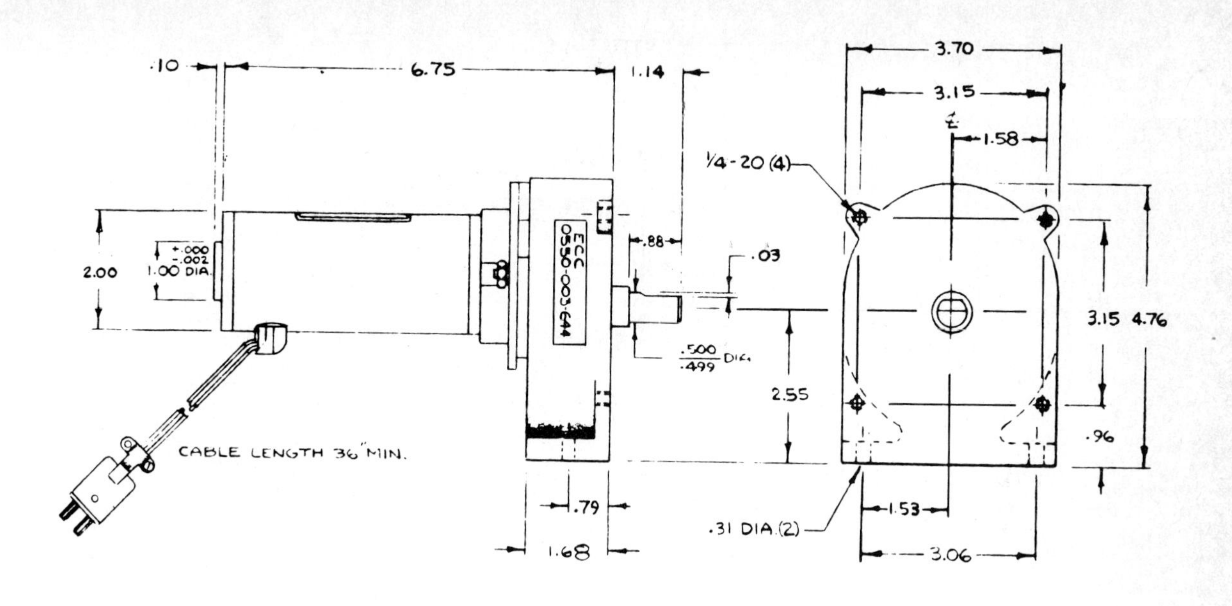

Gearhead Ratio	5.2.1	10:1	20:1	49:1	102:1	193:1	486:1	964:1	5016:1
Maximum Output Torque inch-lbs.	3	6	12	30	60	80	80	80	80

Part Number	Ratio	Model Number	Approx. Shpg. Wt. lbs.	oz.
550-003-640	5.2:1	E550-MGHD	4	12
550-003-643	10:1	E550-MGHD	4	12
550-003-644	20:1	E550-MGHD	4	12
550-003-646	49:1	E550-MGHD	4	12
550-003-647	102:1	E550-MGHD	4	12
550-003-648	193:1	E550-MGHD	4	12
550-003-649	486:1	E550-MGHD	4	12
550-003-650	964:1	E550-MGHD	4	12
550-003-653	5016:1	E550-MGHD	4	12
0550-00-003		E550-MG w/o gearhead	1	8

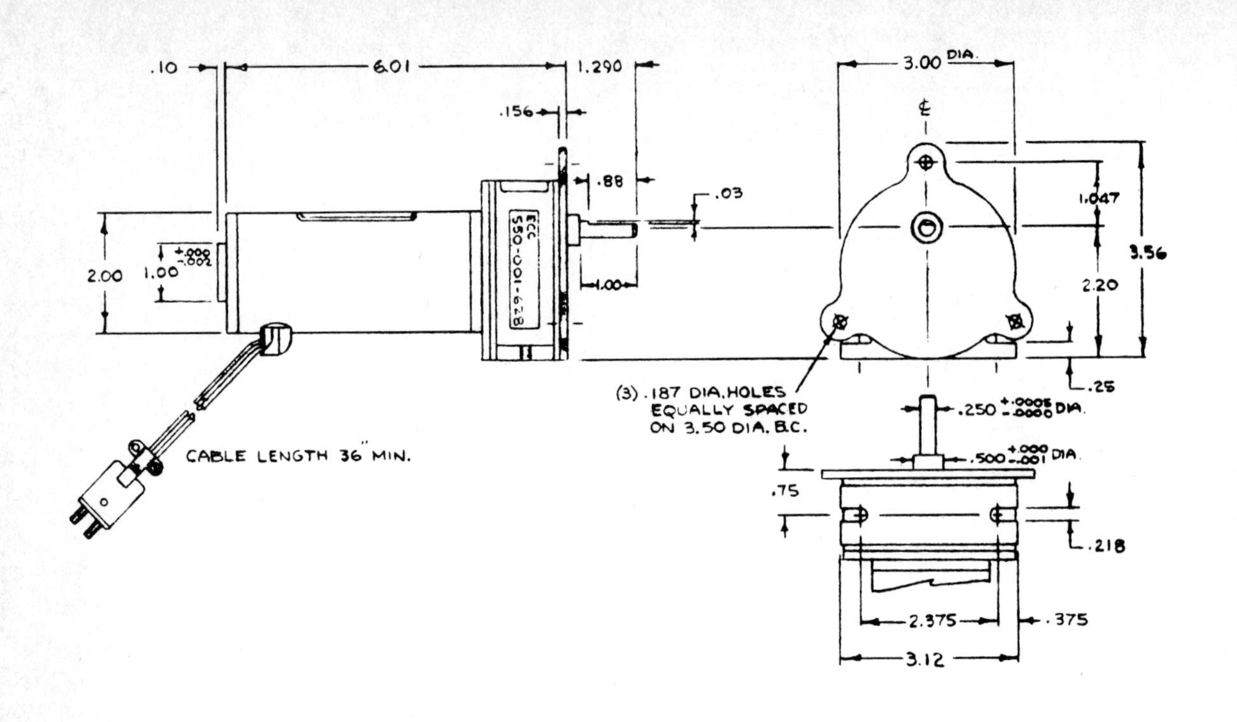

Gearhead Ratio	10:1	30:1	50:1	100:1	500:1
Maximum Output Torque inch-lbs.	3	20	30	50	50

Part Number	Ratio	Model Number	Approx. Shpg. Wt. lbs.	oz.
550-001-625	10:1	E550-MGHP	3	9
550-001-626	30:1	E550-MGHP	3	9
550-001-627	50:1	E550-MGHP	3	9
550-001-628	100:1	E550-MGHP	3	9
550-001-629	500:1	E550-MGHP	3	9
0550-00-001		E550-MG w/o gearhead	2	6

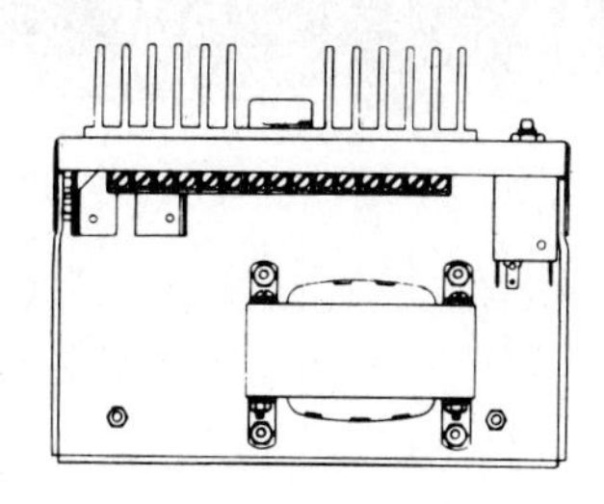

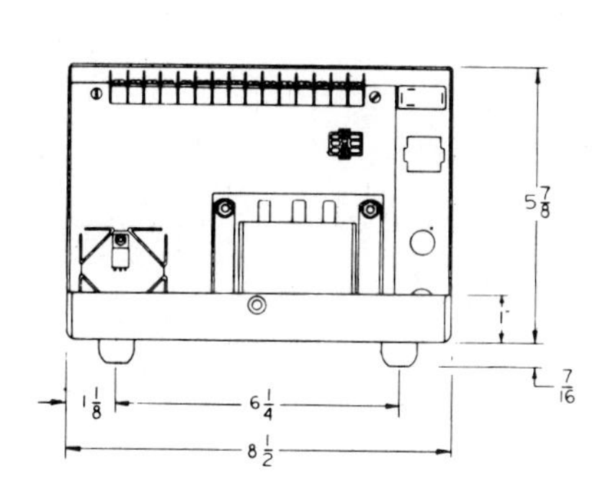

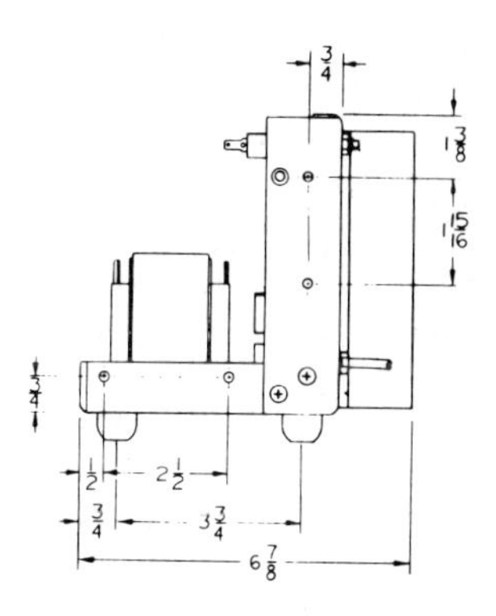

CONTROLLER SPECIFICATIONS E586-0

Input Power

105-125 VAC 50/60 Hz
210-230 VAC

Line Regulation

1%

Maximum Output Voltage No Load

36 VDC

Output Current

3.3 A

Part Number	Model Number	Approx. Shpg. Wt. lbs.	oz.
9078-0012	E586-0	9	6

Options

Speed & Torque Readout 0041-5012		9
Reversing Switch 0041-5005-1		3

Ten-Turn Potentiometer
Use Spectrol Model 534-5K potentiometer and Model 11 dial or Bourns 3507S-5K potentiometer and H492-3 dial.

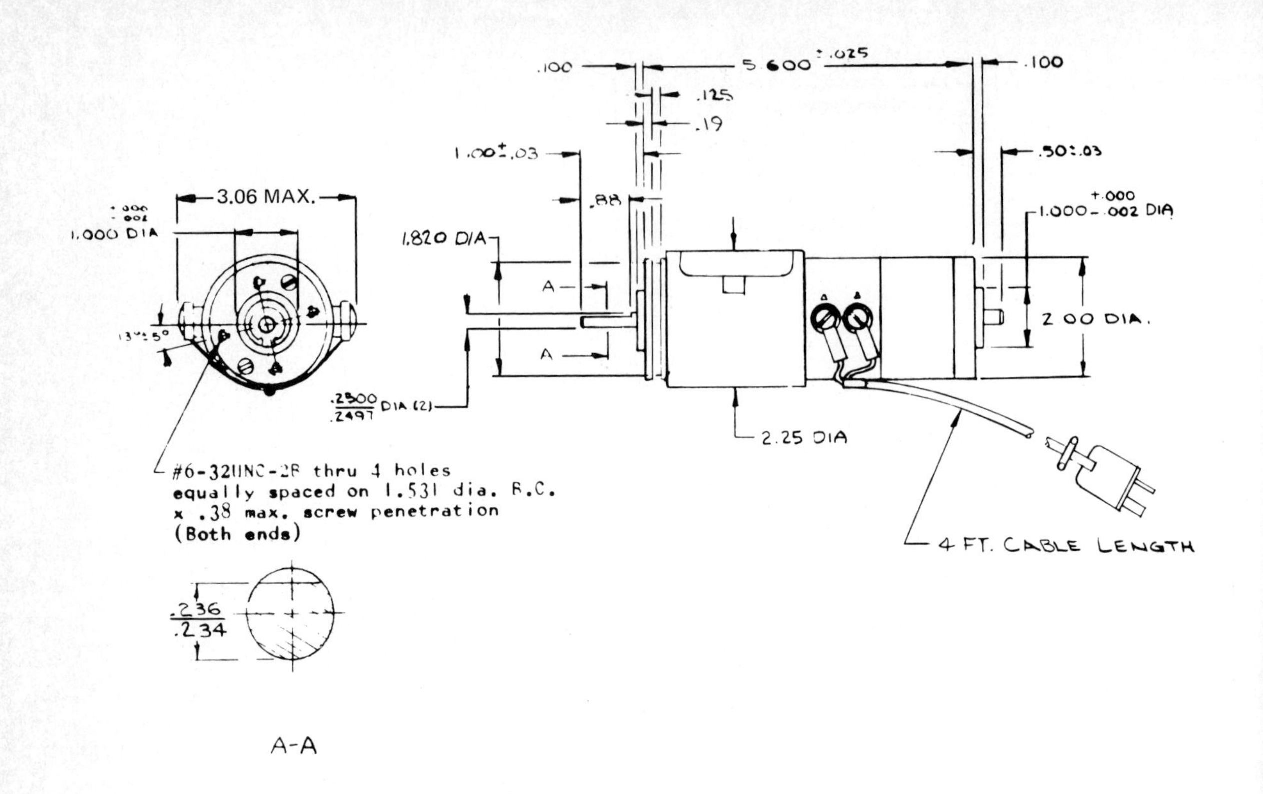

Part Number	Model Number	Approx. Shpt. Wt. lbs.	oz.
0586-00-010	E586-MG	2	13

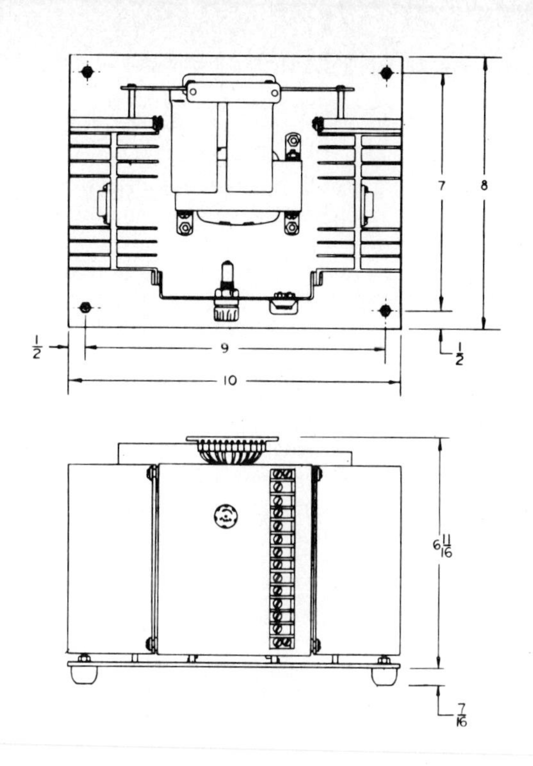

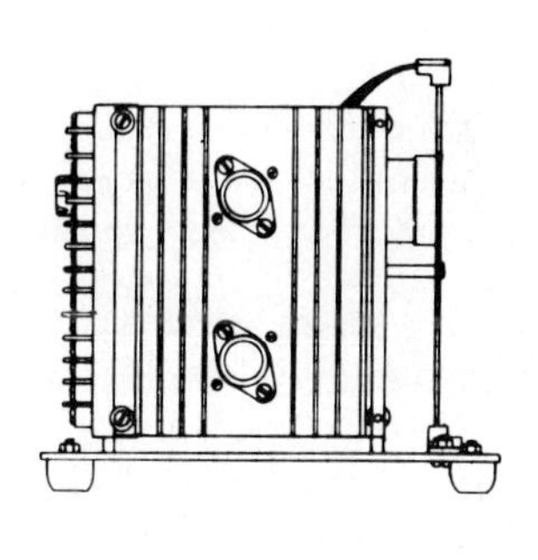

CONTROLLER SPECIFICATIONS E586-BVC, E586-BPC

Input Power
105-125 VAC 50/60 Hz
210-230 VAC

Line Regulation
1%

Maximum Output Voltage No Load
36 VDC

Output Current
3.3 A

Part Number	Model Number	Approx. Shpg. Wt. lbs.	Approx. Shpg. Wt. oz.
9017-0025	E586-BVC (velocity)	11	2
9017-0050	E586-BPC (position)	11	2

(See reverse side for additional information on E586-BPC.)

Options
E586-BVC only – Digital Readout

E586-BPC BI-DIRECTIONAL POSITION CONTROL SYSTEM

Brief Description:

The E586-BPC system is a position control system employing a high-gain, closed-loop electronic control. The input command can be any variable resistance or ungrounded reference DC voltage. The motor angular position will be proportional to the command setting, with the position feedback coming from the angular position of a potentiometer or any other error voltage source. Velocity feedback is employed to control the travel rate while the motor is "homing in" on the desired position.

System Specifications:

The *position accuracy* of the system is limited by two variables: (1) the drift characteristics of the reference power supplies, and (2) the accuracy of the command and feedback signals. Using the potentiometer supplied with the standard E586-BPC, system accuracy is approximately 1% per revolution or 3.6°. Using improved power supply regulation and sensors, the accuracy can easily be improved to .1% per revolution or .36°. For high accuracy systems, please consult an Electro-Craft representative.

The *system travel speed* is controlled by velocity feedback from an integral tachometer and closed-loop servo. The standard E586-BPC controls the motor to run at 5,000 rpm during travel to the "target" position.

Command input can be accomplished in either of two ways: (1) a position setting on a potentiometer, using the E586-BPC reference power supply voltage, or (2) an ungrounded command voltage of 0 to +15 volts DC. Care must be exercised to insure that the input command range is not greater than the feedback sensor output range.

The *feedback input* requirements are similar to the command requirements and can be either a position on a potentiometer, using the feedback reference voltage source, or an ungrounded 0 to −15 volts DC signal.

BLOCK DIAGRAM

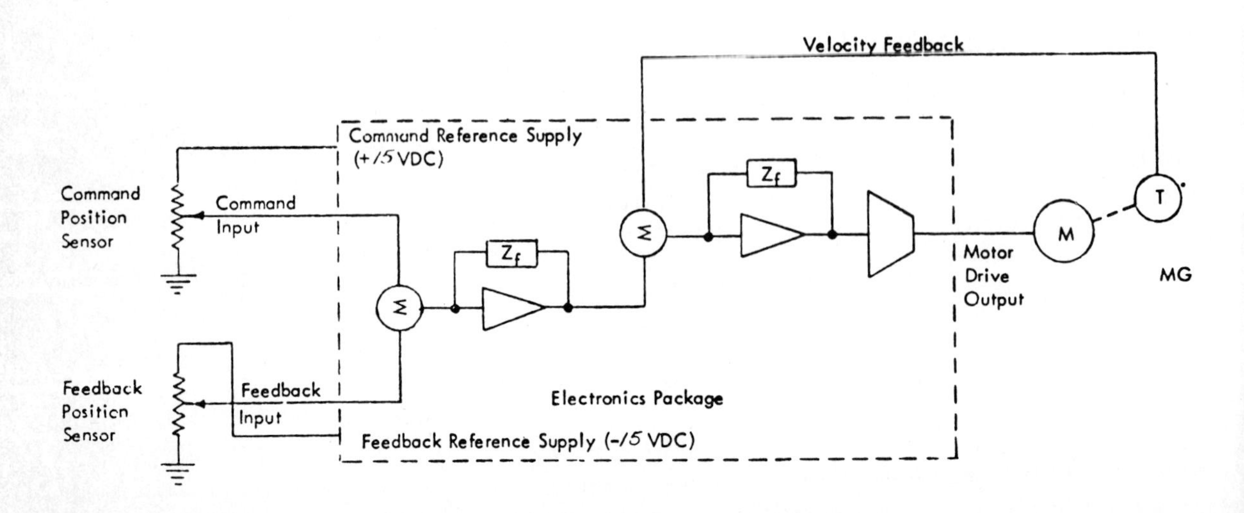

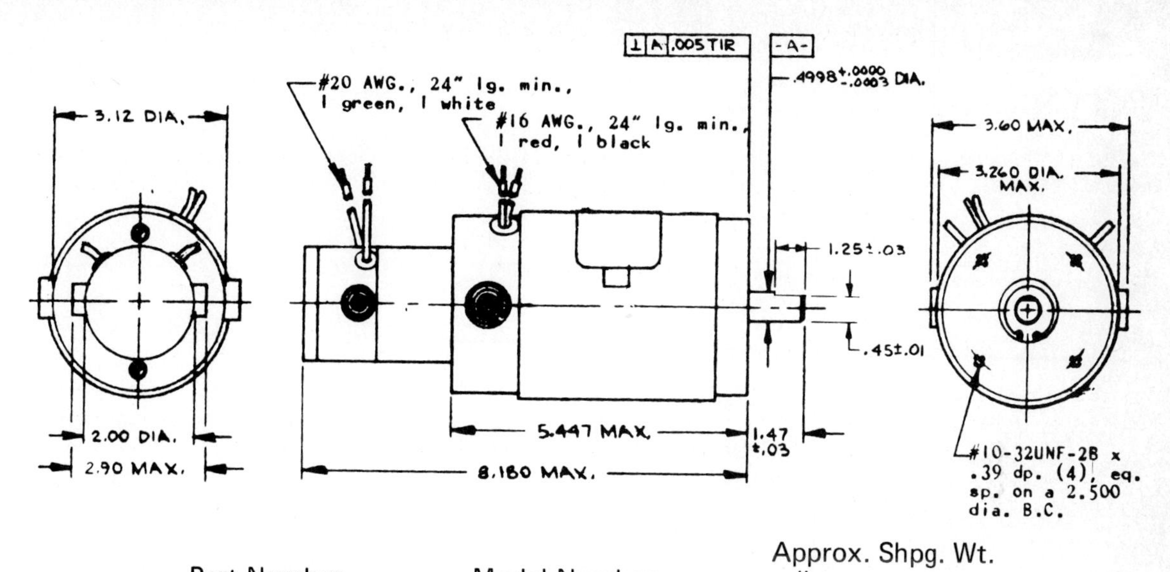

Part Number	Model Number	Approx. Shpg. Wt. lbs.	oz.
0660-08-014	E660/596	7	6
0660-08-015 (42cz flange)	E660/596	7	6

Note: Matching amplifier E586-BVC or E586-BPC only.

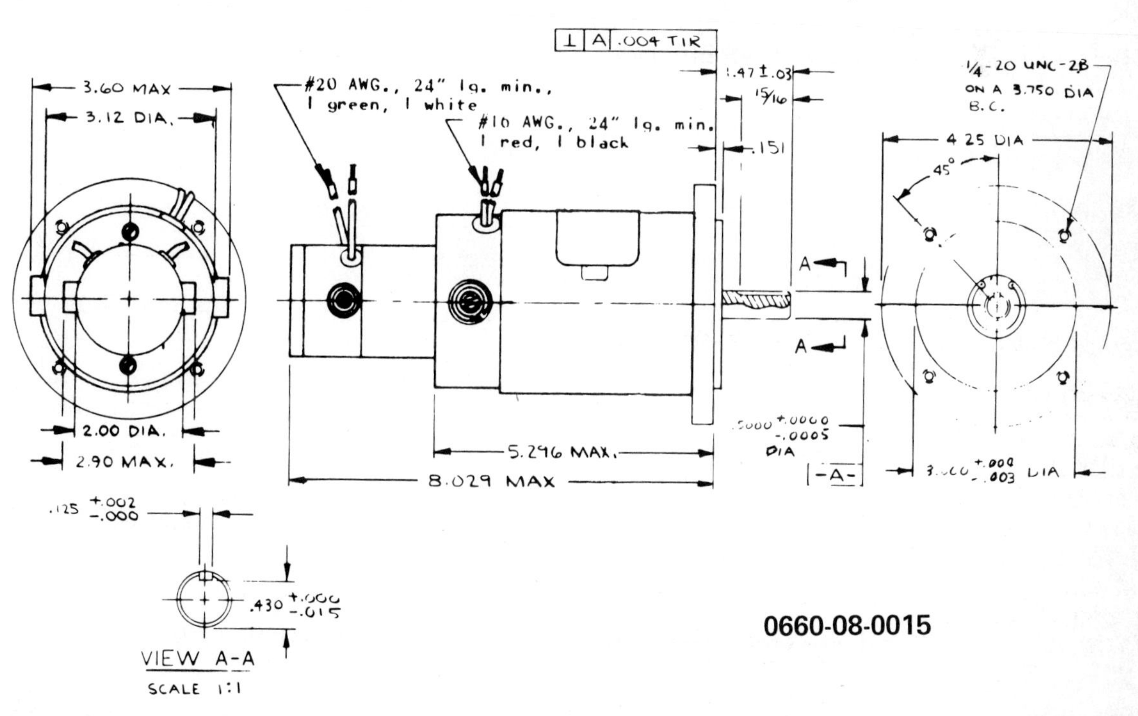

0660-08-0015

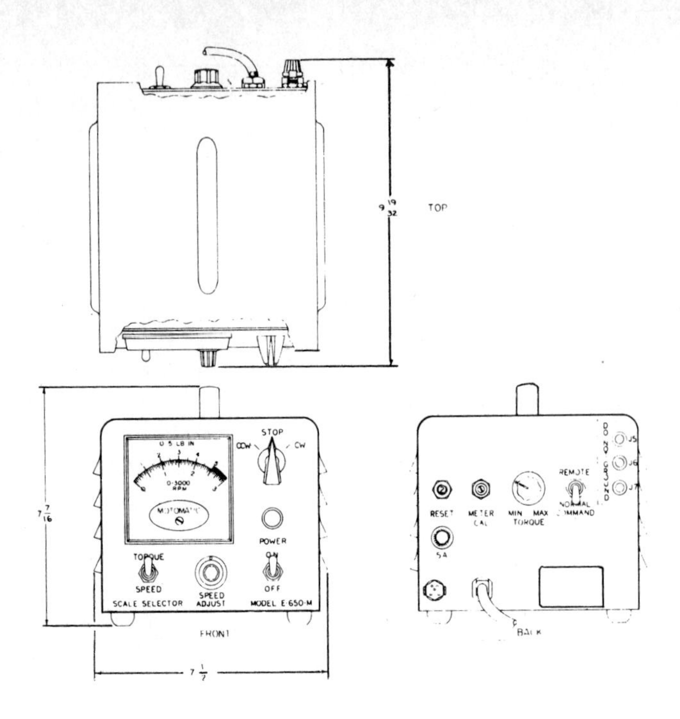

CONTROLLER SPECIFICATIONS E650-M

Input Power

105-125 VAC 50/60 Hz

Line Regulation

1%

Maximum Output Voltage No Load

130 VDC

Output Current

2.2 A

Part Number	Model Number	Approx. Shpg. Wt. lbs.	oz.
9025-0025	E650-M	10	9

Options

Ten-Turn Potentiometer
Use Spectrol Model 534-30K potentiometer and Model 11 dial or Bourns Model 3507S-30K potentiometer and H492-3 dial.

Remote Input
Order schematic 9025-0026 for installation diagram and recommended hardware.

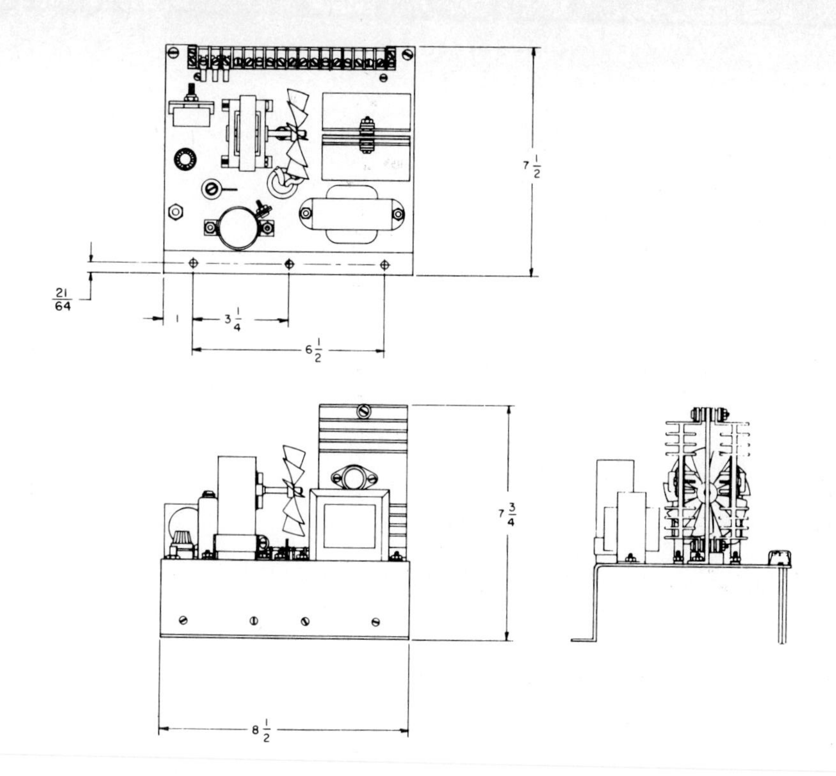

CONTROLLER SPECIFICATIONS E650-O/M

Input Power	Line Regulation
105-125 VAC 50/60 Hz	1%

Maximum Output Voltage No Load	Output Current
130 VDC	2.2 A

Part Number	Model Number	Approx. Shpg. Wt. lbs.	Approx. Shpg. Wt. oz.
9009-0025	E650-O/M	7	12

Options

Speed & Torque Readout
0041-5000-1

Reversing Switch
0041-5002-1

Ten-Turn Potentiometer
Use Spectrol Model 534-30K potentiometer and Model 11 dial or Bourns Model 3507S-30K potentiometer and H492-3 dial.

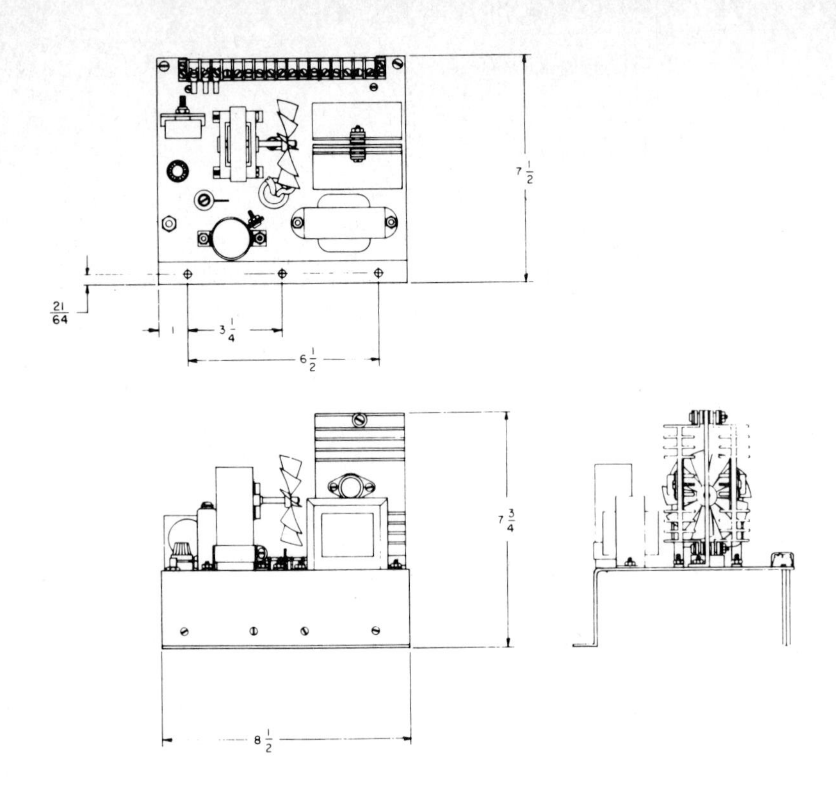

CONTROLLER SPECIFICATIONS E650/0/S

Input Power
105-125 VAC 50/60 Hz

Line Regulation
1%

Maximum Output Voltage No Load
110 VDC

Output Current
1.8 A

Part Number	Model Number	Approx. Shpg. Wt. lbs.	oz.
9010-001-4	E650-O/S	6	14

Options

Speed & Torque Readout
0041-5003-1

Reversing Switch
0041-5002-1

Ten-Turn Potentiometer
Use Spectrol Model 534-30K potentiometer and Model 11 dial or Bourns Model 3507S-30K potentiometer and H492-3 dial.

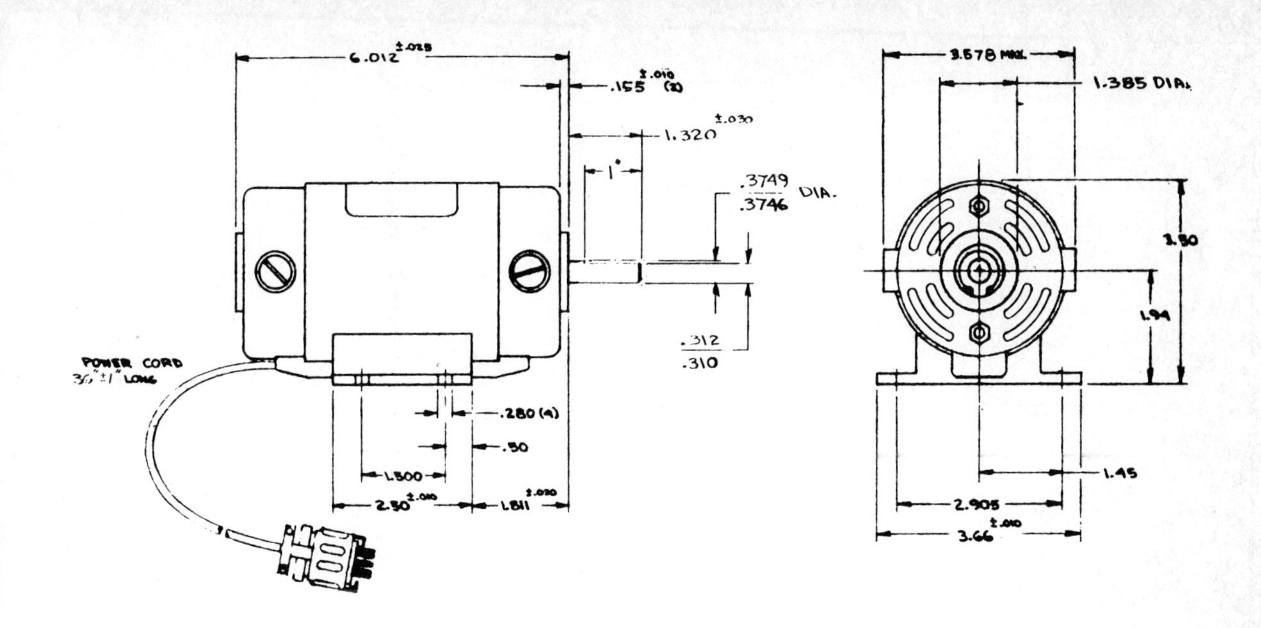

Part Number	Model Number	Approx. Shpg. Wt. lbs.	oz.
0650-00-004	E650-MG	5	10
0650-00-020	E650-MG	5	10

(face mounting — See reverse side for outline drawing.)

Options

Base Mount Conversion Kit for 0650-00-020

Part Number	Model Number	lbs.	oz.
0029-6000-3			6

Interchangeable Replacement Motors for Discontinued E600 Series

Part Number	Model Number	lbs.	oz.
600-005 0650-10-051	E650-MG	5	10
600-003 0650-10-052	E650-MG	5	10

Replacements for Obsolete Part Numbers

Part Number	Replacement
0650-00-003	0650-00-025
0650-00-009	0650-00-020
0650-00-026	0650-00-055

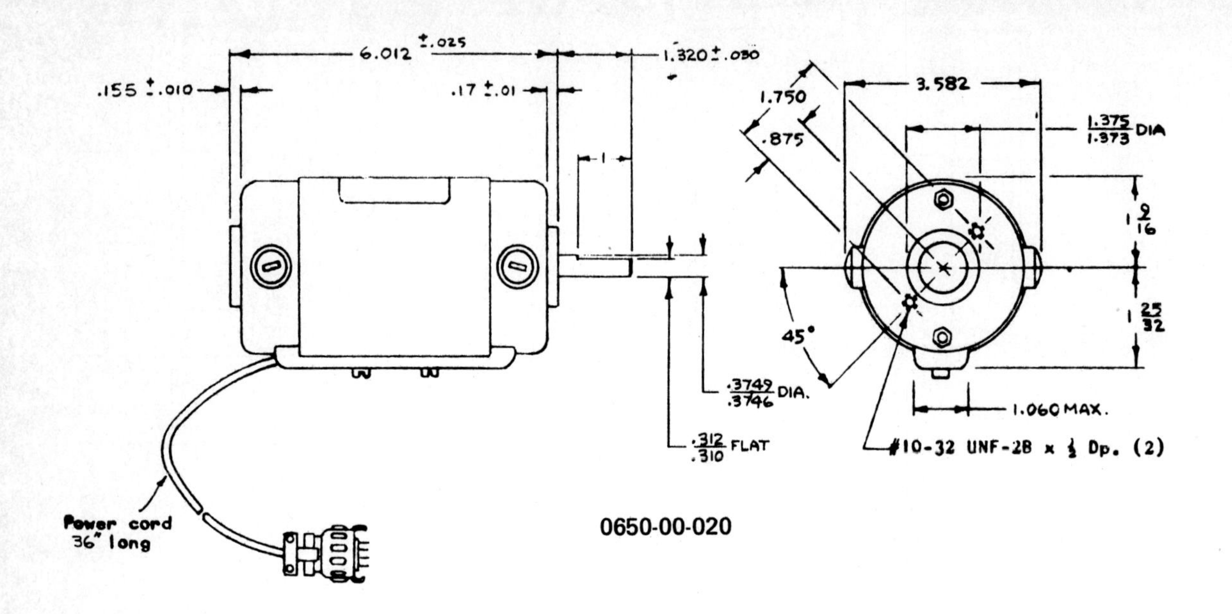
6.012 ±.025
1.320 ±.030
.155 ±.010
.17 ±.01
1
1.750
.875
3.582
1.375 / 1.373 DIA
1 9/16
1 25/32
45°
.3749 / .3746 DIA.
.312 / .310 FLAT
1.060 MAX.
#10-32 UNF-2B x ½ Dp. (2)
Power cord
36" long
0650-00-020

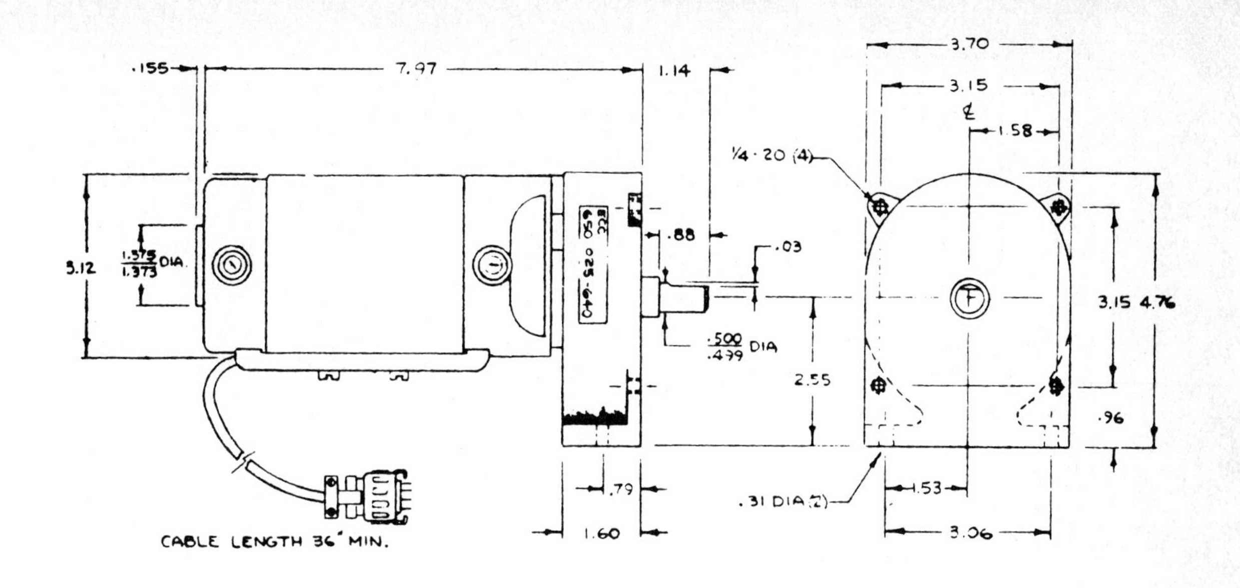

Gearhead Ratio	49:1	102:1	193:1	486:1	964:1
Maximum Output Torque inch-lbs.	80	80	80	80	80

Part Number	Ratio	Model Number	Approx. Shpg. Wt. lbs.	oz.
650-025-646	49:1	E650-MGHS	8	12
650-025-647	102:1	E650-MGHS	8	12
650-025-648	193:1	E650-MGHS	8	12
650-025-649	486:1	E650-MGHS	8	12
650-025-650	964:1	E650-MGHS	8	12
0650-00-025		E650-MG w/o gearhead	5	10

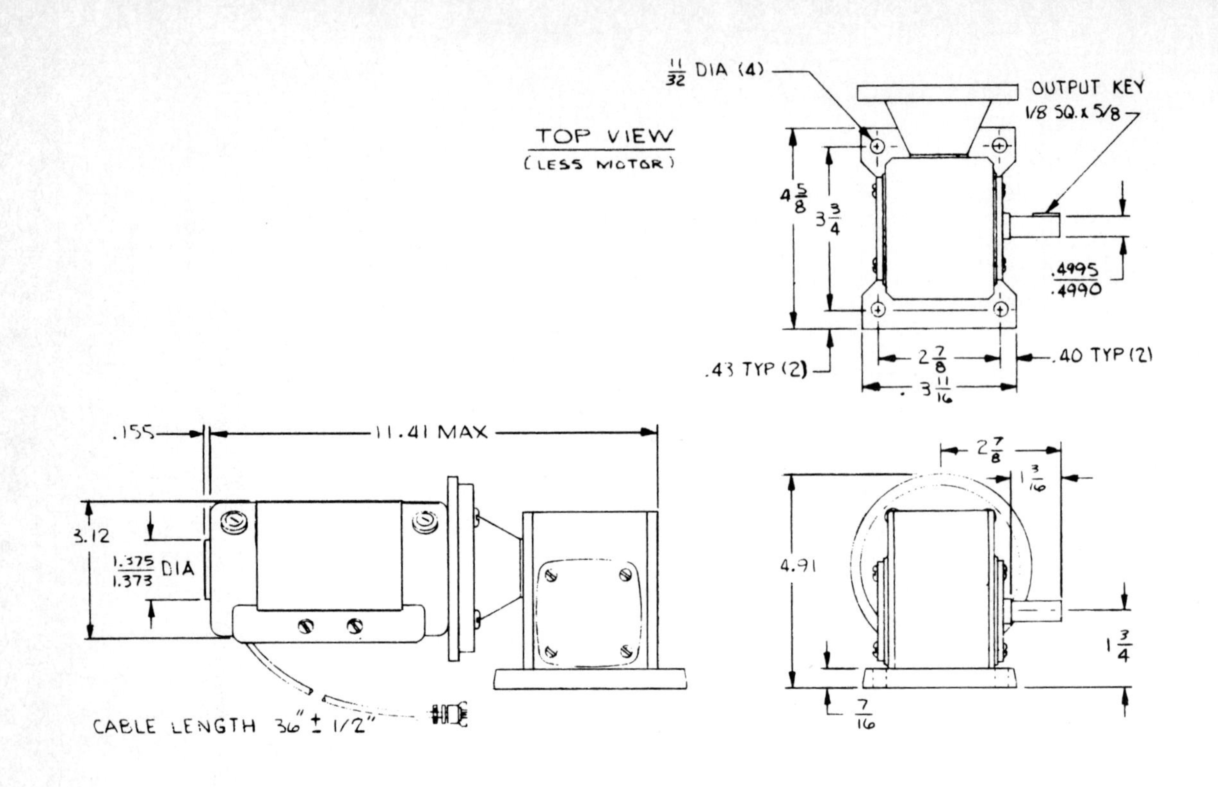

Gearhead Ratio	5:1	10:1	15:1	20:1	30:1	40:1	50:1
Maximum Output Torque inch-lbs.	12	24	36	48	73	97	115

Part Number	Ratio	Model Number	Approx. Shpg. Wt. lbs.	oz.
650-055-701*	5:1	E650-MGHR	12	10
650-055-702*	10:1	E650-MGHR	12	10
650-055-703*	15:1	E650-MGHR	12	10
650-055-704*	20:1	E650-MGHR	12	10
650-055-705*	30:1	E650-MGHR	12	10
650-055-706*	40:1	E650-MGHR	12	10
650-055-707*	50:1	E650-MGHR	12	10
0650-00-055		E650-MG w/o gearhead	5	10

*Shipped oil filled.

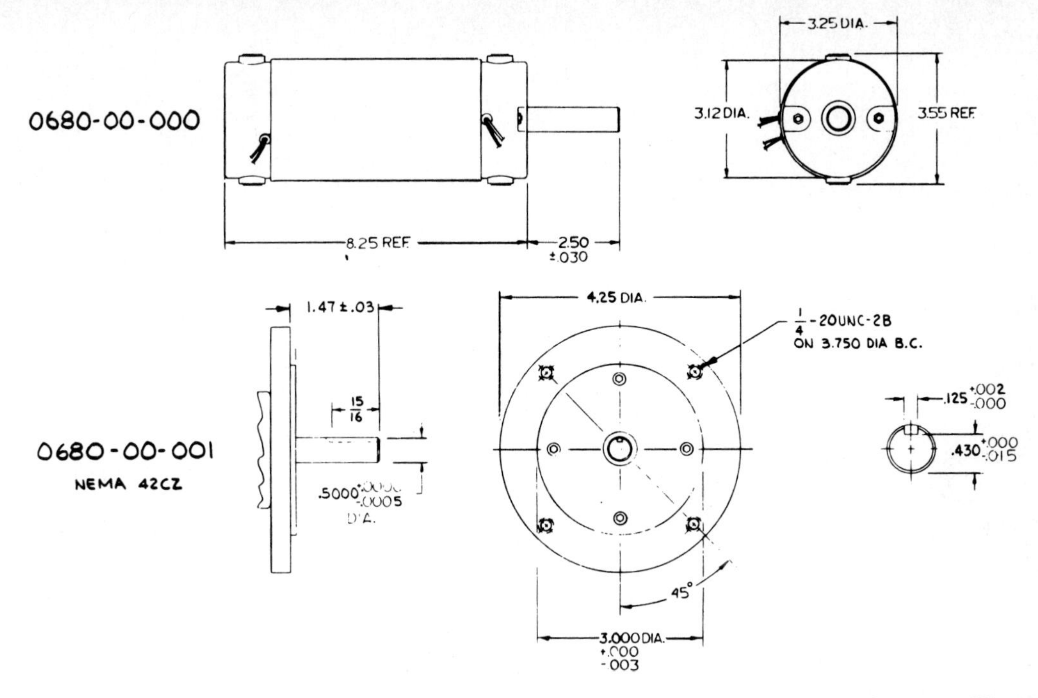

Part Number	Model Number	Approx. Shpg. Wt. lbs.	oz.
0680-00-000	E680-MG	8	12
0680-00-001 (NEMA 42 cz)	E680-MG	8	12

Matching Amplifiers:

Part Number	Model Number	Approx. Shpg. Wt. lbs.	oz.
9025-0035	E680-M	10	9
9009-0035	E680-O/M	7	12

Options:

Ten-Turn Potentiometer
Use Spectrol Model 534-30K potentiometer and Model 11 dial or Bourns Model 3507S-30K potentiometer and H492-3 dial.

Remote Input - E680-M only
Order schematic 9025-0026 for installation diagram and recommended hardware.

Speed & Torque Readout - E680-O/M only
0041-5013 — 9

Reversing Switch - E680-O/M only
0041-5002-1 — 7

For controller specifications and outline dimensions - see E650-M and E650-O/M.

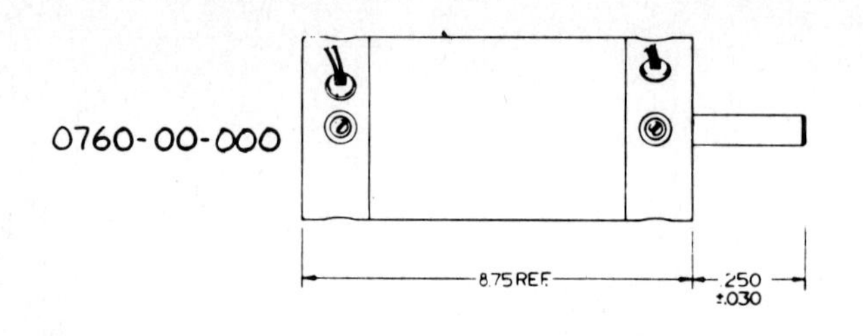

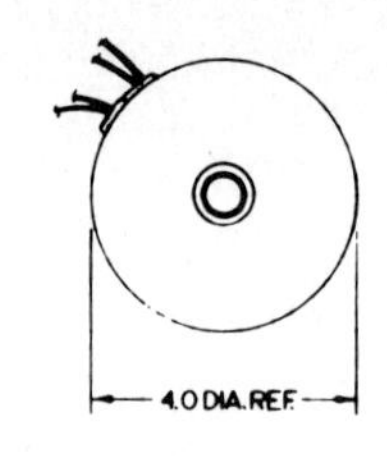

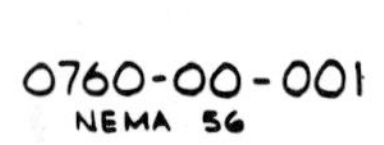

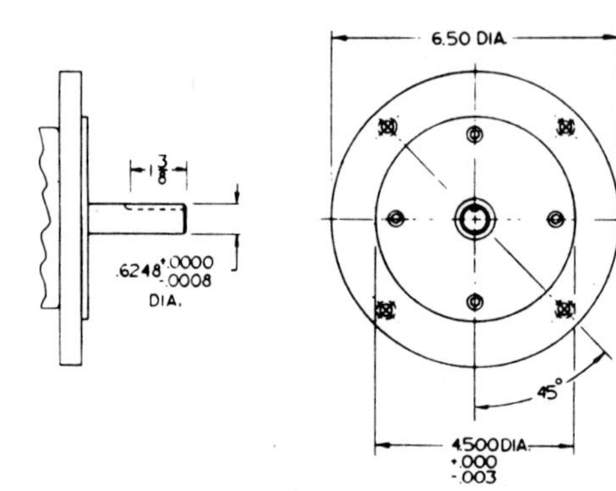

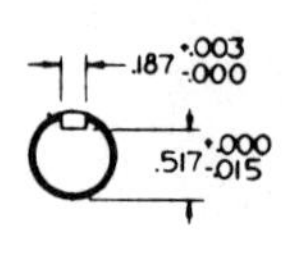

Part Number	Model Number	Approx. Shpg. Wt. lbs.	oz.
0760-00-000	E760-MG	11	10
0760-00-001 (NEMA 56)	E760-MG	11	10

Matching Amplfiers:

Part Number	Model Number	Approx. Shpg. Wt. lbs.	oz.
9025-0045	E760-M	10	9
9009-0045	E760-O/M	7	12

Options:

Ten-Turn Potentiometer
Use Spectrol Model 534-30K potentiometer and Model 11 dial or Bourns Model 3507S-30K potentiometer and H492-3 dial.

Remote Input — E760-M only
Order schematic 9025-0026 for installation diagram and recommended hardware.

Speed & Torque Readout - E760-O/M only
0041-5014 — 9

Reversing Switch - E760-O/M only
0041-5002-1 — 7

For controller specifications and outline dimensions - see E650-M and E650-O/M.

7.6.2. P6000 PULSE WIDTH MODULATED D.C. SERVO MOTOR CONTROLLERS

	P6100	P6200
• Voltage at Rated Current	±60V	±55V
• Rated Current	7A	15A
• Peak Current	12A	20A
• Rated Power	420W	825W
• Peak Power	670W	1000W
• Bandwidth	3000Hz	3000Hz

FEATURES:

Unique Open Modular Construction
Completely self contained
Plug in Standard or Custom Options
Regenerative Energy Protected
Silent Operation
115/230 V AC or DC Power Options
Independent Average and Peak Current Limits
Crowbar **Protected AC Power Supply**

The P6000 series are the fastest response pulse width modulated servo controls competitively available, providing bandwidths at full load exceeding 3000 Hz. The unique* fixed frequency design gives completely silent operation free from annoying audible "chirps". The standard unit, which is available for operation from AC line (model P6000AP) or for operation from a DC power supply (model P6000A), has the following features: three input fully adjustable summing amplifier with independent gain and lead-lag adjustment to enable operation with a wide range of motor and load types; a dual stage current limiter providing independent adjustment of peak and average current limit as well as pulse duration; unique crowbar protection (model P6000AP) which protects the amplifier from damage due to short circuits, motor failure or misuse; space for two 4-5/16" x 2-3/4" option cards which can be selected from a wide range of standard options or custom designed options; logic compatible end of travel limit and clamping inputs; all major subassemblies are mounted on plug-in connectors to enable replacement without the need to change the entire unit; and all major test points are brought out to a 14 pin D.I.P. socket for fast, easy troubleshooting and set up. This, combined with the unique open construction, makes the P6000 series the most accessible and serviceable units of their type available. The P6000 series amplifiers can be readily adapted, by means of plug-in options, for interfacing with an unlimited range of remote inputs, programmers, and computer controls. Large volume users can purchase the basic modules and harnesses for use in applications where space and cost are critical.

The P6000 amplifiers are also available in Electro-Craft's D.P.S. range of Digital Positioning Systems which typically can provide fully programmable step motion drives capable of completing single steps of 1.8° in 8 ms including settling, and slew speeds of 15,000 steps per second with inertial loads of 0.05 oz-in-sec^2 and friction loads of 50 oz-in. The D.P.S. series are available in a wide range of step sizes; consult our factory for further details.

*Patents applied for.

P6100 PWM Servo Control Specifications

OUTPUT RATINGS

Parameter	Conditions		Value
Output Voltage	Maximum		±75 V
	At continuous rated current (nominal)		±60 V
	At continuous rated current (low line)		±53 V
	At peak rated current (nominal)		±56 V
	At peak rated current (low line)		±50 V
Current	Continuous rated (independent of line)		7A
	Peak rated (100 ms maximum) (independent of line)		12A
Power	Maximum continuous	(nominal line)	420 W
		(low line)	370 W
	Maximum peak	(nominal line)	670 W
		(low line)	600 W
Temperature Range	Operating		0-50°C
	Storage		-30 to 80°C
Ripple Current (I_R)	E = Output voltage (Avg)		$I_R = \left[\frac{4900 \cdot E^2}{3080 \cdot I}\right]$ A p-p
	I = Output current (RMS)		
	L = Load inductance (mH)		
Form Factor	Typically < 1.01		$FF = \sqrt{1 + \frac{I_R^2}{12 \cdot L^2}}$

INPUT REQUIREMENTS P6100A

Parameter	Conditions	Value
DC Input Voltage	(High current supply) unipolar	
	Nominal at continuous rated current	+ 70v DC nom
	High line no load	+ 80V DC max
	Low line at continuous rated current	+ 60v DC min
Ripple	At continuous rated current	6v p-p
Fusing Requirements	3AB Type (normal Blo)	8A
Current	Continuous rated	7A
	Peak (for 100 ms)	12A
Low Current		
Supply Voltage	Bi-polar	± 24v DC ± 5v
Current	Continuous	± 75 mA
Ripple	At above current	2v p-p max
Fusing	3AG Type	1/8 A
Fan Supply	AC 50/60 Hz single phase	115v†

INPUT REQUIREMENTS P6100AP

Parameter	Conditions		Value
Voltage	Nominal	(single phase)	115/230v RMS
	High line	(single phase)	125/250v RMS
	Low line	(single phase)	105/210v RMS
Frequency		(single phase)	50/60 Hz
Current	Continuous		6.2 Amps RMS
	Peak (for 100 ms)		10.7 Amps RMS

†230v by special option.

PERFORMANCE

Parameter	Conditions	Value
Audible Noise Generation	All conditions (excludes fan)	None
Modulation Frequency	All conditions (fixed)	22 kHz nom
Frequency Response	Open loop – 3db point	3 kHz nom
Gain	Open loop (adjustable) reference input Open loop (adjustable) velocity tach input Open loop (selectable) auxiliary Optional v/v gain is available; consult factory for details.	0-4000 Amps/volt 0-1475 Amps/volt 0-8000 Amps/volt
Offset Voltage		adjustable to zero
Deadband		negligible
Drift	Maximum gain referred to input, auxiliary input not in use For details consult 741 op-amp data.	20 $\frac{uv}{^{o}C}$ nom

STANDARD FEATURES

Parameter	Conditions	Value
Current Limit Peak Current	Dual stage, independent peak and average limits Range (adjustable)	Average setting to 12 A nom
	Duration (adjustable)	10 to 100 ms nom
Average Current	Range (adjustable)	3 to 7 A nom
Cro-Bar	P6100AP	
Time to Clear Fuse	Amplifier output shorted + to - or to ground	4 ms nom
Time to Reduce Power Input to 10v	Amplifier output shorted + to - or to ground	500 μsec nom
Current Monitor Output Signal		.4v/A
Remote Current Limit Pots	Peak Average	10k Ω 20k Ω
End of Travel Limits and Amplifier Clamp	Enable Inhibit or Enable Inhibit	switch closure switch open TTL low TTL high
External Connections	Amp. Mate-N-Lock pin & socket connectors (Mating connectors are supplied with each unit.)	

*All specifications have been verified by exhaustive testing under operating conditions.

P6200 PWM Servo Control Specifications

OUTPUT RATINGS

Parameter	Conditions	Value
Output Voltage	Maximum At continuous rated current (nominal) At continuous rated current (low line) At peak rated current (nominal) At peak rated current (low line)	± 75 V ± 55 V ± 50 V ± 50 V, ± 47 V,
Output Current	Continuous rated (independent of line) Peak rated (100 ms maximum) (independent of line)	15 A 20 A
Output Power	Maximum continuous (nominal line) (15 A) Maximum continuous (low line) (15A) Maximum peak (nominal line) Maximum peak (low line)	825 W 750 W 1000 W 940 W
Temperature Range	Operating Storage	0-50°C -30 to 80°C
Ripple Current (I_R) Form Factor	E = Output voltage (Avg) I = Output current (RMS) L = Load inductance (mH) Typically $<$ 1.01	$I_R = \left[\frac{5625 \cdot E^2}{3300 \cdot L}\right]$ A p-p $FF = \sqrt{1 + \frac{I_R^2}{12 \cdot L^2}}$

INPUT REQUIREMENTS P6200A

Parameter	Conditions	Value
DC Input Voltage	(High current supply) unipolar Nominal at continuous rated current High line no load Low line at continuous rated current	 + 70v DC nom + 80V DC max + 60v DC min
Ripple	At continuous rated current	6v p-p
Fusing Requirements	3AB Type (normal Blo)	15 A
Current	Continuous rated Peak (for 100 ms)	15 A 25 A
Low Current Supply Voltage Current Ripple Fusing	 Bi-polar Continuous At above current 3AG Type	 ± 24v DC ± 5v ± 75 mA 2v p-p max 1/8 A
Fan Supply	AC 50/60 Hz single phase	115v†

INPUT REQUIREMENTS P6200AP

Parameter	Conditions	Value
Voltage	Nominal (single phase) High line (single phase) Low line (single phase)	115/230v RMS 125/250v RMS 105/210v RMS
Frequency	(single phase)	50/60 Hz
Current	Continuous Peak (for 100 ms)	13 Amps RMS 17 Amps RMS

†230v by special option

PERFORMANCE

Parameter	Conditions	Value
Audible Noise Generation	All conditions (excludes fan)	None
Modulation Frequency	All conditions (fixed)	22 kHz nom
Frequency Response	Open loop – 3db point	3 kHz nom
Gain	Open loop (adjustable) reference input Open loop (adjustable) velocity tach input Open loop (selectable) auxiliary Optional v/v gain is available; consult factory for details.	0-4000 Amps/volt 0-1475 Amps/volt 0-8000 Amps/volt
Offset Voltage		adjustable to zero
Deadband		negligible
Drift	Maximum gain referred to input, auxiliary input not in use For details consult 741 op-amp data.	20 $\frac{uv}{^{o}C}$ nom

STANDARD FEATURES

Parameter	Conditions	Value
Current Limit Peak Current	Dual stage, independent peak and average limits Range (adjustable) Duration (adjustable)	 Average setting to 25 A nom 10 msec to 100 msec nom
Average Current	Range (adjustable)	6 to 15 A nom
Cro-Bar Time to Clear Fuse Time to Reduce Power Input to 10v	(P6200AP only) Amplifier output shorted + to - or to ground Amplifier output shorted + to - or to ground	 4 msec nom 500 μsec nom
Current Monitor Output Signal Remote Current Limit Pots	 Peak Average	.2v/A 10k Ω 20k Ω
End of Travel Limits and Amplifier Clamp	Enable Inhibit • or Enable Inhibit	switch closure switch open TTL low TTL high
External Connections	Amp. Mate-N-Lock pin & socket connectors (Mating connectors are supplied with each unit.)	

* All specifications have been verified by exhaustive testing under operating conditions.

P6300 PWM D.C. Servo Motor Controller

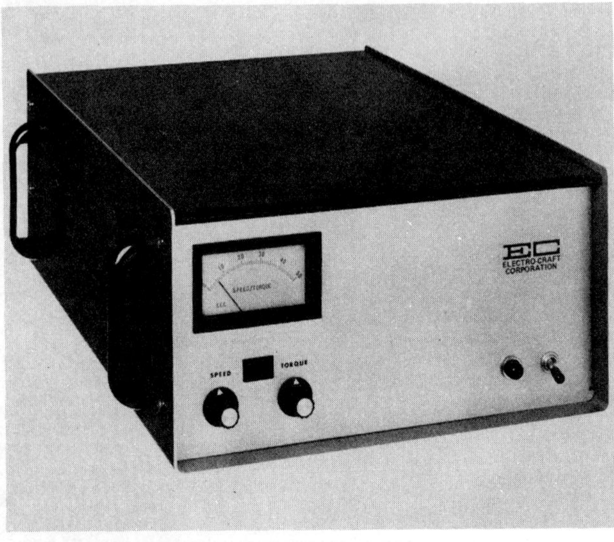

- ***Voltage at Rated Current*** **±55V**
- ***Rated Current*** **15A**
- ***Peak Current*** **25A**
- ***Rated Power*** **825W**
- ***Peak Power*** **1150W**
- ***Bandwidth*** **1000Hz**

FEATURES:

Unique Open Modular Construction
Completely self contained
Plug in Standard or Custom Options
Regenerative Energy Protected

115/230 V AC or DC Power Options
Independent Average and Peak Current Limits
Crowbar Protected AC Power Supply

The P6000 series are the fastest response pulse width modulated servo controls competitively available, providing bandwidths at full load exceeding 1000 Hz. The unique* fixed frequency design gives quiet operation free from annoying audible "chirps". The standard unit, which is available for operation from AC line (model P6000AP) or for operation from a DC power supply (model P6000A), has the following features: three input fully adjustable summing amplifier with independent gain and lead-lag adjustment to enable operation with a wide range of motor and load types; a dual stage current limiter providing independent adjustment of peak and average current limit as well as pulse duration; unique crowbar protection (model P6000AP) which protects the amplifier from damage due to short circuits, motor failure or misuse; space for two 4-5/16 x 2-3/4" option cards which can be selected from a wide range of standard options or custom designed options; logic compatible end of travel limit and clamping inputs; all major subassemblies are mounted on plug-in connectors to enable replacement without the need to change the entire unit; and all major test points are brought out to a 14 pin D.I.P. socket for fast, easy troubleshooting and set up. This, combined with the unique open construction, makes the P6000 series the most accessible and serviceable units of their type available. The P6000 series amplifiers can be readily adapted, by means of plug-in options, for interfacing with an unlimited range of remote inputs, programmers, and computer controls. Large volume users can purchase the basic modules and harnesses for use in applications where space and cost are critical.

The P6000 amplifiers are also available in Electro-Craft's D.P.S. range of Digital Positioning Systems which typically can provide fully programmable step motion drives capable of completing single steps of 1.8° in 8 ms including settling, and slew speeds of 15,000 steps per second with inertial loads of 0.05 oz-in-sec^2 and friction loads of 50 oz-in. The D.P.S. series are available in a wide range of step sizes; consult our factory for further details.

*Patents applied for.

P6300 PWM Servo Control Specifications*

OUTPUT RATINGS

Parameter	Conditions	Value
Output Voltage	Maximum	± 75 V
	At continuous rated current (nominal)	± 55 V
	At continuous rated current (low line)	± 50 V
	At peak rated current (nominal)	± 47 V
	At peak rated current (low line)	± 47 V
Output Current	Continuous rated (independent of line)	15 A
	Peak rated (500 ms maximum) (independent of line)	25 A
Output Power	Maximum continuous (nominal line) (15 A)	825 W
	Maximum continuous (low line) (15A)	750 W
	Maximum peak (nominal line)	1150 W
	Maximum peak (low line)	1075 W
Temperature Range	Operating	0-50°C
	Storage	-30 to 80°C
Ripple Current (I_R)	E = Output voltage (Avg)	$I_R = \left[\frac{5625 - E^2}{3300 \cdot L}\right]$ A p-p
	I = Output current (RMS)	
	L = Load inductance (mH)	
Form Factor	Typically < 1.01	$FF = \sqrt{1 + \frac{I_R^2}{12 \cdot L^2}}$

INPUT REQUIREMENTS P6200A

Parameter	Conditions	Value
DC Input Voltage	(High current supply) unipolar	
	Nominal at continuous rated current	+ 70v DC nom
	High line no load	+ 80V DC max
	Low line at continuous rated current	+ 60v DC min
Ripple	At continuous rated current	6v p-p
Fusing Requirements	3AB Type (normal Blo)	15 A
Current	Continuous rated	15 A
	Peak (for 500 ms)	25 A
Low Current Supply Voltage	Bi-polar	± 24v DC ± 5v
Current	Continuous	± 75 mA
Ripple	At above current	2v p-p max
Fusing	3AG Type	1/8 A
Fan Supply	AC 50/60 Hz single phase	115v†

INPUT REQUIREMENTS P6200AP

Parameter	Conditions		Value
Voltage	Nominal	(single phase)	115/230v RMS
	High line	(single phase)	125/250v RMS
	Low line	(single phase)	105/210v RMS
Frequency		(single phase)	50/60 Hz
Current	Continuous		13 Amps RMS
	Peak (for 500 ms)		22 Amps RMS

†230v by special option.

PERFORMANCE

Parameter	Conditions	Value
Modulation Frequency	All conditions (fixed)	8 kHz nom
Frequency Response	Open loop – 3db point	1 kHz nom
Gain	Open loop (adjustable) reference input Open loop (adjustable) velocity tach input Open loop (selectable) auxiliary Optional v/v gain is available; consult factory for details.	0-4000 Amps/volt 0-1475 Amps/volt 0-8000 Amps/volt
Offset Voltage		adjustable to zero
Deadband		negligible
Drift	Maximum gain referred to input, auxiliary input not in use For details consult 741 op-amp data.	20 $\frac{uv}{^{o}C}$ nom

STANDARD FEATURES

Parameter	Conditions	Value
Current Limit	Dual stage, independent peak and average limits	
Peak Current	Range (adjustable)	Average setting to 25 A nom
	Duration (adjustable)	10 msec to 100 msec nom
Average Current	Range (adjustable)	6 to 25 A nom
Cro-Bar	(P6300AP only)	
Time to Clear Fuse	Amplifier output shorted + to - or to ground	4 msec nom
Time to Reduce Power Input to 10v	Amplifier output shorted + to - or to ground	500 μsec nom
Current Monitor Output Signal		.2v/A
Remote Current Limit Pots	Peak Average	10k Ω 20k Ω
End of Travel Limits and Amplifier Clamp	Enable Inhibit or Enable Inhibit	switch closure switch open TTL low TTL high
External Connections	Amp. Mate-N-Lock pin & socket connectors (Mating connectors are supplied with each unit.)	

* All specifications have been verified by exhaustive testing under operating conditions.

P6000 series servo motor controllers – Outline Dimensions

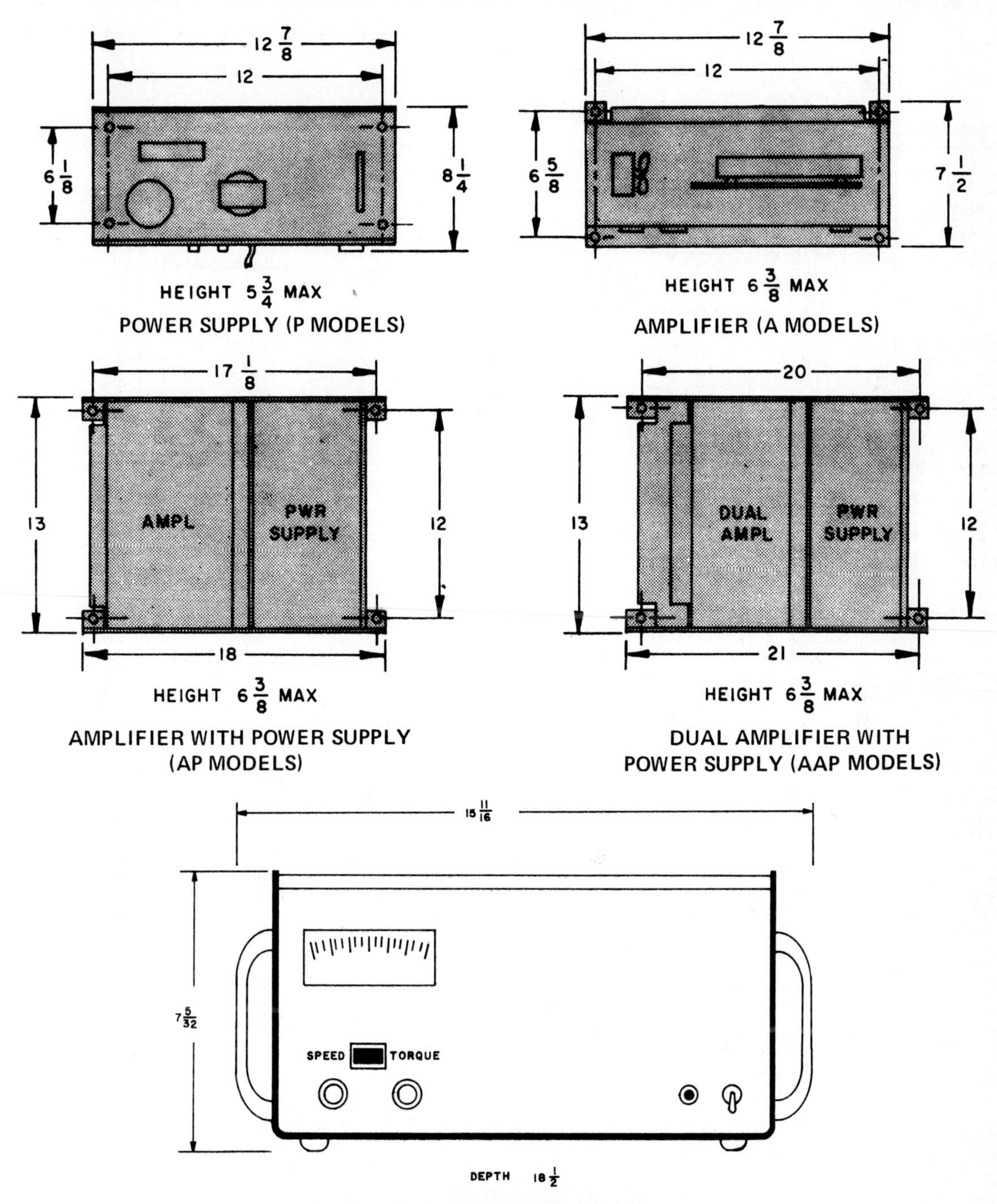

ENCLOSED AMPLIFIER WITH POWER SUPPLY (EAP MODELS)

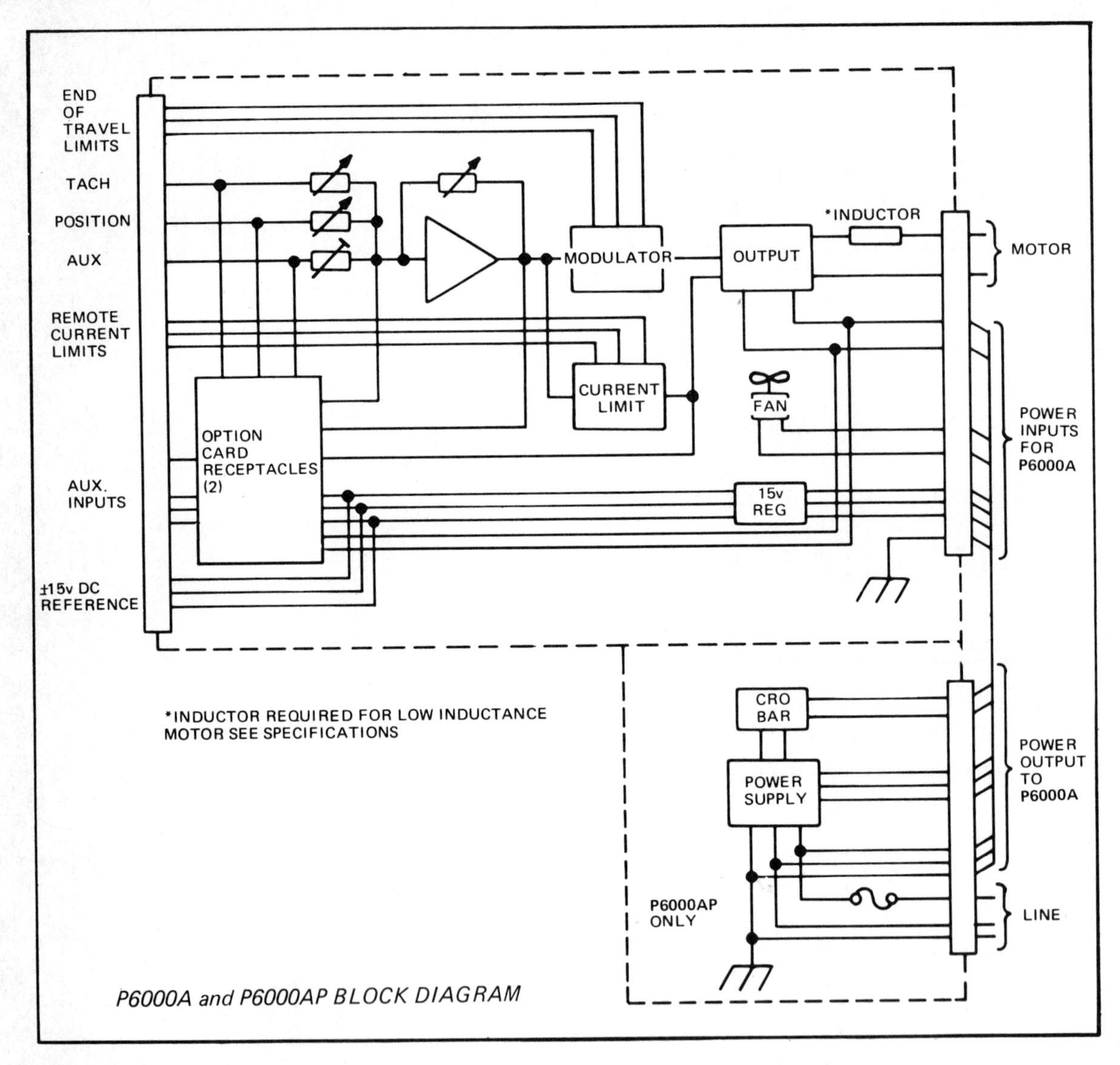

P6000A and P6000AP BLOCK DIAGRAM

Other Quality Products Manufactured by Electro-Craft Corp:

D.C. PERMANENT MAGNET SERVO-MOTORS
M.C.M HIGH PERFORMANCE, LOW INERTIA MOVING COIL MOTORS UP TO 5 H.P.
BRUSHLESS D.C. SERVO MOTORS
MOTOMATIC MOTOR SPEED CONTROL SYSTEMS
D.P.S. DIGITAL POSITIONING SYSTEMS
L5000 SERIES LINEAR D.C. SERVO MOTOR CONTROLS
P7000 HIGH POWER P.W.M. D.C. SERVO MOTOR CONTROLS
CUSTOM DESIGNED SERVO SYSTEMS

THIS PAGE INTENTIONALLY LEFT BLANK

SD6245-00

The SD6245-00 package consists of an Electro-Craft model P6200 Servomotor Controller and a 1450-00 Low Inertia Moving Coil Servomotor. Both the motor and controller are manufactured by Electro-Craft and are matched for optimum performance and reliability.

The controller is a fixed frequency Pulse Width Modulated DC amplifier providing fast response and silent operation. A wide range of standard options enable the system to be used in such applications as Digital Positioning (point to point and contouring), Phase Locked Scanning and other applications where precise speed, position or acceleration control are required. For full dimensions and specifications for the controller consult the appropriate data sheet.

The servomotor utilizes the latest in winding, insulation and high energy magnet technology. Typical motor life is in excess of 15,000 hours when operated at the maximum duty cycle.

For price and other information, contact any Electro-Craft Sales Office.

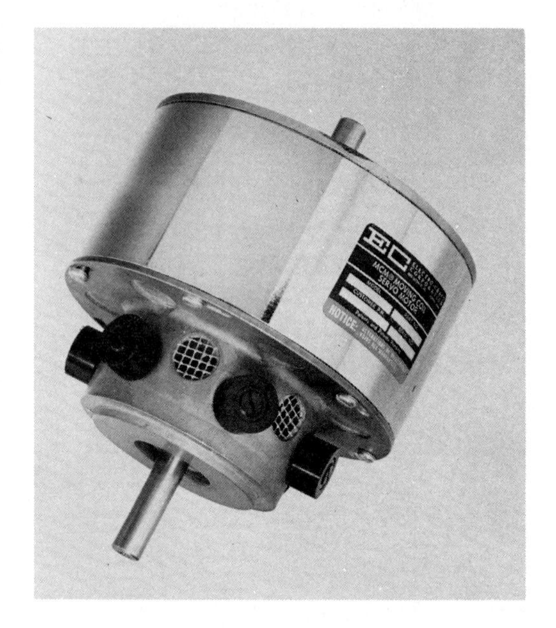

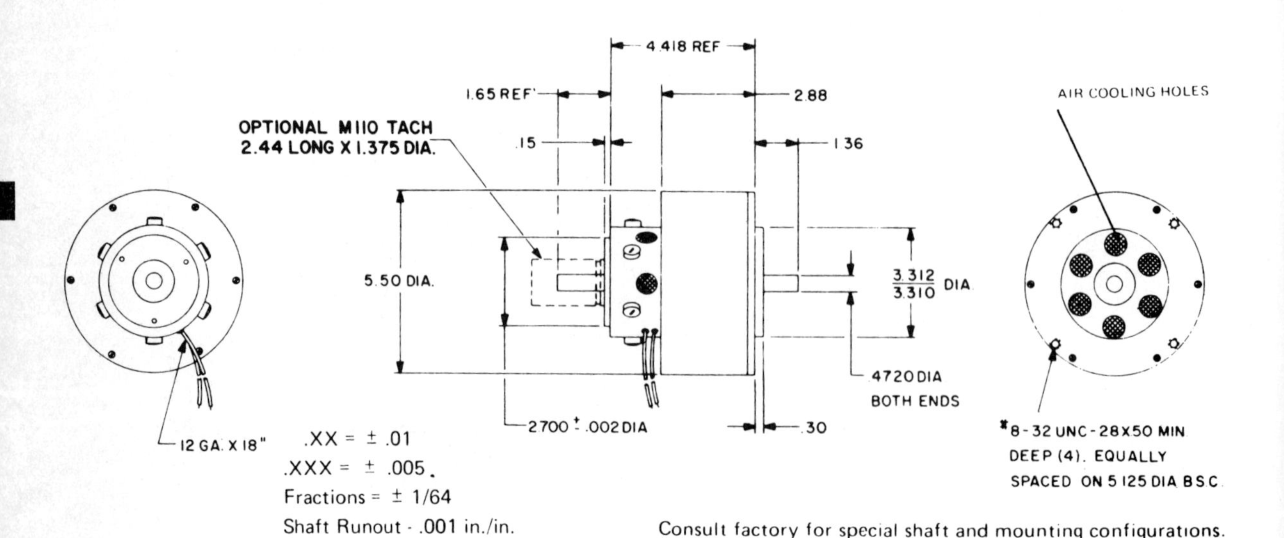

Consult factory for special shaft and mounting configurations.

NOMINAL SPECIFICATIONS (at 25°C) SD6245-00

Maximum Power Output (continuous)	=	290 Watts
Maximum Speed	=	3700 rpm
Maximum Torque Output (continuous)	=	210 oz-in
Maximum Torque Output (pulse operation)	=	490 oz-in
Motor Torque Constant	=	25 oz-in/A
Motor Voltage Constant	=	18.5 V/krpm
Motor Terminal Resistance	=	1.25 ohms
Armature Moment of Inertia	=	0.01 oz-in-s^2
Rotational Loss Torque	=	7.6 oz-in/krpm
Static Friction Torque	=	9 oz-in
Armature to Ambient Thermal Resistance	=	1.0°C/Watt
Maximum Armature Temperature	=	155°C
Armature Inductance	=	< 0.4 mH
Tachometer Voltage Gradient	=	3.0 V/krpm
Tachometer Terminal Resistance	=	150 ohms
Tachometer Armature Inductance	=	4 mH
Tachometer Ripple Amplitude	=	2.0% p-p
Tachometer Ripple Frequency	=	19 cycles/rev.
Tachometer Output Linearity	=	0.2% error
Recommended Load Impedance	=	10,000 ohms
Output Temperature Stability	=	0.01%/°C

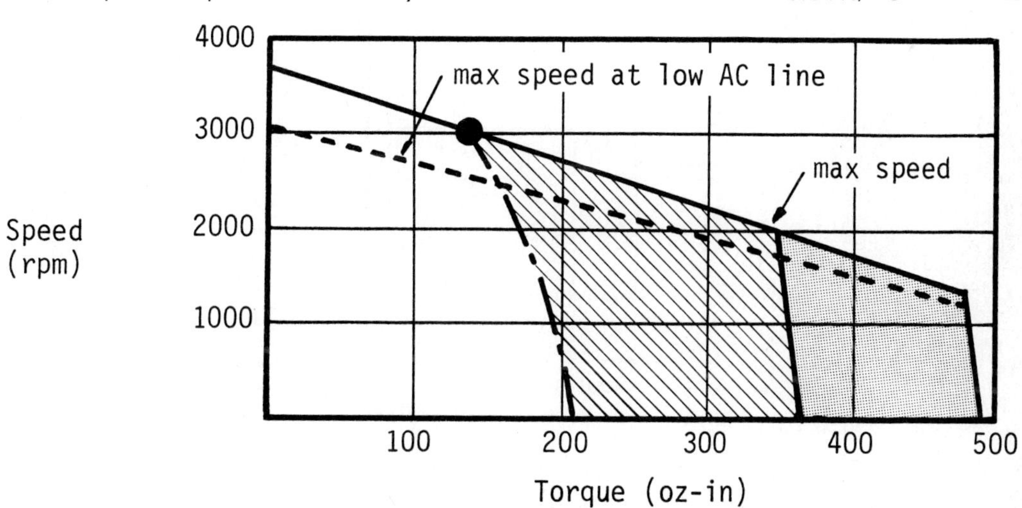

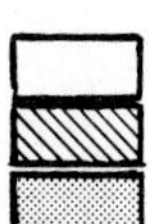

Safe continuous operating area uncooled.
Safe continuous operating area with air cooling or for up to 10 seconds uncooled.
Safe operating area for up to 100 ms.

SD6166-06

The SD6166-06 package consists of an Electro-Craft model P6100 Servomotor Controller and a 660-06-011 High Performance Servomotor. Both the motor and controller are manufactured by Electro-Craft and are matched for optimum performance and reliability.

The controller is a fixed frequency Pulse Width Modulated DC amplifier providing fast response and silent operation. A wide range of standard options enable the system to be used in such applications as Digital Positioning (point to point and contouring), Phase Locked Scanning and other applications where precise speed, position or acceleration control are required. For full dimensions and specifications for the controller consult the appropriate data sheet.

The servomotor utilizes the latest in winding, insulation and high energy magnet technology. Typical motor life is in excess of 15,000 hours when operated at the maximum duty cycle. The mounting face conforms to NEMA 42CZ which enables use with a wide range of readily available gearheads and coupling devices.

Both the motor and control are normally available from stock for 7-14 day delivery; for price and other information, contact any Electro-Craft Sales Office.

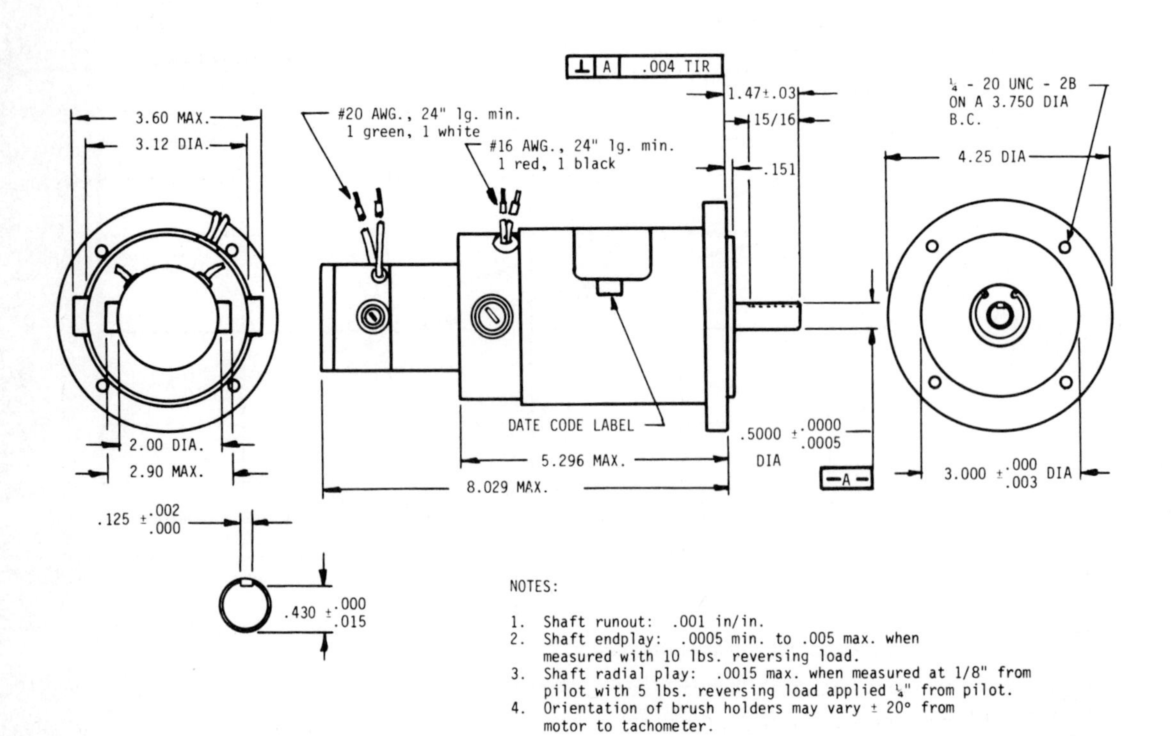

NOTES:

1. Shaft runout: .001 in/in.
2. Shaft endplay: .0005 min. to .005 max. when measured with 10 lbs. reversing load.
3. Shaft radial play: .0015 max. when measured at 1/8" from pilot with 5 lbs. reversing load applied ¼" from pilot.
4. Orientation of brush holders may vary ± 20° from motor to tachometer.

NOMINAL SPECIFICATIONS (at 25°C) SD6166-06

Maximum Power Output (continuous)	=	310 Watts
Maximum Speed	=	4200 rpm
Maximum Torque Output (continuous)	=	127 oz-in
Maximum Torque Output (pulse operation)	=	245 oz-in
Motor Torque Constant	=	21 oz-in/A
Motor Voltage Constant	=	15.5 V/krpm
Motor Terminal Resistance	=	1.15 ohms
Armature Moment of Inertia	=	0.03 oz-in-s^2
Rotational Loss Torque	=	1.0 oz-in/krpm
Static Friction Torque	=	7.0 oz-in
Armature to Ambient Thermal Resistance	=	2.8°C/Watt
Maximum Armature Temperature	=	155°C
Armature Inductance	=	2.0 mH
Tachometer Voltage Gradient	=	14.2 V/krpm
Tachometer Terminal Resistance	=	750 ohms
Tachometer Armature Inductance	=	138 mH
Tachometer Ripple Amplitude	=	5.0% p-p
Tachometer Ripple Frequency	=	11 cycles/rev.
Recommended Load Impedance	=	5000 ohms
Output Temperature Stability	=	-0.05%/°C
Tachometer Output Linearity	=	0.2% error

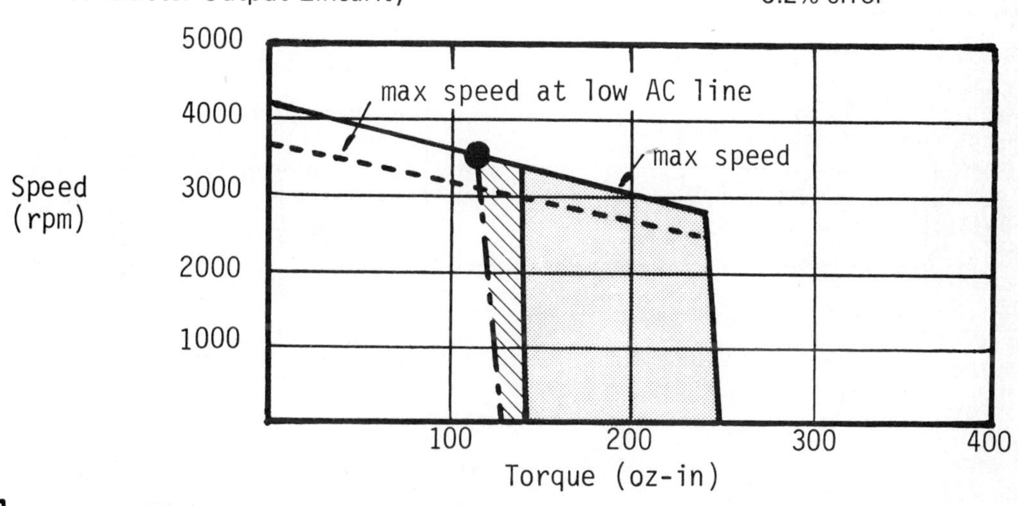

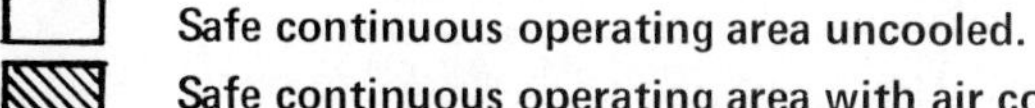

Safe continuous operating area uncooled.

Safe continuous operating area with air cooling or for up to 60 seconds uncooled.

Safe operating area for up to 100 ms.

SD6267-01

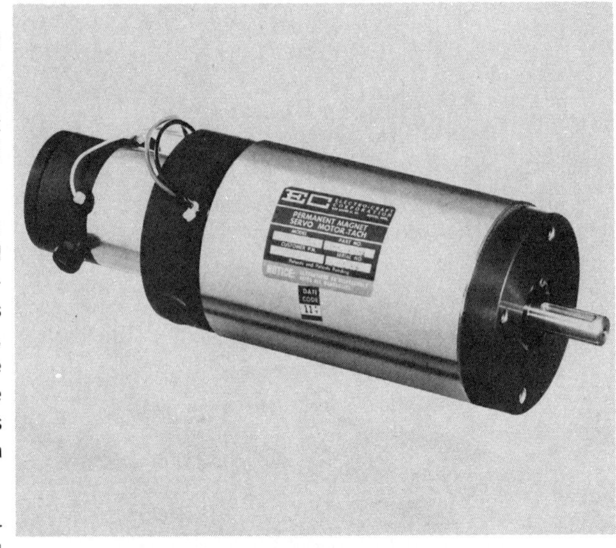

The SD6267-01 package consists of an Electro-Craft model P6200 Servomotor Controller and a 670-01-004 High Performance Servomotor. Both the motor and controller are manufacturered by Electro-Craft and are matched for optimum performance and reliability.

The controller is a fixed frequency Pulse Width Modulated DC amplifier providing fast response and silent operation. A wide range of standard options enable the system to be used in such applications as Digital Positioning (point to point and contouring), Phase Locked Scanning and other applications where precise speed, position or acceleration control are required. For full dimensions and specifications for the controller consult the appropriate data sheet.

The servomotor utilizes the latest in winding, insulation and high energy magnet technology. Typical

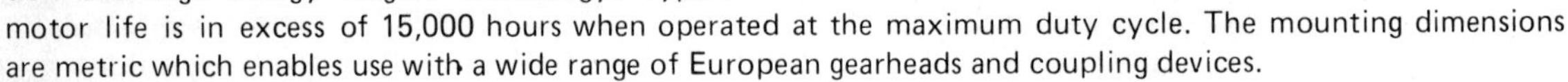

motor life is in excess of 15,000 hours when operated at the maximum duty cycle. The mounting dimensions are metric which enables use with a wide range of European gearheads and coupling devices.

Both the motor and control are normally available from stock for 7-14 day delivery; for price and other information, contact any Electro-Craft Sales Office.

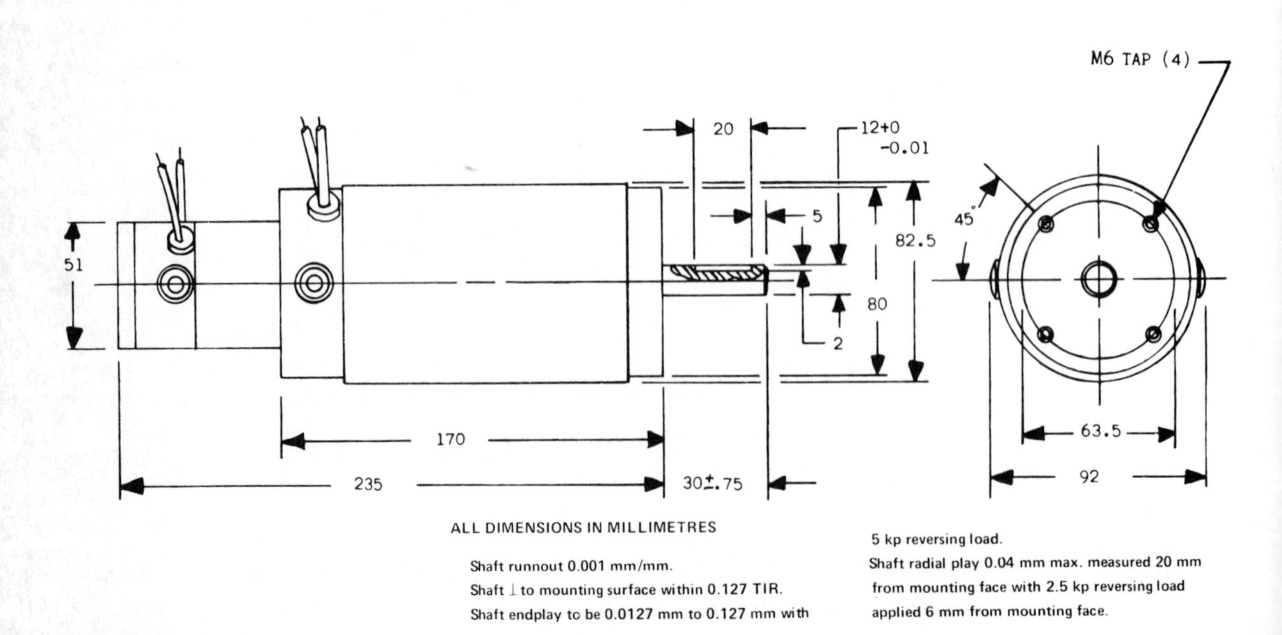

ALL DIMENSIONS IN MILLIMETRES

Shaft runnout 0.001 mm/mm.
Shaft ⊥ to mounting surface within 0.127 TIR.
Shaft endplay to be 0.0127 mm to 0.127 mm with 5 kp reversing load.
Shaft radial play 0.04 mm max. measured 20 mm from mounting face with 2.5 kp reversing load applied 6 mm from mounting face.

NOMINAL SPECIFICATIONS (at 25°C) SD6267-01

Maximum Power Output (continuous)	=	475 Watts
Maximum Speed	=	4950 rpm
Maximum Torque Output (continuous)	=	185 oz-in
Maximum Torque Output (pulse operation)	=	370 oz-in
Motor Torque Constant	=	18.9 oz-in/A
Motor Voltage Constant	=	14.0 V/krpm
Motor Terminal Resistance	=	0.65 ohms
Armature Moment of Inertia	=	0.05 oz-in-s^2
Rotational Loss Torque	=	1.0 oz-in/krpm
Static Friction Torque	=	8.0 oz-in
Armature to Ambient Thermal Resistance	=	2.1 °C/Watt
Maximum Armature Temperature	=	155°C
Armature Inductance	=	1.91 mH
Tachometer Voltage Gradient	=	14.2 V/Krpm
Tachometer Terminal Resistance	=	750 ohms
Tachometer Armature Inductance	=	138 mH
Tachometer Ripple Amplitude	=	5.0% p-p
Tachometer Ripple Frequency	=	11 cycles/rev.
Tachometer Output Linearity	=	0.2% error
Recommended Load Impedance	=	5000 ohms
Output Temperature Stability	=	–0.05%/°C

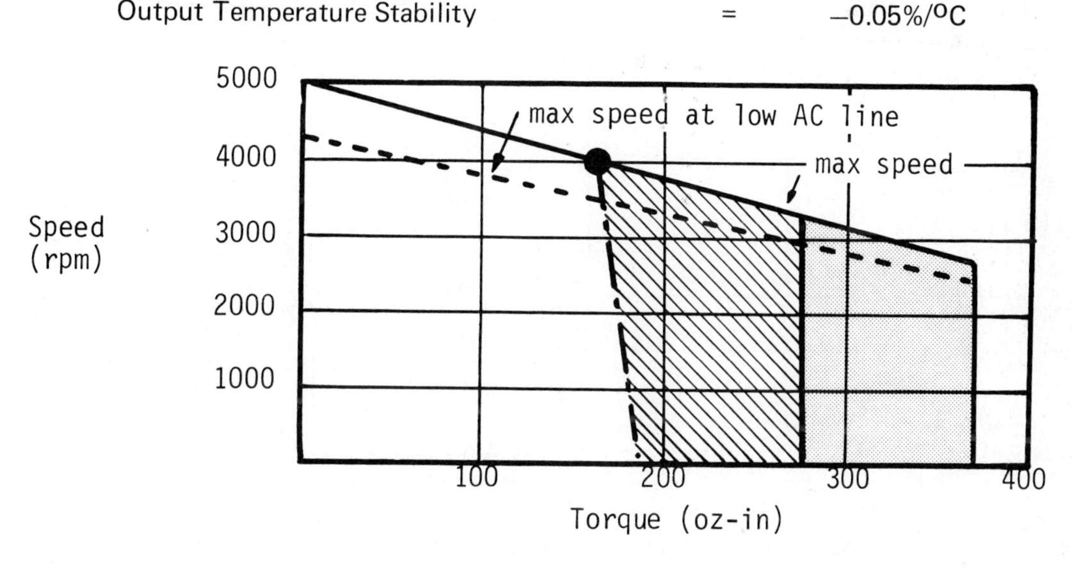

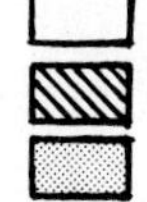

Safe continuous operating area uncooled

Safe continuous operating area with air cooling or for up to 60 seconds uncooled

Safe operating area for up to 100 ms.

SD6272-06

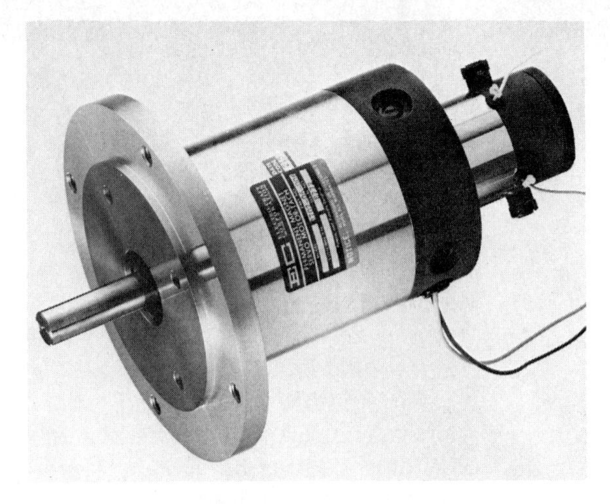

The SD6272-06 package consists of an Electro-Craft model P6200 Servomotor Controller and a 702-06-006 High Performance Servomotor. Both the motor and controller are manufactured by Electro-Craft and are matched for optimum performance and reliability.

The controller is a fixed frequency Pulse Width Modulated DC amplifier providing fast response and silent operation. A wide range of standard options enable the system to be used in such applications as Digital Positioning (point to point and contouring), Phase Locked Scanning and other applications where precise speed, position or acceleration control are required. For full dimensions and specifications for the controller consult the appropriate data sheet.

The servomotor utilizes the latest in winding, insulation and high energy magnet technology. Typical motor life is in excess of 15,000 hours when operated within the maximum duty cycle. The mounting face conforms to NEMA 56c which enables use with a wide range of readily available gearheads and coupling devices.

Both the motor and control are normally available from stock for 7-14 day delivery; for price and other information, contact any Electro-Craft Sales Office.

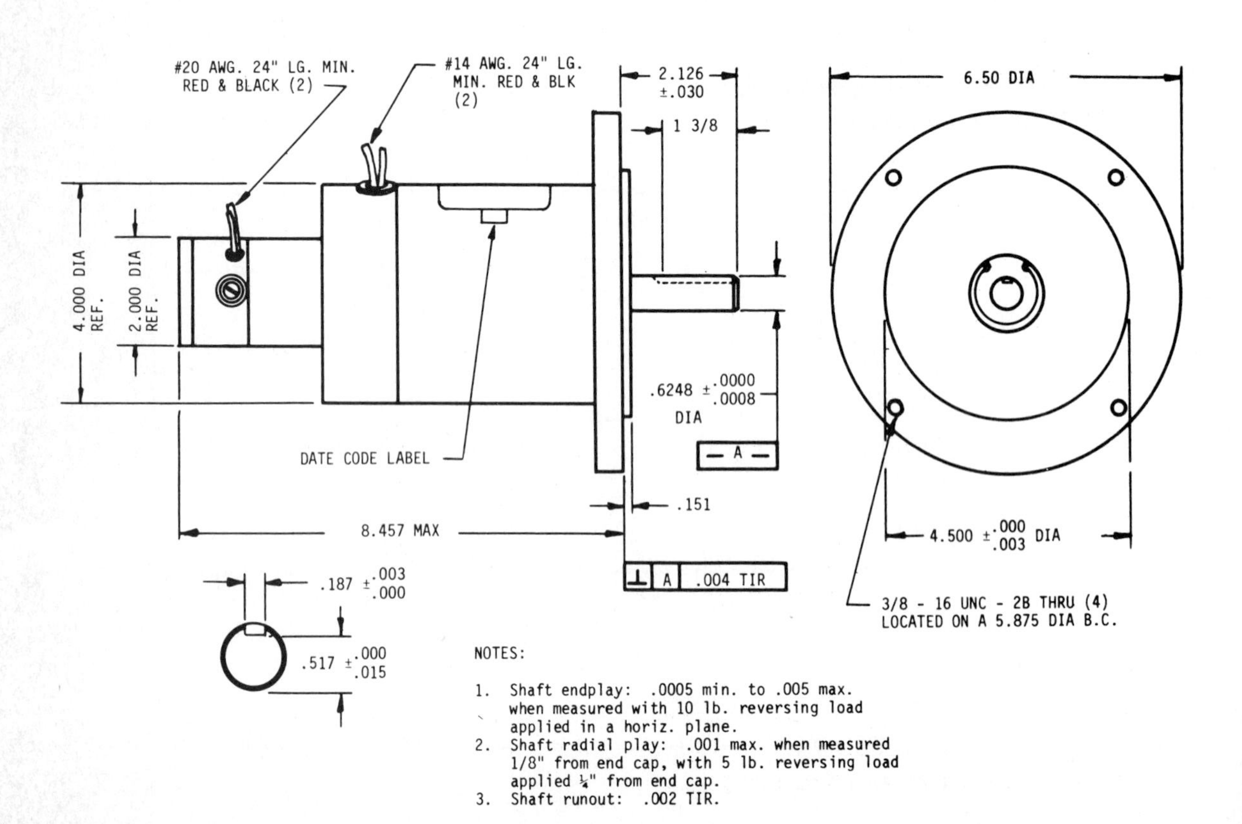

NOMINAL SPECIFICATIONS (at 25°C) SD 6272-06

Maximum Power Output (continuous)	=	230 Watts
Maximum Speed	=	2000 rpm
Maximum Torque Output (continuous)	=	280 oz-in
Maximum Torque Output (pulse operation)	=	460 oz-in
Motor Torque Constant	=	23 oz-in/A
Motor Voltage Constant	=	17 V/krpm
Motor Terminal Resistance	=	0.48 ohms
Armature Moment of Inertia	=	0.15 oz-in-s^2
Rotational Loss Torque	=	7.0 oz-in/krpm
Static Friction Torque	=	10 oz-in
Armature to Ambient Thermal Resistance	=	1.6°C/Watt
Maximum Armature Temperature	=	155°C
Armature Inductance	=	1.5 mH
Tachometer Voltage Gradient	=	14.2 V/krpm
Tachometer Terminal Resistance	=	750 ohms
Tachometer Armature Inductance	=	138 mH
Tachometer Ripple Amplitude	=	5.0% p-p
Tachometer Ripple Frequency	=	11 cycles/rev.
Recommended Load Impedance	=	5000 ohms
Output Temperature Stability	=	–0.05%/°C
Tachometer Output Linearity	=	0.2% error

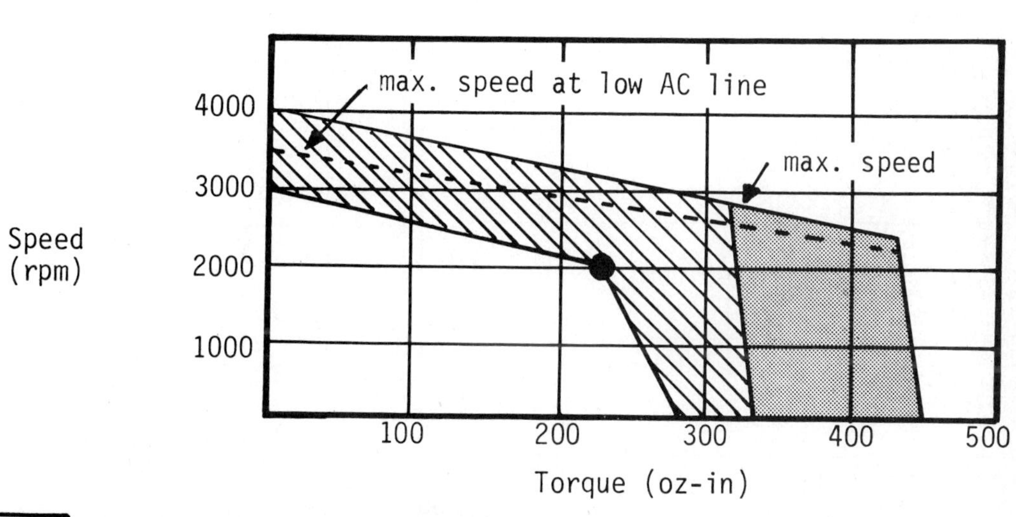

Safe continuous operating area uncooled.

Safe continuous operating area with air cooling or for up to 60 secs. uncooled.

Safe operating area for up to 100 ms.

SD6273-02

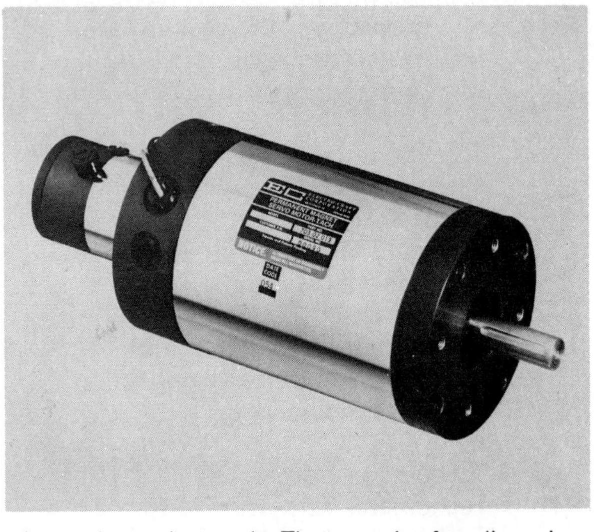

The SD6273-02 package consists of an Electro-Craft model P6200 Servomotor Controller and a 703-02-018 High Performance Servomotor. Both the motor and controller are manufactured by Electro-Craft and are matched for optimum performance and reliability.

The controller is a fixed frequency Pulse Width Modulated DC amplifier providing fast response and silent operation. A wide range of standard options enable the system to be used in such applications as Digital Positioning (point to point and contouring), Phase Locked Scanning and other applications where precise speed, position or acceleration control are required. For full dimensions and specifications for the controller consult the appropriate data sheet.

The servomotor utilizes the latest in winding, insulation and high energy magnet technology. Typical motor life is in excess of 15,000 hours when operated at the maximum duty cycle. The mounting face dimensions are metric which enables use with a wide range of European gearheads and coupling devices.

Both the motor and control are normally available from stock for 7-14 day delivery; for price and other information, contact any Electro-Craft Sales Office.

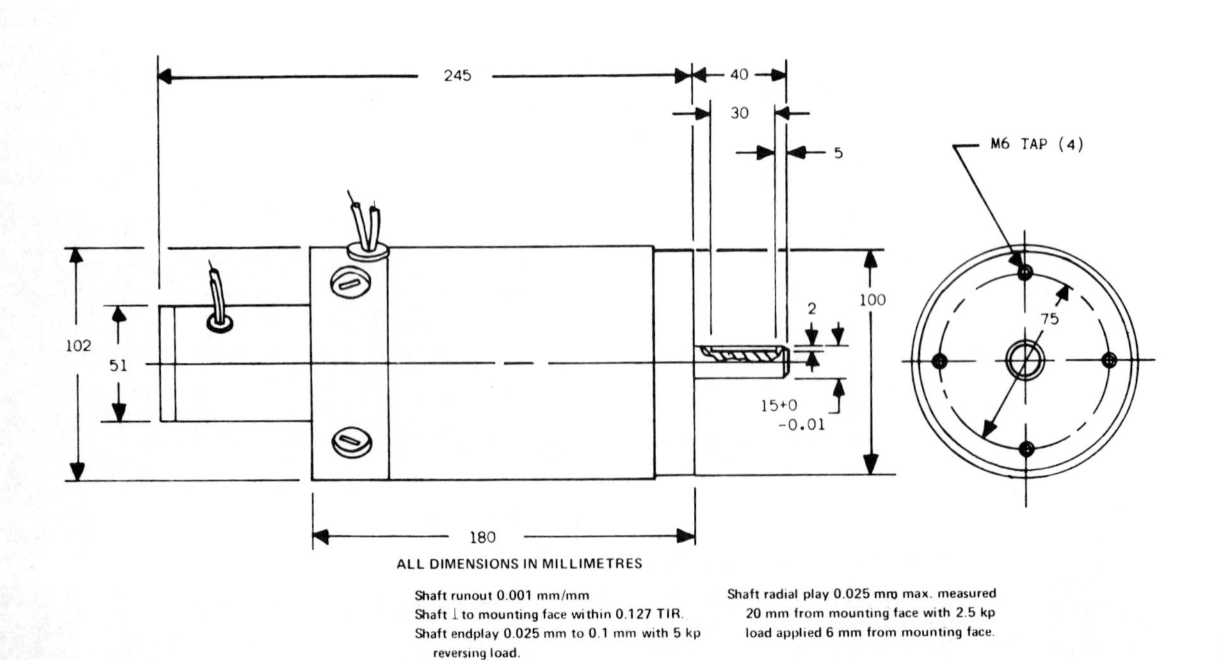

ALL DIMENSIONS IN MILLIMETRES

Shaft runout 0.001 mm/mm
Shaft ⊥ to mounting face within 0.127 TIR.
Shaft endplay 0.025 mm to 0.1 mm with 5 kp reversing load.

Shaft radial play 0.025 mm max. measured 20 mm from mounting face with 2.5 kp load applied 6 mm from mounting face.

NOMINAL SPECIFICATIONS (at 25°C) SD6273-02

Maximum Power Output (continuous)	=	450 Watts
Maximum Speed	=	2000 rpm
Maximum Torque Output (continuous)	=	480 oz-in
Maximum Torque Output (pulse operation)	=	970 oz-in
Motor Torque Constant	=	49 oz-in/A
Motor Voltage Constant	=	36 V/krpm
Motor Terminal Resistance	=	0.85 ohms
Armature Moment of Inertia	=	0.2 oz-in-s^2
Rotational Loss Torque	=	10 oz-in/krpm
Static Friction Torque	=	14 oz-in
Armature to Ambient Thermal Resistance	=	1.2°C/Watt
Maximum Armature Temperature	=	155°C
Armature Inductance	=	4.4 mH
Tachometer Voltage Gradient	=	14.2 V/krpm
Tachometer Terminal Resistance	=	750 ohms
Tachometer Armature Inductance	=	138 mH
Tachometer Ripple Amplitude	=	5% p-p
Tachometer Ripple Frequency	=	11 cycles/rev
Tachometer Output Linearity	=	0.2% error
Recommended Load Impedance	=	5000 ohms
Output Temperature Stability	=	-0.05%/°C

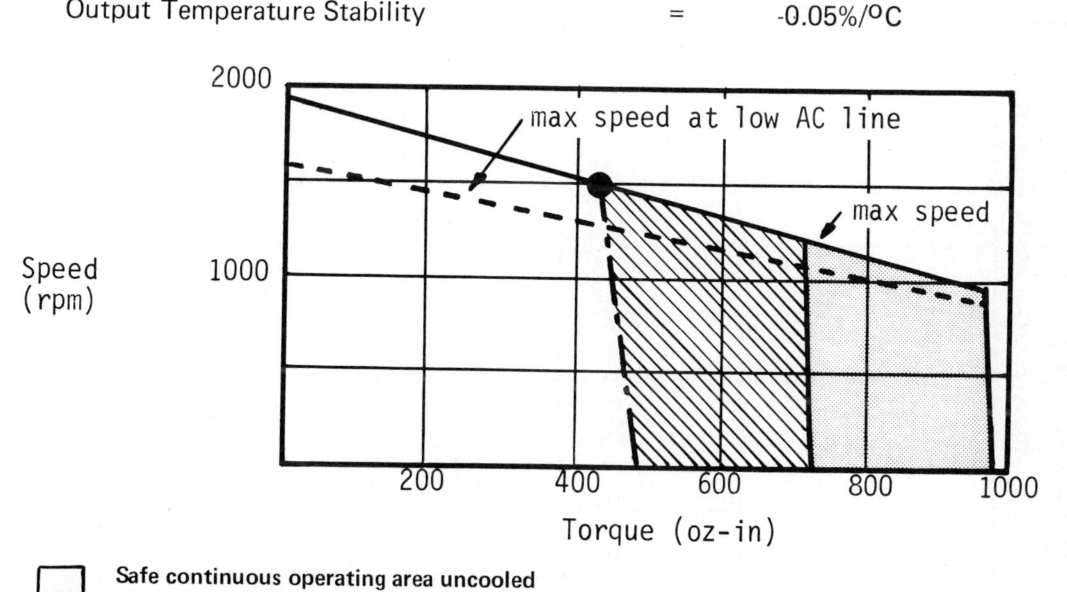

L5100

The L5100 amplifier is a high performance four quadrant (fully regenerative) DC amplifier having a continuous power output of 120 watts and a peak power output of 420 watts. This amplifier, which has an open loop frequency response of 10,000 Hz, is ideal for application where fast response is required. The amplifier is equipped with dual state load line current limiting and can be used with any of Electro-Craft's many standard and custom plug-in options.

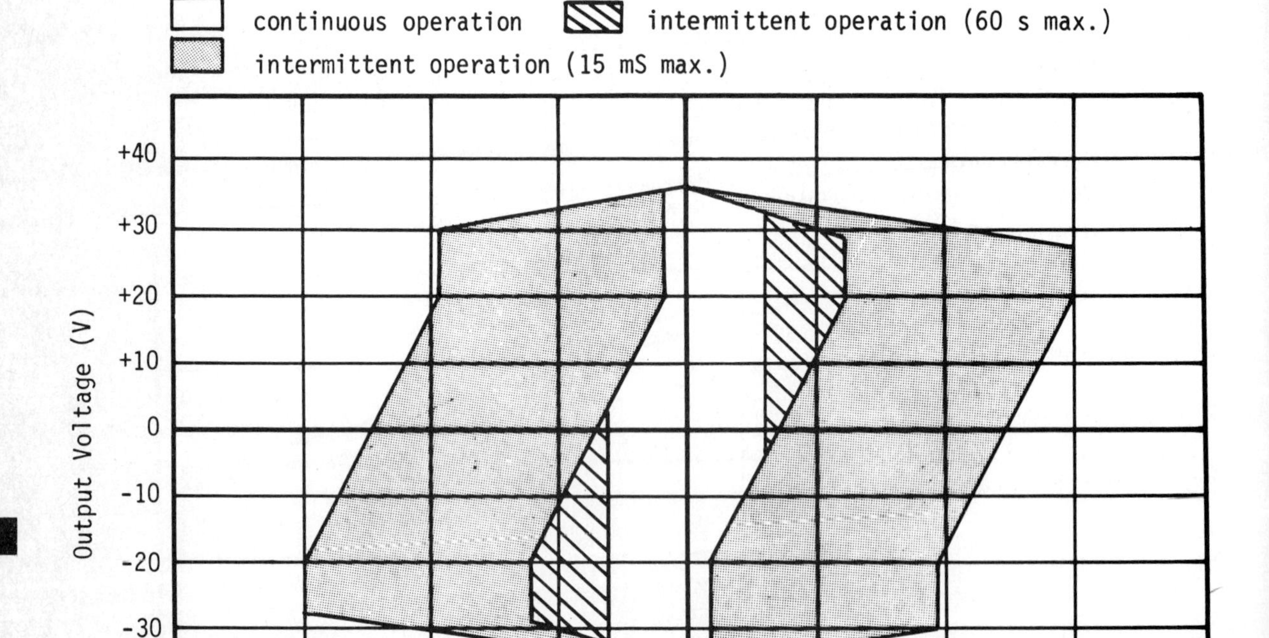

L5100 NOMINAL RATINGS

Maximum Internal Power Dissipation	215w
Maximum Voltage Output	± 35v
Maximum Continuous Current	± 6A
Maximum Peak Current	± 15A
Maximum Continuous Power Output	120w*
Maximum Peak Power Output	420w

FEATURES

— Single ended complementary symmetry output stage.

— ± supply inputs fused.

— Will withstand a continuous motor lead short circuit.

— A short from either output lead to chassis will cause a fuse to blow and the amplifier to inhibit.

— Unit will accept up to 3 plug-in option boards such as a ramp generator or phase lock servo board.

— Independent ± peak current limit controls.

— 2 stage dissipation limiting current limiter.

— A TTL high on the proper input terminal inhibits all output current.

— See separate sheet for physical dimensions.

*Amplifier continuous output is limited to 4A (120w) by power supply rating. Continuous operation at 6A (170w) is possible using a modified power supply. In this case use 60 seconds intermittent limits as continuous limits.

L5200

The L5200 amplifier is a high performance four quadrant (fully regenerative) DC amplifier having a continuous power output of 200 watts and a peak power output of 840 watts. This amplifier, which has an open loop frequency response of 10,000 Hz, is ideal for applications where fast response is required. The amplifier is equipped with dual stage load line current limiting and can be used with any of Electro-Craft's many standard and custom plug-in options.

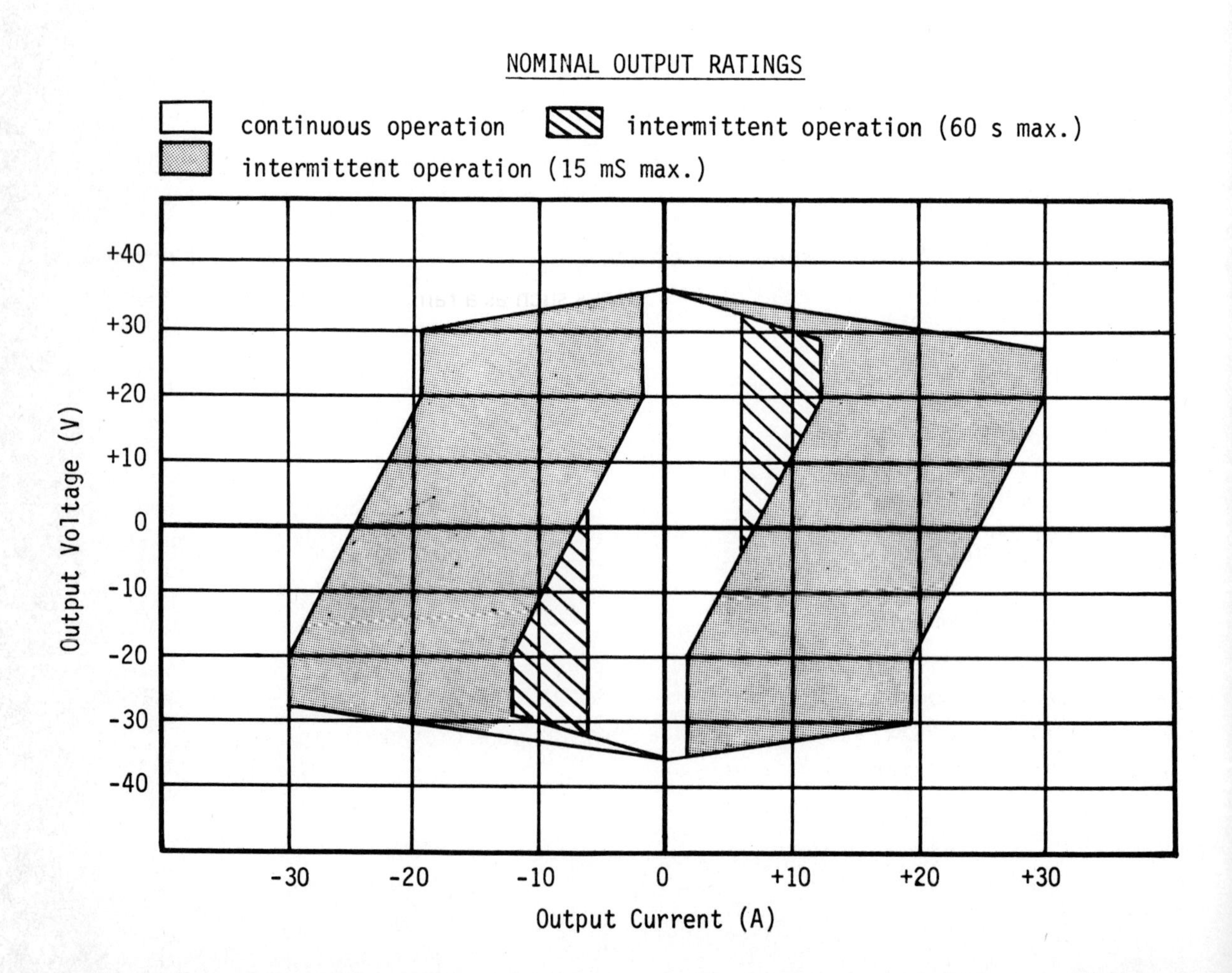

L5200 NOMINAL RATINGS

Maximum Internal Power Dissipation	330w
Maximum Voltage Output	± 35v
Maximum Continuous Current	± 12A
Maximum Peak Current	± 30A
Maximum Continuous Power Output	200w*
Maximum Peak Power Output	840w

FEATURES

— Single ended complementary symmetry output stage.

— ± supply inputs fused.

— Will withstand a continuous motor lead short circuit.

— A short from either output lead to chassis will cause a fuse to blow and the amplifier to inhibit.

— Unit will accept up to 3 plug-in option boards such as a ramp generator or phase lock servo board.

— Independent ± peak current limit controls.

— 2 stage dissipation limiting current limiter.

— A TTL high on the proper input terminal inhibits all output current.

— See separate sheet for physical dimensions.

*Amplifier continuous output is limited to 6 A (200w) by power supply rating. Continuous operation at 12A (340w) is possible using a modified power supply. In this case use 60 second intermittent limits as continuous limits.

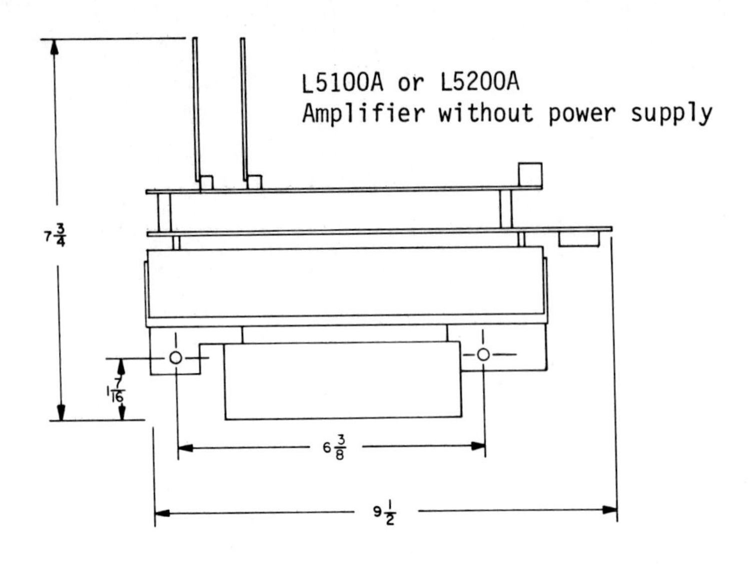

L5100A or L5200A
Amplifier without power supply

HEIGHT $4\frac{3}{4}$ MAX

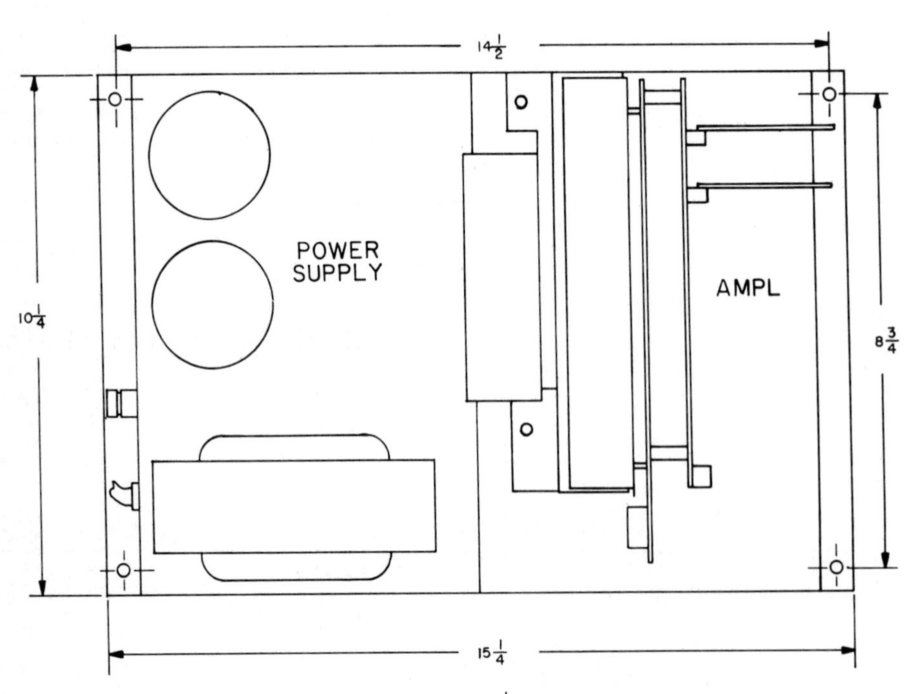

HEIGHT $6\frac{1}{4}$ MAX

L5100AP or L5200AP
Amplifier with power supply

THIS PAGE INTENTIONALLY LEFT BLANK

7.6.5. OPTIONS FOR USE WITH P6000 AND L5000 CONTROLS

The DPS MKII is a digital position control option which can be ordered with any of Electro-Craft's four quadrant servo controllers. The unique design of the DPS option and the fast response of the servo controllers combine to provide a positioning system that can achieve higher accuracy and faster step times than stepper motors, clutch brakes and other widely used systems.

The performance data which follows is for a system consisting of an SD6267-06 servo drive package and the DPS MKII option. Performance data for other combinations is available on request.

A typical 200 step per revolution system will give single steps times of 15 ms including settling and slew speeds of 16,000 steps/second — even with load inertias as high as 1 lb-in^2 — without loss of position.

Positional accuracy is typically in the order of 0.5° even with high disturbing torques. Greater accuracy can be obtained by using systems having higher resolutions without deterioration of high speed slew and large step performance. Holding torques are in the order of 400 in-oz peak, 175 in-oz continuous.

The system accepts parallel binary or BCD step size inputs which enables direct control from thumb wheel switches and direct operation from minicomputers eliminating the need for a translator/indexer or other interface logic.

An 8 bit speed input is provided so that the speed of the motor during stepping can be controlled. This feature, or the alternative analog speed input, enables the DPS 6100 to be used for contouring as well as point to point positioning applications.

For application assistance, detailed specifications and prices, contact Electro-Craft's Applications Engineering Department.

DESCRIPTION OF INPUTS AND OUTPUTS

Inputs	Designation	Description
Step Size	SIZ	20 bits binary on BCD (high = 1)
Direction	DIR	High = CW, low = CCW
Speed	SPD	8 bit binary (high = 1) or 0 to +10v DC analog level
Start	STA	2 μs minimum low level pulse = step in direction DIR for SIZ steps at SPD speed
Initialize (move to reference position)	INT	2 μs minimum low level pulse = initialize
CW End of Travel	CWT	Low = end of travel
CCW End of Travel	CCT	Low = end of travel
Reset (emergency stop)	RST	Low = reset
Slew	SLW	Low = slew at speed SPD

Outputs	Designation	Description
Initialize Required	INR	Low = initialize required
Motion Complete	MCT	Low = motion complete
Encoder Output	EOP	+ve square wave

NOTES:

SIZ and DIR are gated and must be held for 5 μs before and 10 μs after the leading edge of the STA pulse.
Input sink current = 0.23 mA per input.
Output sink current = 5 mA maximum per output.
Standard units are TTL compatible, CMOS, HiNil and switch compatible systems are available.
INT, CWT, CCT, and RST inputs include anti-bounce circuits as standard and can use logic or contact inputs.

Performance of DPS MKII with SD6267-06

TYPICAL MOTION TIMES (including settling)

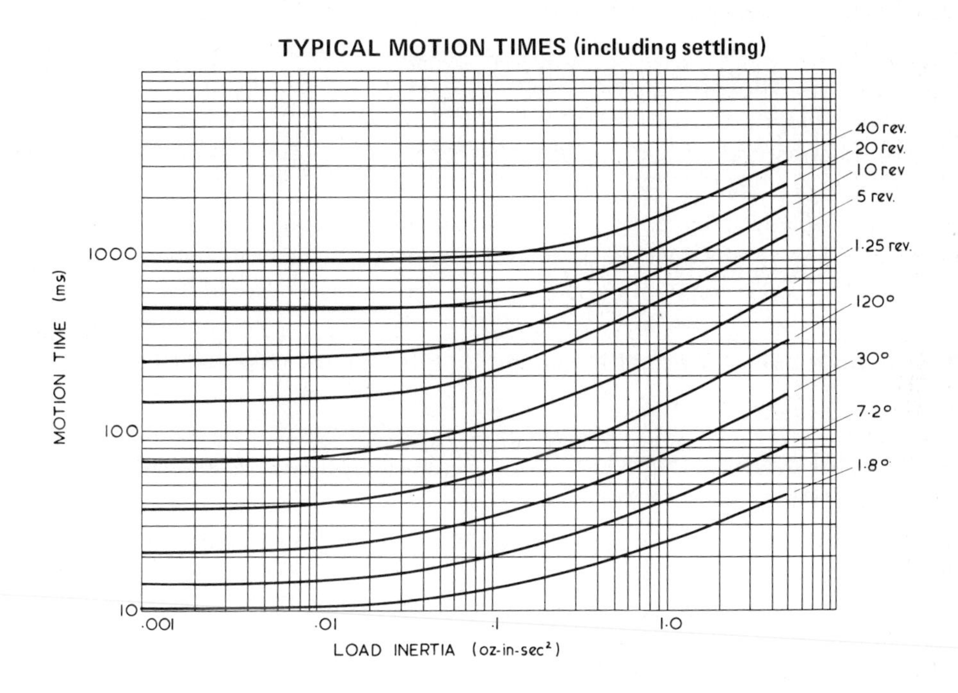

MAXIMUM CONTINUOUS REPETITION RATE

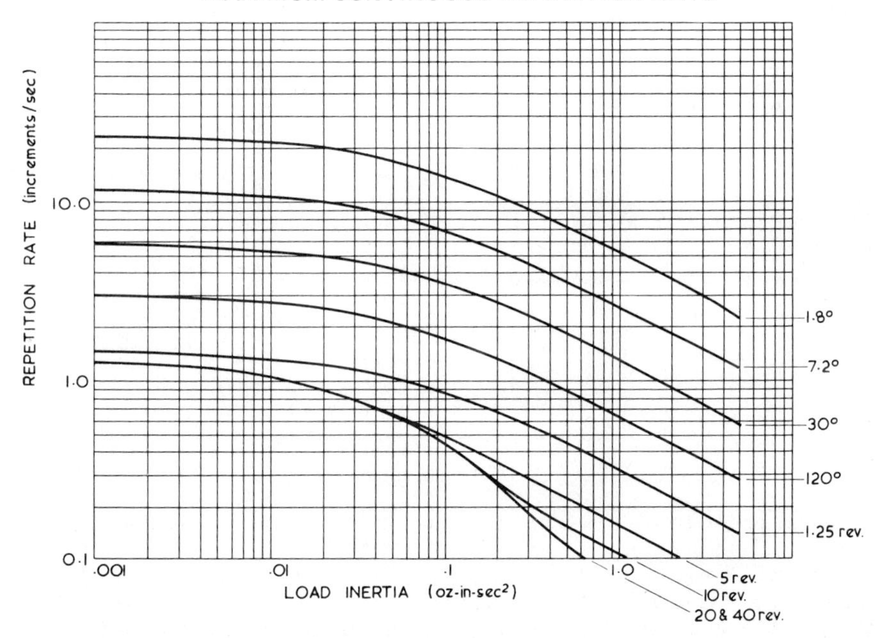

Ramp Generator for Controlled Acceleration

R1000

The R1000 Ramp Generator is a standard plug in option for use with Electro-Craft's L5000 series Servomotor Control. The option permits the selection of one of two preset adjustable acceleration rates and one of two preset adjustable deceleration rates. The adjustments for ramp slopes are independent of each other and the different preset slopes can be selected by switch contacts or logic signals.

The R1000 also permits the selection of one of two preset adjustable motor speeds or a speed set by an external analog signal, thus the option can be used for digital selection of speed.

The Ramp Generator does not affect the closed-loop frequency response of the system to load disturbances.

The option is useful in any application where the acceleration and deceleration need to be controlled in a repeatable manner.

1 Adjustment for speed 1   2 Adjustment for speed 2

a+ Adjustment for +ve slope a   a- Adjustment for -ve slope a

b+ Adjustment for +ve slope b   b- Adjustment for -ve slope b

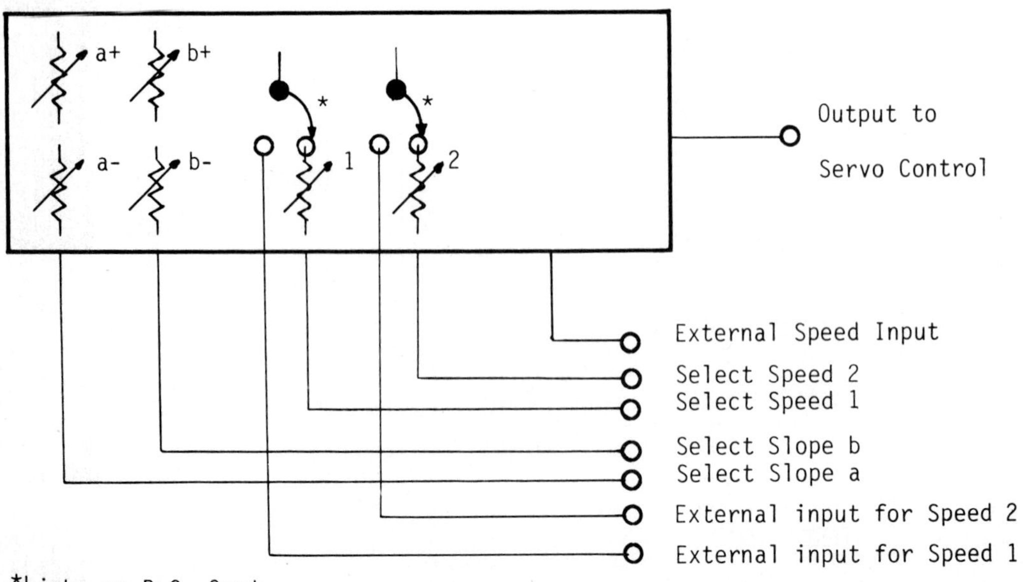

*Links on P.C. Card

R1000 RAMP GENERATOR SPECIFICATIONS

RAMP RATES

Range of potentiometer adjustment	10:1
Maximum rate (standard unit)	100 v/s
Minimum rate (standard unit)	10 v/s
Maximum rate (with component change)	100 v/ms
Minimum rate (with component change)	0.5 v/s
Rate drift with temperature	0.1 %/^{o}C

OUTPUT

Maximum Output	10 v min.
Load Impedance	10 kM min.

INTERFACE

Logic Type	TTL

(To select slope or speed apply 0 v to appropriate terminal)

For further information contact Electro-Craft's Applications Engineering Department.

PL1000

The PL1000 Phase Lock System is a standard option which can be used with Electro-Craft's L5000 and P6000 series Servomotor Controls.

The Phase Lock option enables speed control with zero long term speed drift and synchronization of a motor with any other moving part.

This option is particularly suited to applications where the speed of the motor forms a reference such as chart drives or scanning drives. The system is also ideally suited to component marking applications where a motor driving a marking head has to be accurately synchronized with components on a moving conveyor.

For further information and technical assistance contact Electro-Craft's Applications Engineering Department.

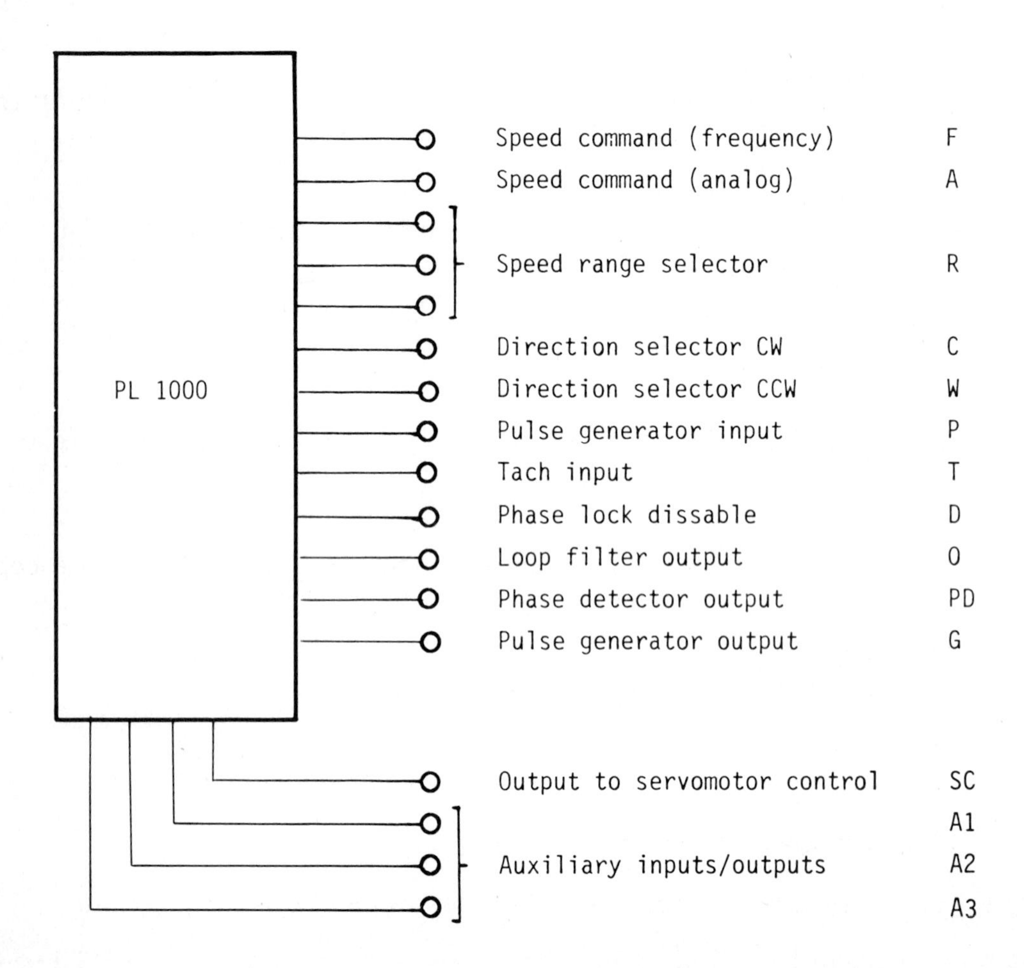

DESCRIPTION OF INPUTS AND OUTPUTS

F *Speed Command Frequency*
Square wave or sine wave 5v minimum peak to peak minimum frequency 1 Hz maximum 10 kHz. Frequency proportional to speed. Also can be used for synchronizing pulse input for synchronizing systems.

A *Speed Command Analog*
Analog voltage proportional to F. Maximum value 10v. Polarity of signal can be used to set direction of rotation. An internal frequency discriminator can be provided to generate this signal from F or a voltage controlled oscillator can be provided to generate the frequency signal from the analog signal.

R *Speed Range Selector*
The overall speed range of the system is approximately 10,000:1. This is divided into three ranges of 20:1 each; a logic zero signal applied to the speed range selector inputs selects the appropriate speed range.

C&W *Direction Selectors*
A logic 0 on the appropriate input selects the direction of rotation. Alternately the polarity F signal A can be used to select direction.

P *Pulse Generator Input*
Pulse signal feedback from the pulse generator mounted on the driver motor.

T *Tach Input*
Analog signal feedback from tachometer mounted on driver motor. This signal is only required on high performance systems.

D *Phase Lock Disable*
Logic signal to disable the phase comparitor to return the system to conventional analog control.

PD *Phase Detector Output*
Analog signal proportional to the phase error between F and P. This and the loop filter output can be used to examine the stress on the servo loop.

O *Loop Filter Output*
The same as PD except filtered to remove harmonics.

G. *Pulse Generator Output*
Same signal as P but squared up by internal circuits.

SC Output to servomotor control.

A1 thru A3 auxiliary inputs and outputs-spare inputs for special applications.

7.7. BRUSHLESS DC MOTORS

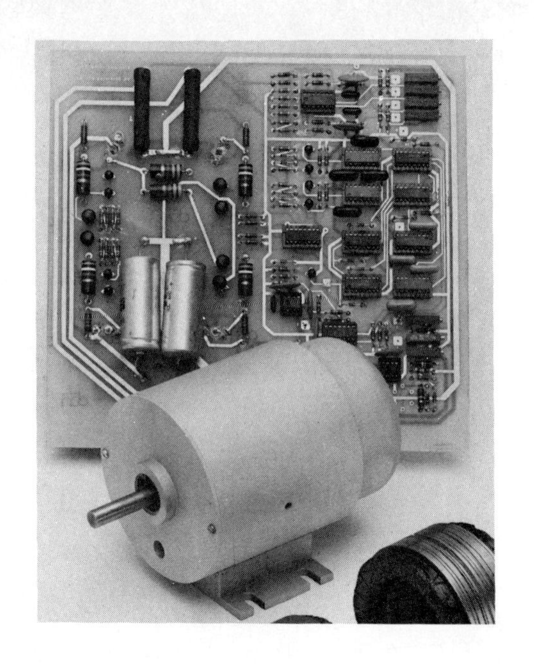

MODEL BLM 340

The Model 34 series BLM is a line of high performance brushless DC servomotors. Their design is suited for a wide range of servo applications including computer peripherals and machine tool drives.

SPECIFICATIONS

MOTOR RATINGS

Peak Torque (stall)	145 oz-in
Rated Load Torque	40 oz-in
Rated Speed	3600 rpm
Rated Output Power	0.144 hp
No Load Speed	4880 rpm
Electrical Time Constant	2.4 ms
Mechanical Time Constant	14.5 ms
Friction Torque	1.3 oz-in
Armature Inertia	0.01 oz-in-sec
Viscous Damping Constant	0.1 oz-in/krpm
Maximum Winding Temperature	155 °C
Ripple Torque (avg. to peak)	7
Thermal Resistance (no heatsink)	2.9 °C/W
Theoretical Acceleration from Stall	14,500 rad/sec^2

WINDING CONSTANTS*

Winding Number	Voltage (Volts)	Peak Pulse Current (Amps)	Rated Current (Amps)	Winding Resistance (Ohms @ 25°C)	Voltage Constant K_E (V/krpm)	Torque Constant K_T (oz-in/Amp)	Winding Inductance (mH)
01	24	24	6.7	0.56	5.1	6.1	1.3
02	36	17	4.8	1.10	7.1	8.6	2.7
03	48	12	3.5	1.97	9.7	11.7	4.7
04	60	10	2.9	3.00	11.8	14.2	7.1
05	80	7	2.1	5.42	16.7	20.1	11.7
06	100	6	1.7	8.60	20.7	24.8	17.1

*All Winding Values are with 2 of 3 Phases energized sequentially by electronic commuata-tion.

SCHEMATIC DIAGRAM

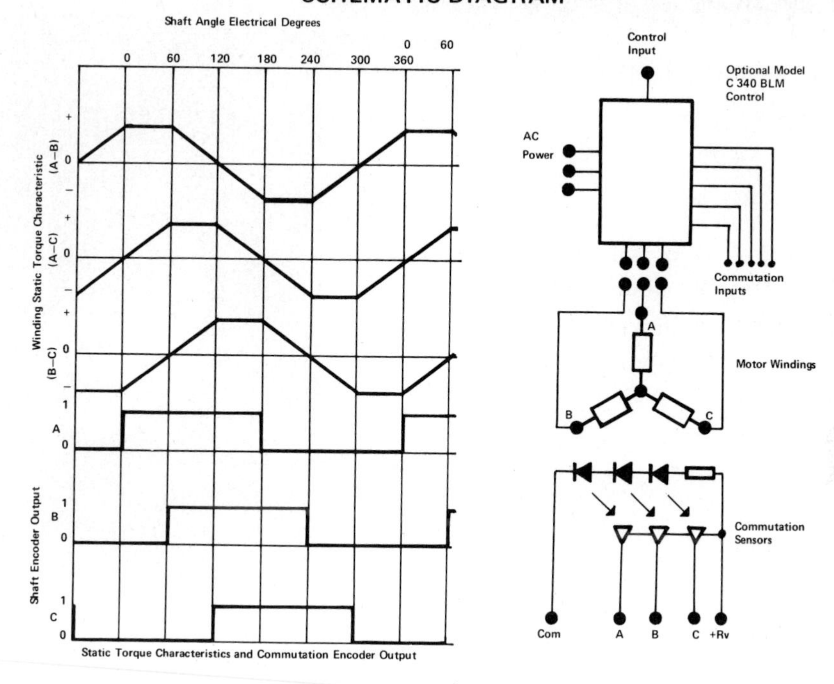

OUTLINE DRAWING

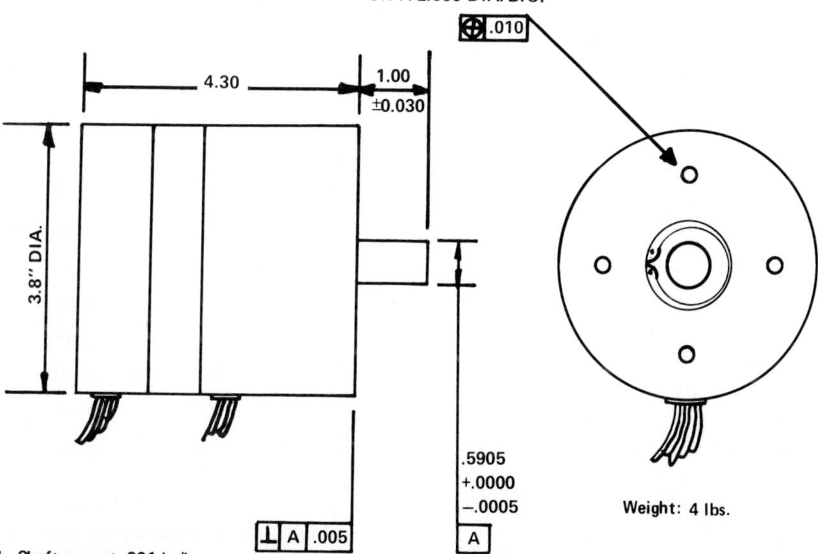

1. Shaft runout .001 in/in
2. Shaft end play .0005 min to .005 max measured with 10 lbs reversing load.
3. Shaft radial play .0015 max measured ¼ in from mtg. face with 5 lbs reversing load ¾ in. from mtg face.

MODEL BLM 341

The Model 34 series BLM is a line of high performance brushless DC servomotors. Their design is suited for a wide range of servo applications including computer peripherals and machine tool drives.

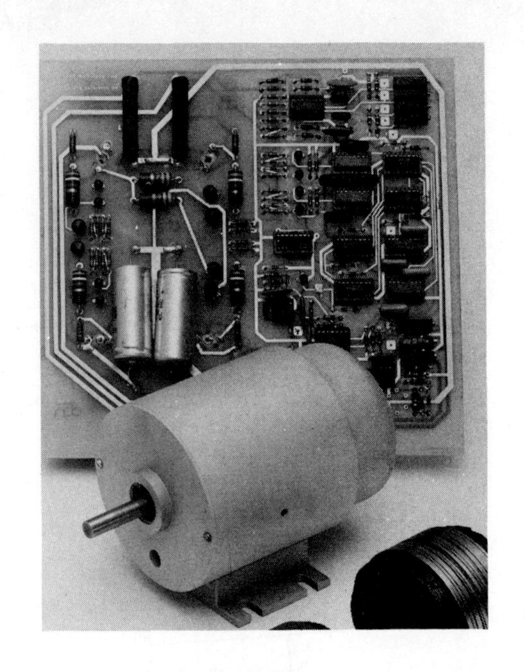

SPECIFICATIONS

MOTOR RATINGS

Peak Torque (stall)	320 oz-in
Rated Load Torque	80 oz-in
Rated Speed	3600 rpm
Rated Output Power	0.28 hp
No Load Speed	4530 rpm
Electrical Time Constant	1.85 ms
Mechanical Time Constant	8.20 ms
Friction Torque	2.50 oz-in
Armature Inertia	0.016 oz-in-sec^2
Viscous Damping Constant	0.1 oz-in/krpm
Maximum Winding Temperature	155 ^{o}C
Ripple Torque (avg. to peak)	7
Thermal Resistance (no heatsink)	1.95 ^{o}C/W
Theoretical Acceleration at Stall	20,000 rad/sec^2

WINDING CONSTANTS*

Winding Number	Voltage (Volts)	Peak Pulse Current (Amps)	Rated Current (Amps)	Winding Resistance (Ohms @ 25°C)	Voltage Constant K_E (V/krpm)	Torque Constant K_T (oz-in/Amp)	Winding Inductance (mH)
01	24	48	12.3	0.31	5.6	6.7	0.5
02	36	34	8.8	0.54	7.8	9.4	1.0
03	48	87	6.8	0.88	10.1	12.1	1.8
04	60	20	5.2	1.38	13.2	15.9	2.7
05	80	16	4.2	2.45	16.4	19.6	4.1
06	100	12	3.1	4.68	22.4	26.9	7.4

*All Winding Values are with 2 of 3 Phases energized sequentially be electronic commutation.

SCHEMATIC DIAGRAM

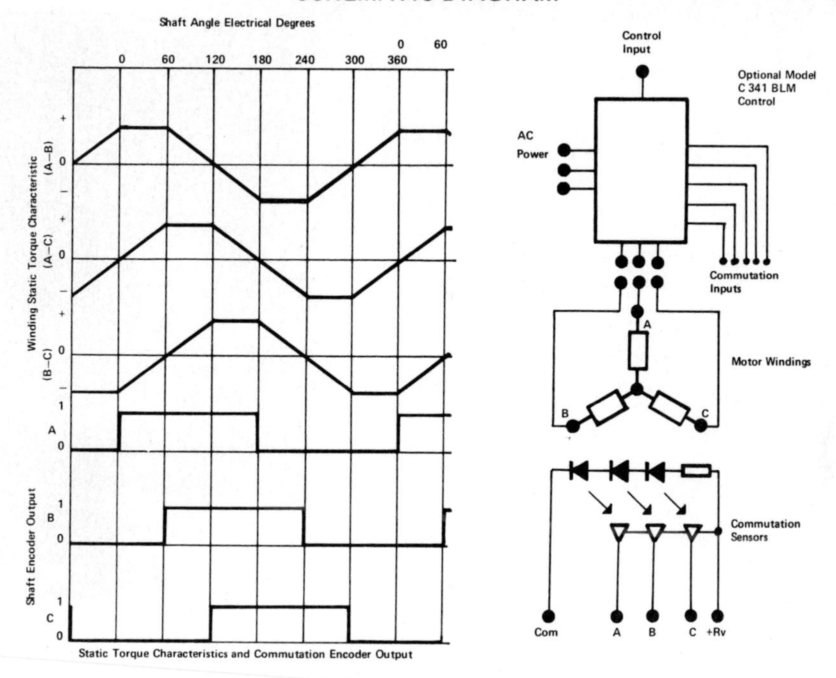

Static Torque Characteristics and Commutation Encoder Output

OUTLINE DRAWING

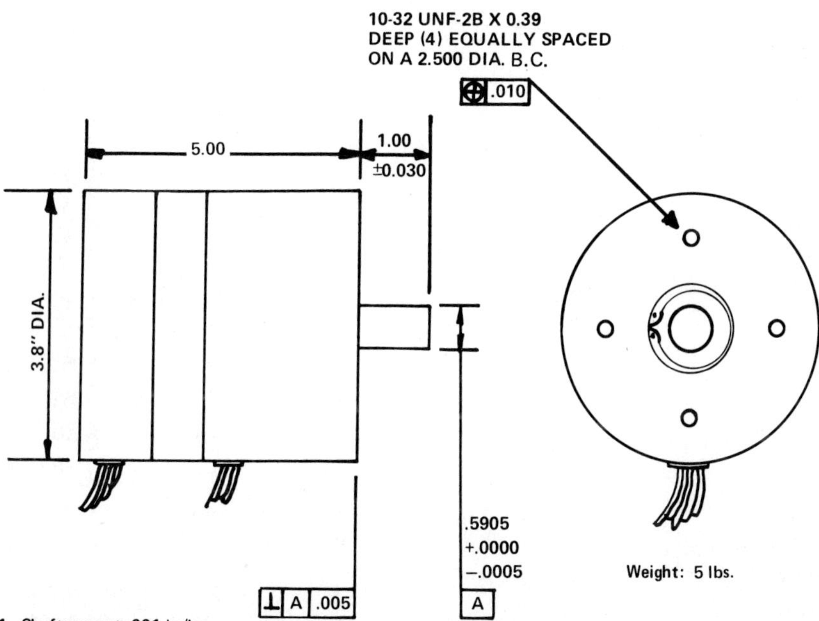

Weight: 5 lbs.

1. Shaft runout .001 in/in
2. Shaft end play .0005 min to .005 max measured with 10 lbs reversing load.
3. Shaft radial play .0015 max measured ¼ in from mtg. face with 5 lbs reversing load ¾ in. from mtg face.

MODEL BLM 342

The Model 34 series BLM is a line of high performance brushless DC servomotors. Their design is suited for a wide range of servo applications including computer peripherals and machine tool drives.

SPECIFICATIONS

MOTOR RATINGS

Peak Torque (stall)	525 oz-in
Rated Load Torque	140 oz-in
Rated Speed	3600 rpm
Rated Output Power	0.51 hp
No Load Speed	4460 rpm
Electrical Time Constant	1.3 ms
Mechanical Time Constant	7.4 ms
Friction Torque	3.5 oz-in
Armature Inertia	0.022 oz-in-sec
Viscous Damping Constant	0.22 oz-in/krpm
Maximum Winding Temperature	155 ^{o}C
Ripple Torque (avg. to peak)	7
Thermal Resistance (no heatsink)	1.2 ^{o}C/W
Theoretical Acceleration from Stall	23,800 rad/sec^2

WINDING CONSTANTS*

Winding Number	Voltage (Volts)	Peak Pulse Current (Amps)	Rated Current (Amps)	Winding Resistance (Ohms @ 25°C)	Voltage Constant K_E (V/krpm)	Torque Constant K_T (oz-in/Amp)	Winding Inductance (mH)
01	36	53	14.3	0.36	8.3	10.0	0.5
02	48	40	10.9	0.60	11.0	13.2	0.8
03	60	31	8.5	0.93	14.0	16.8	1.2
04	80	21	6.8	1.50	18.0	21.6	2.0
05	100	20	5.4	2.31	22.0	26.4	3.0
06	125	16	4.3	3.70	28.0	33.6	4.8

*All Winding Values are with 2 of 3 Phases energized sequentially be electronic commutation.

SCHEMATIC DIAGRAM

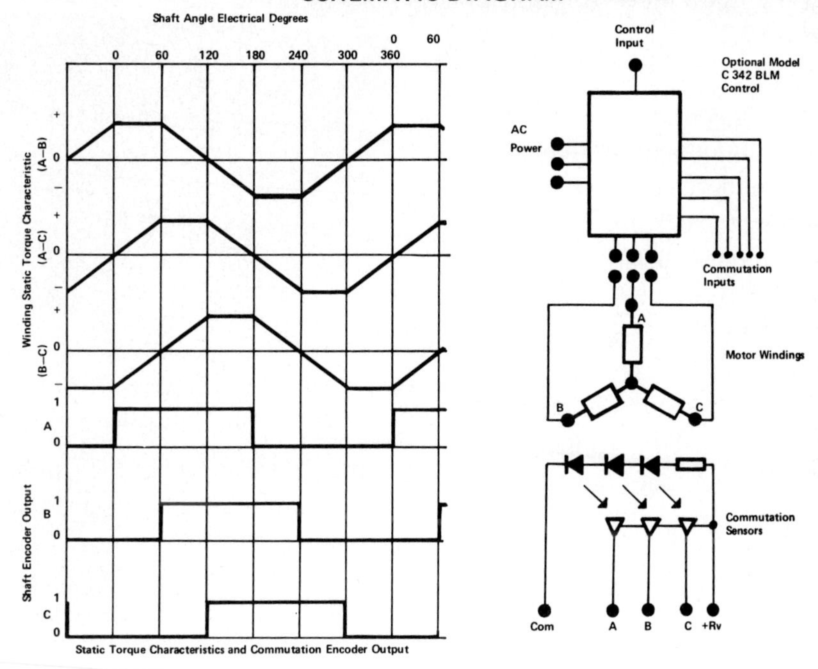

OUTLINE DRAWING

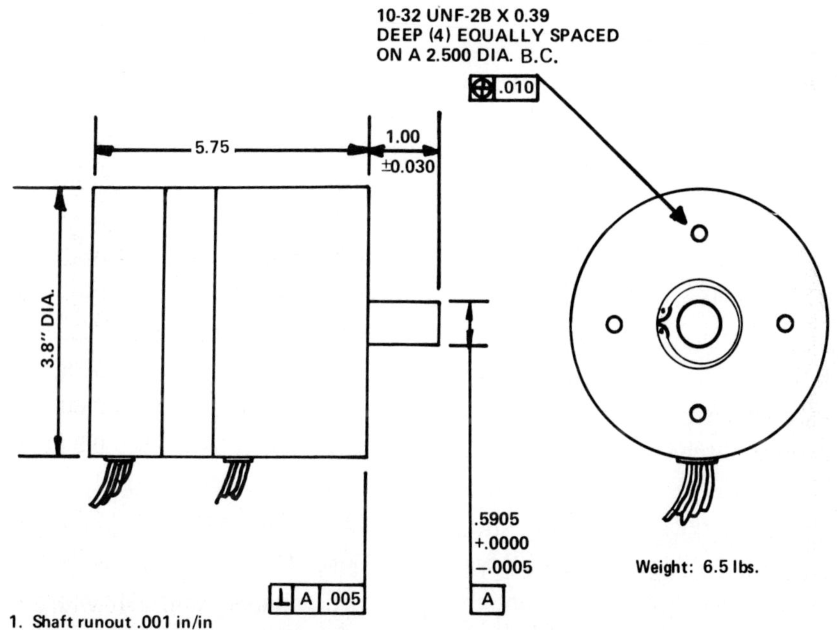

1. Shaft runout .001 in/in
2. Shaft end play .0005 min to .005 max measured with 10 lbs reversing load.
3. Shaft radial play .0015 max measured ¼ in from mtg. face with 5 lbs reversing load ¾ in. from mtg face.

7.8. SERVO SYSTEMS FOR ENGINEERING EDUCATION

This section is intended to inform about special laboratory equipment available from Electro-Craft. The more complex Motomatic Control System Laboratory (Fig. 6.8.1) was developed for senior level university or graduate course instruction. The simpler Motomatic Experimenter (Fig. 6.8.2) was aimed for vocational technical or lower level college instruction. Both are complete hardware-software packages, student-tested in educational and industrial (in-plant training) environments.

MOTOMATIC CONTROL SYSTEM LABORATORY (MCSL)

The Motomatic Control System Laboratory, developed in conjunction with control engineers at the University of Minnesota, is a comprehensive piece of laboratory equipment including both hardware and software suitable for a one-year laboratory course in electro-mechanical control system principles.

Hardware

The Motomatic Control System Laboratory consists of an electronic control chassis, an electro-mechanical chassis, a kit of small additional parts, plus an instrumentation chassis. The electronic chassis is a solid state device which is used to power and compensate the electro-mechanical components in conjunction with simple plug-in cards on the top of the chassis. On the chassis,

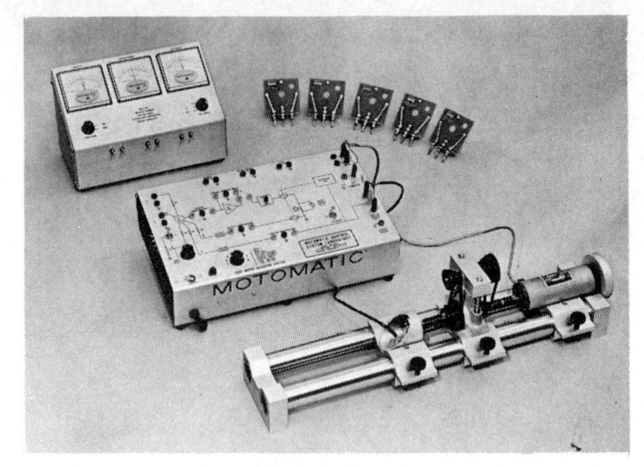

Fig. 6.8.1. The Electro-Craft Motomatic Control System Laboratory (MCSL).

provision is made for control and instrumentation of the electro-mechanical components. The instrumentation chassis is simply a collection of suitable meters to monitor the behavior of the system. The mechanical chassis is, simply stated, the ultimate in "hands on", "student proof" mechanical design of a general purpose electro-mechanical instrument which is available today. There is no chance of shaft misalignment and no opportunity for poor mechanical connection between the components with this device. It is simple to use, and can be connected in seconds. The motor, tachometer, potentiometer and speed reduction unit are the highest industrial quality components.

Software

The laboratory manual was written by Professor Stephen J. Kahne of the University of Minnesota, Department of Electrical Engineering and Director of Hybrid Computer Laboratory. It consists of a complete description of the MCSL, a complete laboratory course of experiments which have been developed especially for this equipment in a senior level, two quarter laboratory course in automatic control and a description of projects in which this equipment may be used. Many of these projects are suitable for senior thesis or basic graduate level projects in electro-mechanical control principles. The emphasis in this equipment and projects is on basic engineering, analysis and design principles and includes all literature references needed for a self-contained laboratory course to accompany a regular senior level control course. Experience has shown that with two or three students per MCSL, this laboratory equipment is ideally suited for this senior level course.

MCSL Manual

Table of Contents

Section I - Introduction

- I-1 Laboratory Philosophy
- I-2 Parts List
- I-3 The Electronic Equipment
- I-4 The Electro-mechanical Equipment
- I-5 The Instrumentation Chassis
- I-6 Miscellaneous Remarks on Operation of MCSL

Section II - The Experiments

- II-0 Format of Experiments
- II-1 Experiment 1 - Amplifiers for Control Systems
- II-2 Experiment 2 - Control Devices and Their Transfer Functions
- II-3 Experiment 3 - Frequency Response
- II-4 Experiment 4 - Transient Response
- II-5 Experiment 5 - Velocity Control
- II-6 Experiment 6 - Position Control
- II-7 Experiment 7 - Adaptive Control

Section III - Special Projects

- III-0 Introduction
- III-1 Pneumatic Loading
- III-2 Temperature Control
- III-3 Illumination Control
- III-4 Weight Measurement
- III-5 Large Speed Reduction
- III-6 Air Foil Control
- III-7 Bang-Bang Control
- III-8 Effects of Backlash
- III-9 Optimal Control - Linear State Feedback
- III-10 Flexible Couplings

Fig. 6.8.2. The Electro-Craft Motomatic Experimenter.

Section IV - Appendices

1. An Introduction to State Variable Characterization of Linear Systems.

2. Bibliography

THE MOTOMATIC EXPERIMENTER

This complete educational kit uses industrial equipment to provide the student with a "hands-on" introduction to the theory and practice of servo control systems. Equipment includes a DC motor-generator, a transistor control unit, and miscellaneous parts necessary to all experiments in the course. The kit was designed to guide the student in a thorough examination of the basic concepts, physical principles, construction, and important parameters of each component and the entire system.

Motomatic Experimenter Manual

Table of Contents

THE MOTOMATIC SERVO CONTROL COURSE MANUAL

The detailed, complete study course was specifically prepared for students in vocational-technical schools. The author, Robert N. Bateson, holds a master degree in electrical engineering and has extensive industrial and teaching experience. He is currently chairman of the Engineering and Technology Division of Anoka-Ramsey State Junior College in Minnesota. The 144-page manual includes 20 student-tested experiments. Each is preceded by a thorough explanation of the physical principles involved and presents detailed description of the necessary operations in the experiment. □

Appendix

SI System and Conversion Factors

A.1. SI SYSTEM OF UNITS (METRIC SYSTEM)

The **International System of Units** (in French: **S**ystème **I**nternational d'Unités - **SI**) was given its name in 1960 by the XI. Conférence Générale des Poids et Mesures. This system covers the whole science and technology and includes as subsystems the MKS (Meter-Kilogram-Second) system of units, which covers mechanics, and the MKSA (Meter-Kilogram-Second-Ampere) system of units, which covers mechanics, electricity, and magnetism.

The former CGS (Centimeter-Gram-Second) system of units has been, until now, sometimes used — especially in magnetism regardless of whether its use has been deprecated by law in nations which are on the metric system. Therefore, in the Conversion Factors Table in the Appendix A.2. some of the CGS magnetic units and their relation to SI and British units are introduced.

The SI system is the coherent system of metric units, which is based on the following six *basic units* and quantities:

QUANTITY	UNIT	SYMBOL
length	meter	m
mass	kilogram	kg
time	second	s
electric current	ampere	A
temperature difference	degree	deg
luminous intensity	candela	cd

From these basic units there is derived successively and systematically a series of *main units* which, together with basic units, form the SI system. These units are coherent; i.e., in their definition equations there are no numerical coefficients at all. Therefore, also in all practical equations there are no numerical coefficients if all quantities are expressed in basic or main units.

Besides this, a series of *secondary units* is used which are derived from basic or main units by means of certain numerical coefficients. These units do not belong to the SI system, but their use has either historical reason (e.g., minute for time, 1 min = 60 s), or practical reason, because some of main units are too small or too great for usual applications (e.g., main unit of pressure, $1\ N/m^2 \approx 1.44 \times 10^{-4}\ lb/in^2$). In the following text each kind of unit will be distinguished by:

(B) for basic units

(M) for main units

(S) for secondary units

In the SI system, the following prefixes are used to indicate multiples or submultiples of units:

MULTIPLE	PREFIX	SYMBOL
10^{12}	tera	T
10^{9}	giga	G
10^{6}	mega	M
10^{3}	kilo	k
10^{2}	hecto	h
10^{1}	deka	da
10^{-1}	deci	d
10^{-2}	centi	c
10^{-3}	milli	m
10^{-6}	micro	μ*
10^{-9}	nano	n
10^{-12}	pico	p
10^{-15}	femto	f
10^{-18}	atto	a

*Lower-case u is frequently used in typing

Examples:

millimeter	1 mm	$= 10^{-3}$ m
kilovolt	1 kV	$= 10^{3}$ V
microampere	1 μA	$= 10^{-6}$ A
nanofarad	1 nF	$= 10^{-9}$ F
femtogram	1 fg	$= 10^{-15}$ g
gigawatt	1 GW	$= 10^{9}$ W

DEFINITION OF BASIC SI UNITS

Length

(B) meter	m
1 m = 1 650 763.73 λ, where λ is the wavelength of radiation of the orange spectral line of the nuclid 86 of krypton in vaccum	

Originally, the meter was defined as $1/(4 \times 10^{7})$ of the earth's meridian at sea level and realized by the length of the metallic international prototype.

Frequently used multiples:

kilometer	1 km	$= 10^{3}$ m
centimeter	1 cm	$= 10^{-2}$ m
millimeter	1 mm	$= 10^{-3}$ m
micrometer	1 μm	$= 10^{-6}$ m

Mass

(B) kilogram	kg
1 kg is the mass of the international mass prototype which is secured by the International Bureau of Weights and Measures at Sèvres, France	

Very accurately, 1 kg is the mass of 1 l (liter) = 1 dm^3 = 10^{-3} m^3 of water at 4 °C and at normal atmospherical pressure.

Frequently used multiples:

ton (metric)	1 t	$= 10^{3}$ kg
gram	1 g	$= 10^{-3}$ kg
milligram	1 mg	$= 10^{-6}$ kg

Time

(B) second	s
1 s = 9 192 631 770 T, where T is the period of oscillations of the atom of cesium 133 in the transition between the levels F = 4, M = 0 and F = 3, M = 0 of the basic state 2 $S_{1/2}$ without influence of external magnetic fields	

Frequently used multiples and secondary units:

(S) day (mean solar day)

1 d = 24 h = 86 400 s

(S) hour

1 h = 3600 s

(S) minute

1 min = 60 s

millisecond

1 ms = 10^{-3} s

microsecond

1 μs = 10^{-6} s

nanosecond

1 ns = 10^{-9} s

Electric current

(B) ampere	A
1 A is the DC current which, being conducted by two straight, parallel and infinitely long conductors with negligible cross-section, spaced in vacuum 1 m apart, causes the force of 2 x 10^{-7} N per meter of length between them	

Frequently used multiples:

kiloampere	1 kA = 10^{3} A
milliampere	1 mA = 10^{-3} A
microampere	1 μA = 10^{-6} A
nanoampere	1 nA = 10^{-9} A
picoampere	1 pA = 10^{-12} A

Temperature difference

(B) degree	deg, ^{o}
1 deg = 1/273.16 of the temperature difference between the temperature of absolute zero (0 K = –273.15 ^{o}C) and the temperature of triple point of water (273.16 K = 0.01 ^{o}C), expressed in the thermodynamical scale	

Temperature is measured and designated in two different scales:

a) absolute temperature	kelvin	K, deg K
b) normal temperature	degree Celsius (centigrade)	^{o}C, deg C

where

1 deg = 1 K = 1 ^{o}C

(magnitude of degrees)

and

0 ^{o}C = 273.15 K, so that

$$\Theta\,[^{o}C] = \Theta\,[K] - 273.15$$

Originally, the degree Celsius was defined as 1/100 of the temperature difference between the boiling and freezing points of water at normal atmospherical pressure .

Luminous intensity

(B) candela	cd
1 cd is the normal (perpendicular) luminous intensity of (1/60) cm^2 of the surface of an absolutely black body at the temperature of solidification of platinum (≈1773.5 °C) and at pressure of 1.01325×10^5 N/m^2	

The *candela* is one of photometric units (like *lumen, lux,* etc.) which are related to optical properties of the human eye. Therefore, in physical radiation theory and measurement only the radiometric units *watt* for radiant flux, etc.) must be used.

REVIEW OF MAIN SI UNITS AND SECONDARY UNITS

This review is limited to mechanics, electricity and magnetism, and contains only the units related to the contents of this book.

The successive method of derivation from basic units is described and the dimension of each unit, expressed by means of basic dimensions (m, kg, s, A) is introduced, e.g., $[V] = m^2\ kg\ s^{-3}\ A^{-1}$

A. MECHANICAL UNITS

Area

(M) square meter	m^2

Frequently used multiples:

square kilometer	$1\ km^2$	$= 10^6\ m^2$
hectare	1 ha	$= 10^4\ m^2$
are	1 a	$= 10^2\ m^2$
square centimeter	$1\ cm^2$	$= 10^{-4}\ m^2$
square millimeter	$1\ mm^2$	$= 10^{-6}\ m^2$

Volume

(M) cubic meter	m^3

Frequently used multiples:

liter	1 l	$= 1\ dm^3 = 10^{-3} m^3$
cubic centimeter	$1\ cm^3$	$= 10^{-6}\ m^3$
cubic millimeter	$1\ mm^3$	$= 10^{-9}\ m^3$

Velocity (linear)

(M) meter per second	$m\ s^{-1}$

Frequently used multiples and secondary unit:

(S) kilometer per hour

$1\ km/h = (1/3.6)\ m\ s^{-1}$

centimeter per second

$1\ cm\ s^{-1} = 10^{-2}\ m\ s^{-1}$

millimeter per second

$1\ mm\ s^{-1} = 10^{-3}\ m\ s^{-1}$

The velocity of electromagnetic radiation in vacuum $c = 2.99793 \times 10^8$ m s^{-1}

Acceleration (linear)

(M) meter per square second	m s^{-2}

The standard (average) gravitational acceleration is internationally stated by the value $g = 9.80665$ m s^{-2} (exactly).

Angular displacement

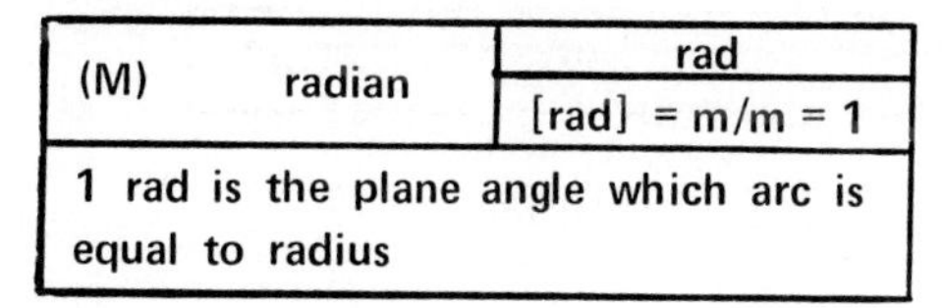

(M) radian	rad
	[rad] = m/m = 1
1 rad is the plane angle which arc is equal to radius	

Secondary units:

(S) angular degree

$1^{\circ} = (2\pi/360)$ rad
$= 0.017453$ rad

(S) revolution

$1\ r = 360^{\circ} = (2\pi)$rad $= 6.2832$ rad

Angular velocity

(M) radian per second	rad s^{-1}
	[rad s^{-1}] = s^{-1}

Secondary units:

(S) angular degree per second

$1^{\circ}\ s^{-1} = 0.017453$ rad s^{-1}

(S) revolution per second

$1\ r\ s^{-1} = 1$ rps
$= 6.2832$ rad s^{-1}

(S) revolution per minute

$1\ r\ min^{-1} = 1$ rpm
$= 0.10472$ rad s^{-1}

Angular acceleration

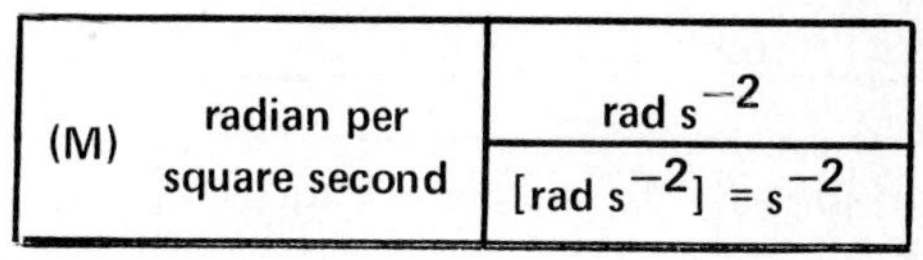

(M) radian per square second	rad s^{-2}
	[rad s^{-2}] = s^{-2}

Secondary unit:

(S) angular degree per square second

$1^{\circ}\ s^{-2} = 0.017453$ rad s^{-2}

Force

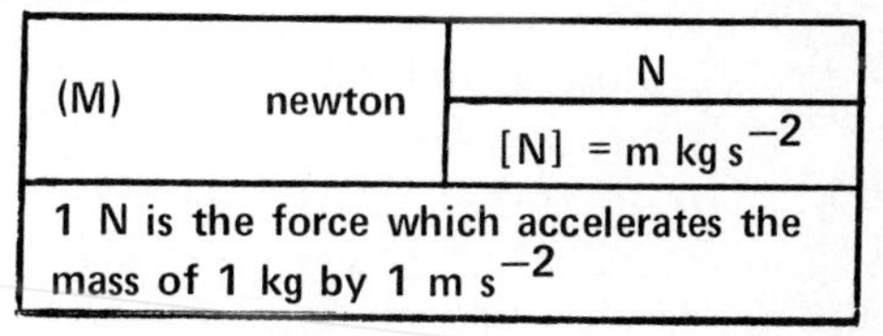

(M) newton	N
	[N] = m kg s^{-2}
1 N is the force which accelerates the mass of 1 kg by 1 m s^{-2}	

Secondary units:

(S) kilopond

1 kp = 9.80665 N

(S) pond

$1\ p = 10^{-3}$ kp
$= 9.80665 \times 10^{-3}$ N

1 kp is the force which accelerates the mass of 1 kg by $g = 9.80665$ m s^{-2} so that 1 kp is the approximate weight (heaviness) of the mass of 1 kg. The unit kp is sometimes called kilogram-force (kgf).

Torque

(M) newtonmeter	Nm
	[Nm] = m^2 kg s^{-2}
1 Nm is the torque produced by the force of 1N acting on the radius of 1 m	

Secondary unit:

(S) kilopondmeter

$$1 \text{ kpm} = 9.80665 \text{ Nm}$$

Moment of inertia

(M) kilogram – square meter	$kg\ m^2$

Moment of inertia of a body (with respect to the given axis of rotation) is given by the volume integral

$$J = \int_m r^2 \, dm$$

where **dm** is the mass element of the body and **r** is its radius with respect to the axis of rotation.

Energy (work)

(M) joule	J
	$[J] = m^2\ kg\ s^{-2}$
1 J is the work of the force of 1 N along the path of 1 m	

Secondary and equivalent units:

newtonmeter

$$1 \text{ Nm} = 1 \text{ J}$$

wattsecond

$$1 \text{ Ws} = 1 \text{ J}$$

(S) kilopondmeter

$$1 \text{ kpm} = 9.80665 \text{ J}$$

(S) kilocalorie

$$1 \text{ kcal} = 4186.8 \text{ J}$$

(amount of heat to increase the temperature of 1 l of water by 1 ^{o}C)

(S) electronvolt

$$1 \text{ eV} = 1.60206 \times 10^{-19} \text{ J}$$

(S) kilowatthour

$$1 \text{ kWh} = 3.6 \times 10^{6} \text{ J}$$

Power

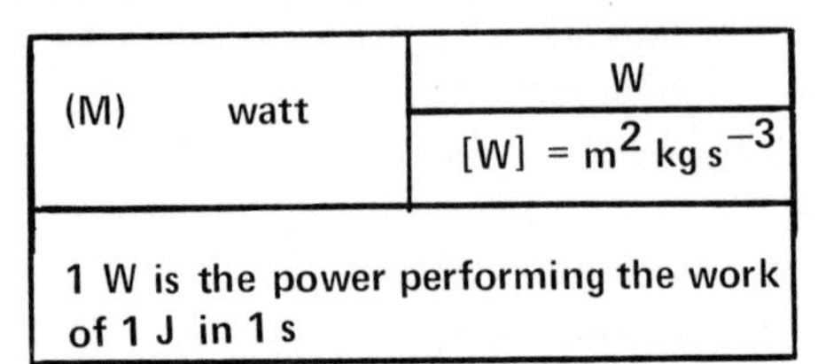

(M) watt	W
	$[W] = m^2\ kg\ s^{-3}$
1 W is the power performing the work of 1 J in 1 s	

Secondary and equivalent units:

newtonmeter per second

$$1 \text{ Nm/s} = 1 \text{ W}$$

joule per second

$$1 \text{ J s}^{-1} = 1\text{W}$$

(S) kilopondmeter per second

$$1 \text{ kpm s}^{-1} = 9.80665 \text{ W}$$

(S) horsepower (metric)

$$1 \text{ hp} = 75 \text{ kpm s}^{-1} = 735.5 \text{ W}$$

Pressure

(M) newton per square meter	$N\ m^{-2}$
	$[N\ m^{-2}] = m^{-1}\ kg\ s^{-2}$

Secondary units:

(S) kilopond per square centimeter

$$1 \text{ kp cm}^{-2} = 9.80665 \times 10^{4} \text{ N m}^{-2}$$

(S) technical atmosphere

$$1 \text{ at} = 1 \text{ kp cm}^{-2} = 9.80665 \times 10^{4} \text{ N m}^{-2}$$

(S) normal (physical) atmosphere

1 atm $= 1.01325 \times 10^5 N m^{-2}$

(S) bar

1 bar $= 10^5 N m^{-2}$

(S) torr

1 torr $= 133.322 N m^{-2}$

1 bar is approximately equal to normal atmospherical pressure at sea level. 1 torr is equal to hydrostatical pressure of 1 mm column of mercury at 0 °C and g = 9.80665 $m s^{-2}$ (1 atm = 760 torr).

B. ELECTRICAL AND MAGNETIC UNITS

Electric charge

(M)	coulomb	C
		[C] = s A
1 C is the charge delivered by constant current of 1 A in 1 s		

The elementary electric charge (charge of proton)

$$e = (1.60206 \pm 0.00003) \times 10^{-19} C$$

Electric flux density

(M)	coulomb per square meter	$C m^{-2}$
		$[C m^{-2}] = m^{-2} s A$

Electric potential, voltage

(M)	volt	V
		$[V] = m^2 kg s^{-3} A^{-1}$
1 V is the voltage (or potential difference) between two points of a conductor conducting the current of 1 A if the power expended in this part of the conductor is 1 W		

Electric field intensity

(M)	volt per meter	$V m^{-1}$
		$[V m^{-1}] = m kg s^{-3} A^{-1}$

Electric field intensity (E) is related to electric flux density (D) by

$$D = \epsilon_r \epsilon_o E$$

where

ϵ_r is relative permittivity, $[\epsilon_r] = 1$

$\epsilon_o = 8.85416 \times 10^{-12} F m^{-1}$ is the permittivity of vacuum

Capacitance

(M)	farad	F
		$[F] = m^{-2} kg^{-1} s^4 A^2$
1 F is the capacitance of a capacitor which, being charged by the charge of 1 C, has the terminal voltage of 1 V		

Magnetic flux

(M)	weber	Wb
		$[Wb] = m^2 kg s^{-2} A^{-1}$
1 Wb is the magnetic flux which, being linearly attenuated to zero in 1 s, induces in surrounding turn an emf of 1 V		

1 Wb = 1 Vs (voltsecond)

Magnetic flux density

(M) tesla	T
	$[T] = kg\ s^{-2}\ A^{-1}$
1 T is the magnetic flux density of a homogenous magnetic field which acts by the force of 1 N per 1 m of a conductor conducting the current of 1 A, if the field direction is perpendicular to the axis of the conductor	

$$1\ T = 1\ Wb\ m^{-2}$$

Magnetic field intensity

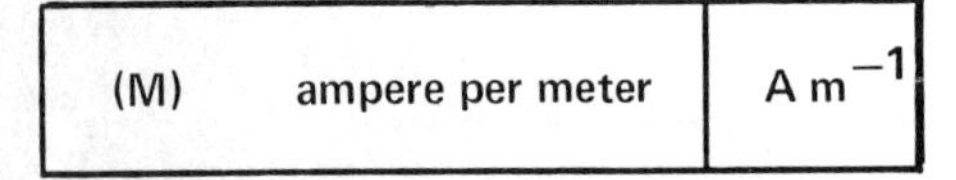

(M) ampere per meter	$A\ m^{-1}$

Magnetic field intensity (H) is defined according to the 1st Maxwell equation by the curve integral

$$\oint_C H dL = \Sigma I$$

where ΣI is the sum of currents inside the closed curve, **C**, and **d***L* is the element of C. It is related to the magnetic field density (B) by

$$B = \mu_r \mu_o H$$

where

μ_r is the relative permeability, $[\mu_r] = 1$

$\mu_o = 1.25664 \times 10^{-6}\ H\ m^{-1}$ is the permeability of vacuum

Permeability and permittivity of vacuum are natural constants related together by

$$c = \frac{1}{\sqrt{\epsilon_o \mu_o}} = 2.99793 \times 10^8\ m\ s^{-1}$$

which is the velocity of electromagnetic radiation in vacuum.

Inductance

(M) henry	H
	$[H] = m^2\ kg\ s^{-2}\ A^{-2}$
1 H is the inductance of a conductor in which the counter emf of 1 V is generated if the current in the conductor is linearly changed by 1 A in 1 s	

Current density

(M) ampere per square meter	$A\ m^{-2}$

For practical use this main unit is too small. Practical unit is ampere per square millimeter

$$1\ A\ mm^{-2} = 10^6\ A\ m^{-2}$$

Resistance

(M) ohm	Ω
	$[\Omega] = m^2\ kg\ s^{-3}\ A^{-2}$
1 Ω is the resistance of a conductor which, being connected to the voltage of 1 V, conducts the current of 1 A	

The same unit is used for reactance and impedance.

Resistivity

(M) ohm-meter	Ωm
	$[\Omega m] = m^3\ kg\ s^{-3}\ A^{-2}$

For practical use this unit is too great. Usually, the resistivity is measured and stated for the sample of a conductor which is 1 m long and has the cross-section of 1 mm^2. Therefore, the practical unit is

$$1\ \Omega mm^2 m^{-1} = 10^{-6}\ \Omega m = 1\ \mu\Omega\ m$$

which is called *microohm-meter.* For example, resistivity of copper is

$$\rho_{Cu} = 0.0172\ \mu\Omega m$$

Resistance of a given conductor is then given by

$$R = \rho \frac{L}{S} \quad [\Omega;\ \mu\Omega m,\ m,\ mm^2]$$

Electric power, reactive power, apparent power

a) power (effective power)

(M) watt	W
	$[W] = m^2\ kg\ s^{-3}$

b) reactive power

(M) var	var
	$[var] = m^2\ kg\ s^{-3}$

c) apparent power

(M) voltampere	VA
	$[VA] = m^2\ kg\ s^{-3}$

The coherency of basic and main SI units and great practical advantage of their use is best demonstrated in the following set of examples:

a) Power, transmitted by rotating shaft, given as the product of torque **T** and angular velocity ω

$$P = T\omega \qquad [W;\ Nm,\ rad\ s^{-1}]$$

b) Centrifugal force acting on the body with the mass **m** which moves by the velocity **v** along the curved path with instantaneous radius **r**

$$F = m\frac{v^2}{r} \qquad [N;\ kg,\ m\ s^{-1},\ m]$$

c) Acceleration torque necessary to accelerate a rotating body with the moment of inertia **J** by angular acceleration a

$$T_a = Ja \qquad [Nm;\ kg\ m^2,\ rad\ s^{-2}]$$

d) Force acting on current-carrying conductor in homogenous magnetic field perpendicular to conductor axis (**I** = current in conductor, L = length of conductor, **B** = magnetic flux density)

$$F = BLI \qquad [N;\ T,\ m,\ A]$$

e) Voltage generated in conductor, moved in homogenous magnetic field

$$E = BLv \qquad [V;\ T,\ m,\ m\ s^{-1}]$$

f) Voltage generated in the turn due to time change of magnetic flux

$$e = -\frac{d\Phi}{dt} \qquad [V;\ Wb,\ s]$$

g) Voltage generated in the coil with an inductance **L** due to time change of the current in the coil

$$v_L = L\frac{di}{dt}$$ [V; H, A, s]

h) Motor power loss expressed in terms of torque and angular velocity for output and armature current and terminal DC voltage for input

$$P_L = P_i - P_o = I_a V - T\omega$$

[W; A, V, Nm, rad s^{-1}]

A.2. UNITS' CONVERSION FACTORS AND TABLES

CONVERSION FACTORS

Unless otherwise specified, the units *oz* and *lb* in the following table are *units of force.*

LENGTH	*SI units* m meter km kilometer cm centimeter mm millimeter μm micrometer nm nanometer Å angstrom	1 m = 10^{-3} km 10^2 cm 10^3 mm 10^6 μm 10^9 nm 10^{10} Å	1 m = 39.370 in 3.2808 ft 1.0936 yd
		1 in = 25.4 mm (exactly) 1 ft = 12 in = 0.3048 m 1 yd = 3 ft = 0.9144 m	
		1 mi (statute mile) = 1609.344 m	
		1 mil = 10^{-3} in = 0.0254 mm	
AREA	*SI units* m^2 square meter cm^2 square centimeter mm^2 square millimeter	1 m^2 = 10^4 cm^2 10^6 mm^2	1 m^2 = 1550.0 in^2 10.764 ft^2
		1 in^2 = 645.16 mm^2 = 6.4516 cm^2 1 in^2 = 6.9444 x 10^{-3} ft^2 1 ft^2 = 144 in^2	
VOLUME	*SI units* m^3 cubic meter dm^3 cubic decimeter l liter (= dm^3) cm^3 cubic centimeter mm^3 cubic millimeter	1 m^3 = 10^3 dm^3 10^3 l 10^6 cm^3 10^9 mm^3	1 m^3 = 6.1024 x 10^4 in^3 35.315 ft^3 1 cm^3 = 6.1024 x 10^{-2} in^3
		1 in^3 = 16.387 cm^3 1 ft^3 = 1728 in^3 = 2.8317 x 10^{-2} m^3	
LINEAR VELOCITY	*SI units* m/s meter per second mm/s millimeter per second km/h kilometer per hour	1 m/s = 10^3 mm/s 3.6 km/h	1 m/s = 39.370 in/s 3.2808 ft/s
		1 in/s = 2.54 x 10^{-2} m/s = 25.4 mm/s 1 mph = 1.6093 km/h	

Quantity	SI units	Conversions
LINEAR ACCELERATION	*SI unit* m/s^2 meter per sqaure second	$1\ m/s^2 = 39.370\ in/s^2$ $3.2808\ ft/s^2$
		$1\ in/s^2 = 2.54 \times 10^{-2}\ m/s^2$
PLANE ANGLE	*SI units* rad radian o angular degree r revolution ′ angular minute ″ angular second	$1\ rad = 57.296^o = (360/2\pi)^o$
		$1^o = 60'$ $3600''$ 1.7453×10^{-2} rad 2.7778×10^{-3} r
		1 r = 360° 6.2832 rad = (2π) rad
ANGULAR VELOCITY	*SI units* rad/s radian per second r/s (rps) revolution per second r/min (rpm) revolution per minute $^o/s$ angular degree per second	1 rad/s = 0.15915 rps 9.5493 rpm
		1 rpm = 10^{-3} krpm 1.6667×10^{-2} rps 6 $^o/s$ 0.10472 rad/s
		1 rps = 60 rpm 360 $^o/s$ 6.2832 rad/s
ANGULAR ACCELERATION	*SI units* rad/s^2 radian per second per second r/s^2 revolution per second per second r/min/s revolution per minute per second $^o/s^2$ angular degree per second per second	$1\ rad/sec^2 = 0.15915\ rps^2$ 9.5493 rpm/s
		1 rpm/s = 10^{-3} krpm/s $1.6667 \times 10^{-2}\ rps^2$ $6\ ^o/s^2$ $0.10422\ rad/s^2$
		$1\ rps^2$ = 60 rpm/s $360\ ^o/s^2$ $6.2832\ rad/s^2$
MASS	*SI units* kg kilogram g gram	1 kg = 10^3 g 35.274 oz (mass) 2.2046 lb (mass) 6.8522×10^{-2} slug
		1 oz (mass) = 28.3495 g
		1 lb (mass) = 16 oz (mass) 0.45359 kg
		1 slug = 14.5939 kg

Quantity	Units	Conversions
FORCE	*SI units* N newton kp kilopond kgf (= kp) kilogram-force P pond	1 N = 0.10197 kp 0.22481 lb 3.5969 oz 7.2330 poundals
		1 kp = 9.80665 N 2.2046 lb 35.274 oz
		1 oz = 0.27801 N 2.83495×10^{-2} kp 28.3495 P
		1 lb = 16 oz 4.4482 N 0.45359 kp
		1 poundal = 0.138255 N
PRESSURE	*SI units* N/m^2 newton per square meter kp/cm^2 kilopond per square centimeter at (=kp/cm^2) technical atmosphere atm normal (physical) atmosphere	1 N/m^2 = 1.0197×10^{-5} kp/cm^2 1.45034×10^{-6} $16/in^2$
		1 at = 1 kp/cm^2 9.80665×10^4 N/m^2 14.223 lb/in^2
		1 atm = 1.01325×10^5 N/m^2 14.696 lb/in^2
		1 lb/in^2 = 0.07031 at
TORQUE	*SI units* Nm newtonmeter kpm kilopondmeter	1 Nm = 0.10197 kpm 0.73756 lb-ft 8.85075 lb-in 141.612 oz-in
		1 kpm = 9.80665 Nm 1.3887×10^3 oz-in
		1 oz-in = 7.0615×10^{-3} Nm 7.2008×10^{-4} kpm
		1 lb-ft = 192 oz-in 1.3558 Nm 0.13825 kpm

	Units	Conversions
MOMENT OF INERTIA	*SI units* kg cm^2 kilogram-square centimeter kg m^2 kilogram-square meter g cm^2 gram-square centimeter	1 kg cm^2 = 0.01416 oz-in-s^2 1 kg m^2 = 10^7 g cm^2 8.85075 lb-in-s^2 141.612 oz-in-s^2 1 oz-in-s^2 = 6.25×10^{-2} lb-in-s^2 7.06155×10^{-3} kg m^2 70.6155 kg cm^2
ENERGY (WORK)	*SI units* J joule Nm newtonmeter Ws wattsecond kpm kilopondmeter kWh kilowatthour kcal kilocalorie cal calorie	1 J = 1 Nm 1 Ws 0.10197 kpm 2.7778×10^{-7} kWh 2.38846×10^{-4} kcal 9.4781×10^{-4} Btu 1 kcal = 10^3 cal 4186.8 J 1.1630×10^{-3} kWh 3.9683 Btu 1 kWh = 3.6×10^6 J 859.845 kcal 3.4121×10^3 Btu 1 Btu = 1055.06 J 2.9307×10^{-4} kWh 0.251997 kcal
POWER	*SI units* W watt kW kilowatt J/s joule per second kpm/s kilopondmeter per second hp horse power (metric)	1 W = 10^{-3} kW 1 J/s 0.10197 kpm/s 0.73756 lb-ft/s 1.3596×10^{-3} hp (metric) 1.3410×10^{-3} hp (British) 1 hp (British) = 550 lb-ft/s 745.7 W 1.0139 hp (metric) 1 hp (metric) = 75 kpm/s 735.5 W 0.98632 hp (British)

Quantity	Units	Conversions
OUTPUT POWER OF MOTOR	expressed as product of torque and angular velocity (speed of rotation)	$P = T\omega$ [W; Nm, rad/s]
		$P = 0.10472\ T\ n$ [W; Nm, rpm]
		$P = 7.3948 \times 10^{-4}\ T\ n$ [W; oz-in, rpm]
		$P \approx 10^{-6}\ T\ n$ [hp; oz-in, rpm]
TEMPERATURE	*SI units* deg — temperature degree ^oC — degree Celsius (centigrade) K — Kelvin	magnitude of degrees 1 deg = 1 ^oC = 1 K = 9/5 ^oF
		0 ^oC = 273.15 K = 32 ^oF
		$\Theta\ [^oC] = (\Theta\ [^oF] - 32) \cdot \frac{5}{9}$
		$\Theta\ [^oF] = \frac{9\ \Theta\ [^oC]}{5} + 32$
		$\Theta\ [K] = \Theta\ [^oC] + 273.15$
MAGNETIC FLUX	*SI units* Wb — weber Vs — voltsecond Mx — maxwell (CGS unit)	1 Wb = 1 Vs 10^8 Mx 10^5 kilolines 10^8 lines
		1 line = 1 Mx = 10^{-8} Wb
MAGNETIC FLUX DENSITY	*SI units* T — tesla G — gauss (CGS unit)	1 T = 1 Wb/m^2 10^4 G 10^4 $lines/cm^2$ 6.4516×10^4 $lines/in^2$
		1 $line/in^2$ = 0.1550 G 1.550×10^{-5} T
MAGNETIC FIELD INTENSITY	*SI units* A/m — ampere per meter Oe — oersted (CGS unit)	1 A/m = 10^{-2} A/cm 1.2566×10^{-2} Oe 2.54×10^{-2} A-turn/in
		1 Oe = 79.577 A/m 2.0213 A-turn/in
		1 A-turn/in = 39.370 A/m 0.49474 Oe

DERIVED MOTOR CONSTANTS		
	TORQUE CONSTANT	1 Nm/A = 0.10197 kpm/A 141.612 oz-in/A
		1 kpm/A = 9.80665 Nm/A 1.3887×10^{3} oz-in/A
		1 oz-in/A = 7.06155×10^{-3} Nm/A 7.2008×10^{-4} kpm/A
	VOLTAGE CONSTANT	1 V/krpm = 9.5493×10^{-3} V/rad s^{-1}
		1 V/rad s^{-1} = 104.72 V/krpm
	DAMPING CONSTANT *VISCOUS DAMPING FACTOR*	1 Nm/rad s^{-1} = 141.612 oz-in/rad s^{-1} 104.72 Nm/krpm
		1 Nm/krpm = 141.612 oz-in/krpm 9.5493×10^{-3} Nm/rad s^{-1}
		1 oz-in/rad s^{-1} = 7.0615×10^{-3} Nm/rad s^{-1} 104.72 oz-in/krpm
		1 oz-in/krpm = 7.0615×10^{-3} Nm/krpm 9.5493×10^{-3} oz-in/rad s^{-1}
	SPEED REGULATION CONSTANT	1 rpm/oz-in = 0.14161 krpm/Nm 1 krpm/Nm = 7.0615 rpm/oz-in

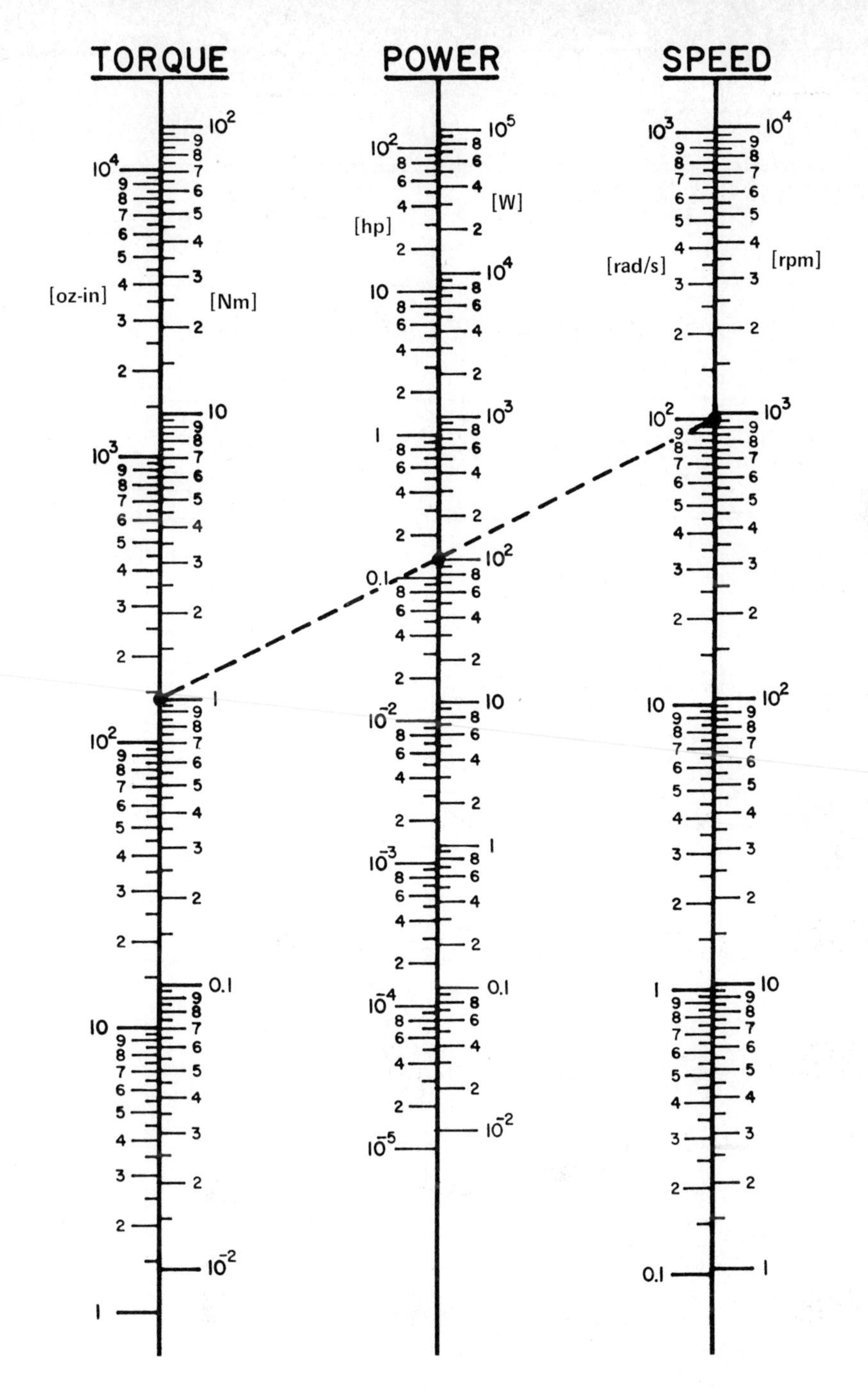

Example shown: 100 W = 1 Nm x 100 rad/s

TORQUE–POWER–SPEED NOMOGRAPH

A \ B	kg m^2	g cm^2	oz-in-s^2	lb-in-s^2	oz-in^2	lb-in^2	lb-ft^2
kg m^2	1	10^7	141.612	8.85075	5.46745×10^4	3.41716×10^3	23.7303
g cm^2	10^{-7}	1	1.41612×10^{-5}	8.85075×10^{-7}	5.46745×10^{-3}	3.41716×10^{-4}	2.37303×10^{-6}
oz-in-s^2	7.06155×10^{-3}	7.06155×10^4	1	6.25×10^{-2}	386.088	24.1305	0.167573
lb-in-s^2	0.112985	1.12985×10^6	16	1	6.17741×10^3	386.088	2.68117
oz-in^2	1.82901×10^{-5}	182.901	2.59008×10^{-3}	1.61880×10^{-4}	1	6.25×10^{-2}	4.34028×10^{-4}
lb-in^2	2.92641×10^{-4}	2.92641×10^3	4.14413×10^{-2}	2.59008×10^{-3}	16	1	6.94444×10^{-3}
lb-ft^2	4.21403×10^{-2}	4.21403×10^5	5.96755	0.372972	2304	144	1

Tab. A.1. Moment of inertia conversion factors

To convert from A to B, multiply by entry in table.

B / A	Nm	kpm	oz-in	lb-in	lb-ft
Nm	1	0.101972	141.612	8.85075	0.737562
kpm	9.80665	1	1.38874×10^{3}	86.7962	7.23301
oz-in	7.06155×10^{-3}	7.20077×10^{-4}	1	6.25×10^{-2}	5.20833×10^{-3}
lb-in	0.112985	1.15212×10^{-2}	16	1	8.33333×10^{-2}
lb-ft	1.35582	0.138255	192	12	1

Tab. A.2. Torque conversion factors

To convert from A to B, multiply by entry in table.

EXPLANATION TO TAB. A.1 AND A.2

The British system of units has two forms:

a) Technical system

which is more common in engineering (and is also used in this book) and which is based on the following basic units:

1 s - for time

1 ft - for length

1 lb (force) - for force

The derived unit of mass is then 1 *slug,* defined as a mass which is accelerated by 1 ft/s^2 upon application of a force of 1 lb.

b) Physical system

which is based on the following basic units:

1 s - for time

1 ft - for length

1 lb - for mass

The derived unit of force is then 1 *poundal,* a force which accelerates the mass of 1 lb by 1 ft/s^2.

The basic equation which relates acceleration torque $\mathbf{T_a}$, moment of inertia **J** and angular acceleration α has therefore two forms, depending on the system of units used:

a) in the British technical system

$$\mathbf{T_a} = \mathbf{J}\alpha \qquad [\text{oz-in; oz-in-s}^2, \text{rad/s}^2]$$

where *oz* is the unit of force, and

b) in the British physical system

$$\mathbf{T_a} = \mathbf{J}\alpha \qquad [\text{poundal-ft; lb-ft}^2, \text{rad/s}^2]$$

where *lb* is the unit of mass.

The unit of torque, *poundal-ft,* is not common and is not used in engineering practice but the moment of inertia, **J**, is sometimes given in units of *lb-ft*^{2} (or, in related units, as *lb-in*^{2} or *oz-in*^{2}) – even in technical literature. Therefore, in Tab. A.1 all necessary conversion factors are introduced.

In British units of torque, which are introduced in Tab. A.2, the *oz* and *lb* are units of force, according to technical form of the British system.

As for the SI (metric) units in both tables, see the Appendix A.1.

SPEED [rpm]	OUTPUT POWER [hp] (British)																			
	1/75	1/60	1/50	1/40	1/35	1/30	1/25	1/20	1/15	1/12	1/10	1/8	1/6	1/4	1/3	1/2	5/8	3/4	1	3/2
850												148	198	296	396	592	740	883	1184	1776
1140												110	149	220	294	440	550	660	880	1320
1200	11.2	14	16.8	21	24	28	33.7	42.1	56.1	70.1	84.1	105	140	210	269	420	525	630	840	1260
1425	9.4	11.8	14.2	17.7	20.2	23.6	28.4	35.5	47.3	59	70.8	90	115	180	235	360	450	540	720	1080
1600	8.4	10.5	12.6	15.8	18.0	21	25.2	31.5	42.1	52.6	63.1	78.8	105	158	210	316	395	474	632	943
1725	7.8	9.8	11.7	14.6	16.7	19.5	23.4	29.3	39	48.8	53.6	72.8	97	146	194	291	364	437	582	874
1800	7.5	9.4	11.2	14	16	18.7	22.4	28	37.4	46.8	56.1	69.9	91	140	182	280	350	418	559	836
2000	6.7	8.4	10.1	12.6	14.4	16.8	20.2	25.3	33.6	42.1	50.5	63.2	84.2	126	168	252	315	378	504	756
2200	6.1	7.6	9.2	11.5	13.1	15.3	18.4	23	30.6	38.2	45.9	57.3	76.4	115	153	230	287	335	460	670
2400	5.6	7.0	8.4	10.5	12.0	14.0	16.8	21.0	28.0	36.0	42.1	52.6	70.1	105	140	210	263	315	420	630
2600	5.2	6.5	7.8	9.7	11.1	12.9	15.5	19.4	25.9	32.4	38.8	48.6	64.7	97.2	129	194	243	291	389	582
2800	4.8	6.0	7.2	9.0	10.8	12.0	14.4	18.0	24.0	30.0	36.0	45.0	60	90	120	180	225	270	360	540
3000	4.5	5.6	6.7	8.4	9.6	11.2	13.5	16.8	22.4	28.1	33.6	42.1	56.1	84.1	112	168	210	252	336	504
3200	4.2	5.3	6.3	7.9	9.0	10.5	12.6	15.8	21.0	26.3	31.5	39.4	52.6	78.8	105	158	197	236	315	472
3450	3.9	4.9	5.8	7.3	8.4	9.8	11.7	14.6	19.5	24.4	29.3	36.4	48.8	72.8	97.7	146	182	218	201	436
3600	3.7	4.7	5.6	7.0	8.0	9.4	11.2	14.0	18.7	23.4	28.1	35.1	46.8	70.1	93.6	140	175	210	280	420
3800	3.5	4.4	5.3	6.6	7.6	8.8	10.6	13.3	17.7	22	26.6	33.2	44.3	66.4	88.6	133	166	199	265	399
4000	3.4	4.2	5.0	0.3	7.2	8.4	10.1	12.6	16.8	21	25.3	31.5	42.1	63.2	84.2	126	158	189	252	378
4500	3	3.7	4.5	5.6	6.4	7.5	9	11.2	15	18.7	22.4	28	37.4	56	74.8	112	140	168	224	336
5000	2.7	3.4	4	5	5.8	6.7	8.1	10.1	13.5	16.8	20.2	25.2	33.7	50.5	67.3	101	126	152	202	303
5500	2.45	3.1	3.7	4.6	5.2	6.1	7.3	9.2	12.2	15.3	18.4	22.9	30.6	45.9	61.2	91	114	137	182	273
6000	2.2	2.8	3.4	4.2	4.8	5.6	6.7	8.4	11.2	14	16.8	21	28	42	56.1	84	105	126	168	252
6500	2.1	2.6	3.1	3.9	4.4	5.2	6.2	7.8	10.4	13.0	15.6	19.4	25.9	38.8	51.8	77	96	116	154	232
7000	1.9	2.4	2.9	3.6	4.1	4.8	5.8	7.2	9.6	12	14.4	18	24	36	48	72	90	108	144	216
	TORQUE [oz-in]																			

Tab. A.3. Relationship of the speed, torque and output power of a motor.

[in]		[mm]
fraction	decimal	
1/64	0.0156	0.397
1/32	0.0312	0.794
3/64	0.0469	1.191
1/16	0.0625	1.588
5/64	0.0781	1.984
3/32	0.0938	2.381
7/64	0.1094	2.778
1/8	0.1250	3.175
9/64	0.1406	3.572
5/32	0.1562	3.969
11/64	0.1719	4.366
3/16	0.1875	4.763
13/64	0.2031	5.159
7/32	0.2188	5.556
15/64	0.2344	5.953
1/4	0.2500	6.350
17/64	0.2656	6.747
9/32	0.2812	7.144
19/64	0.2969	7.541
5/16	0.3125	7.938
21/64	0.3281	8.334
11/32	0.3438	8.731
23/64	0.3594	9.128
3/8	0.3750	9.525
25/64	0.3906	9.922
13/32	0.4062	10.319
27/64	0.4219	10.716
7/16	0.4375	11.113
29/64	0.4531	11.509
15/32	0.4688	11.906
31/64	0.4844	12.303
1/2	0.5000	12.700

[in]		[mm]
fraction	decimal	
33/64	0.5156	13.097
17/32	0.5312	13.494
35/64	0.5469	13.891
9/16	0.5625	14.288
37/64	0.5781	14.684
19/32	0.5938	15.081
39/64	0.6094	15.478
5/8	0.6250	15.875
41/64	0.6406	16.272
21/32	0.6562	16.669
43/64	0.6719	17.066
11/16	0.6875	17.463
45/64	0.7031	17.859
23/32	0.7188	18.256
47/64	0.7344	18.653
3/4	0.7500	19.050
49/64	0.7656	19.447
25/32	0.7812	19.844
51/64	0.7969	20.241
13/16	0.8125	20.638
53/64	0.8281	21.034
27/32	0.8438	21.431
55/64	0.8594	21.828
7/8	0.8750	22.225
57/64	0.8906	22.622
29/32	0.9062	23.019
59/64	0.9219	23.416
15/16	0.9375	23.813
61/64	0.9531	24.209
31/32	0.9688	24.606
63/64	0.9844	25.003
1	1.0000	25.400

Tab. A.4. Inches to millimeters conversion.

°C	°F
–40	–40
–35	–31
–30	–22
–25	–13
–20	– 4
–17.78	0
–15	5
–10	14
– 5	23
0	32
5	41
10	50
15	59
20	68
25	77
30	86
35	95
40	104
45	113
50	122
55	131
60	140
65	149
70	158
75	167

°C	°F
80	176
85	185
90	194
95	203
100	212
105	221
110	230
115	239
120	248
125	257
130	266
135	275
140	284
145	293
150	302
155	311
160	320
165	329
170	338
175	347
180	356
185	365
190	374
195	383
200	392

Tab. A.5. Celsius to Fahrenheit degrees conversion.

Index

PRODUCT DATA INDEX

To use this index, bend book and follow margin index to page with black edge marker.

HETERICK MEMORIAL LIBRARY
621.3132 E38d 1977
onuu
Electro-Craft Corpo/DC motors, speed con
3 5111 00065 9106